Texts in Statistical Science

Graphics for Statistics and Data Analysis with R

Second Edition

CHAPMAN & HALL/CRC
Texts in Statistical Science Series

Recently Published Titles

Extending the Linear Model with R
Generalized Linear, Mixed Effects and Nonparametric Regression Models, Second Edition
J.J. Faraway

Modeling and Analysis of Stochastic Systems, Third Edition
V.G. Kulkarni

Pragmatics of Uncertainty
J.B. Kadane

Stochastic Processes
From Applications to Theory
P.D. Moral and S. Penev

Modern Data Science with R
B.S. Baumer, D.T. Kaplan, and N.J. Horton

Generalized Additive Models
An Introduction with R, Second Edition
S. Wood

Design of Experiments
An Introduction Based on Linear Models
Max Morris

Introduction to Statistical Methods for Financial Models
T.A. Severini

Statistical Regression and Classification
From Linear Models to Machine Learning
N. Matloff

Introduction to Functional Data Analysis
P. Kokoszka and M. Reimherr

Stochastic Processes
An Introduction, Third Edition
P.W. Jones and P. Smith

Theory of Stochastic Objects
Probability, Stochastic Processes and Inference
Athanasios Christou Micheas

For more information about this series, please visit: https://www.crcpress.com/go/textsseries

Texts in Statistical Science

Graphics for Statistics and Data Analysis with R

Second Edition

Kevin J. Keen

CRC Press is an imprint of the
Taylor & Francis Group, an **informa** business

Cover Art

The front cover depicts a stereogram pair of wireframe plots of a two-dimensional kernel density estimate for the joint distribution of length as measured from the tip of the nose to the notch in the tail (cm) and mass (g) for each of 56 perch (*Perca fluviatilis*) caught during a research trawl on a Finnish lake as reported in 1917. Heatmap colors adjusted for improved perception by individuals with anomalous deuteranopic (red-green) color-vision have been draped over the wireframe. This pseudo-color mapping as perceived by individuals with normal color vision has pinkish-purple for the lowest relative frequencies through orange, yellow, and then white for the highest relative frequencies. The axis for length is along the left top edge of the wireframe with values increasing from bottom to top. The axis for mass is along the left bottom edge of the wireframe with values increasing left to right. The major mode in this bimodal distribution corresponds to juvenile fish (prey) while the minor mode corresponds to mature fish (predator). For the optical illusion of depth with the stereogram, cross your eyes until four images appear, then relax gradually to allow the images to converge to a set of three. Focus on the center image. It might take a bit of practice and patience, but the center image will appear to be in three dimensions.

CRC Press
Taylor & Francis Group
6000 Broken Sound Parkway NW, Suite 300
Boca Raton, FL 33487-2742

Printed on acid-free paper
Version Date: 20180411

International Standard Book Number-13: 978-1-4987-7983-8 (Hardback)

Visit the Taylor & Francis Web site at
http://www.taylorandfrancis.com

and the CRC Press Web site at
http://www.crcpress.com

Contents

Preface to the First Edition

This book is intended for those wanting to learn about the basic principles of graphical design as applied to the presentation of data. It is also intended for those well acquainted with the application of these basic principles but who are seeking exposure to a wider palette of graphical displays for the presentation of data. It is assumed that readers have completed a first course in statistics or statistical research methods, ideally with 60 hours of instruction. It is not assumed that students have been previously exposed to any of the graphical displays to be discussed but doubtless there will be some familiarity. Introductory courses typically do not assess graphical displays as to their strengths and weaknesses. This is the point of departure for this book.

By reading selectively, this book can be used to learn the basics of statistical graphics after as few as 30 hours of instruction obtained from an introductory course in statistics. The sections with titles that have been marked with an asterisk require more background in statistical and mathematical theory than could be obtained from such a course. These sections can be omitted with very little loss of continuity in material. These sections are recommended, however, for students with suitable preparation in either the final year of an undergraduate major in statistics or embarking on postgraduate studies in statistics.

Regarding the selection of a statistical software package, there are many packages available with excellent high resolution graphics. The one used in this textbook is R. It is produced by the R Foundation for Statistical Computing (Vienna, Austria) that represents an international consortium of independent software developers, many of whom are statisticians. R is available for a wide variety of operating systems. These include Linux, Windows, and MacOS. R is available as Free Software under the terms of the Free Software Foundation's GNU General Public Licence. See `http://www.r-project.org` for more information.

R has been adopted for this textbook because its graphical capabilities are state-of-the-art. Graphical commands are accessed by scripted function calls. Although this approach has a bit longer learning curve than point-and-click interaction with on-screen images, it does allow the researcher to save drafting commands for future reference. R commands are listed for many of this textbook's examples. This is not done if the length of the R script for a graphical display runs for several lines. However, the scripts for all of the figures in this book rendered by R are available for download from the book's website: `http://www.graphicsforstatistics.com`.

The ideal classroom setting for this book would have all students seated behind computer workstations with the instructor introducing material from a podium workstation onto a projection screen and onto each student's high resolution computer monitor. Instructors with access to less well-endowed teaching environments will likely use a laptop computer projecting onto a single screen in a lecture hall. It is highly recommended that laboratory sessions, in which students work alone or in small groups, be part of the course implementation.

This book has been written to be used in any of three different ways. Firstly, the book can be used as a textbook for a dedicated course in graphical analysis of data. Early drafts of this textbook have been used in this setting. Secondly, the book can be used as a supplementary text throughout a program of study in statistics or data analysis. For this purpose, the textbook chapters have been arranged in such a way to correspond to a progressive program beginning with univariate statistical methods, analysis of contingency tables, linear regression models, and multivariate methods for data analysis. Finally, the book is intended as a handy reference for graphical analysis.

Preface to the Second Edition

The major changes for the second edition are the use of `ggplot2` in addition to the base `graphics` and `lattice` packages of the first edition and the addition of a section on *Learning Outcomes* added to each chapter after the *Introduction*. Major changes include the addition of two appendices: one on human visualization and another on color rendering. Appendix A on human visualization was a standalone chapter in a proposed manuscript for the book but it never made it to print in the first edition. Appendix B on color rendering covers the essentials of how color is rendered on electronic displays and in print. Both appendices contain public domain images from Wikimedia Commons, which are gratefully acknowledged. Wikipedia and the documentation in R were invaluable sources of information on the topic of color spaces when compared to available sources in print. Wikipedia, as does R, relies on the contributions of unpaid volunteer contributors and these are gratefully acknowledged.

Presenting learning outcomes has become a feature of modern textbook design that is a useful student aid before material is presented to help to motivate students and can be used by students after reading the material to review what they have accomplished. For many professions, reporting course learning outcomes is a requirement of the process of review of an accredited educational program. This is true for the discipline of Statistics in those jurisdictions where educational programs of study leading to the designation *Professional Statistician* or *P.Stat.* are accredited. Providing this material in the second edition eases the workload for courses in which this book is adopted as either the sole or auxiliary textbook on the assumption that training in the graphical presentation of data is a requirement.

Although there are many statistical packages available with excellent high-resolution graphics, the second edition is continuing with R, which is produced by the R Foundation for Statistical Computing (Vienna, Austria) and is available as Free Software under the terms of the Free Software Foundation's GNU General Public Licence. See `http://www.r-project.org` for more information. New to the second edition is the use of `ggplot2` in many examples in addition to code in the `base` or `lattice` R graphics packages. Hadley Wickham, one of many independent software developers for R, is the creator of ggplot2, which was released in 2005. ggplot2 is licensed under Version 2 of the GNU General Public Licence.

Consideration was given to adopting ggplot2 solely for the second edition. But this did not happen for two reasons: ggplot2 is a bit slower than lattice (both used the grid graphics package of Paul Murrell), and, more importantly, on February 25th, 2014, Hadley Wickham announced that ggplot2 was shifting into maintenance mode. Nevertheless, when ggplot 2.0.0 was released on December 21st, 2015, it had an extension mechanism allowing developers to add their own code to ggplot2 and incorporate this code in other packages. Two such packages used for the second edition are ggthemes for more graphics themes than available in ggplot2 and ggmosaic for mosaic plots for depicting the joint distribution of two categorical variables as ggplot2 does not produce mosaic plots.

A minor glitch was discovered in release 3.4.1 of R when new figures for the second edition were being drafted. This was determined to be related to the rounding of fractions of inches for dimensions of graphics windows and graphical figures within graphical windows. The pragmatic solution to this problem was to set each dimension for any graphical figure to be 0.05 inch smaller than the corresponding dimension for each outer graphical window. Consequently, none of the original scripts for all of the figures in the first edition, and available for download from the book's website at

https://www.graphicsforstatistics.com,

will produce a graphical figure without these modifications. So a minor glitch became a major problem for the second edition. The solution was to modify all scripts from the first edition retained for the second edition. Since this was required, it was decided to take the code for each graphic and embed it within a function in R and group all the functions together as an R package called graphicsforstatistics. If the data used for a given figure is small, then it is just encoded in the function.

A few of the larger frequently used datasets are coded as functions to be called to create data frames. These data sets can also be downloaded in an R workspace for the second edition from

https://www.graphicsforstatistics.com.

Also available at this website are R scripts containing the functions for each of the figures in the second edition. These are grouped by chapter. Unlike the first edition, there are no text versions of these R scripts for the second edition. However, text versions of the larger frequently used datasets are available on the website for the second edition as for the first.

Acknowledgments

I would like to express my appreciation to a number of colleagues for their assistance. I would very much like to thank Prof. Steven M. Houser, M.D., of the Department of Otolaryngology in the School of Medicine at Case Western Reserve University for permission to use data relating to allergy from clinical studies involving his patients at Cleveland's MetroHealth Medical Center. Steve and I began a productive research collaboration in 1999 that has endured and led to no fewer than three articles, with various co-authors, published in medical journals.

I would like to thank Juha Puranen of the Department of Statistics at the University of Helsinki in Finland for the data from fisheries research trawls conducted on Längelmävesi near Tampere in or before 1917.

I am indebted to Prof. James Hanley of the Department of Epidemiology and Biostatistics at McGill University for rediscovering Galton's original regression data on the heights of family members. The original notebooks were dug up by the staff of the Library at University College London. Jim made the data available to us again in 2004.

I would like to thank Dr. Patricia Tai of the Department of Radiation Oncology at the Allan Blair Cancer Centre of the Saskatchewan Cancer Agency, in Regina, for permission to use data on survival in patients with small-cell cancer.

I would like to thank Dr. Thomas Lietman of the Institute for Global Health at the University of California, San Francisco, for permission to use data on the prevalence of antibiotic resistance in nasospharyngeal *Streptococcus pneumoniae* bacteria between villages treated with topical tetracycline or systemic azithromycin as part of a trachoma control program to prevent blindness in Nepal.

The fourth draft of the first edition was done while with the Department of Epidemiology and Statistics of the School of Medicine at Case Western Reserve University. I am grateful to the Rammelkamp Center for Education and Research at MetroHealth Medical Center in Cleveland, Ohio for the very pleasant physical working environment. Salary support during this period was in part from a supplement to a US Public Health Service Research Grant held by Prof. Robert C. Elston, Professor of Genetic and Molecular Epidemiology,

Case Western Reserve University. I am most grateful to Robert for having offered me this once-in-a-lifetime opportunity.

Final editing and drafting of the first edition were done as an Associate Professor and the entire second edition was done as a Professor in the Department of Mathematics and Statistics at the University of Northern British Columbia. I am very much grateful for having been hired in 2004 by a university that, in the same year, opened its doors to the Northern Medical Program, Canada's most northern medical school. The main campus of the university is situated on a hilltop in the city of Prince George in the midst of a sub-boreal forest, in the foothills of the Cariboo range of the Columbia Mountains and in the shadows of the Rocky Mountains. The university operates four other campuses in an area roughly the size of Germany.

I am most appreciative of Dr. Sheryl Barlett of Health Canada in Ottawa for proofreading and reviewing an earlier manuscript for this book. I am grateful to three referees for their recommendations. I would also like to acknowledge Mr. Mark Pollard, Ms. Stephanie Harding, and Ms. Sharon Taylor of Chapman & Hall in London, for their suggestions for improvement.

I would especially like to thank Mr. Robert Stern of Taylor & Francis for his patience over the years as I worked on the first edition of this book while establishing a research program to support discovery in immune diseases in general, and specifically in the rare rheumatic diseases of unusually high prevalence in northern British Columbia. I am indebted to Ms. Jessica Vakili at Taylor & Francis for editorial assistance on the first edition. I would also like to thank three other individuals at Taylor & Francis: Mr. David Grubbs who took over from Mr. Stern towards the end of preparation of the first edition, Ms. Nadja English, and my Production Editor Ms. Michele Dimont.

I would like to thank again Mr. David Grubbs who was my editor throughout the second edition. I was most lucky to have Ms. Michele Dimont as Production Editor as well for the second edition. Michele is the best. Also at Taylor & Francis, thanks to Ms. Sherry Thomas for additional editorial assistance and Mr. Jonathan Pennell for consultation on the color profiles for the electronic and printed versions of the second edition.

I would like to thank my wife Michelle for her diligent checking of syntax, grammar, and spelling of each draft of both editions. Finally, I would like to thank my daughter Kendra Joanne as a young girl and teenager for understanding why Daddy tended to be preoccupied during the first edition, and later as a young woman during her Bachelor of Science in Biomedical Physiology and then during her Bachelor of Science in Nursing for understanding the value of a book on graphical design for researchers interested in depicting data and statistical summaries clearly.

Part I

Introduction

Chapter 1

The Graphical Display of Information

1.1 Introduction

Graphs and charts are found in almost every nonfiction publication. They are a constant feature in periodicals such as online news outlets, daily newspapers, magazines, technical journals, and annual reports of organizations in the private and public sectors. It has been quite some time since a professional graphic artist or draftsperson has been required to produce publication-quality images with pen and ink at considerable cost with sufficient lead time for the task to be accomplished.

A large number of software packages exist from which to choose to produce high-quality graphical representations of data. Computer software capable of producing images on video displays or paper can be classified into any one of three broad categories according to principal function. These categories are as follows:

- spreadsheet software, typically for accounting applications;
- graphical design software for painting and drawing; and
- statistical software.

Many of these software packages are available at low cost for use on desktop and laptop computers. Consequently, the use of graphical software to display data or the results of statistical analyses is widespread and commonplace. With this has come the proliferation of badly conceived or badly presented graphical material.

The old chestnut that *figures will lie and liars will figure* is no longer solely with respect to numerical figures but also includes graphical figures. The maxim that *a little knowledge is a dangerous thing* in the information age has become *an inexpensive software package is a dangerous thing*. By analogy, just because an individual can afford a car doesn't mean that that individual can sit behind the wheel and drive. Well, actually any individual can. It just doesn't happen that often.

The goal of this book is to provide the basic principles of statistical graphics, the rules-of-the-road for graphical presentation as it were. The intended audience is students and practitioners of social, physical, health, management, statistical, and computational sciences for which information processing is a daily requirement.

With respect to communication with visual media, there are three generally accepted skills that are required. They are

- literacy—the ability to exchange information with letters;
- numeracy—the ability to exchange information with numbers; and
- graphicacy—the ability to exchange information with graphics.

This book attempts to deal explicitly with the latter of the three.

To address concerns as to whether numeracy and graphicacy are really words, a perusal of the Oxford English Dictionary [2, 1] will reveal the following citations of first use.

> **1959** *15 to 18: Rep.Cent.Advisory Council for Educ.* (Eng.) (Min. of Educ.) I.xxv.270 When we say that a historian or a linguist is 'innumerate' we mean that he cannot even begin to understand what scientists and mathematicians are talking about... It is perhaps possible to distinguish two different aspects of numeracy that should concern the Sixth Former.
>
> **1966** *Economist* 22 Jan. 310/2 The need for numeracy today is enormous. Business requires..people who..have grasped the principles of reducing a chaos of information to some kind of order.
>
> **1965** W. G. V. BALCHIN in *Times Educ. Suppl.* 5 Nov. 947/2 Graphicacy..is the communication of relationships that cannot be successfully communicated by words or mathematical notation alone.
>
> **1970** — *Geogr.*iii. 28 Graphicacy is the educated skill that is developed from the visual-spatial ability of intelligence, as distinct from the verbal or numerical abilities.

Taking the definition of graphicacy as that given to us by Balchin, this book is concerned with the communication of data by producing graphical displays of quantitative data or quantitative aspects of qualitative data. In so doing, the following terms will be used more-or-less interchangeably: plot, chart, figure, graph, graphic, and graphical display. These terms will connote a visual representation of either data or a statistical summary of data, or both.

It must be appreciated that the power of statistical graphics comes from the viewer being able to visualize the data. Figure 1.1 is a dot chart of allergy

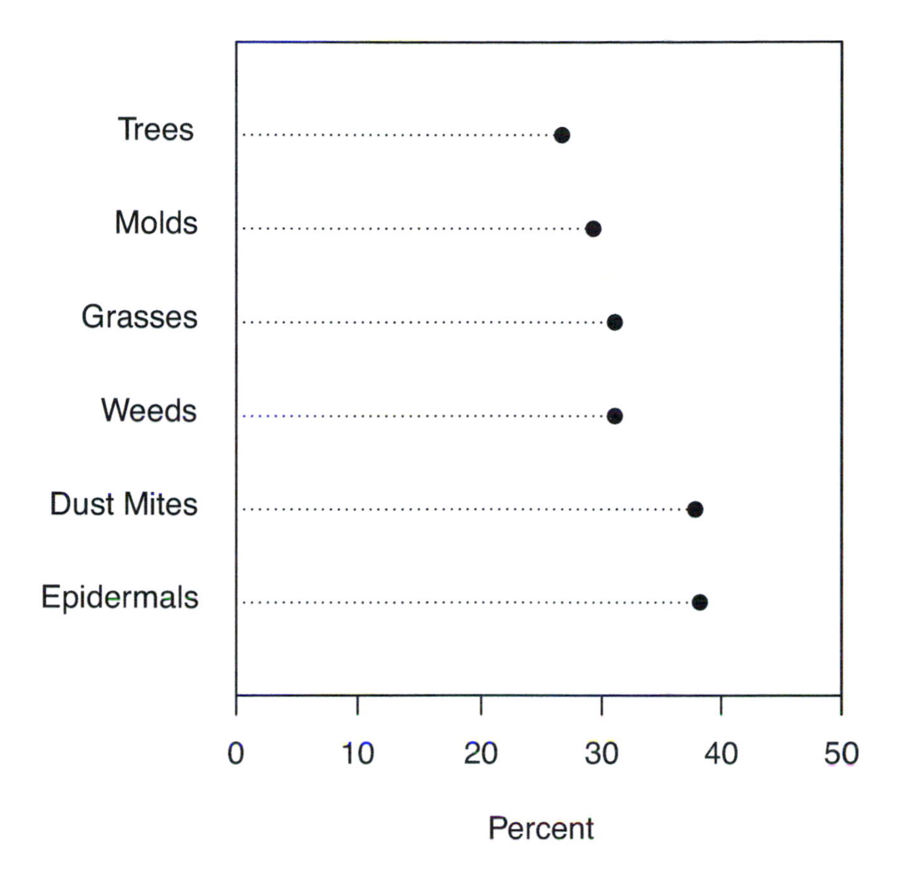

Figure 1.1 *Dot chart of prevalence of allergy in endoscopic sinus surgery patients*

prevalence observed in endoscopic sinus surgery patients. (The term *epidermals* used in the figure refers to epidermal cells shed from cats, dogs, and cockroaches.) The dot chart was developed by Cleveland [21] as a superior alternative to the more familiar and conventional bar chart of the same data in Figure 1.2.

An immediate comparison of the dot chart in Figure 1.1 with the bar chart in Figure 1.2 reveals that the horizontal labels of the dot chart are more easily read than the vertical labels of the bar chart. The horizontal labels of the dot chart allow information concerning the prevalences of the different allergens to be more quickly perceived and compared than with the bar chart.

In this example it is evident that something as simple as the choice of spatial orientation of graphical information can have an impact on the pace of information exchange between the producer and the consumer. There are other

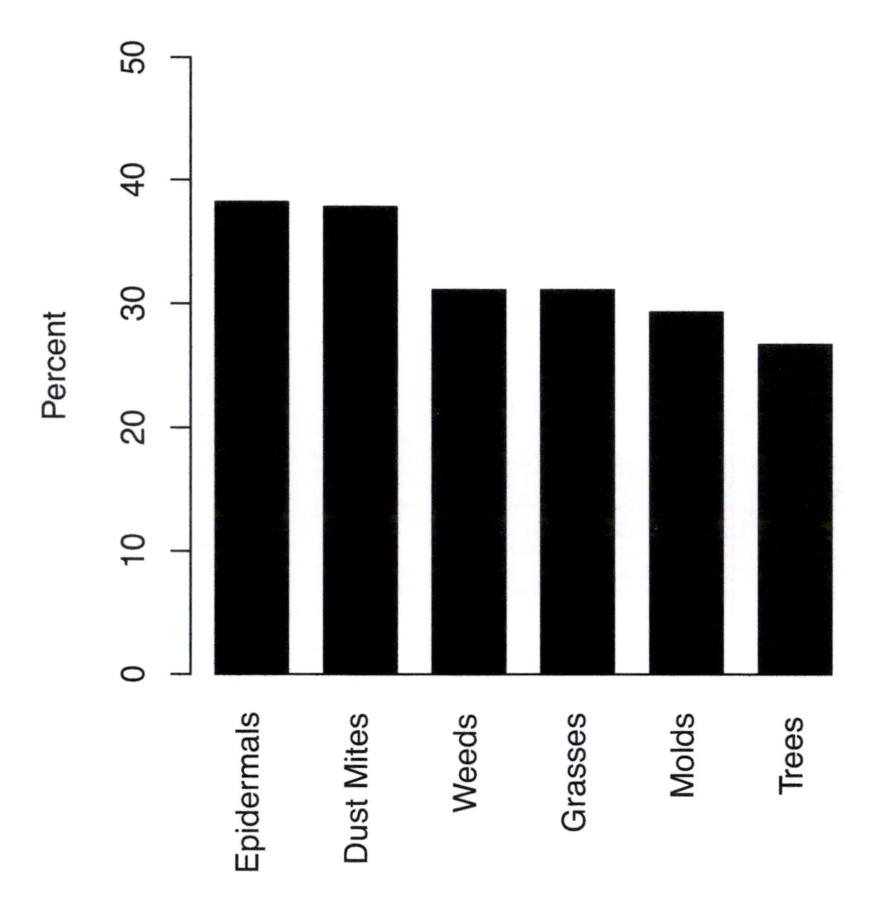

Figure 1.2 *Bar chart of prevalence of allergy in endoscopic sinus surgery patients*

things happening as well that result in the dot chart being a superior choice over the bar chart.

The advantages of the dot chart over the bar chart will be discussed in greater detail in Chapter 2. One impression for most viewers is that the prevalence of allergy being greater with respect to animals than plants and molds is more quickly perceived in the dot chart of Figure 1.1 rather than in the bar chart of Figure 1.2.

The use of spatial intelligence to retrieve data from a graph is an alternative to the language-based intelligence of prose and verbal presentations. For many individuals this skill is acquired through education. Potential users of this book are those new to the process of communicating information by visual displays and experienced researchers interested in exploring ways to better take advantage of the graphical medium.

The use of spatial intelligence affords an opportunity to the viewer of interacting with the data. As a result, the communication process becomes more direct and immediate through graphical displays. But the power of statistical graphics can be thwarted by badly designed graphics, poorly executed graphics, unfamiliar graphics, or some combination of these three factors. This is not to say that the use of unfamiliar graphics, such as the dot chart of Figure 1.1, should necessarily be avoided. Just as with language, with increased familiarity comes increased facility.

1.2 Learning Outcomes

When you complete this chapter, you will be able to do the following.

- Understand the concept of graphicacy and how it relates to but is different from literacy and numeracy.
- Know that the intended audience can impact the selection of a graphic for displaying data or summaries of data.
- Understand the principles underlying the design of an effective graphic.
- Understand the basic factors and processes involved in graphicacy and the design and execution of an effective graphical display.
- Understand that graphics produced by computers require graphical languages, even if coding by a user is not required, and that every graphical language can be described by a grammar.
- Recognize and be able to use a few statistics that can be used to quantify features of a graphic displaying data, a summary of data, or both.

1.3 Know the Intended Audience

It is essential when presenting graphical displays to have a reasonable expectation of the intended audience. Selecting an appropriate graphical display requires recognizing three distinct groups of audiences:

- the public;
- users of data in policy, administrative, legislative positions; and
- professional colleagues.

Most scientists and information professionals, regardless of level of statistical education and training, can count on being required to provide statistical graphics for each of these audiences at various times. The intended audience of scientists will most often be their professional colleagues.

The different target audiences require different kinds of graphical displays for the purpose of communication of the intended message. Communication with the public must not overlook the decorative and entertainment value of graphics. For example, the printed media uses graphs in every issue in these roles. Although decoration may seem anathema to scientists, it plays a useful role in motivating an audience to access important data.

Communication with administrators often busy with workaday issues must be presented in a clear and concise form that motivates the audience to attend to it and must compete effectively for audience time. Therefore broader entertainment values must be adopted without sacrificing credibility, introducing bias, appearing frivolous, or injecting condescension. On the other hand, visual communication with professional colleagues is typically technical with little attention to entertainment values.

For any intended audience, graphs can be used to describe the problem at hand, to identify the point of departure for the study, and to display the findings. Members of the audience ought to have the luxury of time to examine the graphs that are presented and formulate their own questions regarding the data being displayed.

Graphs can be presented to audiences in either a static or dynamic format. Publication of graphs in documents available in print or on webpages are two examples of the static format. The viewer is limited in the degree of interaction with the material being presented. The viewer can formulate questions and conceive of additional graphic displays to address those questions but is left to contact the authors of the document under study or other resources with a response time that is substantially less than immediate.

In a presentation to an audience in a room or in real time over the internet, with suitable statistical graphics software, the presenter and audience can interact dynamically. Audience members can request changes concerning an aspect of presentation or request additional graphical displays not previously prepared. If the presenter is receptive to these requests, the information can be displayed according to the manner of the request. In a collaborative dynamic environment, the distinction between presenter and audience can become sufficiently blurred so that the presenter becomes a facilitator of the graphical presentation and a coordinator of the ensuing discussion.

It should be noted that it is possible to mix elements of the static and dynamic formats. An example of this is the presentation to an audience of graphical material on *PowerPoint* slides. The audience can receive a verbal response from the presenter but visual or graphical interaction is limited to the material prepared in advance. This deficiency can be offset somewhat by the presenter preparing additional graphical material in anticipation of questions from the audience and displaying that material in response to questions but only as required. A much greater degree of interaction of the audience is possible if a link is provided on one of the *PowerPoint* slides to graphics software and the data.

1.4 Principles of Effective Statistical Graphs

A statistical graph is a visual representation of data. This could be the original raw data or a summary of the data produced by statistical analysis, or a combination of the two. Production of an effective statistical graph requires a certain degree of familiarity with quantitative methods from the disciplines of statistics and mathematics as well as a certain degree of exposure to concepts of the visual arts. Conceivably it is possible to assign two marks to any statistical graph: one for technical merit and another for artistic interpretation. It is quite possible for two graphs to display the same statistical data with equal technical merit and the preference between the two to be decided by artistic merit. This section reviews and discusses the components of the technical and artistic aspects of effective statistical graphs.

Inherent to the production of an effective statistical graph is the quality and quantity of the data to be displayed. No amount of statistical expertise or artistic design can rescue a tiny collection of data or a badly fitted model. Given that the number of statistical graphs published (either in print or electronically) worldwide in a given year is estimated to be at least 2 trillion (Tufte [123]), it would not be surprising to find not less than a few attempts to rescue lousy data or statistical models with graphs cleverly designed for maximum visual impact.

Statistical methods typically involve *data reduction*, also known as *data compression.* As a definition by way of example, data compression occurs when one calculates the mean and standard deviation of a random sample of size 30 and does no further analysis with the original observations.

Graphical methods can offset a possible loss of information and lead to *data expression.* Through the use of a plot that presents all the observations of a random sample of size 30, statistical graphics can be used to reveal structures in the data that are missed if only the mean and the standard deviation of the sample are reported in the text of a document. Both data reduction and data expression should be used together to look at data from different perspectives. Indeed, the addition of a graphical representation of the mean and standard deviation to a plot of all observations can simultaneously fulfill the roles of data expression and data compression with a visual check of the representativeness of the form of data compression under consideration.

One of the underlying principles of effective statistical graphics is that there ought to be a formula or prescription for the layout of a figure presenting statistical information.

1.4.1 The Layout of a Graphical Display

According to Burn [16], the elements of form can be separated according to categories that are called *scale* and *graphical frame.* Scale refers to the units of a numerical or categorical system represented along an axis. Considerations with respect to scale include:

- selecting a scale to cover the extremes of the data;
- selecting an appropriate transformation of scale for a quantitative variable, such as logarithmic, exponential, or none at all;
- selecting an ordinal scale for a categorical variable.

Improperly addressing these points regarding scale can lead to misrepresentation or suppression of certain features of the data. This can occur either by an act of omission or commission, that is, incompetence or incapacity with respect to considerations concerning scale, or a deliberate effort to mislead. It goes without saying, the data or curves representing models of the data must be accurately plotted for this is also a potential source of misrepresentation.

The term "frame" is more commonplace among graphic artists in referring to the general aspects of layout of a figure. Among those employing statistical methods in survey sampling the term *frame* denotes a list of members of a population. To avoid confusion, the term *graphical frame* will refer to the layout of a figure. Considerations with respect to graphical frame include:

- deciding on whether to frame the figure region in a box;
- deciding on whether to provide a legend;
- deciding on whether or how to provide a title, a footnote, or a caption;
- deciding on whether or how to provide one or more axis lines;
- deciding on whether or how to label one or more axis lines;
- deciding on whether or how to place tick marks on one or more axis lines;
- deciding on whether or how to provide labels for tick marks;
- deciding on whether or how to provide grid lines;
- deciding on whether or how to provide reference lines;
- deciding on whether or how to provide reference labels for lines, curves, points, or aspects of a display; and
- selecting plotting symbols.

Implicitly included in many of the points above is the choice of font and font size for printed characters, including plotting characters, and line thicknesses.

With regard to the labeling of axes, points, curves, or other aspects of a display including titles and subtitles, the guiding principle should be to avoid ambiguity through clear labeling.

The list of considerations for graphical frame is longer than the list for scale. The considerations for graphical frame tend to be more consistently overlooked than the considerations regarding scale because users of statistical software will typically rely on the default choices rather than decide on a particular form of layout and override the defaults. Blind reliance on software defaults can lead to users ignoring the role of aesthetics in the determination of the graphical frame. Aesthetic choices are to be distinguished from choices of scale for which the omission of data points or an inappropriate choice of scale transformation can be ascribed to the black and white outcome of right or wrong.

Collectively, a list of stipulations for scale and frame for a given figure constitute the *plotting convention*. Typically, researchers don't formally encounter plotting conventions unless preparing graphical material for publication in periodicals. Researchers can also encounter plotting conventions, also called *plotting standards*, for either presentation or publication for corporations, government, or other organizations.

1.4.2 The Design of Graphical Displays

Although beauty is in the eye of the beholder, there are several intrinsic principles that serve to segregate effective statistical graphical displays from ineffective ones. Easy to read labels, plotting only necessary data, and the effective use of color are desirable features of any statistical graph. Burn [16] presents and describes the **ACCENT** rule for the design of graphical displays. It is a simple and elegant mnemonic for six principles of effective graphical communication:

A pprehension;

C larity;

C onsistency;

E fficiency;

N ecessity; and

T ruthfulness.

Each of the words whose first letter is chosen to form the acronym **ACCENT** is in essence a title for a set of questions on one of six topics. The relevant questions for each of the six topics concerned with the form and content of a graphical display are as follows.

Apprehension concerns the ability to correctly perceive relations among the data. One must examine whether the elements of the graph interact to maximize perceptions of the information it contains. A viewer ought to be impressed with the substance of the data and not the artistry of the graphical design or something else.

Clarity concerns the ability to visually distinguish the elements of a graph. One must examine whether the most important elements of the graph are visually prominent and whether it is possible for the viewer to examine the data at different levels of detail ranging from the perspective of an overview to the fine details, or for the viewer to be encouraged to examine the data at different levels of detail.

Consistency concerns the ability to interpret a graph based on exposure to similar graphs in the past or a reasonably universal standard. One must examine whether the graph contains new elements that require elaboration and whether that elaboration is available to the viewer in an accompanying text.

Efficiency concerns the ability of a graph to portray data in as simple a way as possible. One must consider whether another type of graphical display could have been chosen instead. Additional considerations are whether some elements of the graph can serve more than one purpose, or that many numbers can be presented in a small space. An efficient graph ought to present the greatest number of ideas in the shortest time with the least ink in the smallest space.

Necessity concerns the need for the graph or certain of its elements. The graph must serve a clear purpose with respect to either description, exploration, tabulation, or decoration. One must consider whether the information would be better presented in tabular form or not at all. One must make sure that the graph is strongly connected with the verbal discussion and statistical descriptions in the accompanying text. One must consider whether certain artistic aspects of the graphical design are necessary; for example, one must consider whether the figure must be enclosed within a frame or whether the frame itself would just be a source of unnecessary clutter.

Truthfulness concerns whether the data are correctly plotted in a well-defined coordinate system. One must consider whether the positions of the plotted points or curves are correct and also whether the scale is correct. Above all else, it is absolutely necessary to avoid distorting the story that the data can tell.

When producing a graph, it is a worthwhile exercise to consider whether the graph satisfies each of the six principles represented by each letter in the acronym.

The real point of graphic design, which comprises both pictures and text, is clear communication. In viewing graphics produced by statistical software packages one should not be an uncritical consumer. Graphics ought not be merely cosmetic. When they are clear and consistent, they contribute greatly to ease of learning, communication, and understanding. The success of graphic design is measured in terms of the viewer's satisfaction and success in understanding.

1.5 Graphicacy

Graphical displays have the capacity to transmit data at the pace and interest level of the viewer. This does mean in audiovisual presentations that the presenter must make the effort to interact with the audience and to allow them control of the pace.

A potential limitation of graphical display is the portrayal of a facet of reality, the data, on a two-dimensional surface. For the most part, graphical images are depicted on a two-dimensional Cartesian coordinate plane. In a subsequent chapter, we will examine techniques that allow the illusion of three-dimensions on a two-dimensional surface through the use of geometric projections and motion.

While literacy is the ability to convey or receive written information and numeracy is the ability to convey or receive numerical information, graphicacy is the ability to convey or receive graphical information.

At the lowest levels: literacy is concerned with using words to communicate written instructions; numeracy is concerned with using numbers to communicate counts of items in an inventory; and graphicacy is concerned with using graphical symbols to communicate counts of items in an inventory. At higher levels: literacy encompasses criticism and analysis of literary composition; numeracy encompasses algebraic expression and analysis of mathematical and statistical models of the world around us; and graphicacy encompasses the tools of visual imagery to reveal features in data and the fidelity of mathematical, statistical, and even literary models.

In the early stages of formal education, a hierarchy for instruction is present with respect to literacy, numeracy, and graphicacy. Instruction in literacy and numeracy is affected through distinct courses in language, or languages, and arithmetic. Almost seamlessly, literacy and numeracy become interwoven to different degrees in courses in the social, health, and natural sciences. In courses where literacy and numeracy mix, graphical techniques are introduced effortlessly. But for most of the educational process, graphical methods are presented haphazardly in an largely unconscious manner.

Without an apparent discipline-specific home, the basic principles of graphical presentation are typically not formally taught. Perhaps considered to be merely details in the grander scheme of things, the basic principles of graphical display are left to the side. Perhaps it is hoped that the basic principles are absorbed unconsciously through immersion as in second-language instruction or the whole-language approach to literacy. If one is to remain only a receptor of graphical information, then this is likely adequate. When one becomes a provider of graphical information it is rather more important to grasp the basics. Typically, formal exposure to the principles does not occur until the post-secondary level or possibly the postgraduate level of education.

Anecdotally, I recall watching in horror as my masters' thesis supervisor rebuked a distinguished visitor for a particularly bad graph during a presentation. The axes labels were far too small to be legible, a legend for the symbols was needed, and the image was so overlaid with symbols and contours that one couldn't comprehend the main points to be made. The tirade lasted several minutes and included an assault on the presenter's professionalism for presenting such a truly awful slide to colleagues and impressionable postgraduate students who might be led to consider that such reprehensible conduct was acceptable. After a prolonged silence in which the distinguished visitor endured the glares of his colleagues and the stunned silence of students, an apology was proffered by the visitor and the presentation continued with more apologies for even more bad graphics.

On an occasion many years later at a conference, after enduring a series of particularly bad slides, I politely noted the illegibility of axes labels and tiny plotting symbols. Afterward, I was taken aside by a couple of my colleagues and told that I was rude and had missed the big picture. Truly, beauty is in the eye of the beholder.

A few points can made regarding these two anecdotes with regard to graphical principles. Firstly, there is a considerable degree of variation of adherence even among scientists. So it is important that the provider of the graphical display know the intended audience. Secondly, a reviewer can seize on a sloppily drafted figure as an example of professional or intellectual laziness and extrapolate this opinion to be representative of the whole work. For the sake of a single poorly designed or executed figure, it is not worthwhile to have one's intellectual integrity assailed in front of colleagues or superiors. Finally, we all break the rules. But if done in private, you won't get caught.

To realize the full potential of graphical displays, it is necessary to grasp the theoretical framework of graphicacy. One cognitive model for graphicacy depicted in Figure 1.3 involves three distinct processes:

- the production of graphical information;
- the consumption of graphical information; and
- the increase in awareness or knowledge of a subject.

The arc lengths in Figure 1.3 are to be taken to be representative of the ranking of the duration of the time intervals associated with each process of any given cycle of graphical apprehension. Except in the simplest of circumstances, it ought to be appreciated that an audience may have more questions about the data that may require the generation of further charts or graphical figures.

With respect to roles, the producer likely will be the sole member of the audience, at least initially, before the graph is presented to a colleague or a larger audience. Regardless of the assignment of roles, the goal during each cyclical iteration of the process is that the participant or participants ought

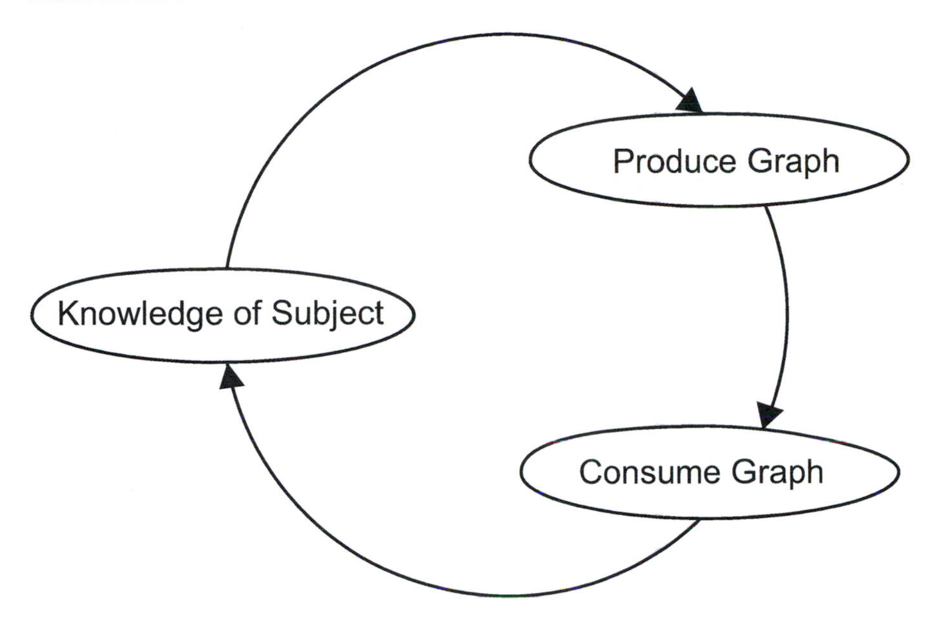

Figure 1.3 *Process of graphicacy*

to have increased their knowledge of the subject of the graph and possibly developed further questions. Within each cycle the producer ought to assemble more data or design additional displays, or both, to present to the consumer.

In an initial scientific investigation, typically the producer and consumer are one and the same person who can take advantage of a dynamic graphics environment to explore a range of different questions concerning the data. In a static-format presentation to managers or administrators, it is imperative to explore possible options with respect to generating additional displays with an eye to keeping the number of cyclic iterations minimized given the time constraints faced by all present and requisite time period until the next meeting or the time periods expected to elapse between subsequent meetings.

The central goal of graphicacy is to increase the base of knowledge about data while minimizing the number of iterations (and, therefore, the time) required to communicate graphically.

The milestones on the road to success in the graphical display of information are:

- the collection of meaningful data;
- the design of displays that reveal the data; and
- the accurate execution of the graphical displays.

Achieving success in graphical display may place requirements on a researcher beyond what the researcher perceives to be as standards of practice if the researcher is unaware of the six principles in the **ACCENT** rule. Repeated drafting and careful evaluation of graphical displays in the time-honored process of trial-and-error are required.

For many researchers there is a need to shift focus from the rigors of data collection and statistical analysis to the data themselves. There is also the need to attend to the mundane but important details of graphical presentation of the data. This can be perceived as less intellectually stimulating and therefore given less attention. But don't expect adversaries or critical colleagues in the audience to let anything slip, especially in settings where the presentation of graphical displays is always under scrutiny.

The data must be sufficiently meaningful to the viewers so that they are motivated to take an active role in the process and look more deeply into the topic. As Tufte [124] wrote: "If the numbers are boring, then you've got the wrong numbers." To this it can be added that if the statistical graphic is boring, then you've got the wrong chart.

For processing a graphical display, the display must be organized into perceptual units that are used in short-term memory. But short-term memory has a very small capacity of about four bits based on research in neuro-physiology. That is, humans have a RAM capacity of 0.5 byte. Simple hand-held calculators have a greater short-term memory capacity. Humans, however, gain the edge over hand-held calculators in processing speed and the capacity for creative cognition.

To process volumes of information larger than 0.5 byte, long-term memory must be used. For analyzing a graphical display, the human equivalent of a hard drive or flash memory stick is needed upon which is stored an algorithm for the analysis of a figure similar to that being viewed.

Conceptual representations for the storage locations of analysis algorithms for different graphical displays include names (for example, bar chart, pie chart, or scatterplot) or thumbnail sketches of representative figures could be used. Such thumbnail sketches will be referred to as *graphical icons*.

Recall of a graphical icon from long-term memory is required to process information from a new graphical display. Current theory about processing data seems to support the proposition that greater volumes of data may be presented more effectively with graphical icons stored in long-term memory. According to Kosslyn [77]: "If one has never seen a display type before, it is a problem to be solved—not a display to be read." Fewer data ought to be graphed when the graphical icon is less familiar to the audience. Consequently, scatterplots of two or more variables are less common in popular publications and the video medium but are widely used in scientific discourse.

A depiction of the retrieval process for graphical displays is given in Figure 1.4.

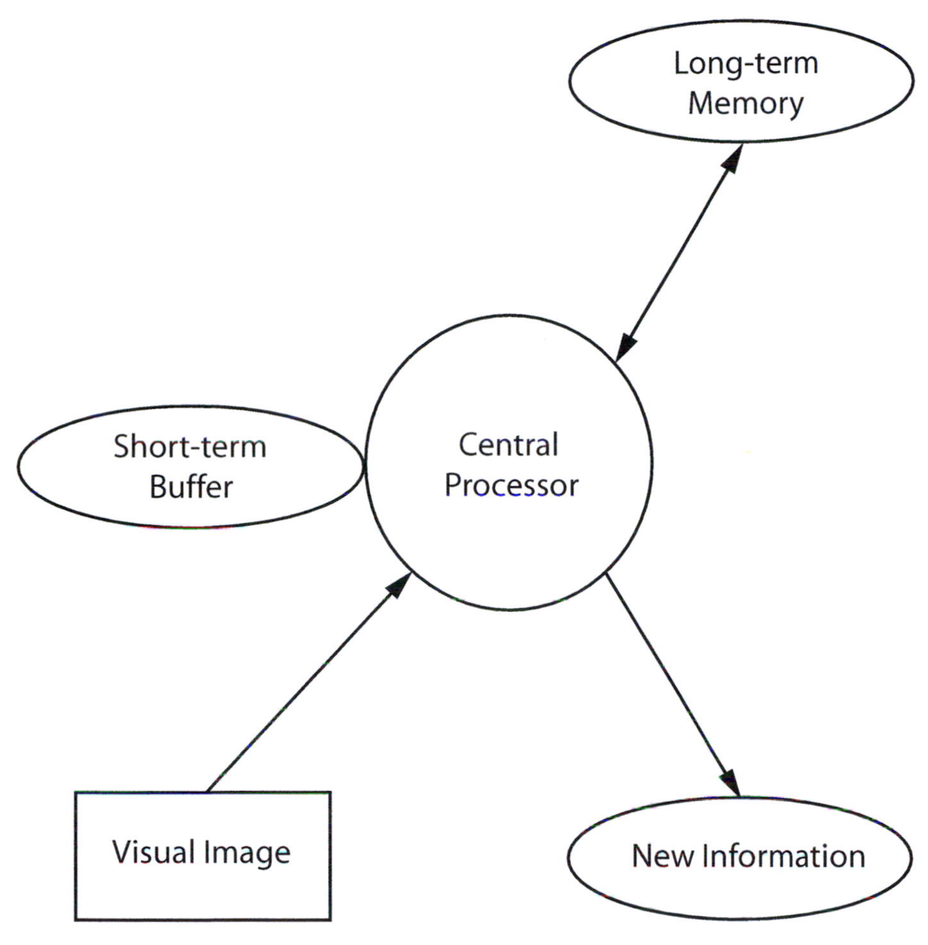

Figure 1.4 *The visual processing of graphical information*

It begins with the viewing of a visual display. An indication of the quick connection between short-term memory and the part of the consciousness of the human mind that functions as a central processing unit for logic and arithmetic operations is visually indicated by the direct connection between the central processor and the short-term memory in the figure. The longer transmission times for conveying information from the visual image and between long-term memory is indicated by the line segments. Note that the connection from the visual image to the central processor is one-way, whereas the connection between long-term memory and the central processor is two-way. An important aspect of any graphical presentation is to accommodate sufficient cycles for processing and understanding the information presented in an image with a pre-existing graphical icon in long-term memory. The conclusion of the process is depicted by a one-way arrow leading, hopefully, to new information about the data under study.

Theory suggests that greater familiarity with abstract graphical forms leads to increased capacity to use graphics with greater sophistication. Exposure to a large number of different types of graphical displays, such as presented in this book, will therefore be doubly beneficial. The odds of encountering a new type of display will be reduced on one hand and the time required to decipher a new type of display never before seen will be reduced on the other.

In summary, two factors underlie graphicacy:

- viewers are capable of retrieving data from well-executed graphs; and
- viewers can learn to use more sophisticated graphics and store their icons in long-term memory for future use.

These are the key considerations for those creating graphical displays either according to pre-existing graphical conventions or developing new ones.

1.6 The Grammar of Graphics

The title of this section is the title of a book by Leland Wilkinson [137] published in 1999. For many interested in the theory of statistical graphics, this book was literally an eye-opener. As noted by Wilkinson, the English word *grammar* stems from the Greek word *γράμμα* for letter. The Greek word for *grammar* is *γραμματική*. The first written grammar, for the Sanskrit language, appeared in the sixth century before the common era (BCE) and is attributed to Yaska. The oldest treatise on the grammar of a European language is acknowledged to be *Τέχνι Γραμματική* by Dionysius Thrax in the second century before the common era. The title *Τέχνι Γραμματική* literally means the *Art of Letters*. In many ways, Wilkinson's book is about the *Art of Graphics*. The English word *writing* in Greek is the word *γραφή*, which is the source of the English word *graph*. Crafting a graph can be considered a parallel to writing a paragraph.

Symbols are often used on a two-dimensional surface in space to construct words, numbers, and graphics. It stands to reason that grammars can be constructed for all three and play an important role in the understanding and teaching of literacy, numeracy, and graphicacy. In linguistics, grammar is concerned with the organization of sounds (phonology), the formation of words (morphology), and the rules governing the composition of sentences (syntax). In his book, Wilkinson [137] writes: "This book is about grammatical rules for creating perceivable graphs, or what I call **graphics**." A reasonable question is whether Wilkinson's rules encompass what can be considered to be graphical equivalents of phonology, morphology, and syntax? Or do Wilkinson's rules focus on equivalents to the production of sounds (phonetics), the meanings of words (semantics), and the contribution of context to meaning (pragmatics).

To continue with the language-graphics analogy, it is important to recognize that computer-drafted graphics rely for the most part on written instructions provided by a user. Point-and-click interfaces do exist in statistical computer systems. One such interface is `Rcmdr` in R [97]. Options in these interfaces, however, are generally limited with respect to what Burn [16] in 1993 referred to as scale and graphical frame, which were described in Section 1.3. For greater control it is necessary to resort to written instructions. So every statistical computer system offers at least one language for conveying instructions on how to construct a graphic. R offers at least three languages. One for each of the `base`, `grid`, and `ggplot2`/index`ggplot2` packages. With these three different languages there is a distinct grammar for each. But as there are common aspects of grammar for all natural languages, it is possible to conceive of a common core of grammar for all graphics. This is what Wilkinson [137] has done.

Authoring a graphic in a graphical language presents the same challenges a writer faces. Unlike a literary composition in which a reader can cope with minor spelling errors, wrong form of a word, or incorrect word order; making these mistakes in communicating with a statistical graphics package generally leads to a puzzling error message and a blank screen for the author. The consequences of an error in graphical grammar are typically more severe.

The language-graphics analogy does break down at the point of view of the individual receiving the communication. In speaking and writing, word order is controlled by the creator. This is not the case for a graphic where the viewer gets to pick and choose where to start with a graphic and where to go next. In written and spoken communication correct syntax is valued by both the author and the recipient. But in graphical communication, the sweating and concentration by the author to get the written syntax correct for the computer is, perhaps unfortunately, not of much concern nor an issue for the recipient. This is quite unlike a natural language among humans.

Looking through Wilkinson's [137] book, one sees a lot of syntax, one block for each figure. The following quotation is taken from page 39 of this book in the section entitled "Notation."

> Each graphic for this book is accompanied by a symbolic specification. As I have indicated, this specification is not a computer language or command system, although the syntax has been designed to fit naturally into a Java string-tokenizer and extended markup language (XML) environment.

An *affix grammar* is essentially a grammar of grammars. It is used to describe the syntax of programming languages, based in part on the grammar of natural languages used by humans for communication. An *extended affix grammar* is a formalization of grammar for both programming and natural languages. Wilkinson [137] has created an affix grammar for graphics. In doing, he has also created a syntax, with its own grammar rules, for describing, or marking up, a graphic.

In the Preface to his book in 1999, Wilkinson [137] notes a collaboration in the 1990s with Daniel Rope and Dan Carr to produce a graphics production library in Java called GPL. Wilkinson, Rope, Carr, and Rubin [139] in an article published in 2000 and entitled "The Language of Graphics," described "a system, called the Graphics Production Library (GPL), that implements a language for quantitative graphics." So is GPL a syntax for an affix grammar for graphical languages, a library of subroutines in Java for executing graphics, or another language for creating graphics? The answer is probably yes to all three questions.

In 2005, the second edition of Wilkinson's [138] book was published with 162 more illustrations than the first edition's 248, 6 new chapters, and 282 pages longer than the first edition's 408 pages. One new chapter is 18 on "Automation," which is devoted to discussing two languages. The first is GPL, which now stands for the Graphics Production Language, and the second is the Visualization Markup Language (ViZml), both of which are now supported by IBM after acquiring the rights to the SPSS statistical software package.

ViZml is an extensible markup language (XML) for describing graphics. XML itself is not a language *per se* but a tree-based specification structure. While XML can remove stylistic details from a language like GPL, it can allow for the specification of a large number of stylistic details in a concise manner. A designer can create design templates in ViZml for a GPL programmer to use.

With the second edition, Wilkinson [138] in Section 2.3 Notation changed the fonts used for the keywords in the syntax of GPL. Yet GPL is not case sensitive except for characters inside double quotation marks.

In an article in 2010, Hadley Wickham [133] proposed an alternative grammar to Wilkinson's based on the idea of building up a graphic from multiple layers. The gg in Wickham's `ggplot2` does not represent the grammar of graphics of Wilkinson but the *layered grammar of graphics* of Wickham [133]. For many, R is not just a statistical analysis system but also a language. When the `UNIX` operating system was created, a programming language for statisticians called `S` and a programming language for computer scientists called `C` were created with it at Bell Labs. As `C+` and `C#` are descendants of C, so `S+` and R are descendants of `S`. `ggplot2` is an open source implementation using `grid` functions in the R language.

Object-oriented design (OOD) is discussed early in both editions of Wilkinson [137, 138]. In a nutshell, OOD in a computing environment treats data, programs, subroutines (functions in `R`) and output as objects to be manipulated. This is quite a change for a programmer writing procedures. Wilkinson sought to treat graphics as objects to be manipulated. Languages such as JAVA and R make OOD easy. Wilkinson chose JAVA for GPL. Wickham chose R for `ggplot2`. Conceivably, a designer could create design templates in ViZml for a `ggplot2` programmer to use.

With respect to the topic of layers, consider the principal components—these can be considered to be keywords—in Wilkinson's affix grammar for graphics. These are as follows.

- SOURCE identifies the source of the data for the graphic.
- DATA identifies the variables in the data used in the graphic.
- TRANS provides information about the functional transformations, if any, of data.
- ELEMENT describes the various layers in the graphic.
- SCALE describes the scale of each axis.
- GUIDE describes the axes themselves and any legends.
- COORD describes the coordinate system (rectangular, polar, or other) for the axes.

Also optionally available in GPL are DO for do-loops and do-while conditional execution, PAGE, and GRAPHICS for starting and ending pages and graphs, respectively. These are not necessary if one elects to rely on defaults.

While statements using keywords of Wilkinson's affix grammar can be used to mark up, that is, describe, any graphic, when they are syntactically correct in GPL they can be used to produce, or reproduce, the graphic when the original data is provided.

The ELEMENT component allows someone to use geometric graphing functions to create or mark up layers of data in a graphic and choose the aesthetics for each layer, such as the choice of symbol or color. The geometric graphing functions allow the creation or mark up of points, lines, smoothed curves, bars, pies, trees, histograms, frequency polygons, maps of the earth, contours, and chloropleths and control or reporting over a host of aesthetics relating to plotting symbol size, curve thickness, and color.

In general, as noted by Wilkinson [138], *aesthetics* concerns attributes such as color, shape, sound, and so on, and the principles for employing same. Aesthetics are a matter of personal taste. Not everyone will agree. Wilkinson's [138] second edition entitles and devotes Chapter 10 to "Aesthetics." The discipline of psychophysics attempts to rationalize aesthetics on a scientific basis. To the extent that this can be done comes down to "the beauty being in the eye of the beholder."

The ELEMENT component also provides for the *faceting* of variables in a dataset. An example of faceting would be using two continuous variables to plot symbols in a Cartesian coordinate plane and depict a third variable by using different symbols or colors, or both. This variable could be categorical or even a continuous variable binned into different ranges to produce different symbols or colors. Alternatively, instead of using different colors, a separate plot in a series of panels could be produced in the same graphic to facet the data. Faceting inherently involves the use of layers but also the coordinate system so the COORD component is also involved.

In any natural language of humans or an artificial language created by humans, word order is important. Many but not all natural languages use subject-verb-object order for statements. In Romance languages, there are grammatical rules for which adjectives precede and which adjectives follow a noun. Even if the word order rules are not followed, a listener or reader can likely understand what is meant. Similarly, if GPL or `ggplot2` are being used to mark up an existing graphic, a reader will be able to understand. But if GPL or `ggplot2` are being used as graphics languages on a computer to produce a graphic, then certain layers must be executed in a specific order. Wilkinson [138] and Wickham [133] do not discuss this in any detail but this is true for any computer graphics language. Implicitly there is a command order syntax for GPL and `ggplot2`, which when used for producing graphics, needs to be acknowledged and reflected when GPL and `ggplot2` are used as markup languages for graphics.

To produce a plot with GPL, there needs first to be a call to GRAPHICS to create an electronic canvas for the graphic, perhaps followed by a call to PAGE. This would be followed by COORD to create the coordinate system in a page on the electronic canvas. This is not synonymous with the creation of lines, tick marks, symbols for tick marks, and a word or words for the labels to name the axes—this can be done by GUIDE. Before anything else can occur, SOURCE must be used to identify the source of the data followed by DATA to identify the variables. Then TRANS needs to be used to identify any transformation of the data followed by SCALE for each axis. After which ELEMENT can be used to add layers of plotted information without strong order restrictions. The point being made is that there are multiple layers, some with a required strong ordering, in any graphic. When drafted by hand, creation of the coordinate system needed to be coincident with laying down the axes and tick marks in order to place the data by hand. This is no longer required in computer graphics, even in the `base` graphics package of `R`, as the default axes can be turned off when plotting the data with custom axes and tick marks added in subsequent calls.

Wickham's [133] article of 2010 states that it "proposes an alternative parameterization of the [Wilkinson's] grammar, based around the idea of building up a graphic from multiple layers of data." Wilkinson's GPL, however, does permit a graphic to be built up from multiple layers of data. Arguably, Wilkinson's grammar was designed not just to do this but also to serve as an affix grammar for statistical graphics. Wickham's grammar does differ from Wilkinson's in its arrangement of components, its hierarchy of defaults and the fact that `ggplot2` is embedded within the statistical language `R`.

Wickham [133] wrote the following.

> By relying on other tools in `R`, `ggplot2` does not need three elements of Wilkinson's grammar: DATA, TRANS, and the algebra.
>
> DATA is no longer needed because the data are stored in R data frames; they do not need to be described as part of the graphic. TRANS can be dropped because variable transformations are already so easy in `R`; they do not need to be part of the grammar. The algebra describes how to reshape data for display and, in `R`, can be replaced by the `reshape` package (Wickham 2005).

Understandably, in developing `ggplot2` as a tool within `R`, Wickham is choosing to not reinvent the wheel. Counterparts to GPL algebra, TRANS, and DATA have not been created in `ggplot2` because they already exist in `R`. This is fine if one considers `ggplot2` to be only a statistical graphics programming language. But if `ggplot2` is to be a grammar of graphics then it is incomplete unless these features from R are also included.

Other structural differences between GPL and `ggplot2` include the following. The SCALE and GUIDE components in GPL are combined into a Scale component in `ggplot2`. The COORD component in GPL is bifurcated into the Coord and Facet components in `ggplot2`. But these changes merely represent a rearrangement of features with choices made on the basis of personal aesthetics.

Aesthetics are just as important to Wickham as they are to Wilkinson. In the second edition of the book *ggplot2 Elegant Graphics for Data Analysis*, Wickham [134] has multiple sub-sections concerned with aesthetics. In `ggplot2`, there is a function called `aes` (for aesthetics) that can be called from within the function `ggplot2` or from one of the functions called within `ggplot2` that perform tasks associated with layers in the ELEMENT component of Wilkinson's affix grammar for graphics. Allowing for setting of aesthetics at the level of the plot or in one of the layers, and creating the necessary algebra for doing so, represents an aesthetic choice.

The statistical computing environment chosen for this book is `R`. There are demonstrations in this book of three graphical packages in `R`: `base`, `lattice` and `ggplot2`. In the background is the R package `grid` developed by Paul Murrell [84]. Part I of Murrell's book, published in 2006, is concerned with the `base` package and Part II is concerned with the `grid` package. This book too has a pragmatic outlook. The choice between the `base` and `lattice` packages will be decided by ease of use. There will be periodic examples where similar graphs will be produced by `ggplot2`. Readers can decide whether the `ggplot2` code is to be preferred. There will be no examples with GPL or ViZml. But readers ought to keep in the back of their mind the utility and completeness of GPL and ViZml as markup languages for the affix grammar of graphics. Perhaps we all look forward to the day of automated translation amongst `base`, `lattice`, `ggplot2`, GPL, and ViZml.

Figure 1.5 *The Golden Section*

1.7 Graphical Statistics

The human eye, vision-corrected if necessary, can make visual distinctions with gaps as small as 0.1 millimeter at a distance of 0.5 meter. Equivalently, the human eye can discern as many as 100 points of intersection in a square 1.0 centimeter on edge consisting of 100 fine lines that are equally spaced and parallel to the edges. Few statistical graphs approach this degree of data density in practice.

Violations of the *Truthfulness* principle of the **ACCENT** rule can be assessed by the following statistic:

$$\textbf{Truth Coefficient} = 1 - \frac{\left|\begin{pmatrix}\text{size of effect}\\ \text{shown in graph}\end{pmatrix} - \begin{pmatrix}\text{size of effect}\\ \text{in data}\end{pmatrix}\right|}{\begin{pmatrix}\text{size of effect}\\ \text{in data}\end{pmatrix}}. \tag{1.1}$$

The goal with respect to the **Truth Coefficient** is to achieve a maximum value of one. Truth Coefficients less than 0.95 are to be considered suspect.

Closely related to the Truth Coefficient is the **Lie Factor** defined by

$$\textbf{Lie Factor} = \frac{\text{size of effect shown in graph}}{\text{size of effect in data}}. \tag{1.2}$$

The goal is to achieve a Lie Factor of one with Lie Factors of less than 0.95 or more than 1.05 indicative of distortion in the presentation of the data by the statistical graph. Compared to the Lie Factor, the Truth Coefficient is insensitive to the direction of the distortion.

If statistical graphs are to be greater in width than height, the issue to be decided is then how much greater the width is to be than the height. One rule for determining the optimum aesthetic rectangular proportions, albeit from the fifth century before the current era, is that of the so-called *Golden Section*, which divides a line such that the lesser line segment is to the greater as the greater is to the whole. From Figure 1.5, the symbolic interpretation of the Golden Section is:

$$\frac{a}{b} = \frac{b}{a+b}. \tag{1.3}$$

Figure 1.6 *Comparison of a rectangle 50% wider than tall (on the left) with a rectangle with dimensions defined by the Golden Section (on the right)*

Upon substitution of $x = b/a$ in the previous equation, one obtains the quadratic equation

$$x^2 - x - 1 = 0, \tag{1.4}$$

the only positive solution of which is

$$\frac{b}{a} = x = \frac{1+\sqrt{5}}{2} \approx 1.61803\ldots. \tag{1.5}$$

Taking h to represent the height of a rectangular figure and w to represent the width, a measure of the fitness of a statistical graph according to the rule of the Golden Section is given by

$$\textbf{Golden Ratio} = \frac{h}{w} = \frac{2}{1+\sqrt{5}}. \tag{1.6}$$

This third graphical statistic is unrelated to the first two. It is a measure of the exterior physical representation of a statistical graphic. Its use assumes as the first premise that the graphical figure is rectangular in shape. Because the human eye is well practiced in ascertaining deviations from the horizontal, the second premise to be accepted is that a properly executed statistical graph ought to take advantage of this effect and, therefore, be wider than its height. The Golden Ratio is inherently an artistic measure. It is a measure relating to both the *Apprehension* and *Efficiency* principles of the **ACCENT** rule.

The goal is to achieve a Golden Ratio equal to one. This statistic is directional as values of the Golden Ratio greater than one are indicative of the width being greater than that required by the rule of the Golden Section. While not desirable, this is tolerable. However, Golden Ratios less than one are more harshly judged.

It is awkward to deal with a height dimension multiplied by an irrational number such as $(1+\sqrt{5})/2$. One simplification would be to prepare a statistical graph that is 50% wider than tall. A comparison of rectangles prepared by this rule and the rule of the Golden Section is given in Figure 1.6. The Golden Ratio is approximately 0.93 when a graph is 50% wider than tall. It is left to the reader to decide whether this simplification is as aesthetically pleasing in comparison with the Golden Section.

1.8 Conclusion

Burn [16] notes the following citation to be found in the *Apple Human Interface Guidelines* [70]: "The real point of graphic design, which comprises both pictures and text, is clear communication. Graphics are not merely cosmetic. When they are clear and consistent, they contribute greatly to ease of learning, communication, and understanding. The success of graphic design is measured in terms of the user's satisfaction and success in understanding..." The context for this quotation is the development guidelines of Apple Computers for its Macintosh (Apple Computers, Inc., Cupertino, California) computer. Apple Computers did more than just introduce a graphical user interface into mass-produced personal computers with the Macintosh computer, it added the acronym GUI, for Graphical User Interface, into the vocabulary of the English language. It is noted that the quotation above is just as applicable to statistical graphics as it is to the MacOS (Apple Computers, Inc., Cupertino, California) operating system and its successors.

In 1983, Tufte [123] wrote: "Graphical excellence is that which gives to the viewer the greatest number of ideas in the shortest time with the least ink in the smallest space." He also noted in the same book: "Each year, the world over, somewhere between 900 billion (9×10^{11}) and 2 trillion (2×10^{12}) images of statistical graphics are printed." Given the passage of time since 1983, and the publication of material on the world-wide web (or internet) and inexpensive laser printers, 2 trillion statistical graphs likely now represents the lower confidence limit at best.

Further information, guidance, and a wealth of examples for researchers to contemplate can be found in the monographs of Tufte [123, 124, 125]. Other sources of equally valuable insight regarding statistical graphics can be found in the books of Cleveland [21, 23] and Wainer [130].

1.9 Exercises

1. For the 5-day period from Monday to Friday, clip with scissors or electronically cut articles from a daily newspaper or online news outlet that have graphs that display data. Review the text that accompanies the graphs.
   (a) List the titles of articles that use graphics.
   (b) Pick out the best example of a graph that illustrates the text.

(c) Analyze the graph in relation to the **ACCENT** rule.
(d) Could the graph stand alone and make the point without the text? Justify your answer.
(e) Explain how the graph visually depicts the information.
(f) Estimate the ratio between height and width of the graph. Was the Rule of the Golden Section apparently used?
(g) Suggest how the graph could be improved.

2. From the set of graphs in your answer to question 1, pick out the most interesting graph.
   (a) What brought the graph to your attention?
   (b) Write down the information that the graph communicates to you.
   (c) What questions do you have after viewing the graph?
3. Pick up a copy of a weekly news magazine and clip out the articles that use graphs.
   (a) List the titles of articles that use graphics.
   (b) How do these graphics compare to the graphics from a newspaper or online media outlet?
   (c) Pick out the best example of a graph that illustrates the text.
   (d) Analyze the graph in relation to the **ACCENT** rule.
   (e) Could the graph stand alone and make the point without the text? Justify your answer.
   (f) Explain how the graph visually depicts the information.
   (g) Estimate the ratio between height and width of the graph. Was the Rule of the Golden Section apparently used?
   (h) Suggest how the graph could be improved.
4. From the set of graphs in your answer to question 3, pick out the most interesting graph.
   (a) What brought the graph to your attention?
   (b) Write down the information that graph communicates to you.
   (c) What questions do you have after viewing the graph?
5. Use a web browser to view content on a news outlet website. The news outlet could be for a television station, radio station, or an internet-only news service.
   (a) Find an article that uses one or more graphics.
   (b) Print the article and its graphics.
   (c) Select and analyze a graph from your answer to part (b) in relation to the **ACCENT** rule.
   (d) Could the graph stand alone and make the point without the text? Justify your answer.
   (e) Explain how the graph visually depicts the information.

(f) Estimate the ratio between height and width of the graph. Was the Rule of the Golden Section apparently used?

(g) Suggest how the graph could be improved.

6. Review one issue of a journal in your field for the graphics that are used.

   (a) How many graphics are used?

   (b) Are they descriptive or analytical?

   (c) Do the graphs appear to contain more or less information than the graphs of the popular press?

   (d) Describe the clarity of titles and labels, purpose of the graphs, and visual appeal.

   (e) Pick out the best example of a graph that illustrates the text.

   (f) Analyze the graph in relation to the **ACCENT** rule.

   (g) Could the graph stand alone and make the point without the text? Justify your answer.

   (h) Explain how the graph visually depicts the information.

   (i) Estimate the ratio between height and width of the graph. Was the Rule of the Golden Section apparently used?

   (j) Suggest how the graph could be improved in ways other than already described.

7. In one issue of a journal in your field, randomly select 20% of the graphics.

   (a) Estimate the ratio between height and width of the graph for your random sample.

   (b) From your answer to part (a) estimate the proportion of graphics compliant with the Rule of the Golden Section. What do you infer regarding the adherence to the Rule of the Golden Section in this journal and in your field?

   (c) Based upon your answers to part (b), what is your opinion regarding the value of the Rule of the Golden Section? Comment.

8. With regard to your random sample of exercise 7 taken of graphics from a journal in your field, do any of the graphics appear to be purely decorative in nature? Are there aspects of decoration with the intent to grab the attention of readers? Comment.

9. Contrast and compare a grammar of graphics and an affix grammar of graphics.

10. With regard to Wilkinson's GPL or Wickham's `ggplot2`, consider the following questions.

    (a) Is GPL an affix grammar of graphics or is it more of a statistical graphics language developed from basic principles of an affix grammar for graphics. Discuss.

    (b) Is `ggplot2` an affix grammar of graphics or is it more of a statistical graphics language developed from basic principles of an affix grammar for graphics. Discuss.

   (c) Which of GPL or `ggplot2` are you more likely to use. State your reasons.

11. By way of assessing what you gained by reading this chapter, consider the following questions.
   (a) Have you encountered the term *graphicacy* before? What did you learn about graphicacy?
   (b) Have you encountered the **ACCENT** rule before? Do you think it likely that you will use it to frame criticisms of graphics presented to you? Do you think it likely that you will use it when you create graphics for yourself and others?
   (c) Discuss any new concepts or considerations regarding statistical graphics that you learned. Comment on whether these will contribute to you being a more critical consumer or producer of graphical displays in the future.

12. In preparation for reading subsequent chapters, consider all the different graphical displays that you already know. (You ought to repeat this exercise at the end of each chapter.)
   (a) Create, from memory only, a table consisting of the names and thumbnail sketches (that is, graphical icons) for all the graphical displays you produce or can recognize. Use only pen and paper for this exercise.
   (b) From your answer to part (a), count the number of different graphical displays you know.

Part II

A Single Discrete Variable

Chapter 2

Basic Charts for the Distribution of a Single Discrete Variable

2.1 Introduction

Charts that present summary data are often used to set the stage for the presentation of the results of a study. Charts are often used as an eye-catching alternative to statistical tables. Perhaps the most frequently encountered statistical chart is that for depicting the distribution of a single discrete variable.

An example of a discrete variable is human eye color, which is classified according to the following comprehensive list: {amber, blue, brown, gray, green, hazel, red, violet}. The eye colors red and violet are observed in albinos. Another example is the number of heads in a simple game of chance consisting of 4 coin tosses. The number of heads in 4 coin tosses can take the value 0, 1, 2, 3, or 4. A *discrete variable* places an individual observation into one of two or more categories.

As a coin can land either heads up or tails up, there are only 5 discrete possible outcomes for the number of heads in 4 tosses. Intermediate values such as 1.5 or π are not possible outcomes in this example. For the example of human eye color, hazel is an intermediate color between brown and green. So the determination of eye-color category would appear to involve a degree of qualitative judgment. By these two examples, discrete variables can be qualitative or quantitative.

Because not all eye colors can be matched with colors in the spectrum of the rainbow, the categories in this example are determined by name. For the example of the coin toss, there is an ordering in the sense that 4 heads are more than 2 heads. In fact, we can judge the interval between these two categories as being equal to 2 heads. But the arithmetic operations of addition and averaging also make sense. In comparison, 4 heads are twice as many as 2 heads. So the scale of measurement is *ratio* for the count of heads in 4 coin tosses. The previous three scales of measurement just considered are *nominal*, *ordinal*, and *interval*, respectively.

A common example of an ordinal scale is personal income if reported to the following five fixed categories: {less than \$25,000, at least \$25,000 but less than \$50,000, at least \$50,000 but less than \$75,000, at least \$75,000 but less than \$100,000, \$100,000 or more}. If this set of five categories were augmented beyond \$100,000 by as many categories, each \$25,000 wide, as needed to accommodate all levels of personal income, then the resulting scale could be considered to be interval.

The number of categories need not be finite. As a variation on the coin-toss example, consider counting the number of tails until the first head appears. If we enumerate the possible outcomes, we come up with the list $\{1, 2, 3, \ldots\}$ that is referred to as the *natural numbers*. As long as the list of categories for a random variable can be put into a one-to-one correspondence with the natural numbers, the random variable is discrete.

The consideration of the distribution of a single nondiscrete random variable is left to Chapters 4 and 5. But in any case, the primary purpose is descriptive for graphical displays that depict distributions of the population of values for a single variable. Such plots can serve several purposes:

- to orient the viewers to the topic or population under study;
- to stress an issue central to the study;
- to provide an overview of the population;
- to expose a commonly held misconception about the population;
- to justify concentration on a subpopulation; or
- to allow viewers to formulate questions that the researcher will address.

A chart can be drafted for just one of these purposes or to accommodate several simultaneously. A sequence of charts can be drafted to step the audience through an argument.

In this chapter we consider three basic charts for plotting the distribution of a single discrete variable and begin with an example.

2.2 Learning Outcomes

When you complete this chapter, you will be able to do the following.

- Know what a dot chart is and be able to create one with either the `graphics` or `ggplot2` packages in R.
- Know what a bar chart is and be able to create one with either the `graphics` or `ggplot2` packages in R.
- Understand the definition of the data-ink ratio and know how to use this statistic to select one graphical display over another.

Item	Amount (United States dollars)
Overall policymaking, direction & coordination	718,555,600
Political affairs	626,069,600
International justice and law	87,269,400
International cooperation for development	398,449,400
Regional cooperation for development	477,145,600
Human rights and humanitarian affairs	259,227,500
Public information	184,000,500
Management and support services	540,204,300
Internal oversight	35,997,700
Jointly financed administrative activities and special expenses	108,470,900
Capital expenditures	58,782,600
Safety and security	197,169,300
Development account	18,651,300
Staff assessment	461,366,000
Total	**4,171,359,700**

Table 2.1 *Appropriations in Resolution 62/327 of the United Nations General Assembly*

- See how the **ACCENT** rule can be used to assess charts for displaying the distribution of a single discrete variable.
- Understand the reasons for avoiding pseudo-three-dimensional bar charts, pie charts, and, even more so, pseudo-three-dimensional pie charts.
- Understand the reasons for preferring the dot chart over other graphic presentations of the distribution of a single discrete variable such as the bar chart and pie chart.

2.3 An Example from the United Nations

The financial operations of the United Nations are guided by a process that produces a budget for a two-year period that is referred to as a *biennium* by this organization of member states.

The budgeting process culminates in a plan produced by the *Fifth Committee* for each biennium that must be passed as a resolution by the General Assembly. Reported in Table 2.1 are the budget appropriations for the 2008–2009 biennium passed in Resolution 62/327 at the 79th Plenary Meeting on the 27th of December 2007. These amounts are reported to the nearest hundred of American dollars (referred to as United States dollars in United Nations documents).

The United Nations leadership would likely prefer to view the process as leading to a biennium budget as deterministic. Given the array of participants, constraints, and issues tugging in one direction or another, it is reasonable to conclude that the numbers presented in Table 2.1 are the result of a complex random process. For each one hundred dollars intended to be spent, there are fourteen possible nonnumerical categorical outcomes. For those responsible for expenditures at the United Nations, the distribution of hundreds of dollars to the expenditure accounts does result in fixed amounts being delivered on account. There is no variation to be reported on distribution of funds.

The items in Table 2.1 are listed in the original order as published in Resolution 62/327. The sequence of items corresponds to budget parts I through XIV. The only amount not included in the table, nor in the original tabulation of the resolution, is the amount of $75,000 appropriated for each year of the biennium from the accumulated income of the Library Endowment Fund for the purchase of books, periodicals, maps, library equipment, or other expenses of the library at the Palais des Nations in Geneva.

We see in Table 2.1 that the overall expenditures exceed four billion dollars. This is about sixty-six cents for every man, woman, and child on the planet. This is not a lot of money per person. With a little effort, it can be noted that the big-ticket item is the category entitled *overall policy making, direction and coordination*. The cost of political affairs is not that far behind.

The development account attracts the least amount of expenditure at just over 18.6 million dollars but this is dwarfed by the combined spending of $875,595,000 for international and regional development. To attempt to extract more information than this by just reading the table would appear to be a questionable effort. This leads to consideration of several different graphical means of depicting the data in Table 2.1 as follows.

2.4 The Dot Chart

An example of a *dot chart* with the spending estimates of Table 2.1 is given in Figure 2.1. Cleveland and McGill [25] refer to the introduction of the dot chart in a Bell Laboratories memorandum of 1983.

From the graphical examples of Cleveland and McGill [25], the plotting standard for the dot chart is inferred to be as follows. The categories, or values, are listed in a vertical column with labels to the left. A horizontal dotted line is adjacent to the category with the value indicated by a large dot. The horizontal and vertical axes are framed in a rectangular box with the category labels and the horizontal coordinates, with tick marks, displayed outside the framing box. There are no tick marks for the vertical categories, presumably because this is not required due to the horizontal lines for the categories within the box.

Figure 2.1 has been plotted using the `dotchart` function that is in the

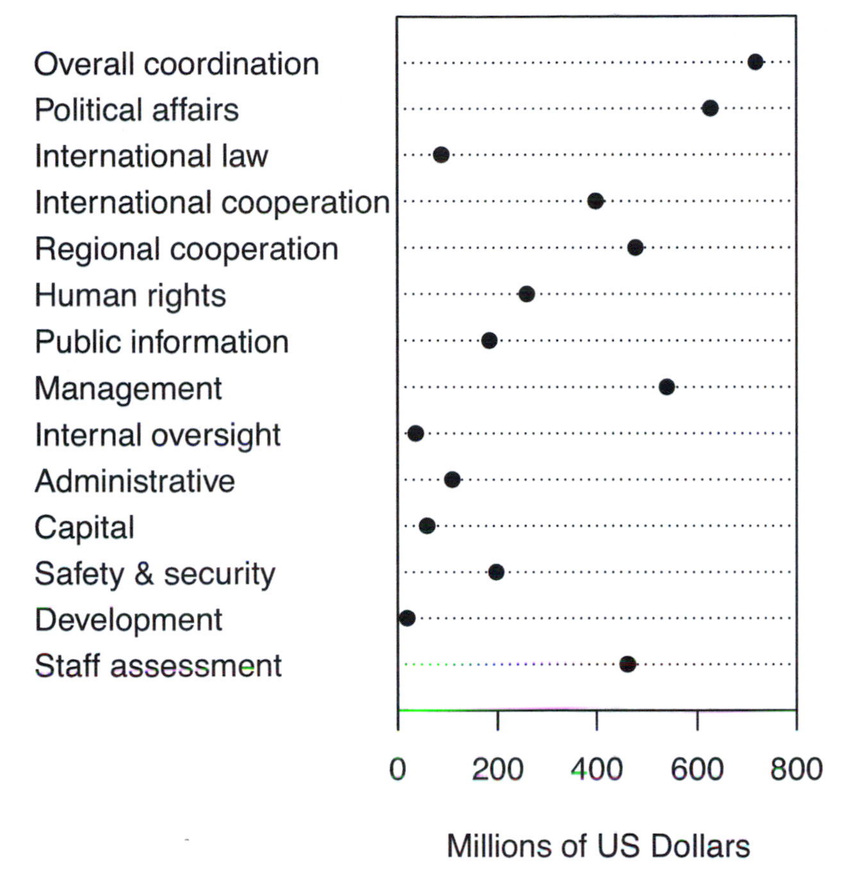

Figure 2.1 *Dot chart of the United Nations budget for 2008–2009*

`graphics` package of the statistical software package R [97]. The R code for producing Figure 2.1 is in the following six lines.

```
graphics.off()
windows(width=4.5,height=4.5,pointsize=12)
par(fin=c(4.45,4.45),pin=c(4.45,4.45),
mai=c(0.875,0.5,0.0,0.25),xaxs="i")
dotchart(x=amount,labels=item,lcolor="black",col="black",
bg="black",xlim=c(0,800),xlab="Millions of US Dollars")
```

The purpose of the call to the R function `graphics.off()` is to shut down all (previously) open graphics devices. This wipes the slate clean. The call to the R function `windows` opens a plotting window on the current display with dimensions of 4.5 inches wide and 4.5 inches tall being set explicitly. Also passed in the call of the function `windows` is the point size of plotted text. The default point size is set to 12 points, which just happens to be the default. A point is a unit of measure used by printing presses in the days when a printer was a human being who ran a press. There are 72.27 *points* to the inch and there are 72 *big points* to the inch—which can be a source of confusion if the two units are confused. For most devices used by R, the actual dimension is a big point. It is recommended that `pointsize` be specified in integers rather than relying on rounding to an integer by functions within R.

One of the purposes of the call to the R function `par` is to set the dimensions of the figure using the variable `fin`. There can be multiple plots within a figure in R. But in this example, there is only one so the dimensions of the plot set by `pin` are the same as the figure. Notice that both `fin` and `pin` set dimensions for 4.45 inches for each of width and height, respectively, which are slightly smaller than 4.5 inches for the width and height of the display window set by the call to the function `windows`. In writing the second edition, it was found that this was necessary due to how numbers for plot dimensions are rounded. By trial and error, it was found that figures will consistently execute if the values for the `fin` dimensions are 0.05 less than the dimensions supplied on the call to the function `windows`. (This was not required for the version of R used for the first edition and for quite a few releases of R afterward.) The variable `mai` sets the dimensions of the margins beginning at the bottom and thence clockwise for the other three margins. Note that the default unit in R for the dimensions of graphical displays is the inch.

Setting `xaxs="i"` (here `"i"` means *internal*) just finds a horizontal axis with pretty labels that fits within the original data range. The only other implemented option for axis style in R, at the time of writing, is `"r"` (regular), which first extends the data range by 4 percent and then finds an axis with pretty labels that fits within the range. The value `"r"` is rarely, if ever, used when plotting graphical displays for this textbook as the results can be less than satisfactory.

The amounts for each category have been stored in the variable `amount` and the corresponding labels in the variable `item` and the names of these variables have been supplied in the call to the plotting function `dotchart`.

In examples of R code to follow, the function calls to `graphics.off`, `windows`, and `par` will be omitted. This will be done with few exceptions as the details of these calls are in the script files that can be downloaded from the textbook's website (`http://www.graphicsforstatistics.com`) for each graphical display produced by R in this book.

A couple of liberties have been taken by the authors of the R function `dotchart`. Figure 2.2 is drafted following the inferred plotting standard of Cleveland and McGill [25]. This was done by creating a script using lower-level graphical functions available in the `graphics` package in R. The script is as follows.

```
plot(amount,1:length(item),type="n",xaxt="n",yaxt="n",
```

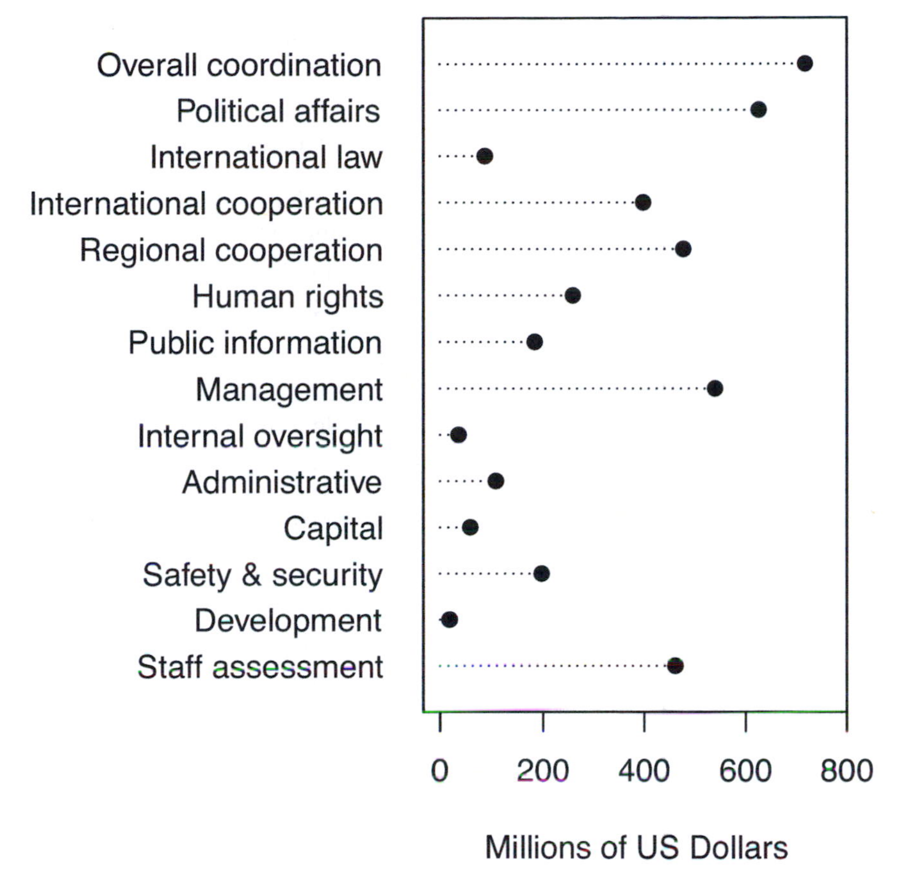

Figure 2.2 *Dot chart in the style of Cleveland and McGill [25] of the United Nations budget for 2008–2009*

```
xlim=c(0,800/1.04),ylim=c(0,length(item)+1),
xlab='Millions of US Dollars',ylab='',xaxs="r",yaxs="i")
for (i in 1:1:length(item)) lines(x=c(0,amount[i]),
y=c(i,i),lty=3)
points(x=amount,y=1:length(item),pch=19,cex=1.0)
axis(1,at=200*(0:4),labels=TRUE,tick=TRUE,outer=FALSE)
axis(2,at=1:14+0.25,labels=item,tick=FALSE,outer=FALSE,
las=2,hadj=1,padj=1)
```

The R function `plot` is a higher-level plot function but setting `type="n"` turns off plotting of lines and plots. Setting `xaxt="n"` and `yaxt="n"` suppresses the plotting of the horizontal and vertical axes, respectively. The R function `lines` adds the dotted line segments. The R function `points` adds the terminating dot for the dotted line segment for each category in the budget. The two separate calls to the function `axis` adds in the missing axes and their labels. Generally, calls to these lower-level routines in R can be used to produce any desired graphical display.

A comparison of Figures 2.1 and 2.2 reveals two differences:

- the dotted lines end at the plotted point in the original design, as in Figure 2.2, rather than continuing to the right margin, as in Figure 2.1; and
- the labels for the categories are right justified in the original, as in Figure 2.2, rather than being left justified, as in Figure 2.1 produced by the `dotchart` function.

With respect to the grammar of graphics and the R code to produce Figures 2.1 and 2.2, the following observations are made with respect to the concept of multiple layers. After calls to the functions `graphics.off`, `windows`, and `par` to create a blank canvas, Figure 2.1 is produced by a call to just one graphics function: `dotchart`. With just one call to a single function and reliance on its plotting defaults, it is hard to perceive any multiple layers in Figure 2.1. But in examining the R code that produced Figure 2.2 according to the inferred plotting standard of Cleveland and McGill [25], the multiple layers inherent in any dot chart become apparent. The call to the function `plot` plots no data nor any axes but does set the coordinate system and extents of the horizontal and vertical axes while otherwise leaving the canvas blank except for the bounding box and the label of the horizontal axis that are placed on the canvas. The label for the vertical axis is set to nothing. The call to `plot` must occur first before any other subsequent plotting or labeling instructions. The next two lines of code for Figure 2.2 are a `for` loop that plots the dotted lines for each category of expenditure. This is followed by a call to the `points` function that plots the larger terminating dots over top of the dotted lines. The final two lines of code are calls to the function `axis` to produce the horizontal and vertical axes, respectively. Tick marks are added outside the bounding box for the horizontal axis together with amounts in millions of US dollars in intervals of 200 beginning at zero and ending at 800. No tick marks are added to the bounding box for the vertical axis but the categorical labels are added with some positional adjustments.

Order is an important feature of a grammar for graphics just as it is for a natural human language. With respect to the grammar for the R code for Figure 2.2, the function `plot` must be called first, but the remaining calls to plot the data and axes can occur in any order. The language of the `graphics` package in R is somewhat more flexible than a human language in this regard. So there is flexibility in how the multiple layers are added to a plot using at least one statistical graphics language.

The quickest way to get something resembling a dot chart using the `ggplot2` package is the line chart in Figure 2.3 that was produced using the following R code.

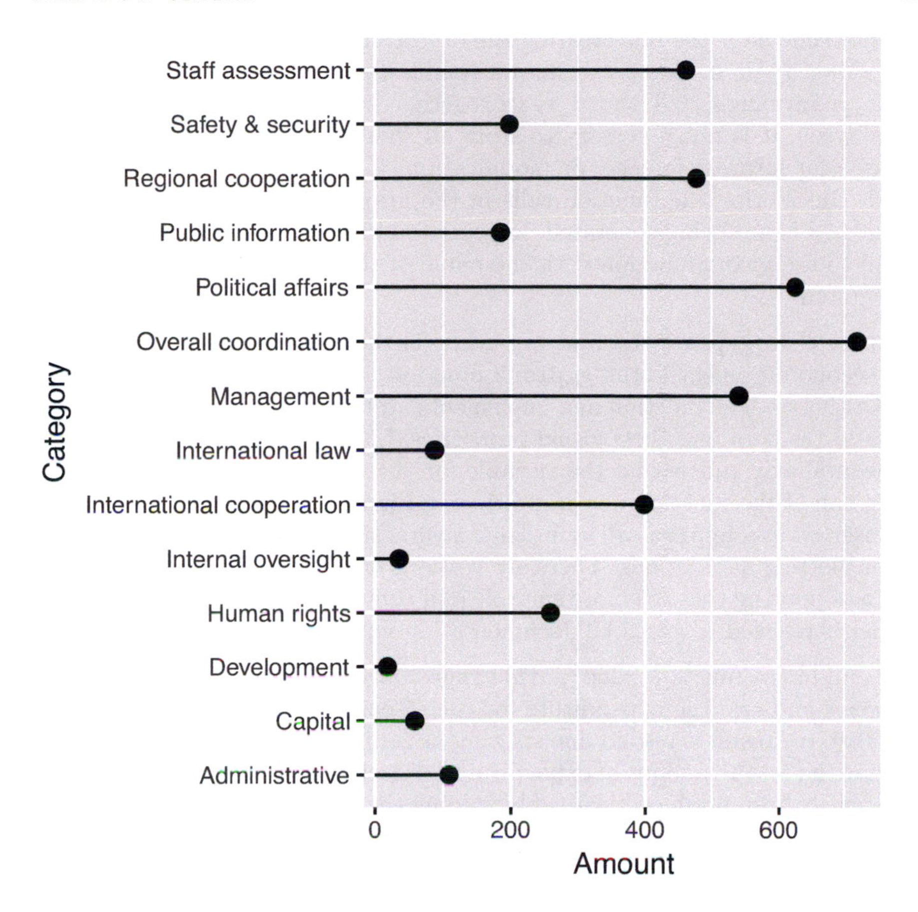

Figure 2.3 *Line chart of the United Nations budget for 2008–2009 using* `ggplot2`

```
Category<-item
Amount<-amount
un_fiscal<-data.frame(Category,Amount)

figure<-ggplot(un_fiscal,aes(Category,Amount)) +
geom_pointrange(ymin=0,ymax=Amount) + coord_flip()

print(figure)
```

The `un_fiscal` object in the code above is a data frame created by the function `data.frame` in the R `base` package. The `un_fiscal` object contains the categorical variable named `Category` and the continuous variable named `Amount`. The variable `Category` consists of the expenditure items. The named `Amount` contains the expenditures for each item. The `ggplot2` package works with variables from within data frames in R. The first two non-empty lines of code after the line creating the object `un_fiscal` creates another object called `figure`. These two lines assemble the `ggplot2` functions `ggplot`, `geom_pointrange`, and `coord_flip` together in a sentence using plus signs in between. It is not necessary to store all these function calls in the object `figure` for future reference, or modification, if only a graphic output is desired. But storing the function calls in the graphical object `figure` will not result in a figure being generated. In order to obtain the output on the screen, a call to the `print` function with the name of the graphical object to appear is required.

In the call to `ggplot`, the first argument is the data frame `un_fiscal` but the second argument is the `ggplot2` function `aes`, which is an abbreviation for aesthetics. The function `aes` implements aesthetic mappings between variables in the data and their visual properties. Listing first the variable for the horizontal axis and second the variable for the vertical axis results in a very basic call of the function `aes` in producing Figure 2.3. The function `aes` can be inserted in a `ggplot` call or in later layers, such as `geom_pointrange` and other plotting instructions. There are versions of the function `aes` called `aes_` and `aes_string` that are more suitable and convenient for programming—the reader is referred to `ggplot2` documentation for more details.

The call to the function `geom_pointrange` conveys instructions to draw lines between zero and the expenditure for each item. Relying on the defaults for this call, the result is a solid line starting at zero and terminating with a large dot for the actual amount of expenditure for each line item. A dot chart could have easily been produced instead by adding the argument `linetype=3` to the call of `geom_pointrange`.

Examining Figure 2.3 reveals a couple of other defaults. The item names have been sorted in increasing alphabetical order outward from the origin. The data are plotted over a tinted grid with wide white gaps every 200 million, narrow white gaps every 100 million. Each axis has black tick marks and there is a white horizontal gap in the grid for each item. The horizontal axis is extended slightly to the left of the origin and there is no label for 800 million dollars. The labels for the axes are the names of the variables themselves in the data frame. There was, however, one default that was overridden. If it were not for the call `coord_flip()`, this would have been a vertical line plot.

It can be argued that the graphics package `ggplot2` lends itself to fast prototyping. That is, if one is willing to accept the plotting defaults, one can see data quickly. But if one is willing to put a little more effort in, a graphic according to a different plotting standard can be obtained. Figure 2.4 was produced according to the inferred plotting standard of Cleveland and McGill [25] with the following `ggplot2` code.

```
un_fiscal<-data.frame(item,amount)

figure<-ggplot(un_fiscal,aes(item,amount)) + theme_base()
```

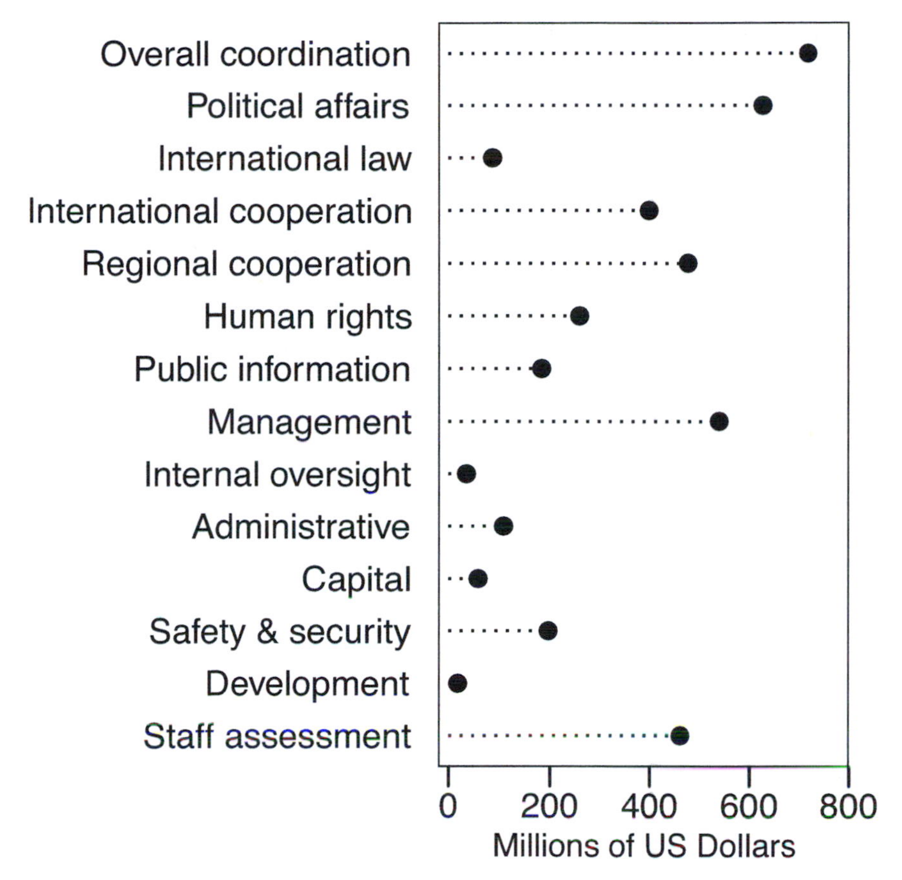

Figure 2.4 *Dot chart in the style of Cleveland and McGill [25] of the United Nations budget for 2008–2009 using* ggplot2

```
figure<-figure+theme(plot.background=
element_rect(fill=NULL,colour = "white",
linetype="solid"))

figure<-figure+theme(axis.title.x=
element_text(size=unit(12,"pt")),
axis.ticks.y=element_blank(),
plot.margin=unit(c(1,5,3,1), "mm")) +
scale_x_discrete(limits=un_fiscal$item) +
scale_y_continuous(expand=c(0,0),breaks=200*(0:5),
limits=c(-20,800))+
geom_pointrange(ymin=0,ymax=amount,linetype=3)+
```

```
coord_flip() + labs(x=NULL,y="Millions of US Dollars")

print(figure)
```

The same data frame `un_fiscal` is used for Figure 2.4 and the arguments to the function `ggplot` are the same as for Figure 2.3 but then there are changes. In addition to the package `ggplot2`, the R code for Figure 2.4 requires an additional package `ggthemes`. Themes in the package `ggplot2` refer to the aspects of a graphic that are described by plotting standards or so-called style guides. This would include whether there is a grid, the background color for the grid, the thicknesses of the axes, font sizes for axis labels, and so on. A number of themes are available in `ggplot2`. One theme missing is the theme implicit in the base `graphics` package in R. There are others missing as well. The package `ggthemes` contains extra themes, scales, and geometric features not in `ggplot2`. The function `theme_base` in the package `ggthemes` provides the theme for base `graphics` package in R but it places a black box around any figure drafted by `ggplot2`. This is not part of the graphing standard, so in the code for Figure 2.4 after the call `theme_base()` there is a call to the `ggplot2` function `theme` to repaint white the original black lines of the figure's bounding box.

With the argument `limits=un_fiscal$item` passed in the function call of `scale_x_discrete` in the code above, the default alphabetical sorting of item names is overridden by the actual sequence order in the variable `un_fiscal$item`. The argument `linetype=3` in the call of `geom_pointrange` is required to obtain the dot chart. The function `coord_flip()` is still used to flip the coordinates while the remainder of the changes relate to controlling other aspects of the axes and their labels. It is left to the user to decide whether they prefer the base `graphics` code of Figure 2.2 or the `ggplot2` code of Figure 2.4 for producing a dot chart in the inferred plotting standard of Cleveland and McGill [25].

Cleveland and McGill [25] argue on the basis of clarity and aesthetics that the dot chart ought to be preferred over other graphical presentations of the distribution of a single discrete variable. But the literature, including Cleveland and McGill [25], is apparently devoid of any discussion or experimental results regarding the use of the dot chart or its advantages over its competitors, which are next discussed.

2.5 The Bar Chart

2.5.1 Definition

The *bar chart* is the most frequently used graphical display for summarizing data with respect to a single discrete variable. See Figure 2.5 for an example with the United Nations proposed expenditure data. The bar chart depicts

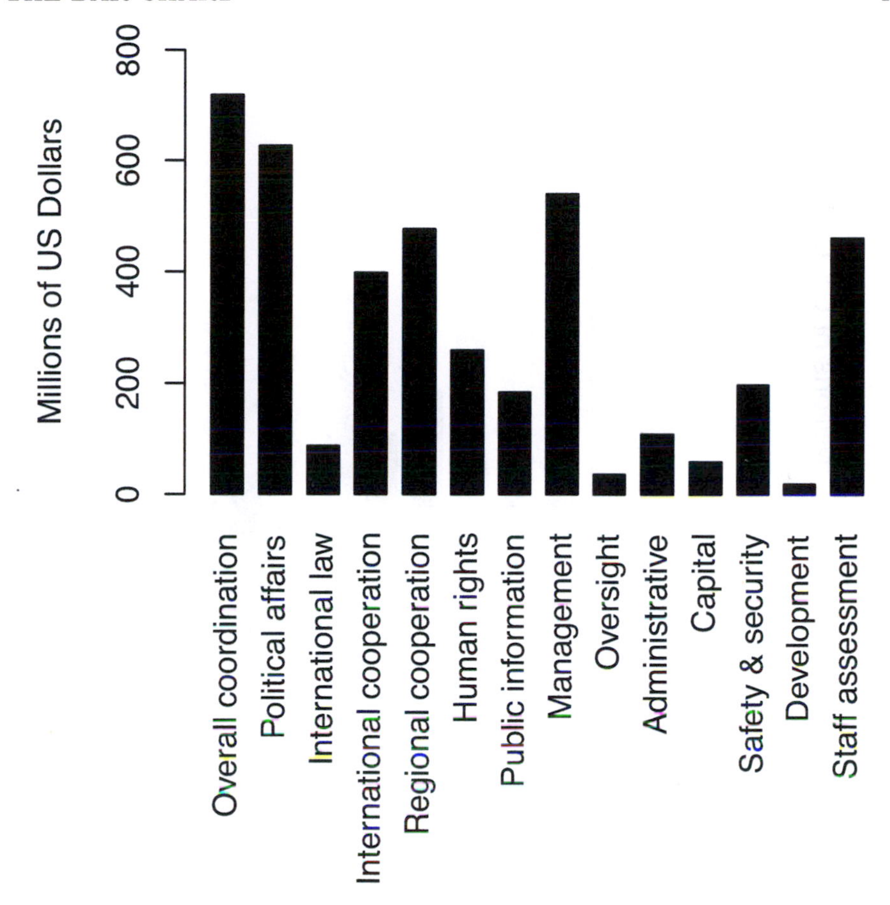

Figure 2.5 *Vertical bar chart of the United Nations budget for 2008–2009*

the distribution of a categorical variable with the lengths of bars scaled accordingly. Figure 2.5 has a vertical orientation for the bars, which arguably is the most frequently encountered orientation for the bar chart.

One problem with the vertical orientation, which is immediately apparent, is that the category names are awkward to read. This awkwardness can be ameliorated by rotating the category labels 45 degrees as in Figure 2.6. It can be eliminated altogether if a horizontal orientation is chosen as in Figure 2.7.

The R script used to produce Figure 2.7 is as follows.

```
barplot(amount,space=0.5,names.arg=NULL,horiz=TRUE,
axes=FALSE,xlim=c(0,800),
xlab="Millions of US Dollars",col="black")
yy<-1.5*(1:length(amount)-0.33)
axis(1,tick=TRUE,yaxp=c(0,800,200))
```

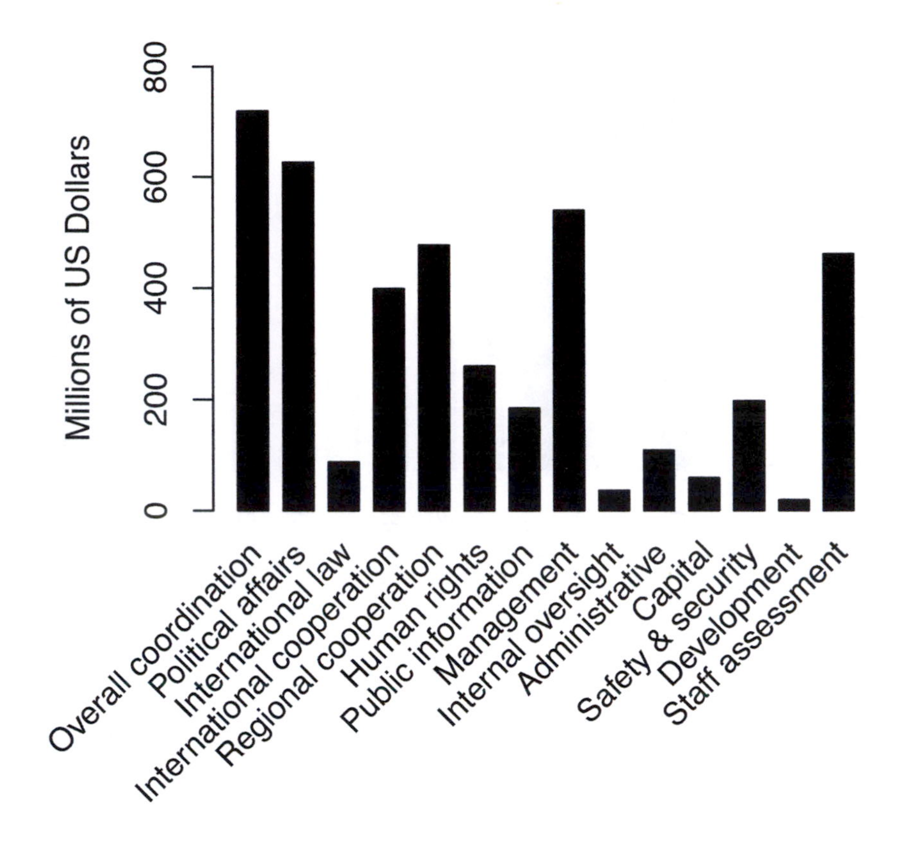

Figure 2.6 *Vertical bar chart of the United Nations budget for 2008–2009 with category labels at an angle of 45 degrees*

```
yy<-1.5*(1:length(amount)-.375)
axis(2,at=yy+0.2,labels=item1,tck=0,tcl=0,col=0,las=2,
hadj=0,outer=TRUE,line=-2)
```

The R function `barplot` is used to produce the bar charts in Figures 2.5, 2.6, and 2.7. But it produces axes that are visually inadequate. The call to `barplot` in the R script for Figure 2.7 switches off the two axes and their labels. These are restored by two calls to the lower-level function `axis`.

A worthwhile question is whether there is any peer-reviewed research in spatial perception that demonstrates the superiority of the horizontal compared to the vertical orientation of the bars, or vice versa. The short answer is that there is no empirical evidence.

The `barplot` function in R, which was used to produce Figures 2.5, 2.6, and

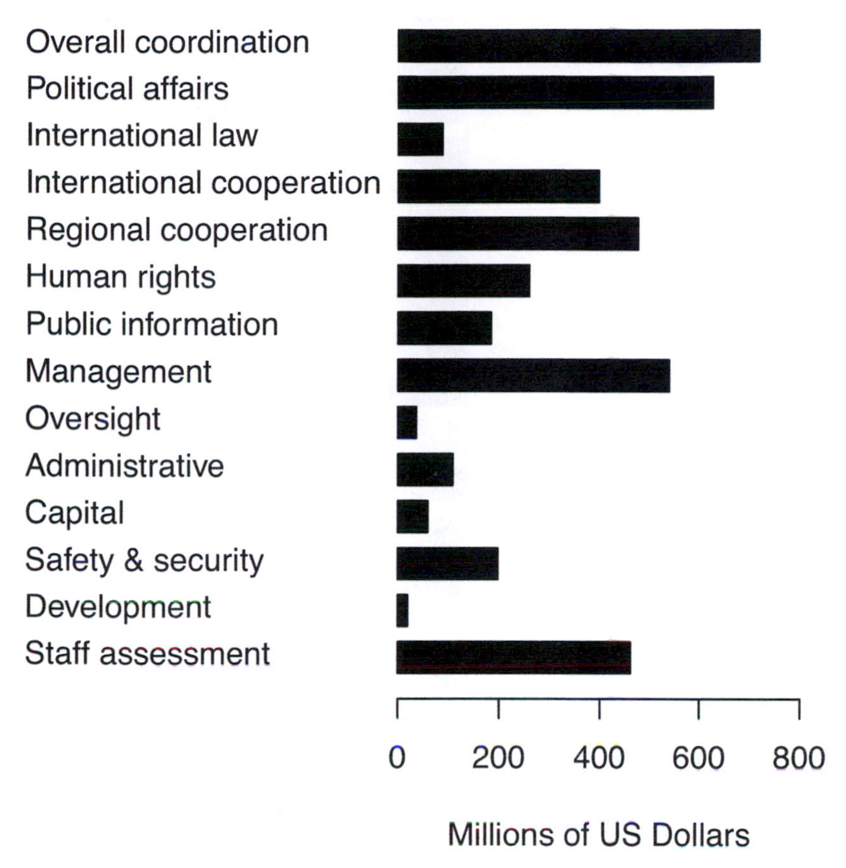

Figure 2.7 *Horizontal bar chart of the United Nations budget for 2008–2009*

2.7, offers a choice of vertical or horizontal orientations. But the `dotchart` function that was used to produce Figure 2.1 offers no such option and will only produce a horizontal orientation for the dotted lines. It is noted that although Cleveland [21] provides illustrations using bar charts with both vertical and horizontal orientations, his dot charts have only the horizontal orientation.

Continuing with a comparison of the dot chart of Figure 2.2 and the bar chart of Figure 2.7, it is noted that both feature a horizontal orientation. Both share the same scale for millions of US dollars on the horizontal axis and the same tick marks at intervals of 200 beginning at zero and ending at 800. The order of the categories in both charts is identical to that of Table 2.1 and is the order published by the United Nations. In Figures 2.1 and 2.7, the category labels are left justified.

The most visually striking difference between the dot chart of Figure 2.2 and the bar chart of Figure 2.7 is the choice of icon that is drawn for each category.

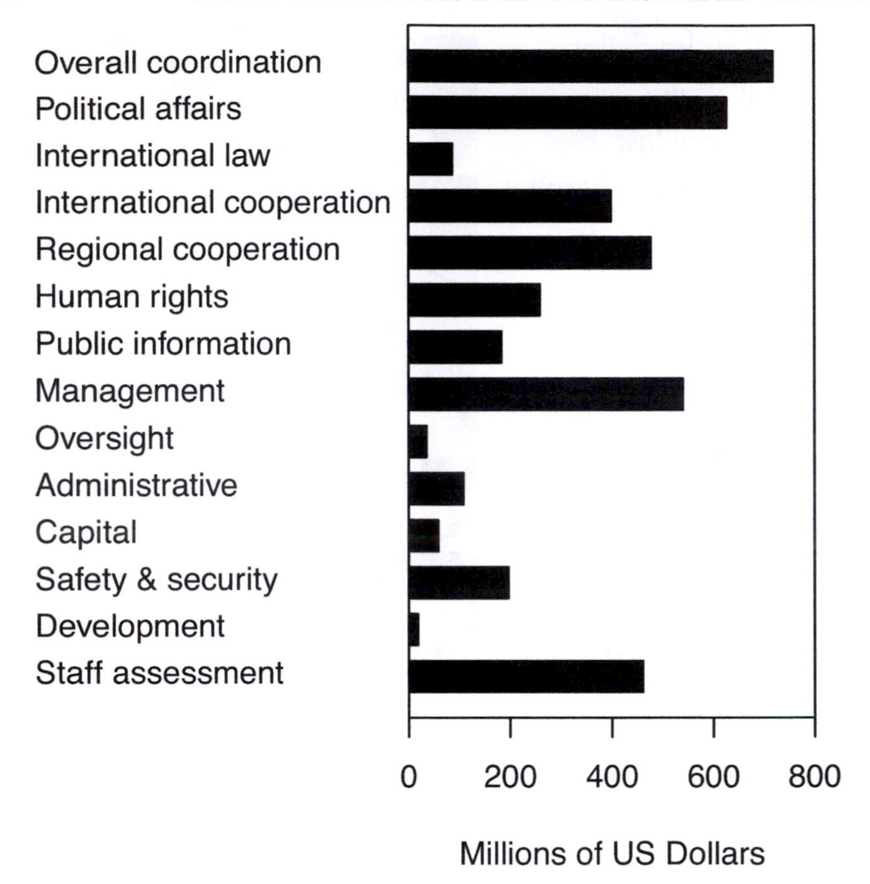

Figure 2.8 *Horizontal bar chart of the United Nations budget for 2008–2009 with framed axes*

In either chart, the viewer is invited to make categorical comparisons based upon the length of the icon. In the case of the classical dot chart of Cleveland and McGill [25] given in Figure 2.2, we have a dotted line starting at zero and ending in a large dot. The bar chart of Figure 2.7 substitutes a bar in the place of this icon.

There is another difference that is more subtle. The axes of the dot chart of Figure 2.2 are fully framed while there is no framing for the bar chart of Figure 2.7. Framing can be easily added to a plotting script in R. A framed version of Figure 2.7 is given in the framed bar chart of Figure 2.8.

A frame can be useful by providing visual cues for locating the end of the bars in a Cartesian coordinate plane. With the single line segment on the bottom for the horizontal axis in Figure 2.7, the further away from the bottom of the chart, the more difficult it is to discern the length of the bar. While the

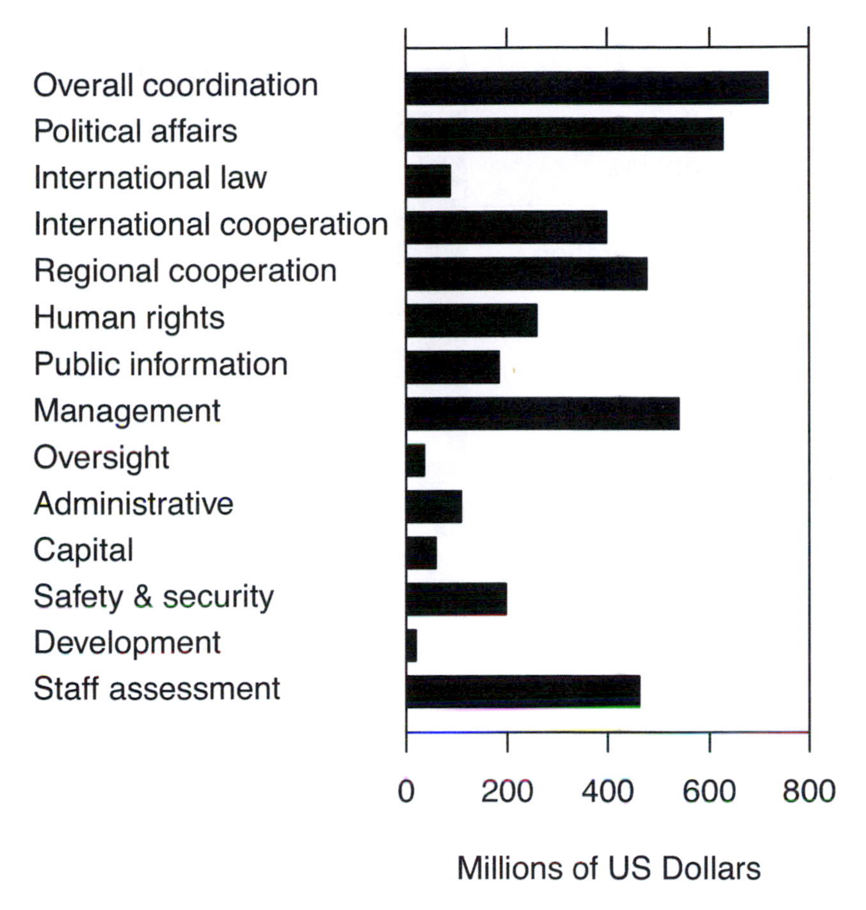

Figure 2.9 *Horizontal bar chart of the United Nations budget for 2008–2009 with framed axes and tick marks on both horizontal axes*

addition of a top line segment parallel to the horizontal axis is helpful, it could be argued that the addition of tick marks to the top line segment, as in Figure 2.9, is even more helpful.

Cleveland [21] recommends parallel axis line segments and with few exceptions, his illustrations with pairs of parallel axes include ticks on both axes. Notable exceptions are the dot charts of Cleveland and McGill [25] and Cleveland [21] in which ticks are only to be found along the lower horizontal axis. As a plotting convention for this book, the advice of Cleveland [21] to place the ticks outside the plotting frame is accepted because this avoids obscuring any data plotted within the frame formed by the two sets of parallel line segments.

It is not uncommon to see bar charts in publications, such as in Figure 2.10, in

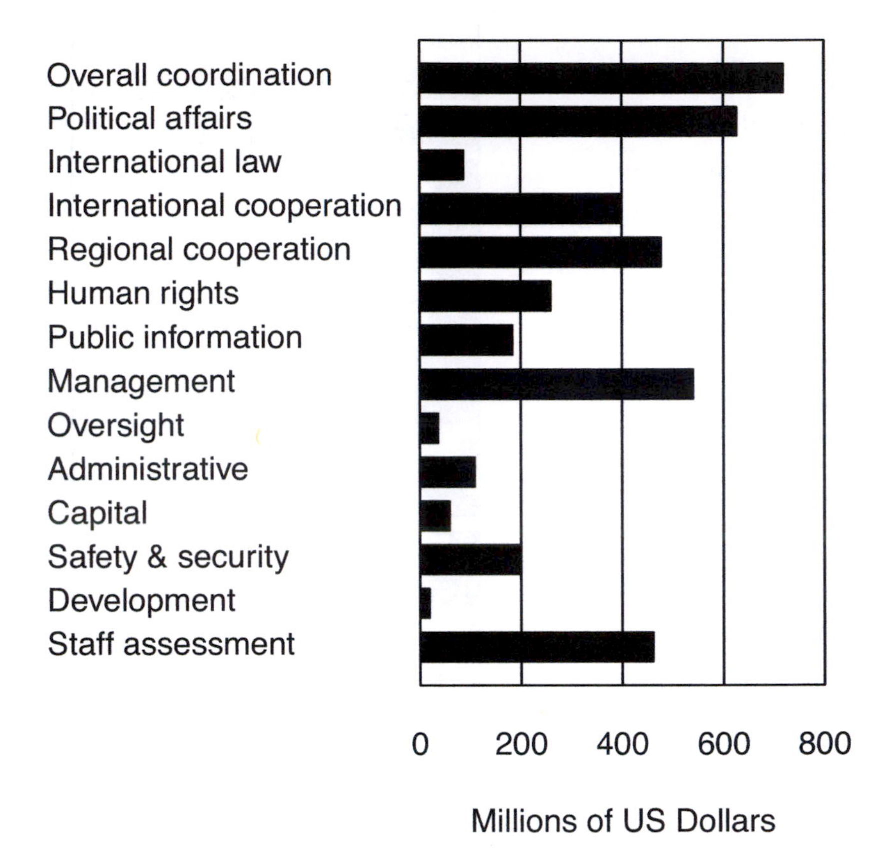

Figure 2.10 *Horizontal bar chart of the United Nations budget for 2008–2009 with framed axes and vertical reference lines*

which the tick marks have been substituted by a system of parallel reference lines. A *grid* can be formed by perpendicular sets of parallel reference lines. Dark grid lines are considered by Tufte [123] to be an example of *chartjunk*: unnecessary graphical decoration.

Tufte [123] developed a descriptive statistic to assess chartjunk. This statistic is defined as follows. *Data-ink* is the nonerasable core of a graphic display, the nonredundant ink arranged in response to variation in the numbers represented. One then considers the total amount of ink used in the graphical display. The resulting statistic is the

$$\textbf{data-ink ratio} = \frac{\text{data-ink}}{\text{total ink used to print the graphic}}.$$

The data-ink ratio is equal to the proportion of a graphic's ink devoted to the nonredundant display of information.

The data-ink ratio for the bar chart in Figure 2.10 with the framed axes and vertical reference lines is evidently less than the data-ink ratio for the bar chart of Figure 2.8 that has framed axes only. With the data-ink ratio, the problem of choosing the best graphical display is reduced to an optimization problem. Difficult optimization algorithms are often solved through iterative algorithms. This provides a mathematical analogy for the process of creating a visually effective graphic display.

The bar chart given in Figure 2.9 was not created at the outset but was obtained after four previous attempts were deemed to be unsatisfactory. The process of using a bar plot to depict planned United Nations expenditures began with the conventional vertical orientation of Figure 2.5. The vertical orientation is the default with the R function `barplot`. An attempt to fix the problem of reading the categorical labels led to Figure 2.6 with the labels at a 45-degree angle. This change was helpful but not fully satisfactory.

Figure 2.7 was produced with a horizontal orientation with the categorical labels now easily read. A frame was added in Figure 2.8 to assist viewers in assessing the length of the bars. Tick marks were added outside the upper horizontal line in Figure 2.9 as a further refinement.

There is a successive lowering in the data-ink ratio in progressing through Figures 2.7, 2.8, and 2.9. This ought not be viewed as counter-productive as it pays to use a bit of common sense. A frame and tick marks on the upper axis that match those on the lower axis have been added to assist the viewer assessing the lengths of the bars. With the addition of vertical reference lines at the tick marks in Figure 2.10, it can be argued that one has gone too far.

Figure 2.11 was created using `qplot` in the package `ggplot2` after a bit of housekeeping code to put the data in a form convenient for `qplot`.

```
Category<-rep(item[1],round(amount[1]))
for (i in 2:length(item))Category<-c(Category,
rep(item[i],round(amount[i])))

figure<-qplot(Category,geom="bar") + coord_flip()

print(figure)
```

The chartjunk in Figure 2.11 is due to the shaded background that adversely impacts the data-to-ink ratio. But what of a comparison between the dot chart and the bar chart on the basis of the data-ink ratio? Deciding to compare two figures with equal nondata ink leads to comparison of the dot chart of Figure 2.2 with the bar chart of Figure 2.8. Because dotted lines ending in a large eye-catching dot for the dot chart require less ink than the solid bars of

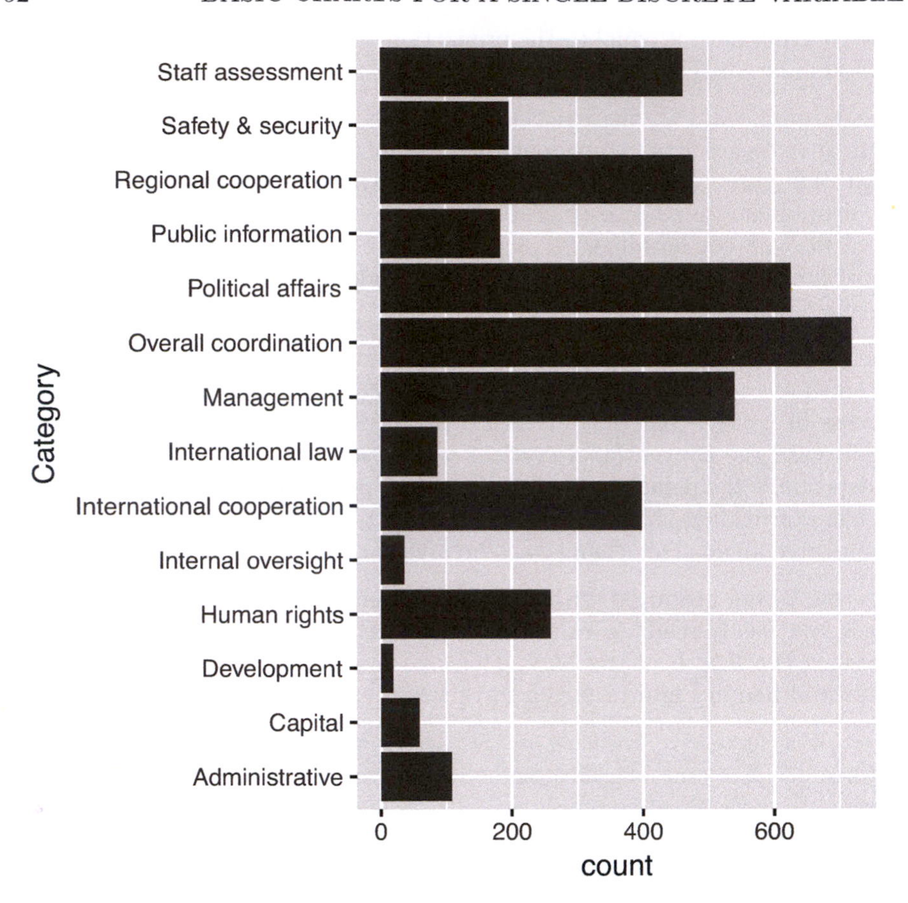

Figure 2.11 *Horizontal bar chart of the United Nations budget for 2008–2009 with grid lines using* `ggplot2`

Figure 2.8, it is clear that the dot chart of Figure 2.2 requires less nondata ink. On the basis of the data-ink ratio, the dot chart is the clear winner over the bar chart.

But has something been overlooked? The amount of black ink used in the frameless bar chart of Figure 2.7 can be easily reduced. Figure 2.12 replaces the black ink in the interior of the bars of Figure 2.7 with the default gray of R when color is not specified.

It can be argued that Figure 2.12 has reduced the amount of ink used and improved the data-ink ratio. Going further, Figure 2.13 replaces the black ink in the interior of the bars of Figure 2.7 with white space. The R code for producing Figure 2.13 is as follows.

```
barplot(amount,space=0.5,
```

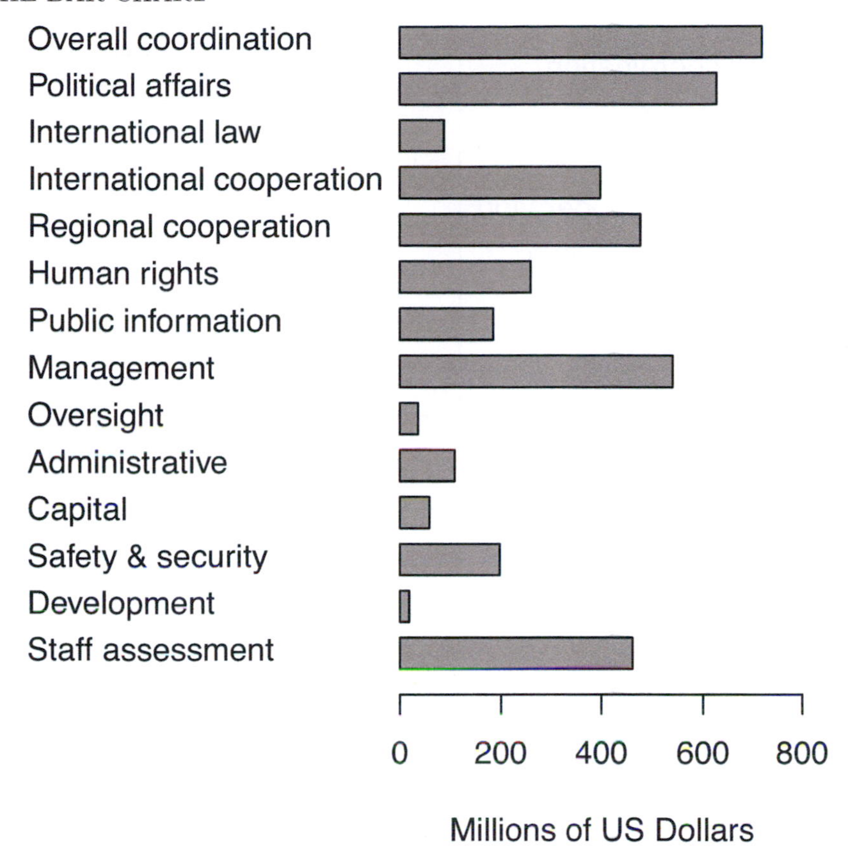

Figure 2.12 *Horizontal bar chart of the United Nations budget for 2008–2009 with gray bar fill*

```
names.arg=NULL,horiz=TRUE,axes=FALSE,xlim=c(0,800),
xlab="Millions of US Dollars",col="white")

yy<-1.5*(1:length(amount)-0.33)
axis(1,tick=TRUE,yaxp=c(0,800,200))
yy<-1.5*(1:length(amount)-.375)
axis(2,at=yy+0.2,labels=item,tck=0,tcl=0,
col=0,las=2,hadj=0,outer=TRUE,line=-2)
```

Ignoring the differences in framing between the dot chart of Figure 2.2 and the white fill bar chart of Figure 2.13, the data-ink ratio is still higher for the dot chart compared to the bar plot.

Figure 2.14 is a version of Figure 2.13 produced using `ggplot2`. Although Figure 2.14 produced by `ggplot2` is not a perfect reproduction of Figure 2.13

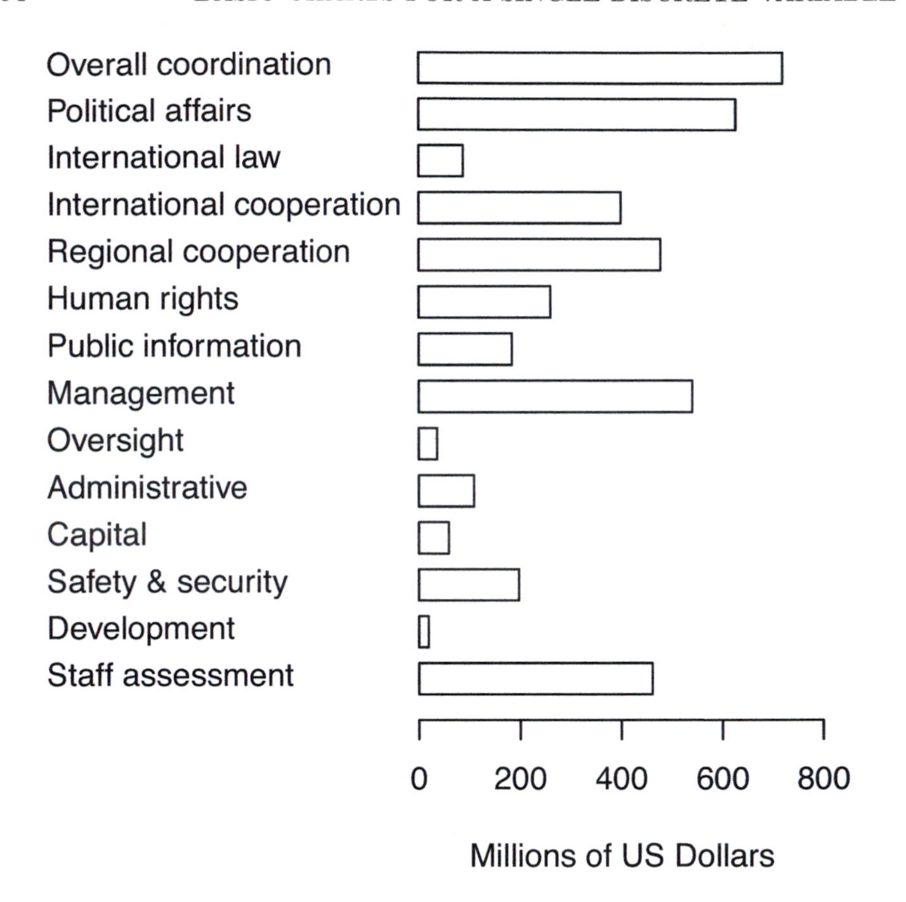

Figure 2.13 *Horizontal bar chart of the United Nations budget for 2008–2009 with white bar fill*

produced by the function **barplot** in the **graphics** package, the differences are ignorable. The R code for producing Figure 2.14 is as follows.

```
figure<-ggplot(un_fiscal2,aes(category)) + theme_classic()

figure<-figure + geom_bar(fill="white",color="black",width=0.7)

>igure<-figure + theme(
axis.title.x=element_text(size=unit(12,"pt")),
axis.ticks.y=element_blank(),
plot.margin=unit(c(1,5,3,1), "mm"),
axis.line.y=element_line(color="white"),
axis.line.x=element_line(color="black"),
axis.ticks.x=element_line(color="black"),
```

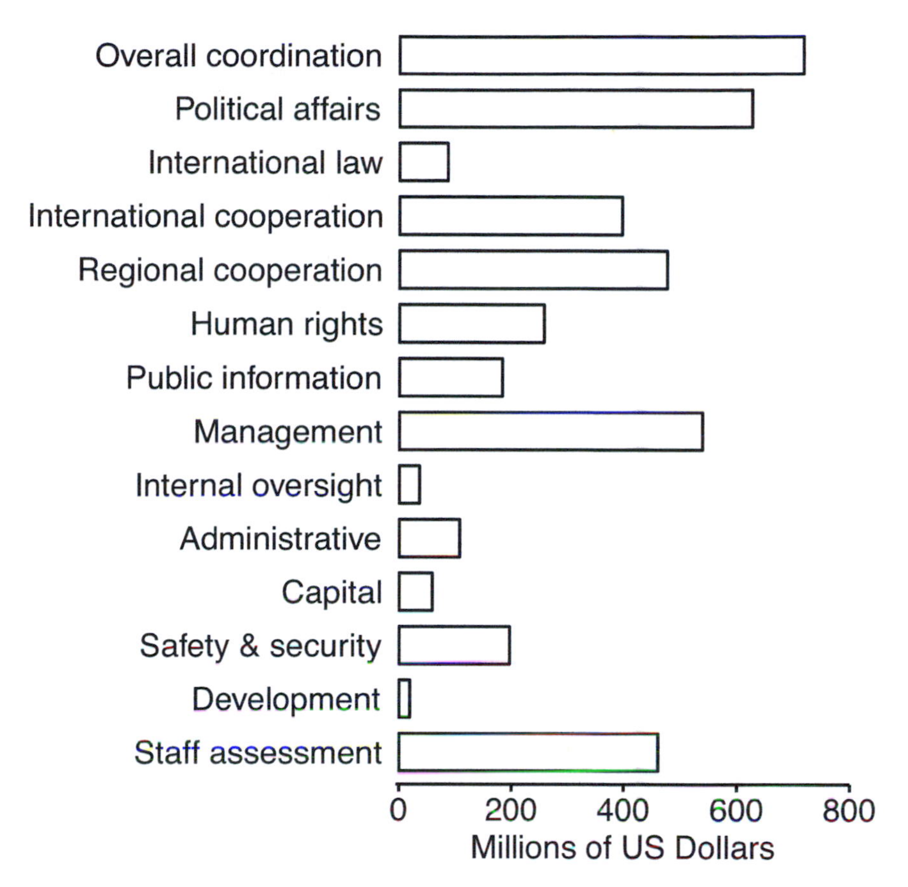

Figure 2.14 *Horizontal bar chart of the United Nations budget for 2008–2009 with white bar fill using* ggplot2

```
axis.text.y=element_text(color="black",
size=unit(12,"pt")),
axis.text.x=element_text(color="black",
size=unit(12,"pt"))) +
scale_x_discrete(limits=item[length(item):1])+
scale_y_continuous(expand=c(0,0),
breaks=200*(0:5),limits=c(-5,800))
figure<-figure+coord_flip() +
labs(x=NULL,y="Millions of US Dollars")

print(figure)
```

Comparing R code and not counting spaces, there are 680 characters of code needed to produce Figure 2.14 with the ggplot2 package compared to just

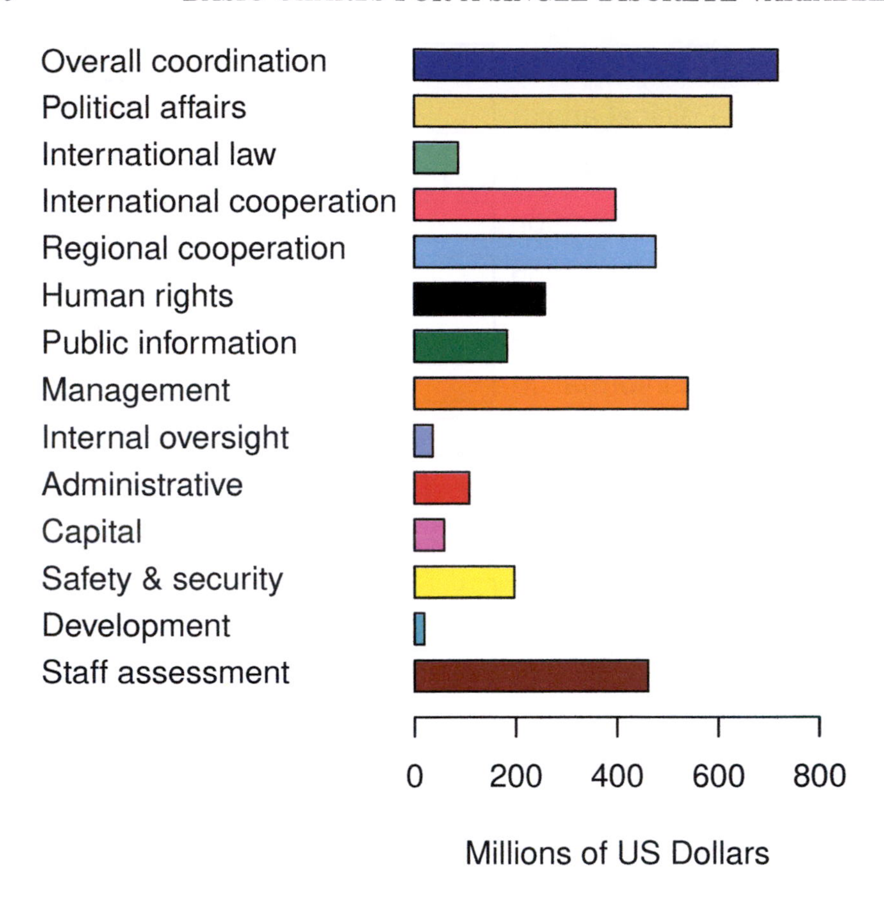

Figure 2.15 *Horizontal bar chart of the United Nations budget for 2008–2009 with color fill for the bars*

292 characters for Figure 2.13 using the `graphics` package. This more than a two-fold difference in extra effort for `ggplot2` may be important to some.

If not yet convinced of the superiority of the dot chart over the bar chart, a more vibrant version of the horizontal bar chart can be viewed in color in Figure 2.15. The use of a distinct color for each category serves to highlight the fact that each bar represents a different category. Nevertheless, the principal purpose for the use of color in a graphic is to catch the eyes of viewers and draw them into the story. It is for this reason that color graphics, such as Figure 2.15, are used by the popular media and in live presentations.

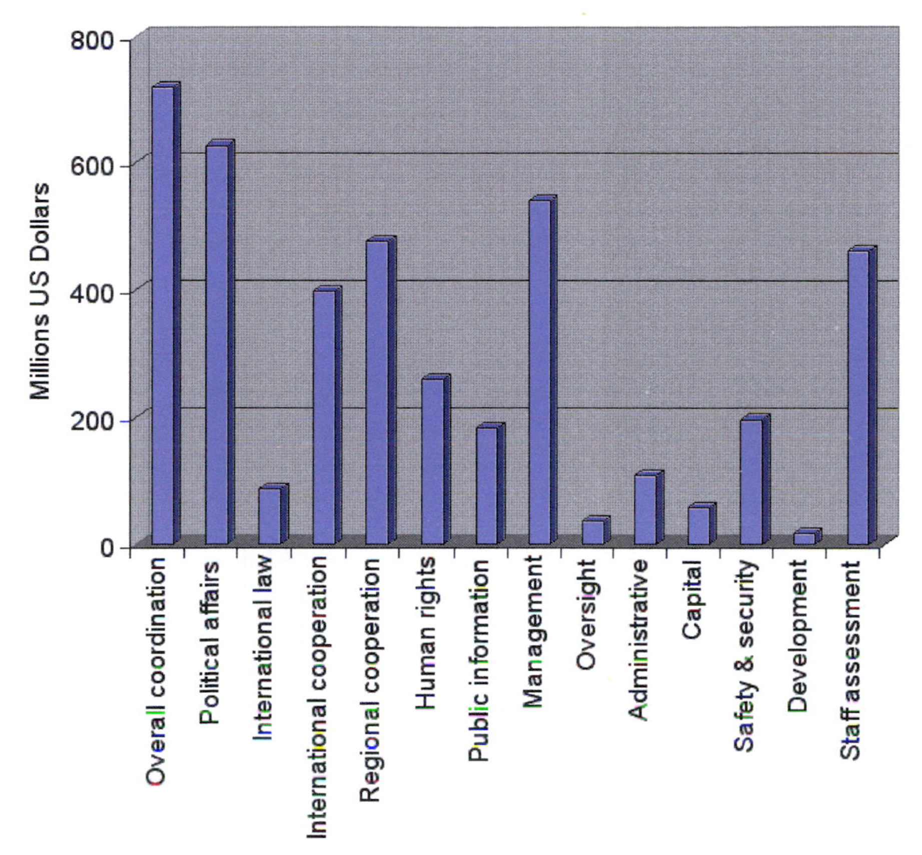

Figure 2.16 *Pseudo-three-dimensional bar chart of the United Nations budget for 2008–2009 with color fill for the bars*

2.5.2 Pseudo-Three-Dimensional Bar Chart

Figure 2.16 is an example of a *pseudo-three-dimensional bar chart.* This color figure was produced using Microsoft (Microsoft Corporation, Redmond, Washington) Office Excel so it is rather frequently encountered. The artistic use of the three-dimensional effect created on a two-dimensional surface serves only to obscure the data rather than reveal its features.

Baird [7] reported on a large number of experiments, by himself and other experimental psychologists, which revealed that judgments concerning lengths tend to be unbiased, whereas, there is distortion for judgments of areas and even more for judgments of volumes.

There is a vertical axis labeled with respect to millions of US dollars. So there appears to be an intent for viewers to compare the lengths, or rather the heights, of the bars. But in the perspective plot of Figure 2.16, the horizontal plane for the tops of the bars is not horizontal. To compensate, Excel has added horizontal references offset from the tick marks of the vertical axis.

Reference lines are in the back plane and are obscured by the opaque pseudo-three-dimensional bars. Further, there is visual evidence at the base of the three-dimensional bars that would suggest that the back plane is further displaced from the backs of the three-dimensional bars. This makes it very hard for the viewer to judge the heights of the three-dimensional bars with respect to the vertical scale. That is, if the viewer takes the time.

It is quite easy for a viewer to only partially use the three-dimensional perspective and incorrectly use the reference lines to judge the foreground heights of the rectangular cross-section of each three-dimensional bar.

While use of color in the two-dimensional bar chart of Figure 2.15 serves a purpose other than to just grab attention, the bar chart of Figure 2.16 only offers a monochromatic alternative to gray. Additionally, the patterning of the background for Figure 2.16 has a shimmering appearance due to the presence of the *moiré effect* that gives an unintentional optical illusion of motion in a graphic. This is a further distraction to the data. A better example of a graphic with a moiré effect will be given in the following section.

Finally in this assessment of the pseudo-three-dimensional bar chart in Excel, it is quite evident that the two-dimensional bar chart of Figure 2.8 is superior to the pseudo-three-dimensional bar chart of Figure 2.16 with respect to the data-ink ratio. It appears that the dot chart remains our first choice for depicting categorical distributions among the graphics so far encountered. In the next section we encounter a graphic as familiar as the bar chart.

2.6 The Pie Chart

2.6.1 Definition

An example of the *pie chart* for United Nations expenditures planned for 2008–2009 is given in Figure 2.17. The distribution of planned expenditures is conveyed visually by the different sizes of the slices for each category. The slices in a pie chart are adjacent, so it is essential that the slices are differentiated from each other. The use of a different color for each slice in Figure 2.17 achieves this. Compare this with the different colors of the nonadjacent bars of Figure 2.15 in which the use of color is not essential but decorative.

It is good practice to avoid putting a region in yellow adjacent to a region in blue to avoid confusion for individuals who are blue-yellow color blind (tritanopia). This has not been done in Figure 2.17. The yellow sector is adjacent to a shade of blue: aquamarine. The prevalence of tritanopia is about 0.001 percent in males and 0.03 percent in females. Approximately 0.01 percent

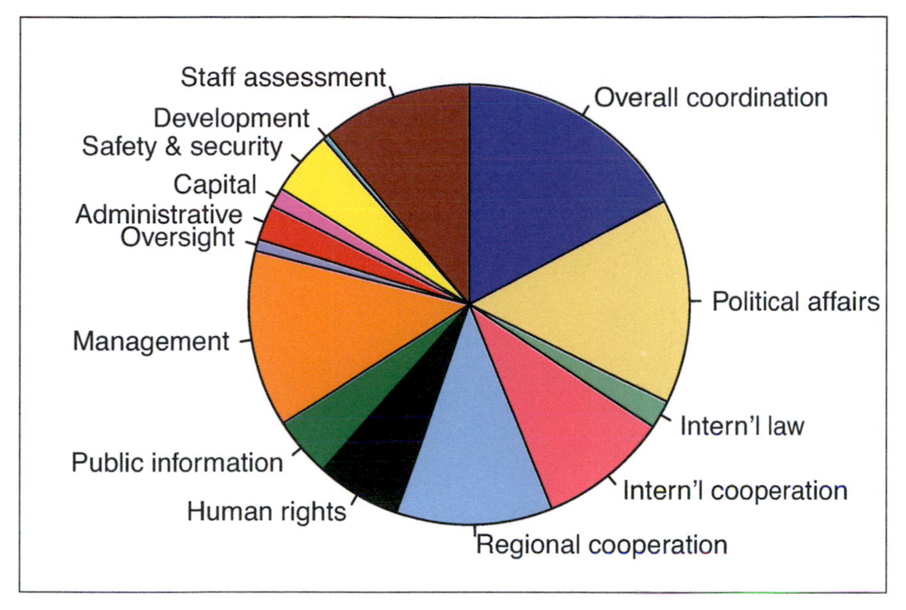

Figure 2.17 *Pie chart of the United Nations budget for 2008–2009 with color fill*

in each gender are deficient to some degree with respect to the color blue (tritanomaly).

It is good practice to avoid putting a region in red adjacent to a region in green to avoid confusion for individuals who are red-green color blind (protanopia and deuteranopia). This has not been done in Figure 2.17. The light green sector of international law in Figure 2.17 is adjacent to the hot pink sector of international cooperation. The prevalence of both forms of red-green color blindness is about 2.3 percent in males and 0.03 percent in females. Approximately 1.3 percent of males and 0.02 percent of females are red deficient (protanomaly). Approximately 5.0 percent of males and 0.35 percent of females are green deficient (deuteranomaly). Complete color blindness (achromatopsia) is rare: occurring once in every 10,000,000 males or 10,000,000 females.

Figure 2.17 was drafted using the `pie` function in R. This function allows for the substitution of color by grayscale shading, as in Figure 2.18. There was a certain amount of trial and error that finally resulted in Figure 2.18. A bit of experimentation was done with different intensities of grayscale to find a selection that avoids virtually indistinguishable wedges as a result of overly dark shades. The use of grayscale is one way around the problem of color blindness among viewers. See Appendix A for more discussion concerning human color perception.

Hatching is the use of close parallel lines to give the effect of shading. *Cross-*

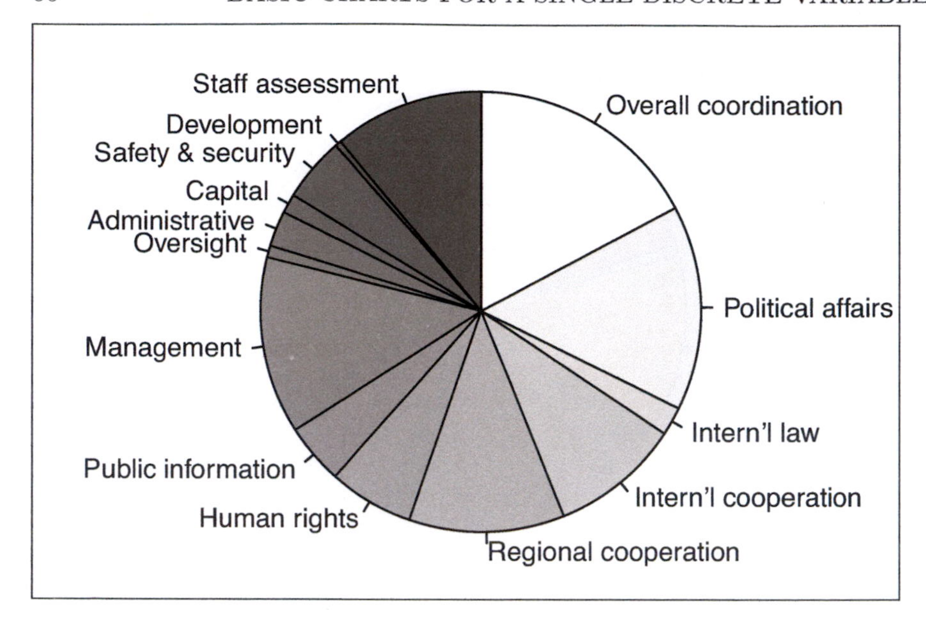

Figure 2.18 *Pie chart of the United Nations budget for 2008–2009 with grayscale fill*

hatching uses intersecting sets of parallel lines to the same effect. Although cross-hatching is not an option available in the R function `pie`, hatching is. Figure 2.19 is a black-and-white version of the pie chart with hatching resulting from several steps of trial-and-error with different combinations for the density of parallel lines and the angle of these lines. This effort was wasted.

Hatching or cross-hatching, or both, can unintentionally add the illusion of motion in a graphic due to the moiré effect. This effect is present in Figure 2.19. After some experimentation to avoid an awkward appearance resulting from using even numbers for line density and the appearance of lines continuing beyond a given sector, it was decided to select the first fourteen prime numbers starting from the number two as the line density for the United Nations expenditure categories in Figure 2.19.

Some experimentation was required with respect to the size of the pie in Figure 2.17, 2.18, and 2.19 as well as the abbreviations of the category names to obtain a visually pleasing layout without overcrowding the category names. A good question is: how many categories are too many for a pie chart? One rule of thumb is that the upper limit for the number of categories is six. The fourteen categories of planned expenditure by the United Nation for 2008–2009 are clearly greater than six.

A question worth considering is how do viewers extract information from a pie chart. Do viewers rely on comparison of areas, arc length, chord length,

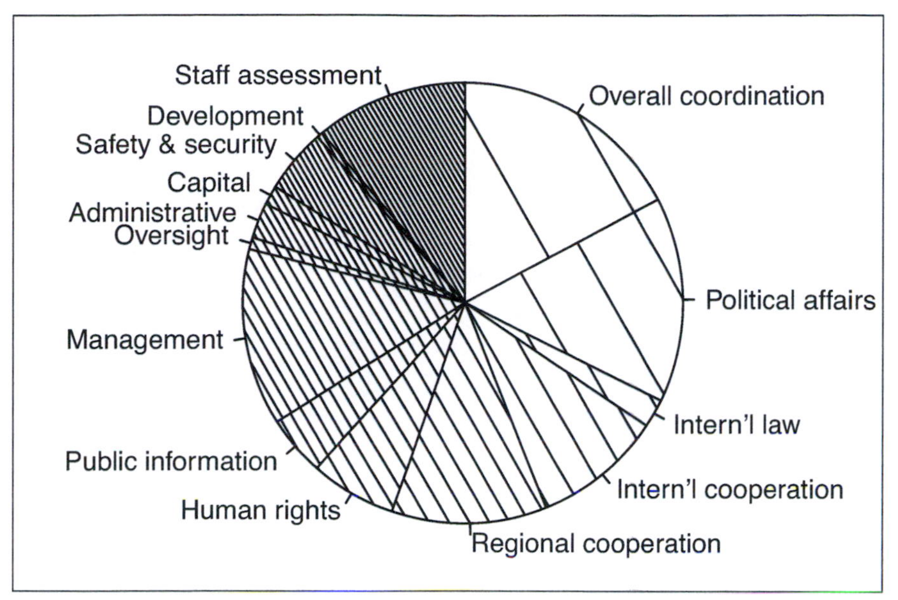

Figure 2.19 *Pie chart of the United Nations budget for 2008–2009 with hatching resulting in a moiré effect*

or central angles of the wedges? Do all viewers uniformly use just one of these comparisons?

With respect to process in estimating the size of pie-chart wedges, Eells [33] conducted an experiment with 94 participants. He found: 51% of subjects used arc length; 25% used area; 23% used central angles; and 1% used chord length. In reviewing Eells' [33] experiment, Kosslyn [77] noted: "Such variability does not bode well for effective communication of quantitative information using pie charts (Eells 1926)." I take this to be a polite understatement by Kosslyn.

Wainer [130] (p. 87) noted that the ability of a child to judge a slice representing one-third of a pie as being larger than another slice representing one-fourth as being a tribute to the graphical sense of humans for being able to make judgments about data from a chart as bad as a pie chart. Wilkinson [137] (p. 21) likewise concurred.

Wainer [130] attributed the origin of the pie chart to Playfair [96]. We cannot hold Playfair [96] responsible for the dubious distinction of conceiving the version of the pie chart discussed in the next subsection.

2.6.2 Pseudo-Three-Dimensional Pie Chart

An all too popular version of the pie chart is the *pseudo-three-dimensional pie chart*, an example of which is given in Figure 2.20 for the estimates of

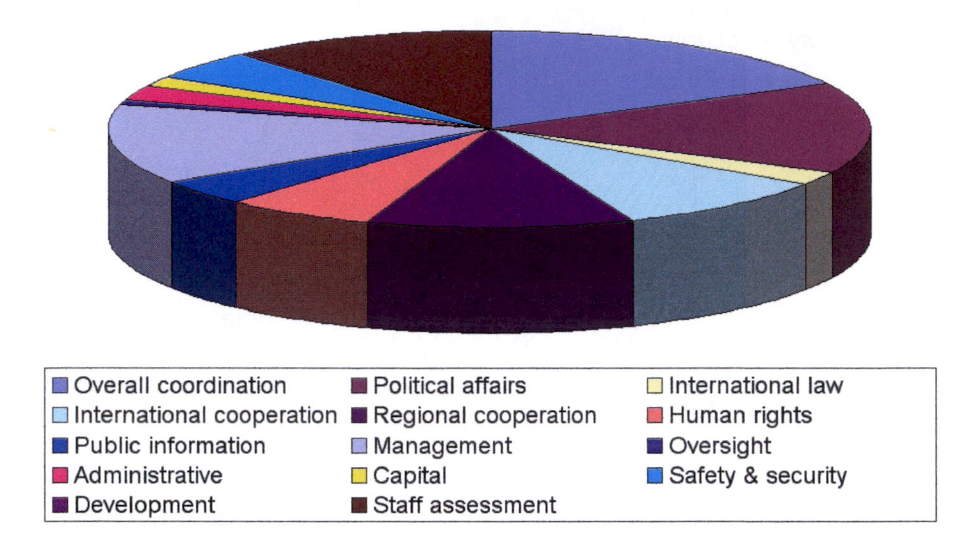

Figure 2.20 *Pseudo-three-dimensional pie chart of the United Nations budget for 2008–2009*

planned expenditure. The pseudo-three-dimensional pie chart suffers from the same deficiencies previously explored with the pseudo-three-dimensional bar chart in Figure 2.16, and then some.

Figure 2.21 attempts to correct the problem in Figure 2.20 that some of the pie wedges are viewed in three-dimensional perspective, while others are not. This is done by separating all the wedges in a so-called exploded pseudo-three-dimensional pie chart. But this attempt is only partially successful with some wedges in Figure 2.21 still remaining obscured.

Figures 2.20 and 2.21 also share a common problem that interferes with a viewer's apprehension of the data. Unlike the two-dimensional pie charts in Figures 2.17, 2.18, and 2.19, all of which have category labels connected to wedges by line segments, the pseudo-three-dimensional pie charts resort to using a color-coded key in a frame beneath the pie in each case. The key, also known as a legend, lists the category name with its corresponding color in a small square. Excel leaves the viewer to their own devices to decide how to read the key.

With a bit of effort, the viewer taking the time can determine that the wedges begin at 12 o'clock and proceed clockwise. The key is to be deciphered as beginning with the top left entry and is to be read one line across and then moving down to the next line until finished.

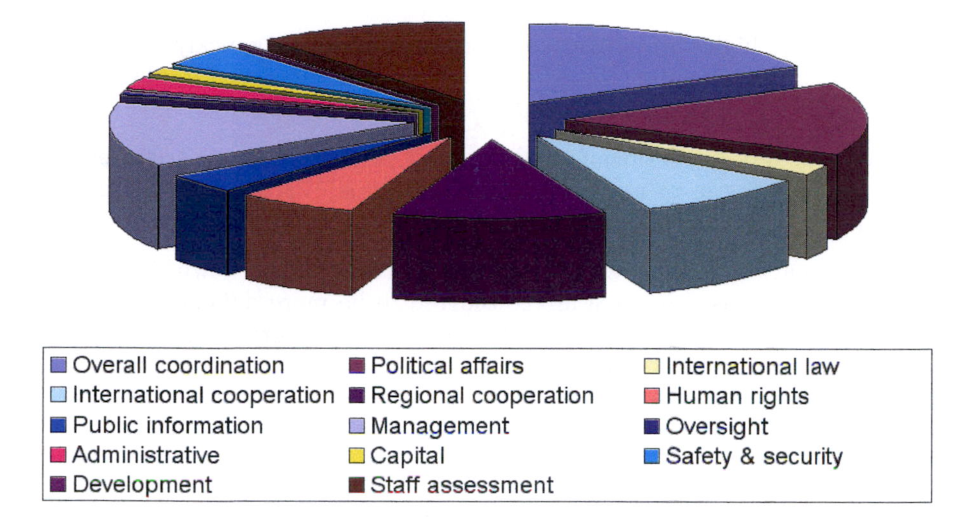

Figure 2.21 *Pseudo-three-dimensional exploded pie chart of the United Nations budget for 2008–2009*

In the previous section, reference to Baird [7] was made with respect to the distortion caused by the graphical rendering on a two-dimensional surface of three-dimensional perspective in the three-dimensional bar chart. This concern also relates to the two-dimensional and three-dimensional pie charts. It is worthwhile to pause here and briefly cover a few points regarding visual perception.

From an experiment with 550 human participants viewing 40 diagrams distributed amongst 9 sets, Croxton and Stein [30] reported in the statistical literature the following ordering of perceptual tasks in increasing order of difficulty further down the list:

- comparing bars of fixed width but varying length;
- comparing squares or circles of different area;
- comparing cubes of different volume.

Tasks that are higher in the list are more accurately performed. For tasks on the same line, there was insufficient data to establish which is preferable for graphical design.

In a large number of psychological experiments reviewed by Baird [7], a theoretical model for the ordering of the degree of difficulty of certain perceptual tasks in increasing order of difficulty further down the list is:

- perceiving amounts or differences in length, direction, or angle;
- perceiving amounts or differences in area; and
- perceiving amounts or differences in volume, or curvature.

Tasks that are higher in the list are more accurately and quickly performed. For tasks on the same line, there has been insufficient data to establish which is preferable for graphical design.

All the wedges of a two-dimensional pie chart can be compared based on arc length, chord length, or angle. Each comparison is of similar discriminating difficulty according to the previous list.

Wedges can also be compared based on area. Area is farther down the list and so comparisons based on area are considered to be more difficult perceptual tasks than comparisons based on either arc length, chord length, or angle. This would place the 25% of viewers, based on the estimate by experiment of Eells [33], at greater disadvantage in making comparisons. So if one insists on drafting a pie chart, one must be aware that about one-quarter of the audience will have greater difficulty analyzing the figure than the rest of the audience.

Alternatively, a pseudo-three-dimensional pie chart could be drafted. This would presumably place everyone at the same degree of difficulty by making comparisons based only upon volume. This assumes, in the absence of experimental results to the contrary, that viewers did not try to make comparisons of either arc length, central angle, chord length, or area. As we see from the previous list ordering tasks of visual perception, comparison of volumes is considered more difficult than comparison of area, and again more difficult than comparisons of either arc length, chord length, or angle. There is the additional consideration that comparisons of each of arc length, chord length, or angle are made more difficult by the distortion introduced by the artificial creation of perspective in three-dimensional pie charts. Comparison by area or volume is also not immune from this effect.

A worthwhile experiment to conduct would be to ask viewers of a pseudo-three-dimensional pie chart whether comparisons are made on the basis of perceived volume, top-surface area, central angle, arc length, or chord length. Notice that in the example of the pseudo-three-dimensional pie chart given in Figure 2.20 the top surface is elliptical not circular. The effect this distortion has on the judging of differences in the sizes of the wedges is a concern. Another aspect valuable to explore would be the effect of the angle of perspective on judging differences in wedge sizes by each of arc length, chord length, angle, top-surface area, or volume.

Figure 2.21 depicts a pseudo-three-dimensional pie chart with all the wedges exploded in an attempt to compensate for the loss of three-dimensional perspective information for the wedges in the rear of the pie of Figure 2.20. One does wonder whether pulling out wedges only increases the degree of difficulty in comparing the size of the wedges rather than achieving the desired effect of simplifying the task.

It is noted that some software packages automatically pull out the largest wedge from the rest of the pie regardless of whether that is the wedge that the researcher might intentionally want to emphasize or not. In other packages the researcher can control which wedge is exploded. In either case, one wonders what distortion this introduces into the process of comparing sizes of wedges.

2.6.3 Recommendations Concerning the Pie Chart

Seven recommendations for drafting a pie chart are listed below. For each recommendation there is a brief justification with respect to the **ACCENT** rule.

1. Do not draft pseudo-three-dimensional pie charts because of the perceptual difficulties in assessing or comparing the sizes of pie segments. The relevant axioms are those of *Apprehension*, *Clarity*, *Efficiency*, and *Truthfulness*.
2. Do not use a pie chart if the number of categories exceeds six because of perceptual difficulty related to the axioms of *Apprehension*, *Clarity*, and *Efficiency*.
3. Do not use hatching or cross-hatching as either will likely result in the perception of a moiré effect in the graphic by viewers. The moiré effect interferes with *Apprehension*, *Clarity*, and *Efficiency*.
4. Do not choose graytone shading or colors that could lead to difficulty in identifying distinct pie segments. To do so would lead to a conflict with the axioms of *Apprehension*, *Clarity*, and *Efficiency*.
5. Do not use a legend with a color key to identify categories with segments because this will require the viewer to take more time to study the graph. This would violate the *Efficiency* and *Necessity* axioms and interfere with *Apprehension*.
6. Do not place labels adjacent to pie segments in such a manner that the identification of the referenced pie segment is ambiguous. This would lead to a conflict with *Apprehension*, *Clarity*, and *Efficiency*.
7. Do not use pullouts in pie charts, or so-called *exploded wedges* if using pseudo-three-dimensional pie charts. To do so interferes with the process of comparing wedge sizes and leads to conflicts with *Apprehension* and *Efficiency*.

Probably the best recommendation is to use the dot chart instead of the pie chart. But even with the passage of time since the introduction of the dot chart by Cleveland and McGill [25] in the statistical literature in 1984, use of the pie chart has persisted. Given the widespread exposure of the pie chart, it is more familiar than the dot chart.

Figure 2.22 *Color dot chart of the United Nations budget for 2008–2009*

With the pie chart being more readily apprehended by the public on the basis of familiarity, there does remain a temptation to use it despite concerns with *Apprehension*, *Clarity*, and *Efficiency*. There is no denying that the dynamic use of colors in a pie chart can yield an attention grabbing advantage over a dot chart produced in black on a white background. Perhaps the solution is to be more creative with the use of color in the dot chart as in Figure 2.22.

2.7 Conclusion

Both Tufte [123] and Henry [64] cite Playfair [96] as presenting the first bar chart in 1786. Playfair's [96] original design was of a bar chart with horizontal bars.

Had William Playfair stopped at the bar chart as a means of depicting the distribution of a single categorical variable then all would have been well and good. But as noted by Wainer [130], the origin of the pie chart can be attributed as well to Playfair.

Eells [33] provided a spirited defense of the use of circles as compared to bars in representing component parts. He began his 1926 paper with a detailed review of condemnation of the pie chart by no less than five authors. Eells [33] lists five more authors who are likewise critical but does not go into detail because "Lack of space forbids the quotation of similar criticisms...."

It is not until 1984 that one finds an experimental comparison of the bar chart with the pie chart. Cleveland and McGill [25] reported the results of an experiment involving comparison of the pie chart with the vertical bar chart. In the experiment there were 51 participants with usable data. This was after discarding 4 participants after an assessment of the answers indicated that these participants had not followed instructions.

Cleveland and McGill [25] do not appear to have used order randomization and did not use a matched pair analysis. Segment magnitudes were selected essentially at random, subject to certain constraints.

Cleveland and McGill [25] reported that judgments with the bar chart were 1.96 times as accurate as with the pie chart and that this was statistically significant. They also found that large errors in estimating central angles were more than twice as likely as large errors in estimating lengths of bars. Cleveland and McGill [25] stated: "A pie chart can always be replaced by a bar chart, thus replacing angle judgments by position judgments."

Opinion and experiment does seem to strongly favor the choice of the simple bar chart over the pie chart as a visual means of presenting information concerning the distribution of a single qualitative variable. But the debate is still not fully closed. Wilkinson [137] (p. 21) asserts that the humble pie chart has been unjustifiably reviled by statisticians while being unjustifiably adored by managers.

Henry [64] does not heap high praise for the lowly pie chart. Indeed, the seven pitfalls noted in this chapter in the use of pie charts are taken from Henry [64] (1995, p. 44). Avoiding these pitfalls is sound advice indeed.

Tufte [123] (pp. 178) is considerably more blunt: "A table is nearly always better than a dumb pie chart; the only worse design than a pie chart is several of them, for then the viewer is asked to compare quantities located in spatial disarray both within and between pies...."

Wilkinson [137] (p. 21) notes that a five-year-old can look at a slice of pie and be a fairly good judge of proportion. Given the weight of evidence, it would appear that the comments by Wilkinson [137] in support of the pie chart represent a dissenting minority view.

A worthwhile recommendation to consider from Cleveland and McGill [25] is that a sensible thing to do when a bar chart is used to replace a pie chart is to make the scale go from 0% to 100% so that the viewer can readily appreciate that each bar represents a fraction of 100%. They also suggest that 0% to 25% or 50% are also reasonable alternative choices.

The prevalence of the pseudo-three-dimensional bar and pie charts in the media and scientific presentations is intriguing. Kosslyn [77] cites the probable source of the misconception of the usefulness of these deplorable graphics as being given by Schmid [104].

Shockingly, the version of the pseudo-three-dimensional pie chart proposed by Schmid [104] entails drawing a circular pie chart and then adding a front lip to the pie to give the illusion of three-dimensional perspective. Following these instructions of Schmid [104] leads to the optical illusion that the circular pie is elliptical. This further compounds the inherent difficulty of the visualization process for the pseudo-three-dimensional pie chart as the viewer receives a distorted version of the frontal area. Schmid [104] gives incredibly bad advice in this regard. Ever since Playfair's [96] introduction of the pie chart in 1786, we have taken the bad with the good in the field of statistical graphics.

There are no experiments comparing the dot chart with the bar chart. This is unfortunate. Cleveland and McGill [25] state their preference for the dot chart over the bar chart. Justification for their choice must necessarily be based upon theoretical considerations. If the **ACCENT** rule of the previous chapter is applied in this situation, the matter is decided by the *Efficiency* principle. This comes down to choosing the dot chart over the bar chart on the basis of the lower data-ink ratio for the dot chart.

In the next chapter, more sophisticated versions of the dot chart and bar chart will be presented. A variation on this chapter's plotting convention for the bar chart will also be given. There will also be further consideration of the relative merits of the pie chart in the next chapter. No further development of the pie chart will be given. Any discussion of a plotting convention for the pie chart is wholly contained in this chapter.

2.8 Exercises

1. Figure 2.23 provides an example of the form of the bar chart recommended on page 128 of Tufte [123] using the expenditures planned by the United Nations for the 2008–2009 biennium.
   (a) Compare Tufte's [123] recommended bar chart in Figure 2.23 with the dot chart given in Figure 2.1 with respect to the data-ink ratio and comment.
   (b) Is Tufte's [123] recommended vertical bar chart in Figure 2.23 superior to the horizontal dot chart of Figure 2.1? Discuss.
   (c) Discuss Figure 2.23 with respect to the principles of the **ACCENT** rule.

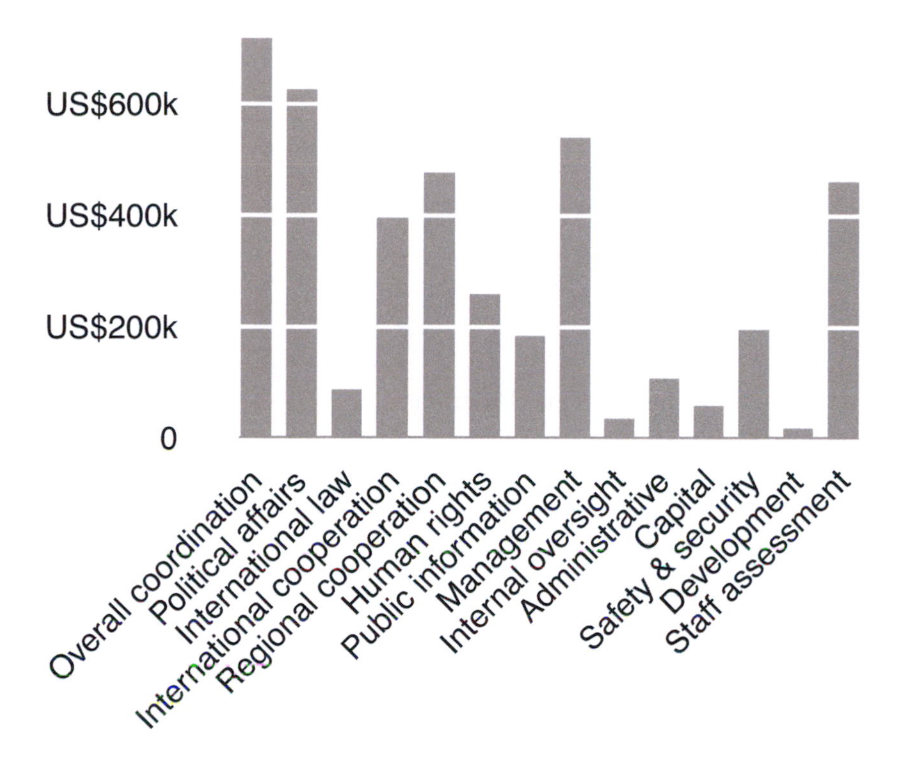

Figure 2.23 *Tufte-style bar chart of the United Nations budget for 2008–2009 with category labels at an angle of 45 degrees*

2. Figure 2.23 provides an example of the form of the bar chart recommended on page 128 of Tufte [123] using the expenditures planned by the United Nations for the 2008–2009 biennium.
   (a) Compare Tufte's [123] recommended bar chart in Figure 2.23 with the dot chart in the style of Cleveland and McGill [25] given in Figure 2.2 with respect to the data-ink ratio and comment.
   (b) Compare Figures 2.2 and 2.23 on a point-by-point basis with the six principles of the **ACCENT** rule.
   (c) In their respective books on statistical graphics, Cleveland [21] advocates the dot chart as the preferred graphic for depicting distributions of a single discrete variable while Tufte [123] recommends a form of the bar chart as illustrated in the example of Figure 2.23. Which one do you prefer? State your reasoning.

Agency	Amount (Billions of Dollars)
Agriculture	20.8
Commerce	8.2
Defense	515.4
Education	59.2
Energy	25.0
Health and Human Services	70.4
Homeland Security	37.6
Housing and Urban Development	38.5
Interior	10.6
Justice	20.3
Labor	10.5
State and Other International Programs	38.3
Transportation	11.5
Treasury	12.5
Veterans Affairs	44.8
Corps of Engineers	4.7
Environmental Protection Agency	7.1
Executive Office of the President	0.4
Judicial Branch	6.3
Legislative Branch	4.7
National Aeronautics and Space Administration	17.6
National Science Foundation	6.9
Small Business Administration	0.7
Social Security Administration	8.4
Other Agencies	7.2
Total	**987.6**

Table 2.2 *Requested discretionary funding by major agencies of the US government for the fiscal year 2009*

3. Figure 2.23 provides an example of the form of the bar chart recommended on page 128 of Tufte [123] using the expenditures planned by the United Nations for the 2008–2009 biennium.
   (a) Compare Tufte's [123] recommended horizontal bar chart in Figure 2.23 with the horizontal bar chart of Figure 2.7 with respect to the data-ink ratio and comment.
   (b) Compare Figures 2.6 and 2.23 on a point-by-point basis with the six principles of the **ACCENT** rule.
   (c) Is Tufte's [123] recommended bar chart in Figure 2.23 superior to the horizontal bar chart of Figure 2.7? Discuss.
4. Table 2.2 reports discretionary funding requested by various US govern-

ment agencies under the budget for the fiscal year 2009. Use the order of categories as listed in the table.

(a) Is the ordering of categories in Table 2.2 alphabetical or something else?

(b) Draft a dot chart for the data.

(c) Produce a pie chart for the requested discretionary funding by agency.

5. Table 2.2 reports discretionary funding requested by various US government agencies under the budget for the fiscal year 2009. Use the order of categories as listed in the table and the `graphics` package in R.

(a) Generate a horizontal bar chart for the data.

(b) Change the orientation of the bar chart to vertical.

(c) Which way is the chart more eye-appealing? Discuss.

6. Table 2.2 reports discretionary funding requested by various US government agencies under the budget for the fiscal year 2009. Use the order of categories as listed in the table and the `ggplot2` package for R.

(a) Generate a horizontal bar chart for the data.

(b) Change the orientation of the bar chart to vertical.

(c) Comment on the ease of use of `ggplot2` for the two tasks above.

7. Governments of foreign nations monitor US government spending and budget plans as reported in the public domain to American taxpayers. This is an example of overt and legal foreign intelligence gathering done by allies and adversaries alike. Assume you are a public servant of a nation located in the eastern hemisphere and that you have been tasked to report on the contents of Table 2.2.

(a) In this context, choose a suitable ordering of agencies and draft an appropriate graphic.

(b) Justify your choice of categorical ordering for your chosen graphic.

(c) Justify your choice of graphic.

(d) What would be your response if your superior requested that you produce a pie chart depicting the contents of Table 2.2?

8. Enter the data of Table 2.3 on crude oil imports into the United States in 2007 into a spreadsheet or statistical graphics package of your choice.

(a) What sort of ordering is used in Table 2.3?

(b) To depict the data in Table 2.3, what sort of ordering do you prefer? Justify your preference.

(c) Produce a vertical bar chart. Did you encounter any difficulties?

9. Enter the data of Table 2.3 on crude oil imports into the United States in 2007 into an R data frame.

(a) Re-order the data in Table 2.3 so that the ranking is from least to greatest in terms of oil imports.

Country	Thousand Barrels per Day
Canada	1,848
Saudi Arabia	1,382
Mexico	1,398
Nigeria	1,085
Venezuela	1,301
Iraq	433
Angola	504
Kuwait	165
Colombia	106
Algeria	474
Ecuador	226
Brazil	156
Congo (Brazzaville)	48
Chad	78
Russia	40

Table 2.3 *Top fifteen crude oil importers into the United States in 2007*

(b) Produce a horizontal dot chart for the data as re-ordered in part (a) using the `graphics` package. If the nation with the greatest oil imports into the US is not located at the bottom of the dot chart, then re-arrange the data as necessary to achieve this effect.

(c) From the dot chart produced for part (b), what can be inferred about the volume of oil imports from the nearest neighbors to the United States?

10. Instead of using the `graphics` package in part (b) of Exercise 9, use `ggplot2` to depict the data in a horizontal dot chart. Does the syntax of `ggplot2` give a better appreciation for the elements of aesthetics in the chart than the syntax of the `graphics` package? Discuss.

11. Consider the data of Table 2.3 on crude oil imports into the United States in 2007.

    (a) Re-order the data in Table 2.3 so that the ranking is from least to greatest in terms of oil imports.

    (b) Produce a horizontal bar chart for the data as re-ordered in part (a) using `ggplot2`. If the nation with the greatest oil imports into the US is not located at the bottom of the dot chart, then re-arrange the data as necessary to achieve this effect.

    (c) Produce a pie chart for the data of Table 2.3 for oil exporting nations arranged clockwise in alphabetical order.

    (d) Which do you prefer: the bar chart of part (b) or the pie chart of part (c)? Justify your answer.

12. You are preparing a media story on crude oil imports into the United States.

Design and execute a graph to provide context in the introduction using the data from Table 2.3 on crude oil imports into the United States in 2007.

Chapter 3

Advanced Charts for the Distribution of a Single Discrete Variable

3.1 Introduction

In the previous chapter, the dot chart, the bar chart, and the pie chart were introduced. The first example chosen for these charts in the previous chapter was the United Nations budget for the years 2008–2009 with fourteen different categories of expenditures. There was no need to depict variation in the number of US Dollars for each category of expenditure. This is a rare example not involving a random sample. Advanced versions of the dot chart and the bar chart are introduced in this chapter that illustrate the degree of variation associated with the proportion illustrated for each value of a discrete random variable.

The pie chart has a complex nature due to its reliance on either the angles, arcs, or areas of sectors for comparisons with regard to the distribution of categories or discrete values. This is unlike either the dot chart or bar chart that rely on length. A variant of the pie chart for incorporating an illustration of variation for the estimate of proportion for each category has not emerged to stand the test of time.

Although a viewer can make comparisons between values of a single discrete variable based on area with the bar chart, it is best to rely on length with respect to a common axis. The building block for the bar chart is rectangles laid adjacent to each other to form bars. The building block for a dot chart is a dot separated by enough space to be individually distinguishable to depict the count for a category or discrete value. Symbols other than rectangles and dots can be used. This chapter discusses how this can be done to minimize distortion of data.

Options with respect to orientation, fill, order, and details with respect to axes were considered in the previous chapter with respect to the dot, bar, and pie charts. In this chapter, order is reviewed again with added discussion regarding the framing of statistical graphics and the merits of using grid lines.

This chapter begins with a straightforward variant of the bar chart using the first example of the previous chapter. The relative merits of both versions of the bar chart are discussed.

3.2 Learning Outcomes

When you complete this chapter, you will be able to do the following.

- Know what a stacked bar chart is and be able to create either a vertical or horizontal orientation with either the `graphics` or `ggplot2` packages in R.
- Know what a pictograph is and be aware of the pitfalls in using a pictograph that can lead to telling lies with data.
- Know how to use either the `graphics` or `ggplot2` packages in R to produce dot and bar charts with and without error bars.
- Know how to produce or suppress frames and grid lines in either the `graphics` or `ggplot2` packages in R.
- Know how to control the order of categorical variables in either the `graphics` and `ggplot2` packages in R.

3.3 The Stacked Bar Chart

3.3.1 Definition

The *stacked bar chart* is a version of the bar chart in which the bars for each value of the discrete variable are stacked on top of one other rather than displayed side-by-side. An example of a stacked bar chart in vertical orientation is given in Figure 3.1 for the planned United Nations budget for 2008–2009.

Figure 3.1 was generated in part by a call to the R function `barplot` in the code excerpt below.

```
colr<-c("royalblue","khaki","lightgreen",
"hotpink","powderblue","black","green",
"orange","lightsteelblue","red","violet",
"yellow","aquamarine","brown")

item1<-rev(item)
amount1<-matrix(rev(amount),ncol=1,
nrow=length(amount))

barplot(height=amount1,space=c(0,4),
horiz=FALSE,axes=FALSE,beside=FALSE,
```

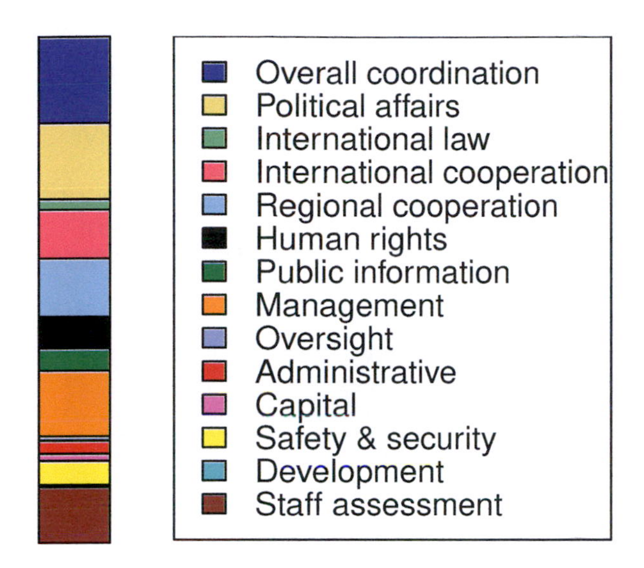

Figure 3.1 *Vertical stacked bar chart of the United Nations budget for 2008–2009*

```
width=rep(0.4,14),xlim=c(0,4),legend.text=FALSE,
col=colr[length(colr):1])

legend(x="top",legend=item,fill=colr,y.intersp=0.86)
```

The colors in the vector variable `colr` were carefully chosen to ensure clear contrast among the 14 categories of planned expenditure. The amounts of expenditure were converted into the form of a matrix consisting of 14 rows and 1 column. This was needed to force `barplot` to produce a stacked bar chart. The legend was not plotted in the call to `barplot` because the default position was poor in this instance. Instead the R function `legend` was called because it resulted in a better position for the legend relative to the stacked bar.

Note that there is no scale provided in Figure 3.1. But neither was a scale provided for the pie chart in Figure 2.18. With the pie chart, the only objective is comparison of the size of the various slices. In comparison with Figure 3.1, the bar chart of Figure 2.7 has a scale and allows the viewer the opportunity to estimate the values of planned expenditure for one or more categories of interest.

Figure 3.2 illustrates a stacked bar in horizontal orientation. Cleveland and McGill [25] refer to the stacked bar chart as the *divided bar chart*. Clearly in the horizontal orientation there is no stacking in the vertical sense. In either

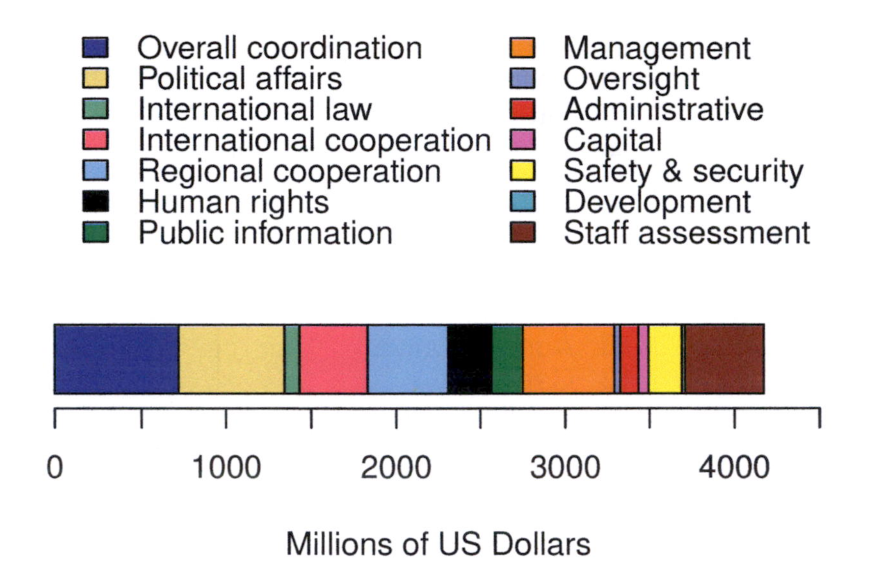

Figure 3.2 *Horizontal stacked bar chart of the United Nations budget for 2008–2009 without framing around color key*

orientation a single bar has been subdivided to depict the proportions of the various categories. Figure 3.2 was generated in part by calls to the R functions `barplot`, `axis`, and `legend` in the code excerpt below.

```
barplot(height=amount1,space=c(0,6),horiz=TRUE,
axes=FALSE,beside=FALSE,width=rep(0.75,14),
ylim=c(0,4),legend.text=FALSE,col=colr,
xlim=c(0,4500),xlab="Millions of US Dollars")

axis(1,at=(0:9)*500,labels=c("0","","1000","",
"2000","","3000","","4000",""))

legend(x="topleft",legend=item[1:7],
fill=colr[1:7],y.intersp=0.8,bty="n")

legend(x="topright",legend=item[8:14],
fill=colr[8:14],y.intersp=0.8,bty="n")
```

The stacked bar charts of Figure 3.1 and 3.2 require viewers to use a legend to decode the colors associated with the expenditure categories. To save vertical space in the horizontal bar chart of Figure 3.2, the legend consists of two columns for the expenditure categories.

The ordering within the color-coding key of the legend is the same as the ordering of the colored bars in the stack in the vertical stacked bar charts of Figure 3.1 and 3.2. For the horizontal stacked bar chart of Figure 3.2, one reads the key from top to bottom of the left column, and then the right, and the bars from left to right.

Despite the color key in the legends of the stacked bar charts of Figure 3.1 and 3.2, both displays still require considerably more effort than reading the names for the bars of the horizontal bar chart of Figure 2.7.

The scale missing from Figure 3.1 has been added to Figure 3.2. But unlike the horizontal bar chart of Figure 2.7, there is no common alignment for the start of each category in the horizontal stacked bar chart of Figure 3.2. The task of estimating bar lengths with a scale is still much more easily accomplished with the bar chart of Figure 2.7 than with the stacked bar chart of Figure 3.2. Despite this difficulty, the stacked bar chart does appear occasionally for a single discrete variable. The stacked bar chart will be encountered again in Chapter 8 that deals with two or more populations. In any case, the stacked bar chart is typically used only with a small number of categories.

A `ggplot2` version of the horizontal bar chart is given in Figure 3.3. The code for producing Figure 3.3 is given below.

```
un_fiscal<-data.frame(item=factor(item,levels=item,
ordered=TRUE),amount=amount)

colr<-c("royalblue","khaki","lightgreen",
"hotpink","powderblue","black","green",
"orange","lightsteelblue","red","violet",
"yellow","aquamarine","brown")

figure<-ggplot(un_fiscal,
aes(1,amount,fill=factor(item,levels=rev(item),
ordered=TRUE))) +
scale_fill_manual(values=rev(colr)) +
theme(axis.ticks.y=element_blank(),
axis.text.y=element_blank(),
legend.title=element_blank(),legend.position="top",) +
guides(fill=guide_legend(ncol=2,byrow=FALSE,reverse=TRUE)) +
geom_bar(stat="identity")+coord_flip() +
labs(x=NULL,y="Millions of US Dollars")

print(figure)
```

Because coding in `ggplot2` requires using R data frames, the first two lines of the preceding code are used to create the data frame for the United Nations budget data. The default ordering for categorical random variables in `ggplot2`

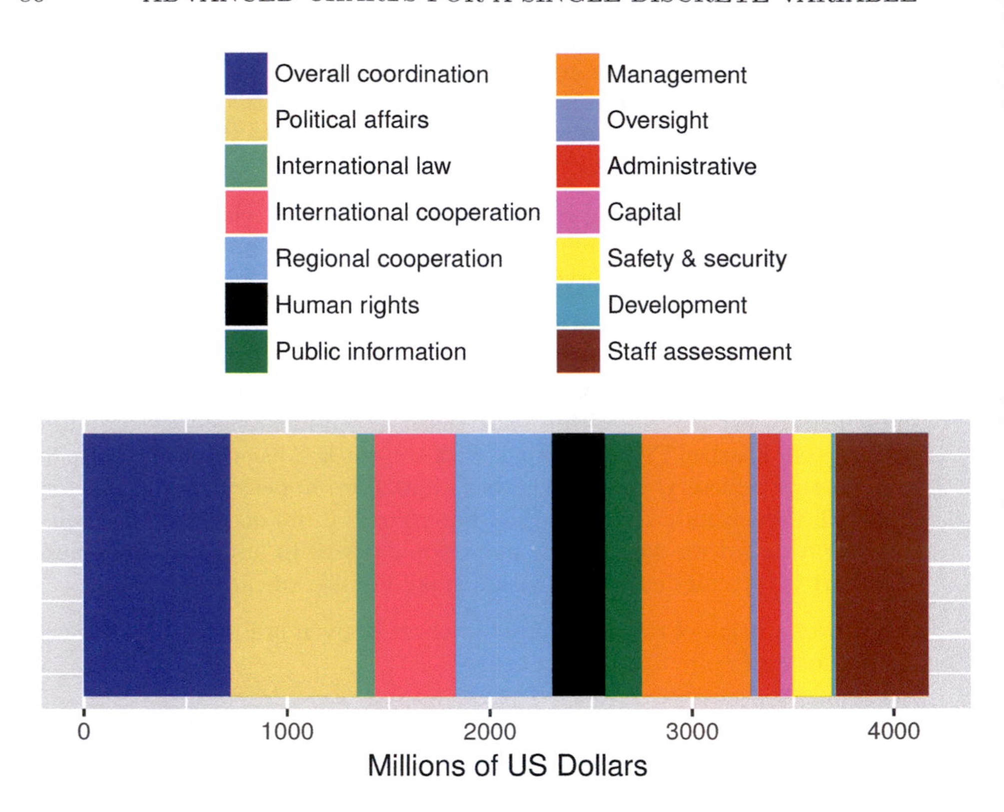

Figure 3.3 *Horizontal stacked bar chart of the United Nations budget for 2008–2009 using* ggplot2 *without framing around the color key*

is alphabetical, so using the R function `factor` to convert the items to a factor vector from a character vector with the argument `ordered=TRUE` will override the default ordering. The same color scheme has been used in all three Figures 3.1 through 3.3. The default graphics `theme` in `ggplot2` has been used so a light blue background with white grid lines appears around the horizontal bar chart in Figure 3.3. To obtain the ordering of color from left to right, it has been necessary to make the call `scale_fill_manual(values=rev(colr))` with the order of the colors reversed. So that the order of colors in the legend matches correctly, the preceding code must also include `guides(fill=guide_legend(ncol=2,byrow=FALSE,reverse=TRUE))`. The default in `ggplot2` is to produce vertical bar charts so the call `coord_flip()` is needed to obtain the horizontal bar chart of Figure 3.3.

In all three Figures 3.1 through 3.3, code has been written to obtain only an axis for amounts in millions of US dollars. This has been done in Figures 3.1 and 3.2 by shutting off all the axes and just adding the axis for the amounts, whereas `axis.text.y=element_blank()` has been passed when calling the function `theme` of `ggplot2` to eliminate the axis for the item names in Figure 3.3. Excluding spaces and the comment symbol `#`, the R code for Figure 3.2 required 679 characters, which is 62 more than the 617 characters for Figure 3.3. The R code for Figure 3.2 uses 57 characters to explicitly code for the numerical axis labels, whereas the `ggplot2` default provided these.

3.3.2 The Stacked Bar Plot versus the Bar Chart and the Pie Chart

Cleveland and McGill [25] reported perceptual experiments involving common graphical displays. Their first experiment compared vertical stacked bar charts with vertical bar charts in a completely randomized design with 55 human participants. They referred to this experiment as the position-length experiment. Their second experiment has already been discussed in Chapter 2. Their second experiment compared pie charts with vertical bar charts in a completely randomized design with 54 human participants. They referred to their second experiment as the position-angle experiment.

In the position-length experiment, the judgments of four people were deleted because it was clear from their answers that they had not followed instructions. As reported in the previous chapter for the position-angle experiment, the judgments of three subjects were deleted for the same reason. For both experiments, 51 participants remained in the analysis set. Most of the participants in the position-length experiment also participated in the position-angle experiment.

The statistical analysis of these two experiments showed that the accuracies of judgments by bar chart, pie chart, and stacked bar chart were different. The ordering from least absolute error to most was vertical bar chart, pie chart, and stacked bar chart.

Fifty-eight years before the publication of the article by Cleveland and McGill [25], Eells [33] in 1926 sparked a lively debate in the *Journal of the American Statistical Association* as to whether pie charts, or *circle diagrams* as Eells referred to them, were better than stacked bar charts. His review of the literature at the time revealed a preference for stacked bar charts over pie charts. Eells [33] found that this preference, while grounded in theory, was without experimental support.

With nearly 100 students in a class in general psychology in Whitman College, Eells [33] ran an experiment comparing pie charts and bar charts. What is sometimes overlooked in later reviews of this work is that Eells [33] compared pie charts with horizontal stacked bar charts. He did not compare pie charts with bar charts.

Eells [33] gave a page containing 15 pie charts to 97 members of a class in general psychology. The participants were instructed to write in each sector their best estimate of the percentage of the whole represented by that sector.

Thirteen minutes were allowed for this work with students instructed to mark the circle on which they were working at the five minute mark.

Three days later, he gave 15 horizontal divided bar charts (each with the same proportion for a corresponding pie chart of the previous test) to 94 of these students. There is no evidence of difference in rapidity of judgment between pie charts and stacked bar charts in the analysis Eells [33] published. He found the pie charts to be read more accurately than the stacked bar charts.

There are flaws with his experimental design. To be fair to Eells [33], Fisher's [40] *Statistical Methods for Research Workers* was just published in the previous year, 1925. Eells [33] failed to randomize the order in which the students received the pie chart test and the stacked bar chart test. He did not conduct a matched pairs analysis nor did he conduct a t-test.

If one disregards the matched pairs procedure, which would have reduced noise, a Student's t-test for comparing the means of two populations with an alternative hypothesis of a difference in accuracy yields a P-value of 0.04 with Eells' [33] data. The difference is in favor of the pie chart over the horizontal stacked bar chart. This result is not inconsistent with the experiments done by Cleveland and McGill [25].

Eells' [33] affront to the conventional wisdom that the stacked bar chart was better than the pie chart did not go unchallenged. Within a period of 9 months, von Huhn [129] and then Croxton [29] replied in rebuttal.

von Huhn [129] stuck to philosophical considerations. He noted the absence of scale in Eells' [33] experiment. He also noted the absence of labels for the sectors. von Huhn [129] took issue with the fact that Eells' [33] experiment was not set up to assess the relative merits of pie charts or horizontal stacked bar charts in the context of comparisons for two or more populations.

Croxton [29] represented the results of an experiment similar to Eells' [33]. Croxton's [29] previously unpublished experiment had been done six years earlier with 287 participants.

Croxton's [29] experiment consisted of an evaluation of two pie charts and two stacked bar charts depicting two categories, each in a ratio of either 1:1.5 or 1:4. Croxton [29] did not randomize the order in which participants received the charts, nor did he conduct a matched pairs analysis.

Croxton [29] reported that the horizontal stacked bar chart appeared to be much superior to the pie chart. Eighty-three years later, using Fisher's exact test with the two contingency tables reported by Croxton [29], P-values are calculated to be 0.002 and less than 0.0001 for the ratios of 1:1.5 and 1:4, respectively. These values cast doubt on Eells' [33] conclusions.

There are important differences between the experiments of Croxton [29] and Eells [33]. Croxton's [29] sample size is much larger but Eells' [33] tests included 30 charts compared to the 4 charts in Croxton's [29] experiment.

Country	Thousand Barrels per Day
Canada	2,460
Mexico	1,538
Saudi Arabia	1,394
Venezuela	1,273
Nigeria	1,120

Table 3.1 *Top five countries exporting petroleum to the United States in 2007*

Eells [33] asked participants to report percentages while Croxton [29] requested ratios. These differences were reported by Croxton [29].

Croxton and Stryker [31], nine months after the publication of the article by Croxton [29], reported a large experiment intended to correct the inadequacies of the experiments of Eells [33] and Croxton [29]. This large experiment consisted of 807 participants viewing 27 diagrams. There were 14 pie charts and 13 horizontal stacked bar charts. Unfortunately, randomization and matched pair analysis were again not done.

The results reported by Croxton and Stryker [31] appear to support in most, but not all instances, that the circle was significantly superior to the bar. But one must take note here that the bar chart under consideration was in fact a horizontal stacked bar chart.

Eells, Croxton, and Stryker failed to go the extra step and add to their comparison the lowly bar chart. We have from the experiments of Cleveland and McGill [25], that the bar chart is the better of the three choices. By application of the **ACCENT** rule, the dot chart is the preferred among four possible choices, considered so far, for depicting the distribution of a single discrete variable.

In the next section, a generalization is given of dot charts and bar charts.

3.4 The Pictograph

3.4.1 Definition

A *pictograph* takes a pictorial symbol representing a unit of measure and then replicates the symbol in a row, or column, for each category with consistent spacing between each symbol. Table 3.1 reports the volume of petroleum, including refined products and crude oil, imported daily into the United States in 2007. Using the illustration of an oil barrel to represent an import of 0.5 million barrels of petroleum per day, Figure 3.4 depicts the data of Table 3.1. The source of this data is the Energy Information Administration, which provides official energy statistics for the US government. The United States imports approximately 60% of its petroleum needs from abroad.

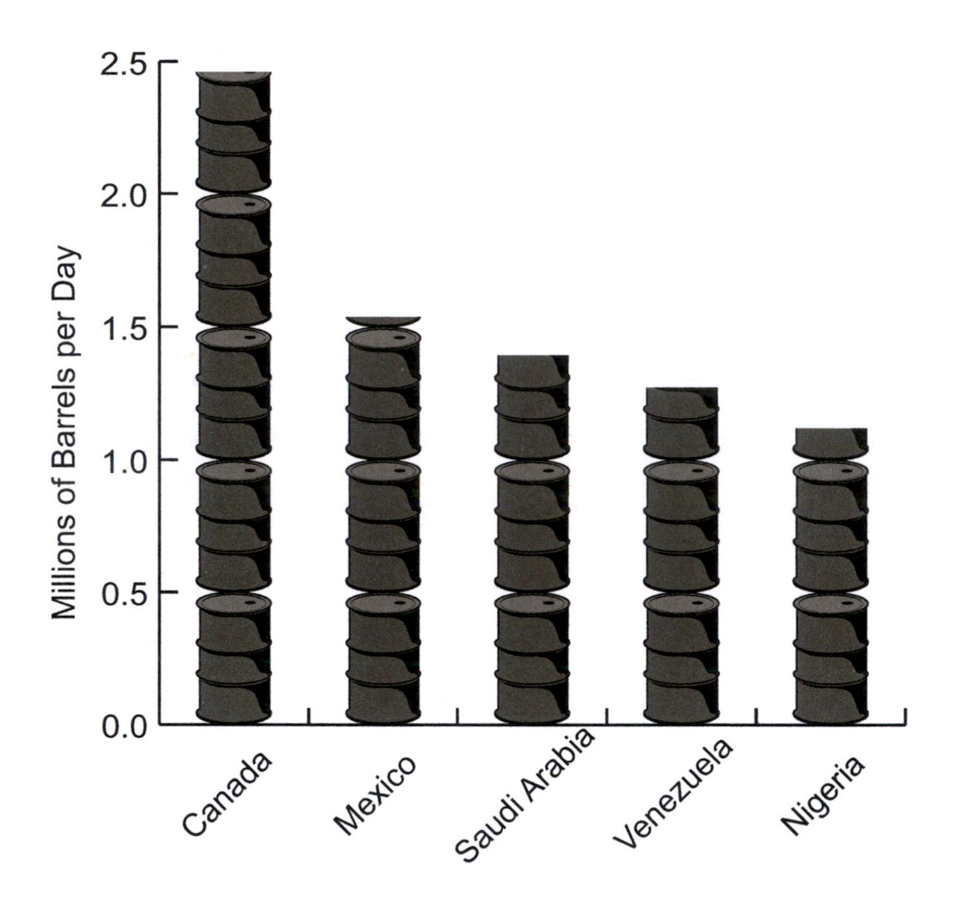

Figure 3.4 *Repeating pictograph of the top five countries exporting petroleum to the United States*

A synonym for pictograph is *pictogram.* Although less familiar, the term *iconograph* is also used because the word icon is synonymous with the term pictorial symbol.

As should be clear from Figure 3.4, each oil barrel represents a daily average import of 0.5 million barrels of crude oil or refined products. Clearly at 2.46 million barrels per day, Canada is the leading exporter of petroleum to the United States. Notice in Figure 3.4 that the topmost barrel for Canada has been chopped because 0.46 million barrels per day is less than a multiple of 0.5. The columns for the other four nations have been similarly affected.

Because the oil barrel icon is pseudo-three-dimensional, the vertical chopping comes across to the viewer as a bit crude. This does serve to educate that the chopping mechanism in the Adobe Illustrator (Adobe Systems Incorporated,

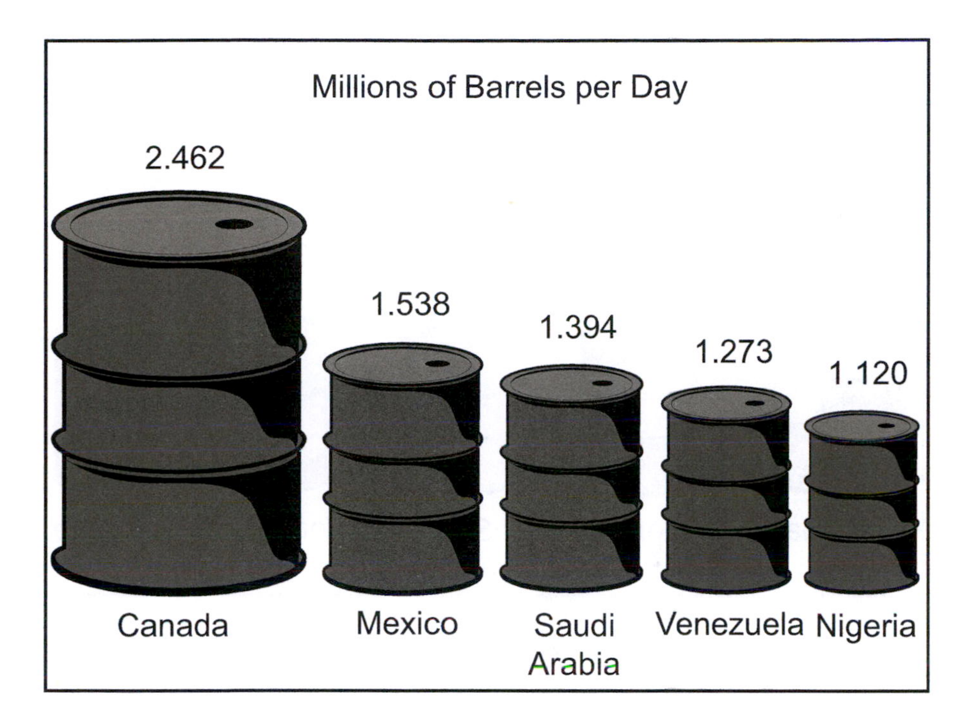

Figure 3.5 *Pictograph of top five countries exporting petroleum to the United States (with icon size proportional to height)*

San Jose, California) software used to draft Figure 3.4 performs its job in just two dimensions not three. Had an oil barrel been drafted in two-dimensional cross section rather than three-dimensional perspective, the result would have been cleaner. But there was a reason for choosing the three-dimensional perspective for the oil barrel, as seen in the pictograph of Figure 3.5.

One oil barrel is illustrated for each country in Figure 3.5. Each oil barrel is intended to represent the volume of daily imports. Illustrator can readily draft such a figure. However, to get separation between the oil barrels, manual features in Illustrator were used rather than relying on the built-in graphing function as was used for the repeating pictograph of Figure 3.4. Either way, a version of Figure 3.5 can be quickly generated by Illustrator.

Figure 3.5 purports to portray the same information in Figure 3.4 but does so in the worst possible way. The height and width of the oil barrels have been uniformly scaled. Coupled with the illusion of three-dimensional perspective, the viewer is left with the impression in Figure 3.5 that the imports from each of the other four countries are much less than the imports from Canada.

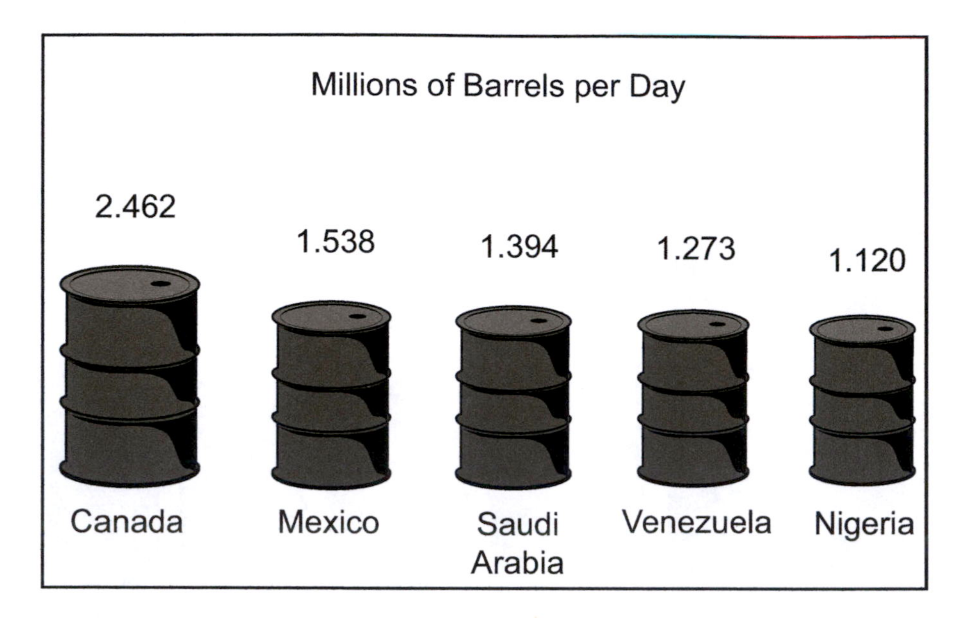

Figure 3.6 *Pictograph of top five countries exporting petroleum to the United States (with icon size proportional to volume)*

Truthfulness of a graphic can be assessed by the following statistic:

$$\textbf{Lie Factor} = \frac{\text{size of effect shown in graph}}{\text{size of effect in data}}.$$

Tufte [123] suggests that the goal is to achieve a Lie Factor of one. He suggests that Lie Factors of less than 0.95 or more than 1.05 are indicative of substantial distortion in the presentation of the data by the graphic.

In Table 3.1, the ratio of Canadian imports to Nigerian imports is 2.128:1. With uniform scaling implicit in all three dimensions in Figure 3.5 the ratio depicted is 2.128^3:1 or 10.622:1. This leads us to conclude that for Figure 3.5 the

$$\textbf{Lie Factor} = \frac{10.662}{2.128} = 4.833.$$

As far as lies go, Figure 3.5 tells a whopper.

The pictograph of Figure 3.6 is an attempt to improve on Figure 3.5. Figure 3.6 does this by achieving a Lie Factor equal to one. But without a careful time-consuming comparison of the oil barrels in Figure 3.6, say of Canada and Nigeria, it is not easily discerned that imports of Canadian petroleum products are slightly more than twice those of Nigerian petroleum.

The scaled pictographs of Figures 3.5 and 3.6 both dispense with a vertical axis and scale. But both figures report the volume of petroleum imports directly above the oil barrel for each country. It is not unusual to see this done. From the perspective of the data-ink ratio, reporting numbers above an icon together with a vertical scale is wasteful. It could be argued in terms of the data-ink ratio that reporting numbers only on a graphic is more efficient than adding and labeling a vertical axis.

Added to Figures 3.5 and 3.6 is a full outside frame for each. There are two purposes for doing this. One is merely artistic. The other is to provide horizontal and vertical references for comparing the sizes of the oil barrels.

3.4.2 The Pictograph versus the Dot Chart and the Bar Chart

Researchers need to be just as concerned with ambiguous graphics as they are concerned with ambiguous questionnaire items. Both can lead to distortion of the real picture of the population. Pictographs ought to be avoided because the viewer is drawn to comparison of areas or volumes when the intended comparison is typically that of lengths along a common axis.

Figure 3.7 presents a bar chart of the top five countries exporting petroleum to the United States. In comparison with the repeating pictograph of Figure 3.4, the bar chart of Figure 3.7 offers a lower data-ink ratio. Had a solid fill color been used for the bar chart, it could be argued that the repeating pictograph has the lower data-ink ratio. It can be argued that the Lie Factor for each of the repeating pictograph and the bar chart is one. Because the bars are of constant width in Figure 3.7, the Lie Factor is identical for the viewer making comparisons based upon areas as opposed to lengths of the horizontal bars.

Because Figure 3.7 allows viewers to choose to make comparisons based upon either area or length of the bars, the dot chart of Figure 3.8 is promoted as the superior alternative to either the bar chart or pictograph. Note that the dot chart is really just a version of the pictograph with a dot being used as a repeating icon. The dot chart of Figure 3.8 has the additional embellishment of the final terminating dot being of much larger size.

The bar chart is also just a version of the pictograph in which each icon is of constant width and is scaled by length alone. It could be splitting hairs to discuss the merits of whether dot charts and bar charts are best considered special cases of the pictograph. Choosing a chart that depicts the data as free of distortion as possible is the main issue.

Careful comparison of Figures 3.7 and 3.8 will reveal that something else is happening with the display of information. The most obvious difference is that Figure 3.7 is a bar chart and that Figure 3.8 is a dot chart. But there is another key difference between the two. The ordering of the categories is different.

The lengths of the bars increase in length in the bar chart of Figure 3.7 with

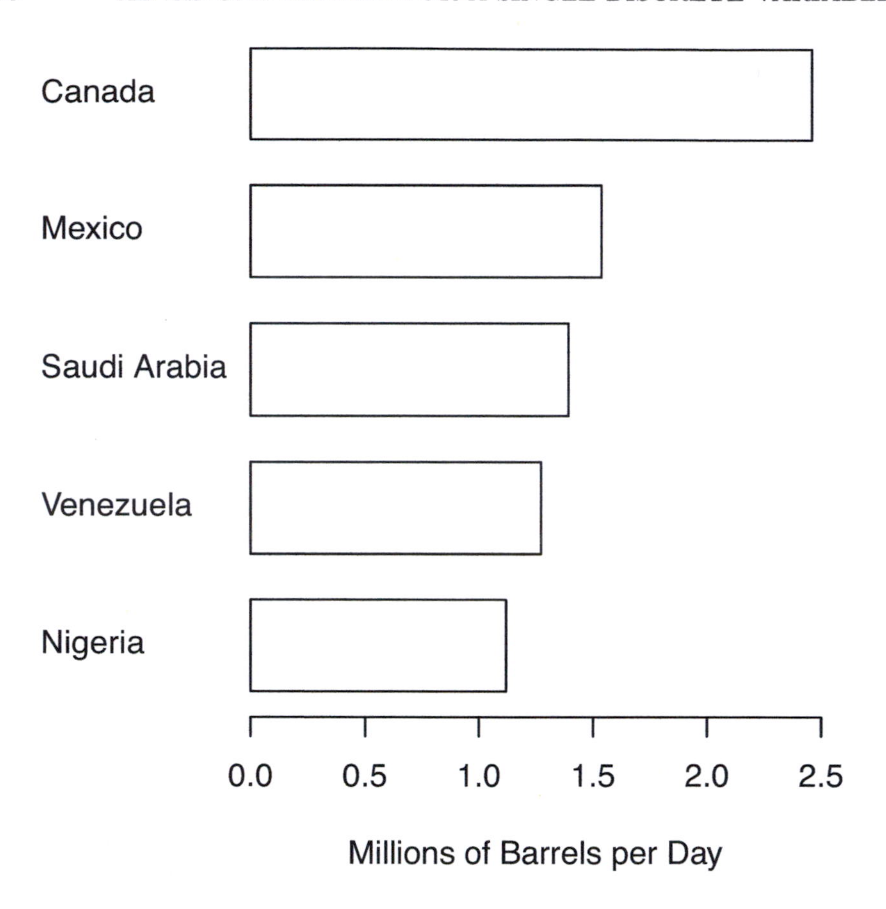

Figure 3.7 *Bar chart of top five countries exporting petroleum to the United States*

increasing distance from the horizontal axis. On the other hand, the lengths of the dotted line segments decrease in length in the dot chart of Figure 3.8 with increasing distance from the horizontal axis. In either figure, the viewer can quickly discern the relative ranking of the countries exporting petroleum products into the United States. More work is required to come to the same conclusion with the dot chart of Figure 3.9 in which the countries are listed in alphabetical order.

Order with respect to a graphic refers to the use of size or the alphabet to achieve a rational sorting of categories. Other than position, the selection of color or graytone of fill can also be used to convey a sense of ordering. With ascending or descending order, the viewer obtains ordinal information about the categories. Ordering can be used to imprint a pattern in the mind of the viewer.

The viewer can quickly discern from the dot chart of Figure 3.8, or the bar

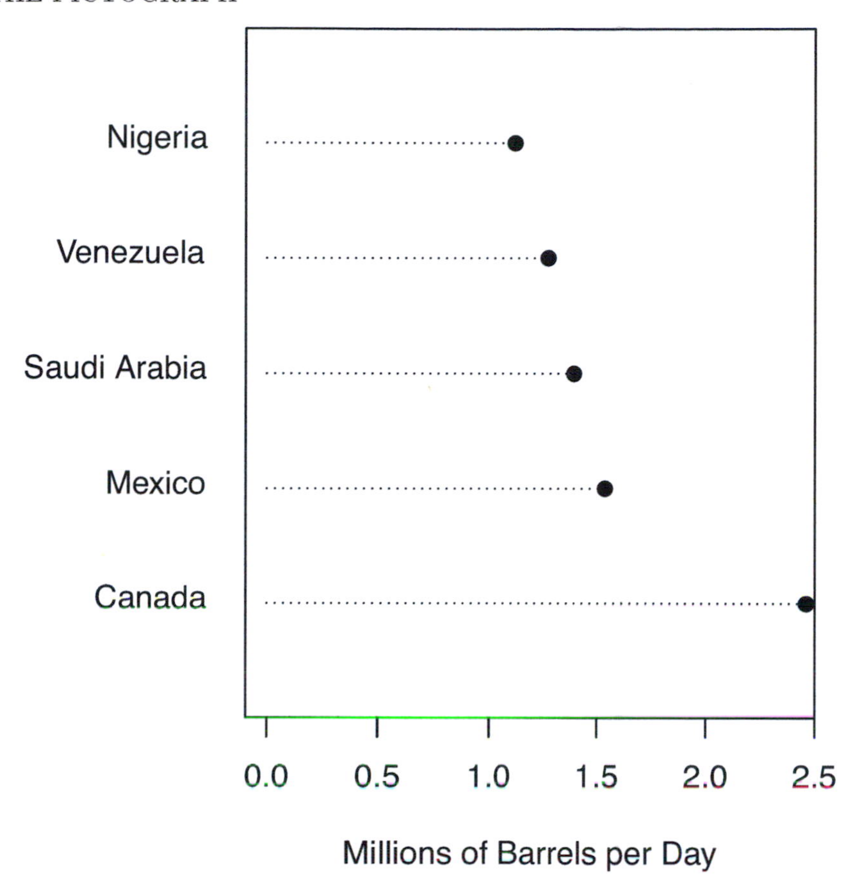

Figure 3.8 *Dot chart of top five countries exporting petroleum to the United States*

chart of Figure 3.7, that the key source of crude oil and petroleum products for the United States is not the Middle East but rather the other two member nations of the North American Free Trade Agreement, namely, Canada to the north and Mexico to the south.

Rarely does the model of ordering according to alphabetic ordering of categorical names prove to be valid. Typically, the use of the alphabetic ordering tends to do little more than imprint upon the viewer that the person who produced the chart was cutting corners. Volume of imports was used in ordering the bar chart of Figure 3.7 and the dot chart of Figure 3.8.

The subtle difference between the two orderings in Figures 3.7 and 3.8 is that each is the reverse of the other. In the dot chart the ordering of the line segments' lengths gives the impression of shorter line segments stacked on top of larger ones. It is the opposite for the bar chart, which gives the viewer an impression of being top heavy.

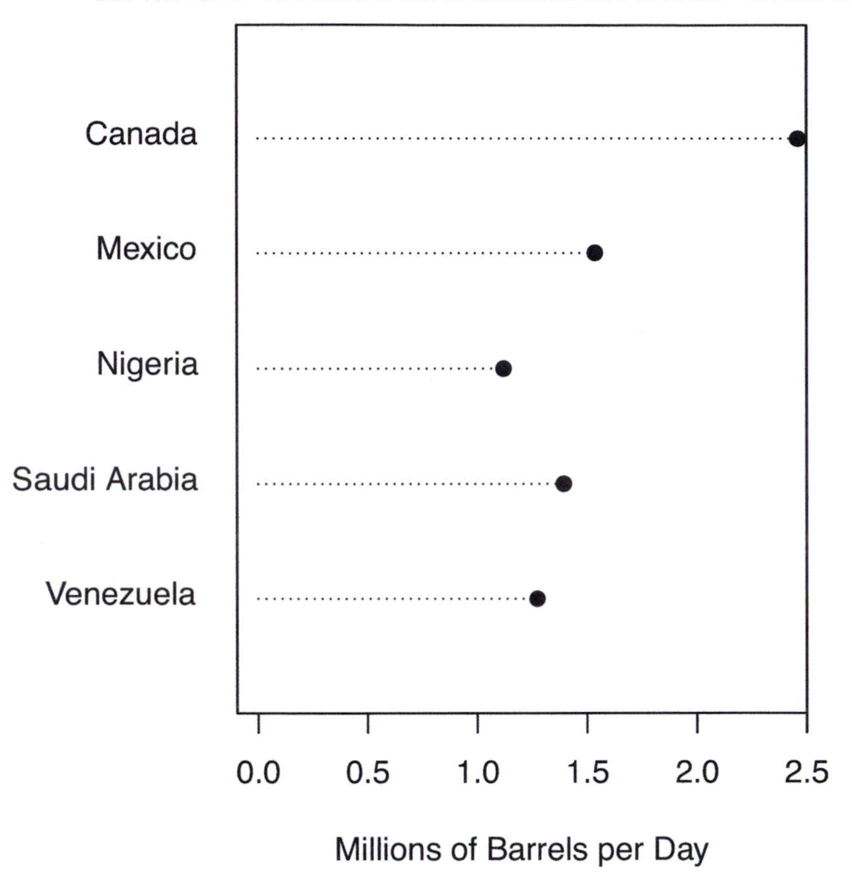

Figure 3.9 *Dot chart of top five countries exporting petroleum into the United States (in alphabetical order)*

It is because some viewers might apprehend an illusion of toppling over in the bar chart of Figure 3.7, that the bottom-heavy approach ought to be preferred, as illustrated in the dot chart of Figure 3.8. Without experimental results to support this conclusion, the choice of ascending versus descending ordering with respect to frequency is a matter of taste for the graphic designer.

Pictographs are often used by the mass media to grab the attention of the viewer. A recommended alternative practice is the use of a dot chart with a picture on the side, as in the example of Figure 3.10.

This sort of production was formerly restricted to newspapers and magazines because the services of a graphic artist were required. But now, with many desktop publishing and word processing software packages, anyone can do this for presentation in print, or any other media for that matter. Figure 3.10 was produced by importing a dot chart produced by R into Illustrator and then

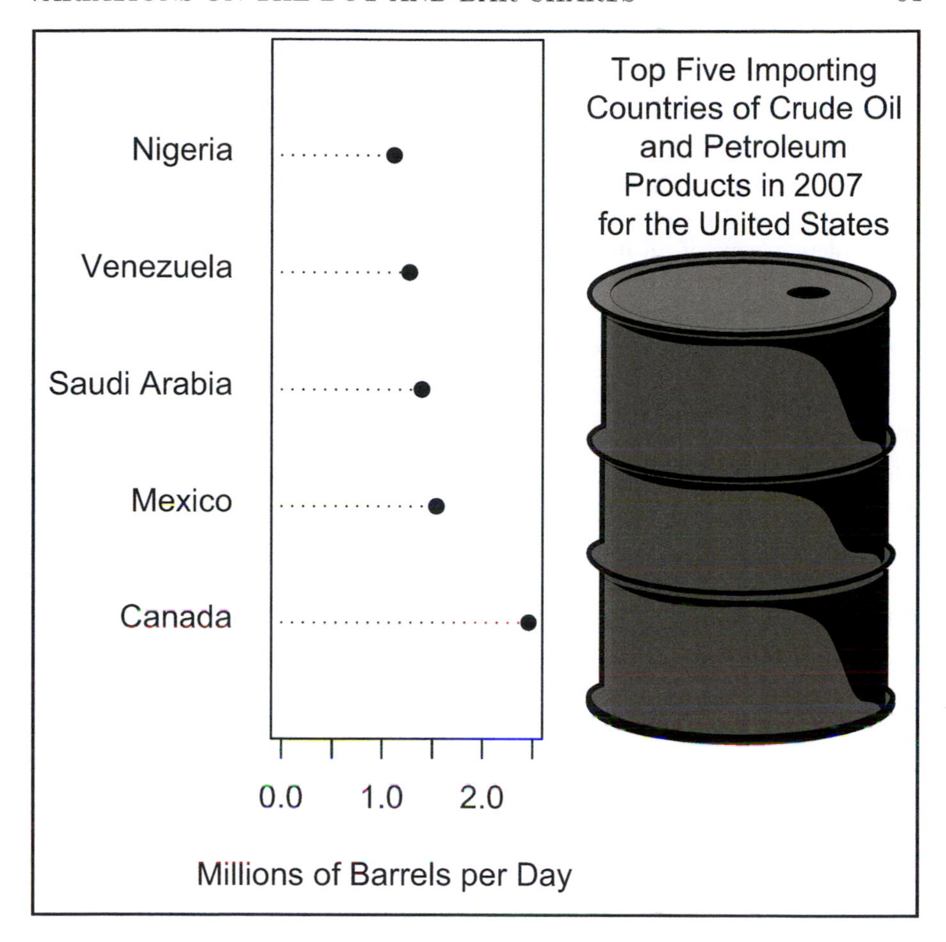

Figure 3.10 *Example of dot chart with artwork*

using Illustrator to add the grayscale depiction of an oil barrel, the frame, and the internal label.

3.5 Variations on the Dot and Bar Charts

For the example of the budget appropriations for the 2008–2009 biennium as passed by the General Assembly of the United Nations, the data do not represent the results of a sample taken of a population. Hence, there is no uncertainty associated with these figures. With respect to the planned expenditures, Figures 2.1 through 2.23, inclusive, are accurate in depicting no uncertainty.

When it comes to graphing actual expenditures, reported expenditure values will be the result of the application of survey sampling methods used by the accountants doing the audit. Any graphic depicting actual expenditure should have some way of illustrating this sampling variability.

Figures 3.4 through 3.10, inclusive, depict the volume of petroleum imports per day reaching the United States from abroad. It is quite unlikely that these figures supplied by the Energy Information Administration are the result of a complete census of all incoming petroleum products. Indeed, the imported volumes are reported to the nearest thousands of barrels per day. Estimates of the sampling variability would be useful. But these are not readily available and that is the excuse for not providing them in Figures 3.4 through 3.10.

Estimates of variation for the allergy prevalences among endoscope sinus surgery patients in Figures 1.1 and 1.2 are available and should be depicted.

The calculation of the standard error for some ith category and its observed proportion $\hat{p}_i$ is given by

$$s_{\hat{p}_i} = \sqrt{\frac{\hat{p}_i(1-\hat{p}_i)}{n}} \tag{3.1}$$

where n is the number of observations for all categories.

It is a major error to present data without an indication of error or discussion of potential sources of error. To do so is to potentially fall prey to finding patterns in the data that the data themselves do not support. This consideration leads to the following presentation of variations on the dot and bar charts that depict, in addition to the category sample-proportions $\{\hat{p}_i\}$, the associated standard errors $\{s_{\hat{p}_i}\}$.

3.5.1 The Bar-Whisker Chart

The *bar-whisker chart* takes the basic bar chart and adds whiskers to inform the viewer about the variation associated with the placement of the ends of the bars. Figure 3.11 is one example of a bar-whisker chart with the length of each whisker on either side of the end of the bar equal to the standard error of the proportion of each category.

Variations do appear in whether the whisker is portrayed as $\pm s_{\hat{p}_i}$, $\pm 2s_{\hat{p}_i}$, $\pm 1.96s_{\hat{p}_i}$, or some similar variation. It is also possible instead to use a whisker to depict the maximum possible standard error for any proportion. For n observations for all categories, the maximum possible standard error for any proportion is given by

$$s_m = \frac{1}{2}\sqrt{\frac{1}{n}}. \tag{3.2}$$

As a viewer of a bar-whisker plot, always check a legend, caption, or accompanying text for explanation of how the whisker length was determined. As a producer, always supply this information.

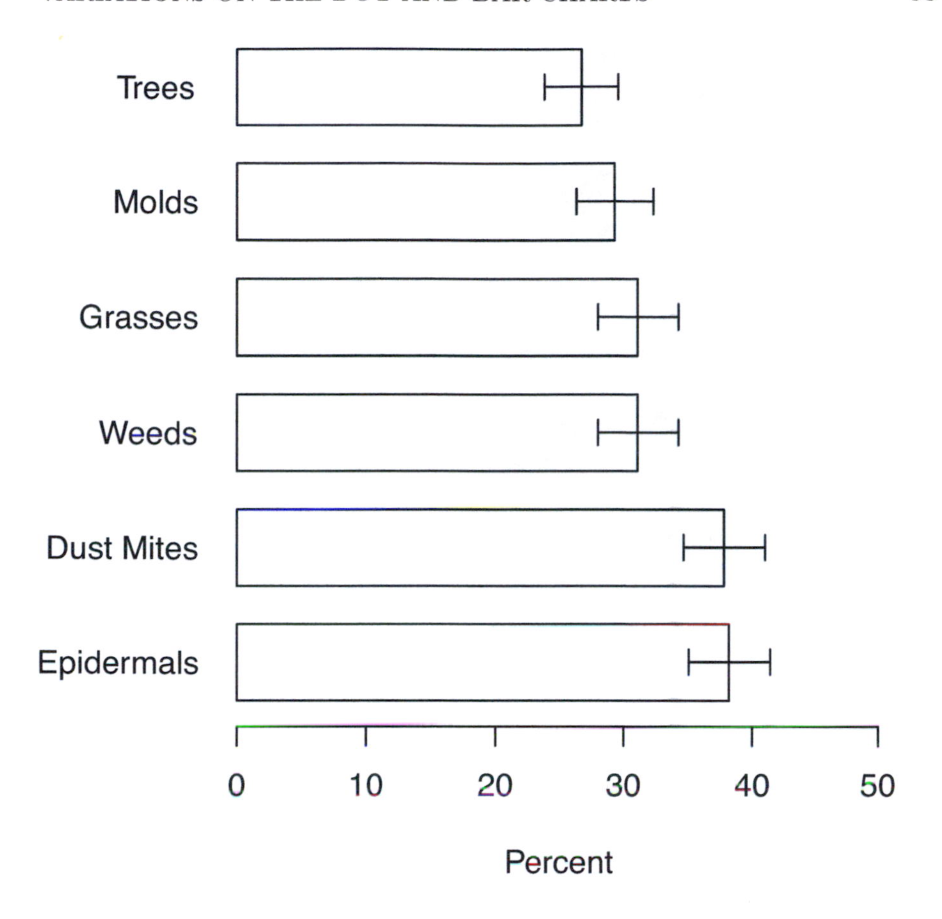

Figure 3.11 *Bar-whisker chart of allergy prevalence with standard errors in endoscopic sinus surgery patients*

A popular version of the bar-whisker chart with a single whisker is depicted in Figure 3.12. The R code for producing this chart is as follows.

```
barplot(prevs,space=0.5,names.arg=NULL,horiz=TRUE,
axes=FALSE,xlim=c(0,50),xlab="Per Cent",col=1)
yy<-1.5*(1:length(prevs)-0.33)
for (i in 1:6) lines(x=c(prevs[i],prevs[i]+se[i]),
y=rep(yy[i],2))
axis(1,tick=TRUE,xaxp=c(0,50,5))
yyy<-1.5*(1:length(prevs)-.5)
axis(2,at=yyy+0.1,labels=names,font=1,tck=0,tcl=0,
las=1,padj=0,col=0)
```

The bars are plotted with the R function `barplot`. The R function `lines` is

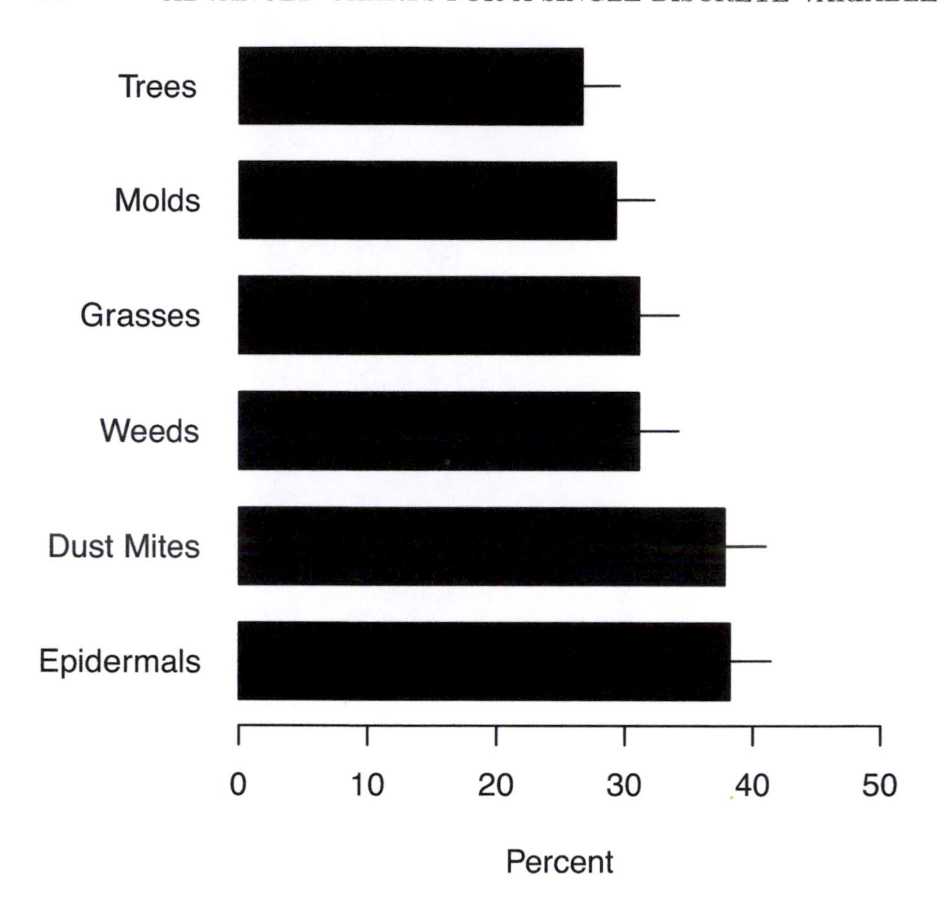

Figure 3.12 *Bar and single whisker chart of allergy prevalence with standard errors in endoscopic sinus surgery patients*

used to add the whiskers with the standard errors stored in the vector variable `se`. The axes are suppressed when calling `barplot` and then added with two separate calls to `axis`. This is done because `axis` gives the user more control over where the ticks are for the horizontal axis and the location of the labels for the vertical axis.

Because black fill is used, a whisker inside each bar would not be visible. But one does see versions with white fill for the bars and a single whisker outside. Both versions appear frequently in biomedical journals. The use of terminating line segments at the ends of the whiskers, as in Figure 3.11, appears to be one of personal preference or journal plotting standards. With respect to maximizing the data-ink ratio, the terminating line segments at the ends of the whiskers are probably best omitted.

Personal preference or publication conventions are known to overrule the ax-

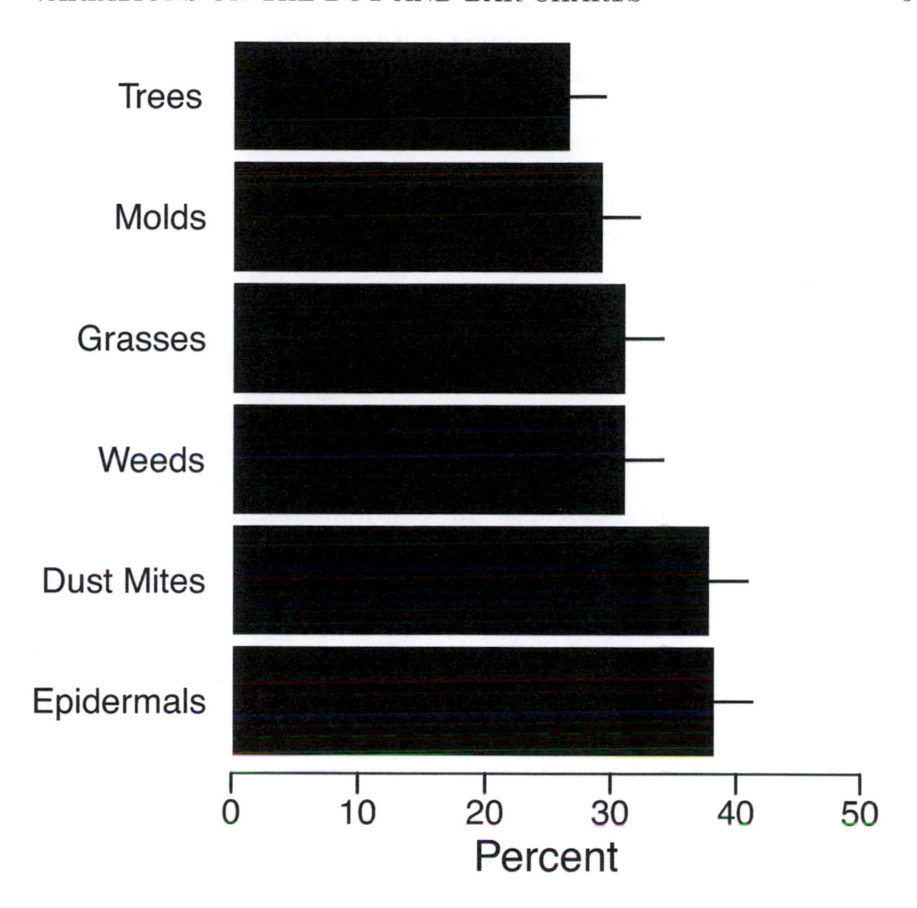

Figure 3.13 *Bar and single whisker chart of allergy prevalence with standard errors in endoscopic sinus surgery patients using* `ggplot2`

iom of *Efficiency* that rejects the two-sided whisker in Figure 3.11 as portraying redundant information given the symmetry of the standard error about the estimate of proportion.

Figure 3.13 uses `ggplot2` to reproduce the horizontal bar-whisker chart of Figure 3.12. The code for producing Figure 3.13 is as follows.

```
figure<-ggplot(allergen, aes(name, prevs)) +
theme_base() + theme(plot.background =
element_rect(fill=NULL, color="white",
linetype="solid"),
panel.border=element_rect(fill=NULL,color="white",
linetype="solid"),
axis.ticks.y=element_blank(),
axis.line.x=element_line(),
```

```
plot.margin=margin(0,14,7,0,unit="pt")) +
scale_y_continuous(expand=c(0,0),
breaks=10*(0:5),limits=c(0,50)) +
geom_col(fill="black") +
geom_errorbar(aes(ymin=prevs,ymax=prevs+se),
width=0) + coord_flip() +
labs(x=NULL,y="Percent")

print(figure)
```

Many biomedical journals have plotting conventions that are parsimonious with respect to the use of grayscale or color, so Figure 3.13 is plotted using `theme_base()` from the `ggthemes` package to mimic the implicit black-on-white background theme in the base `graphics` package in R. This necessitates the call to `element_rect` to paint white the black bounding box for the entire figure. Next is a call to `panel.border` to paint white the framing box for the axis with a subsequent call to `axis.line.x` to produce a solid black line for the horizontal axis. The `ggplot2` coding required 486 characters compared to 316 for the base `graphics` coding as a consequence.

3.5.2 Dot-Whisker Chart

The *dot-whisker chart* takes the basic dot chart and adds visualization of the standard error. An example is given in Figure 3.14. From the perspective of maximizing the data-ink ratio, the full dotted lines are unnecessary in this figure and have been dispensed with entirely in the dot-whisker chart of Figure 3.15.

A purist seeking to maximize efficiency in data presentation should be attracted to the dot-whisker chart of Figure 3.16 that eliminates the terminating line segments at the ends of the whiskers.

Cleveland [21] introduced *two-tiered error bars* to convey sampling variation. Figure 3.17 depicts 68% and 95% confidence intervals for the prevalence estimates with vertical line segments denoting the limits of the 68% confidence interval. The original figure in Cleveland [21], in fact, depicts 50% and 95% confidence intervals. The 68% and 95% confidence intervals in Figure 3.17 are approximately equal to one and two standard errors in either direction from the point estimate of prevalence. This was the basis for choosing the 68% and 95% confidence coefficients.

The R script for producing the two-tiered dot-whisker chart of Figure 3.17 is as follows.

```
dotchart(prevs,labels=names,lcolor="white",col="black",
bg="black",xlim=c(0,50),xlab="Per Cent")
```

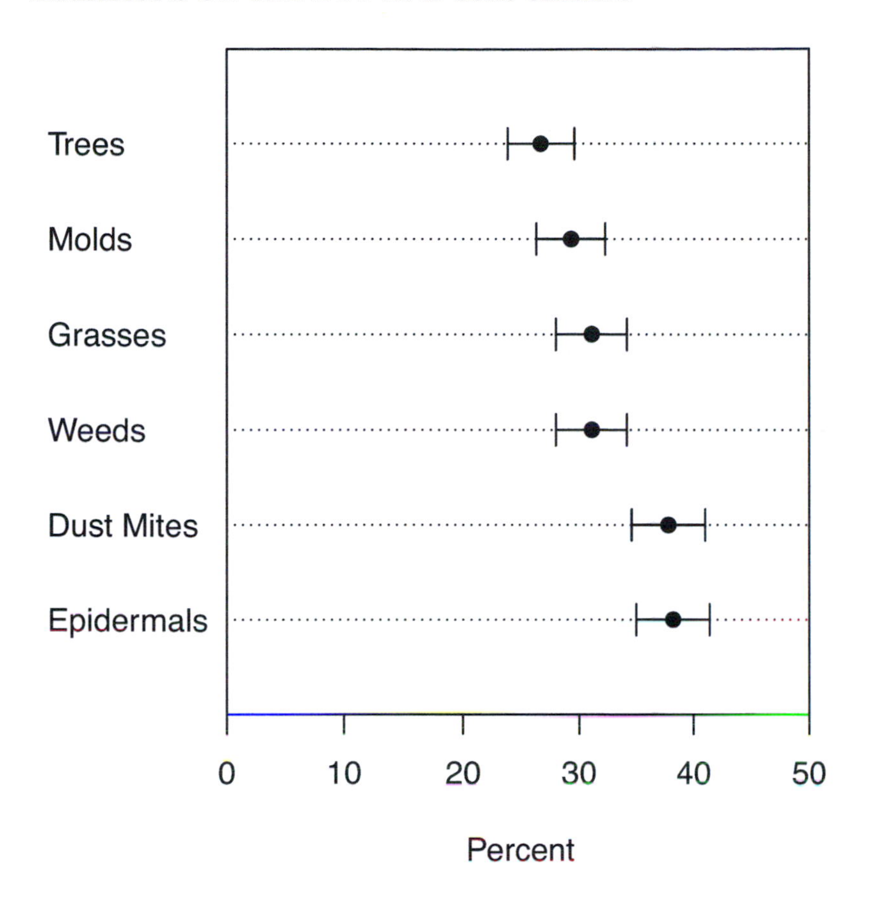

Figure 3.14 *Dot-whisker chart of allergy prevalence in endoscopic sinus surgery patients (with intervals depicting standard errors)*

```
yy<-1:length(prevs)
for (i in 1:6) {
lines(x=c(prevs[i]-1.96*se[i],prevs[i]+1.96*se[i]),
y=rep(yy[i],2))
lines(x=rep(prevs[i]-se[i],2),y=c(yy[i]+0.15/2,
yy[i]-0.15/2))
lines(x=rep(prevs[i]+se[i],2),y=c(yy[i]+0.15/2,
yy[i]-0.15/2))
}
```

The axes, their labels, and the terminating dots were plotted by a call to the R function `dotchart`. The small dots from the left vertical axis to the

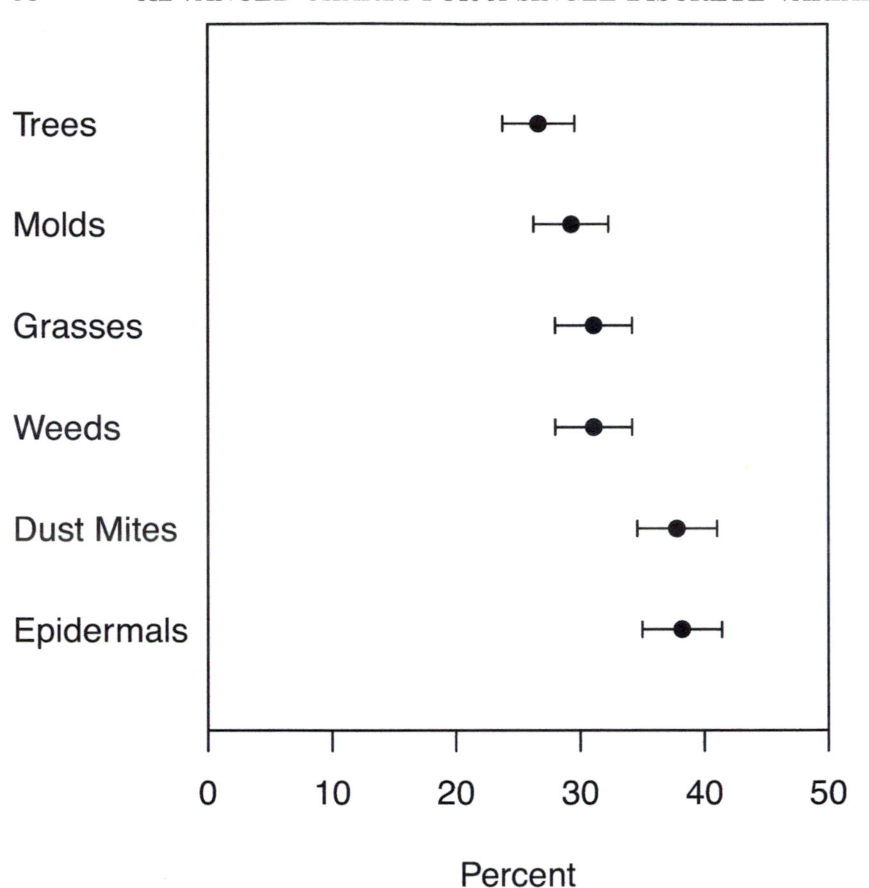

Figure 3.15 *Dot-whisker chart of allergy prevalence in endoscopic sinus surgery patients (with intervals depicting standard error)*

terminating dots were suppressed by setting their color to white with the argument `lcolor="white"`. The horizontal line segments for the whiskers and the vertical line segments identifying the 68% confidence intervals were plotted within a `for` loop by three calls to the R plotting function `lines`.

Figure 3.18 is a facsimile of Figure 3.17 produced by the following code.

```
figure<-ggplot(allergen, aes(name, prevs)) +
theme_base() + theme(plot.background=
element_rect(fill=NULL,color="white",
linetype="solid"),
axis.ticks.y=element_blank(),
plot.margin=margin(7,14,7,0,unit="pt")) +
scale_y_continuous(expand=c(0,0),
```

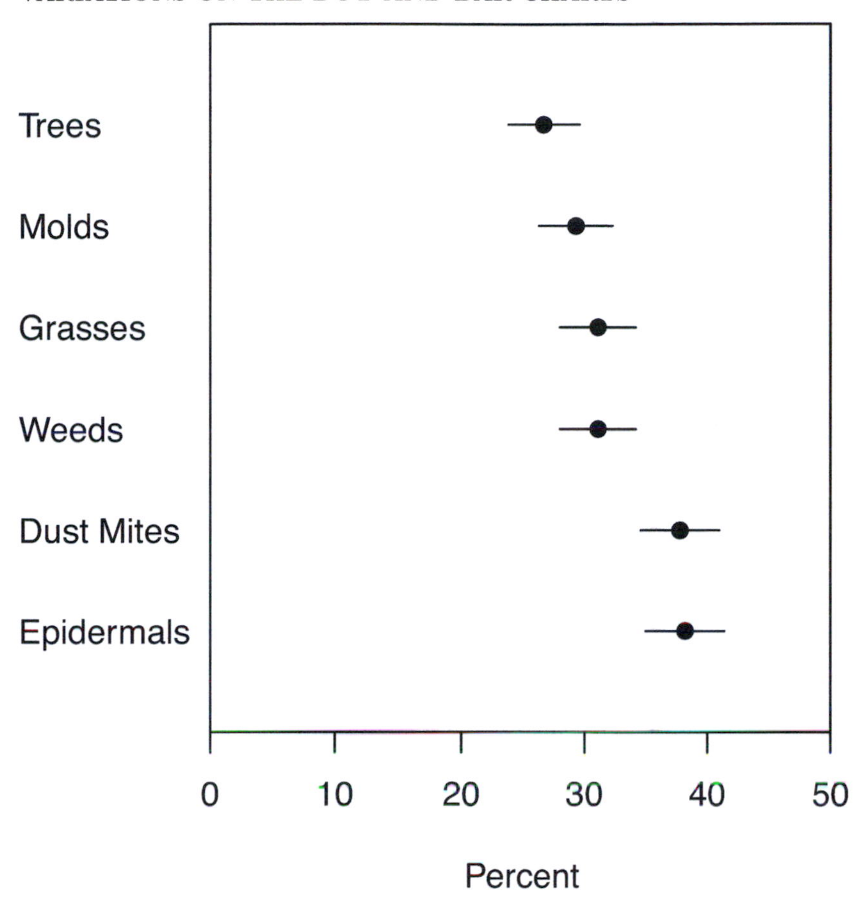

Figure 3.16 *Dot-whisker chart of allergy prevalence in endoscopic sinus surgery patients (with intervals depicting standard error)*

```
breaks=10*(0:5),limits=c(0,50)) +
geom_point(fill="black",size=2) +
geom_errorbar(aes(ymin=prevs-se,ymax=prevs+se),
width=0.1) +
geom_errorbar(aes(ymin=prevs-1.96*se,ymax=prevs+1.96*se),
width=0) + coord_flip() +
labs(x=NULL,y="Percent")

print(figure)
```

Figure 3.18 was produced using the `theme_base()` function in the package `ggthemes` to mimic the base `graphics` package in R. Notice in the code the two calls of the `ggplot2` function `geom_errorbar`: the first to place the whiskers

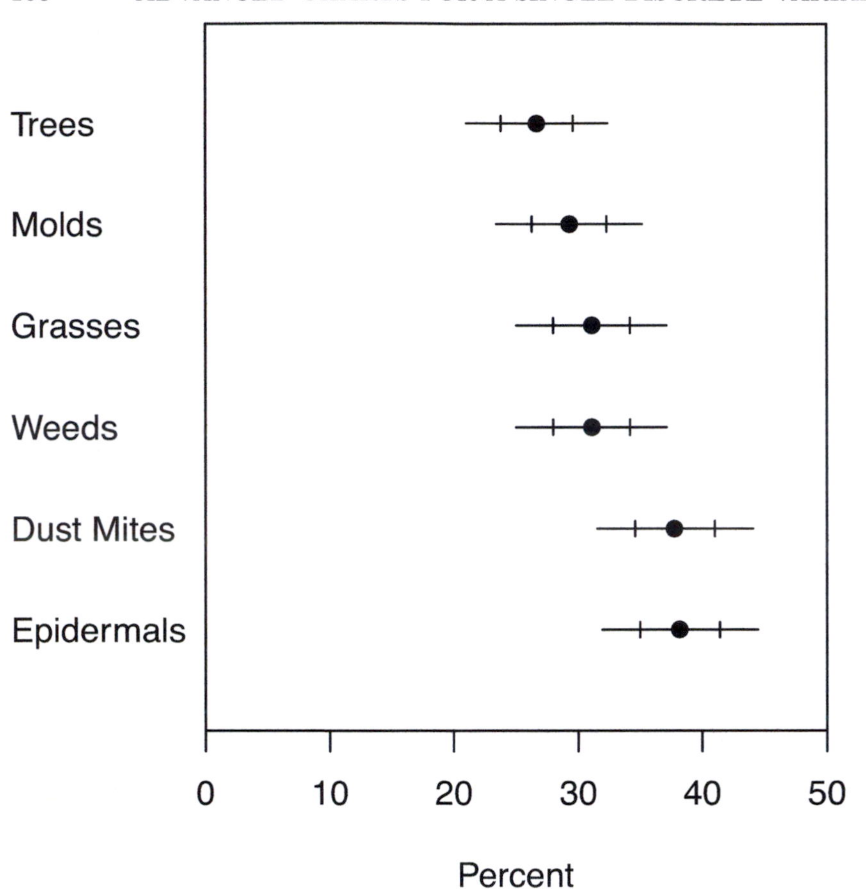

Figure 3.17 *Two-tiered dot-whisker chart of allergy prevalence in endoscopic sinus surgery patients (68% and 95% confidence intervals depicted)*

with an end bar at one standard error and the second to extend from one standard error to 1.96 standard errors without an error bar. The `ggplot2` package code consists of 471 characters without spaces compared to the `graphics` package code of 313 characters.

3.6 Frames, Grid Lines, and Order

The topics of frames, grid lines, and order have been encountered in this and the previous chapter. They will be encountered in other chapters as well. A more detailed consideration for these topics is given here.

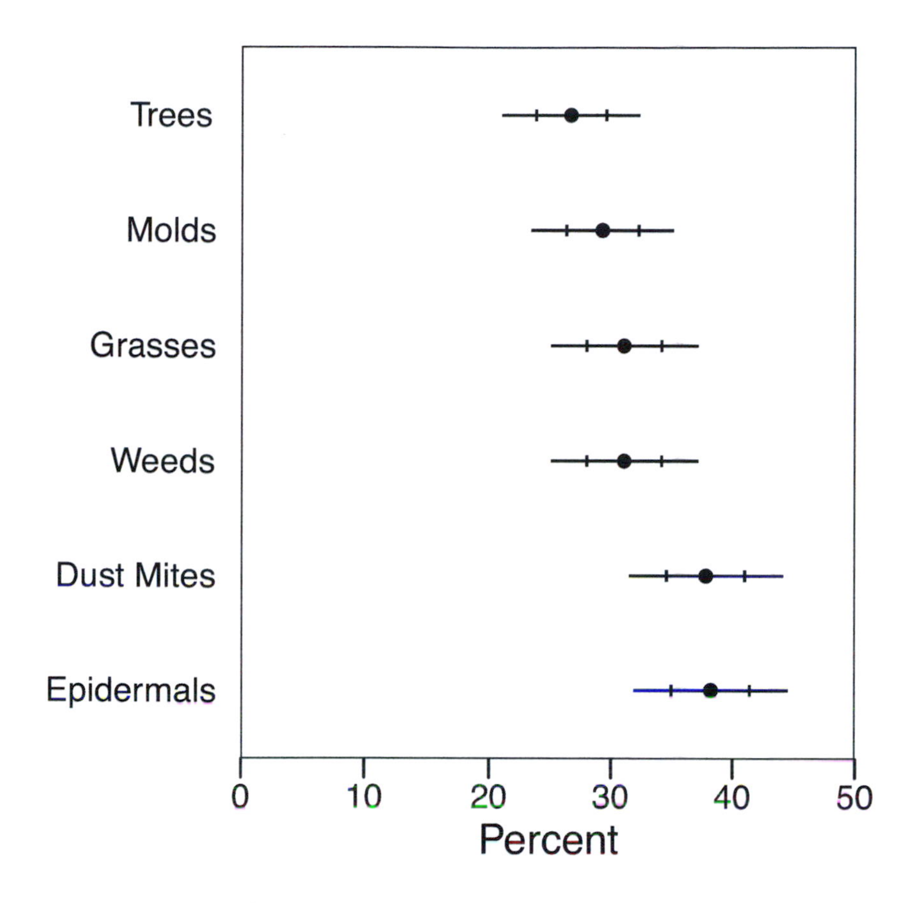

Figure 3.18 *Two-tiered dot-whisker chart of allergy prevalence in endoscopic sinus surgery patients (68% and 95% confidence intervals depicted) with* `gglot2`

3.6.1 Frame

A frame provides a rectangular boundary for a chart with line segments parallel to and of the same length as the horizontal and vertical axes, or by an outer box containing the entire plot, or possibly even both. The dot chart of Figure 2.22 has both.

Two perpendicular axes by themselves are considered by Tufte [123] to provide an adequate frame. Cleveland [21] prefers two perpendicular pairs of parallel axes. Framing any chart, including a bar chart, can be a contentious issue. A frame outside and bounding a statistical graphic is useful for calling attention to a plot or serving to highlight or to separate material from other material on a page or an image projected in a room to an audience. The R function for this purpose is `box`. This function was used for artistic purposes in Figures 2.18 and 2.22.

In the vertically stacked bar chart of Figure 3.1, the `legend` function automatically drew a framing box around the color key. This is a default that can be turned off. There is no framing box around the two-column color key for the horizontal stacked bar chart of Figure 3.2.

A frame can be useful by providing visual cues for locating the position of points in a Cartesian coordinate plane. This is done by using an inner frame so that the two perpendicular axes are set in a completed box. The plotting conventions of many, but not all, scientific periodicals and publishers require a frame for all graphics. It is not a bad practice for projected images for an audience.

Keystoning can occur with projected images. When this happens to a square, the angles are no longer right angles, and the bottom line segment is narrower than the top line segment—the square takes on the appearance of a keystone. An advantage of using frames for graphics in a projected presentation is if there is any distortion in projections, say to due to keystoning, this will be evident.

Many statistical software packages, anticipating a requirement for framing, will always provide a frame. With few exceptions, the use of a frame is optional with the graphic functions of R.

3.6.2 Grid Lines

Grid lines are lines corresponding to specified constant values of a quantitative variable and are typically equally spaced. Grid lines are used for the quantitative variable in the bar chart of Figure 2.10. Obviously, grid lines running parallel to the bars are unnecessary in a bar chart.

The use of grid lines is controversial and a matter of debate not just on technical grounds but artistic as well. Some researchers believe that grid lines, as depicted in Figure 2.10, assist the viewer in better judging the length of the bars with respect to each other and are helpful in interpolating the length of each bar with respect to the scale reported on the quantitative axis. There is some merit to this in the setting of a time-constrained audio-visual presentation in which interpolation can be an alternative to discussion of a table of figures.

The use of grid lines can also be considered for presentations if it is felt that the additional assistance in interpretation is needed for less graphically oriented or adept individuals. If the medium isn't getting the message out, then it is worthwhile to consider adapting the medium. In this context then, it ought to come as no surprise that some authors refer to grid lines as *secondary reference lines*.

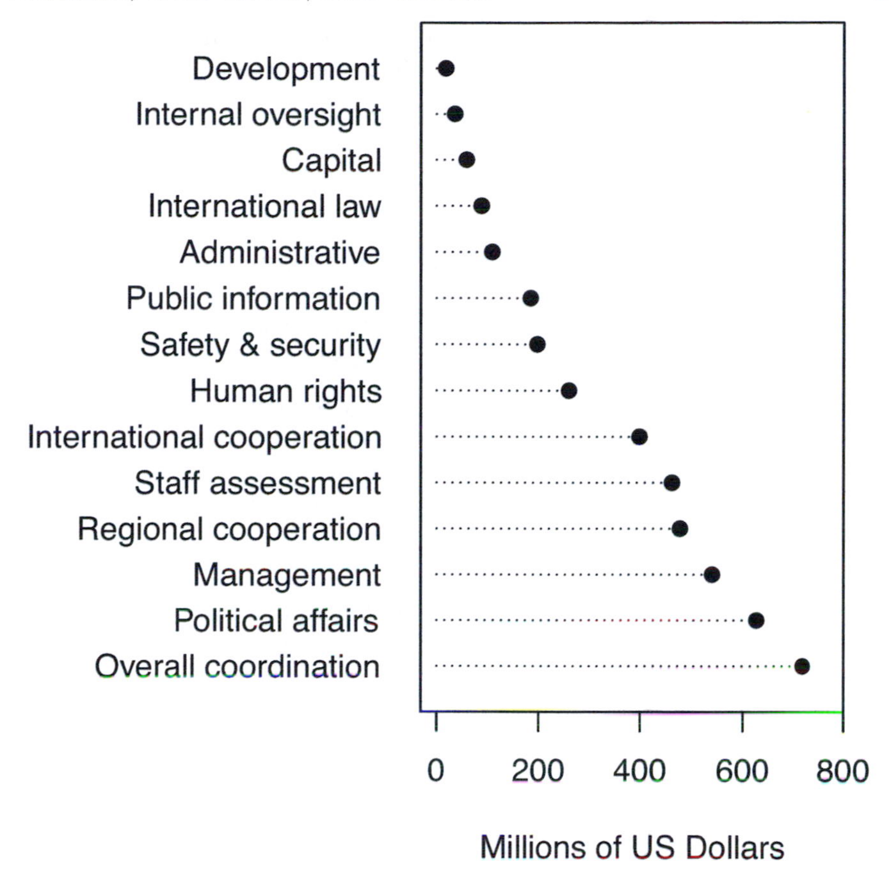

Figure 3.19 *Dot chart in the style of Cleveland and McGill [25] of the United Nations budget for 2008–2009 in ascending order*

3.6.3 *Order*

For a dot chart, the order of the dotted line segments can be ascending, descending, or based on an alphabetical or some other rational sorting of categories. Ordering imprints a pattern in the mind of the viewer. With ascending or descending order, the viewer obtains ordinal information about the categories. Rarely does the model of ordering according to alphabetic ordering of categorical names prove to be important.

The dot chart of Figure 2.2 for planned expenditures by the United Nations lists the categories in the order printed in Resolution 62/327. While this ordering may be relevant in some instances for the United Nations' staff, the ascending order of expenditures in Figure 3.19 is likely of more use for someone studying the planned expenditures for patterns.

With a bar chart, selection of the color or graytone of the fill of the bars can also be used to convey a sense of ordering. This can be to reinforce the order of the bars or to add a second layer of ordering.

3.7 Conclusion

There are good reasons for not drafting pie charts, including documented evidence from large-scale trials comparing pie charts with bar charts. But one does continue to see pie charts. Hence the lengthy discussion of the literature and presentation of examples of pie charts in this book.

Pie charts continue to be very popular especially to depict sources of income and destinations of expenditure for corporations and governments for shareholders and taxpayers, respectively. Despite all their pitfalls and biases, pie charts are popular for communicating with the public and also with administrators and managers.

If asked in the course of employment to draft a piechart, and if time permits, present an alternative using the dot chart and the arguments previously given in favor of the dot chart. If ordered to do a pie, then carefully draft the pie chart using the guidelines listed in the previous chapter.

Dot charts, bar charts, stacked bar charts, and pie charts can be produced with design variations that merit consideration and should not be left to software default settings. Many software packages will produce bar charts when supplied with the data by the user in an automatic manner. For some software packages, what you see is what you get.

For other software packages, what you see is caused by relying on the default settings. By checking the documentation for the particular software package being used, the user may discover that it is possible to control aspects of the display such as:

- Orientation;
- Fill;
- Axes;
- Frame;
- Tick marks on axes;
- Grid lines; and
- Order of the bars.

But what of the stacked bar chart? This is rarely executed to depict the distribution of a single discrete variable. This is a good thing given the experimental results demonstrating the stacked bar chart to be the poorest of the lot.

With regard to pictographs, the only negative concern is that pictures or sketches of objects in 3-D perspective ought to be avoided. Kosslyn [77] notes that "Unfortunately, even in the best of circumstances, humans are not good at estimating volumes of objects and are even worse with pictures of objects (see Teghtsoonian [120] 1965)."

Darrell Huff [68] devoted Chapter 6, pages 66–73, of his book *How to Lie with Statistics* to a discussion of the pitfalls of 3-D perspective in pictographs. This book was first published in 1954 and has been reprinted many times since. His recommendation is that a vertical bar chart is a better option. But this predates by three decades the publication in 1984 of Cleveland and McGill [25].

If one sticks with two-dimensional icons in pictographs, then there ought to be no difficulties. The pictograph has one advantage: namely, steering the viewer toward counts. For pictographs portraying actual counts this is a reinforcing feature for the viewer.

Putting together the comparisons, actual or inferred, as reported from the literature, the hierarchy of preference for depicting distributions of a single discrete variable from most preferred to least is:

1. dot chart;
2. bar chart;
3. pictograph;
4. pie chart;
5. stacked bar chart;
6. pseudo-three-dimensional bar chart; and
7. pseudo-three-dimensional pie chart.

The hierarchical listing of the accuracy of performing tasks in Cleveland and McGill [25] lists length, area, and volume in descending order of accuracy and just plain common sense would indicate that pseudo-three-dimensional charts are to be avoided.

Fortunately, there is no pie chart counterpart to either the bar-whisker chart or the dot-whisker chart if it is desired or necessary to depict variation in estimate of frequency for the distribution of a single discrete variable. Based on perceptual considerations in the absence of experiment, the dot-whisker chart ought to be the preferred choice.

3.8 Exercises

1. Figure 2.23 provides an example of the form of the bar chart recommended on page 128 of Tufte [123] using the expenditures planned by the United Nations for the 2008–2009 biennium.

Political Party	Per Cent
Conservative Party	42
Labour Party	32
Liberal Democrats	14
Scottish & Welsh Nationalist Parties	4
Green Parties	3
United Kingdom Independence Party	1
Other Parties	4

Table 3.2 *Support for political parties in the United Kingdom based upon a telephone poll by Ipsos MORI between March 13–15, 2009, of 1,007 individuals certain to vote*

(a) Compare Tufte's [123] recommended bar chart in Figure 2.23 with the stacked bar chart given in Figure 3.1 with respect to the data-ink ratio and comment.

(b) Is Tufte's [123] recommended vertical bar chart in Figure 3.4 superior to the vertically stacked bar chart of Figure 3.1? Discuss.

(c) Consider Figure 3.1 with respect to the principles of the **ACCENT** rule.

2. Using the `ggplot2` package, produce a vertically stacked bar chart of expenditures planned by the United Nations for the 2008–2009 biennium given in Figure 3.1.

3. Using the **ACCENT** rule and data-ink ratio, compare and contrast the stacked horizontal bar chart of Figure 3.2 produced by the `graphics` package with Figure 3.3 produced by the `ggplot2` package. Which do you prefer?

4. Using the `ggplot2` package, produce a dot chart of the top five countries exporting petroleum into the United States and compare with the dot chart of Figure 3.9.

5. Table 2.2 reports discretionary funding requested by various US government agencies under the budget for the fiscal year 2009. Re-order the categories from the least requested for discretionary funding to the most. Ignore the spending request from the Department of Defense because it is an extreme outlier.

   (a) Draft a horizontal dot chart for the requested funding.

   (b) Produce a pie chart for the requested discretionary funding by agency.

   (c) Draft a horizontal divided bar chart for the requested funding.

   (d) Using the dollar sign $ to represent ten billion dollars, create a pictograph for the spending requests in Table 2.2.

   (e) Which of the four plots produced in parts (a) through (d), inclusive, do you prefer? Justify your preference.

6. Consider the results in Table 3.2 of a public opinion survey in the United Kingdom conducted by the polling firm Ipsos MORI between March 13

and 15, 2009. The results were released on March 17, 2009. The figures in Table 3.2 report the support for political parties in the United Kingdom among 1,007 individuals reached by telephone. Participants were asked: "How would you vote if there were a General Election tomorrow?" Those who were undecided or refused were asked a follow-up question: "Which party are you most inclined to support?" Although a complex survey sampling design was used, assuming simple random sampling for the purpose of estimating random sampling error will lead to a close approximation.

(a) Create a box-whisker chart for the data in Table 3.2.

(b) Create a dot-whisker chart for the data in Table 3.2.

(c) Compare Table 3.2, the box-whisker chart, and the dot-whisker chart. Which of the three does a better job of conveying the results of the survey? Justify your answer.

(d) Which chart would you use if asked to draft a chart for a video media presentation for the general public? Justify your answer.

7. For the country of your choice, find on the internet the results of the latest public opinion survey regarding party preference in advance of the next general election.

(a) Create a pie chart for preference among the respondents of the survey.

(b) Create a dot-whisker chart with two-tiered error bars for the survey results. Depict 68% and 95% confidence intervals.

(c) Would you use the pie chart from part (a) or the dot-whisker chart from part (b) to present the results of the survey as a party insider to the executive committee of your choice of political party? Justify your answer.

8. For a 5-day period from Monday to Friday, clip or cut electronically the articles from a daily newspaper or online news outlet that have charts that display the distribution of a single discrete variable.

(a) Comment on any mistakes or violations of the **ACCENT** rule in the charts.

(b) Pick out the best example of a chart. Justify your selection.

(c) Pick out the worst example of a chart. Suggest how it could be improved. Re-draft the chart using your suggestions. Use numbers from the accompanying article or estimate values from the chart in the article.

9. Review one issue of the *New England Journal of Medicine* for graphics that are used to depict the distribution of a single discrete variable.

(a) Comment on any mistakes or violations of the **ACCENT** rule in the graphics.

(b) Pick out the best example of a graphic. Justify your selection.

(c) Pick out the worst example of a graphic. Suggest how it could be improved. Re-draft the graphic using your suggestions. Use numbers from the accompanying article or estimate values from the graphic.

10. Review one issue of a journal in your field for the charts that are used to depict the distribution of a single discrete variable. (If your field is medicine or an allied health science, choose another journal different from the *New England Journal of Medicine*.)
    (a) Determine the proportion of dot, bar, and pie charts in the issue.
    (b) Pick out the best example of a chart. Justify your selection.
    (c) Pick out the worst example of a chart. Suggest how it could be improved. Re-draft the chart using your suggestions. Use numbers from the accompanying article or estimate values from the chart in the article.
11. Based upon your answers to questions 8 through 9, compare and contrast how distributions of single discrete variables are depicted in the popular media and the *New England Journal of Medicine*.
12. Based upon your answers to questions 9 and 10, compare and contrast how distributions of single discrete variables are depicted in the two different journals.

Part III

A Single Continuous Variable

Chapter 4

Exploratory Plots for the Distribution of a Single Continuous Variable

4.1 Introduction

There is quite a variety of graphical displays for exploring distributions of a single continuous variable. Arguably, not all of them are suitable for a public audience and perhaps ought to be avoided. Others are suitable but less familiar to a public audience. An unfamiliar graphic must be introduced and explained to an audience if the information conveyed by the graphic is to be received. The plot of the next section is one such graphic.

The convention of this book is to refer to graphical displays for more-or-less continuous variables as *plots*. Although not previously stated, the convention of this book is to refer to graphical displays for discrete (categorical) variables as *charts*. This nomenclature is by-and-large conventional with occasional departures. The next section adapts the dot chart of the previous chapter from the setting of a discrete variable to that of a continuous variable.

4.2 Learning Outcomes

When you complete this chapter, you will be able to do the following.

- Know what a dot chart is and be able to create it or one of its variations with either the `graphics` or `ggplot2` packages in R.
- Understand how a stemplot is constructed and be aware of its variations.
- Know what a boxplot is and how to use either the `graphics` or `ggplot2` packages in R to produce a few of its variations.
- Know how to use either the `graphics` or `ggplot2` packages in R to plot an empirical distribution function.

Mass (g)							
5.9	32.0	40.0	51.5	70.0	100.0	78.0	80.0
85.0	85.0	110.0	115.0	125.0	130.0	120.0	120.0
130.0	135.0	110.0	130.0	150.0	145.0	150.0	170.0
225.0	145.0	188.0	180.0	197.0	218.0	300.0	260.0
265.0	250.0	250.0	300.0	320.0	514.0	556.0	840.0
685.0	700.0	700.0	690.0	900.0	650.0	820.0	850.0
900.0	1,015.0	820.0	1,100.0	1,000.0	1,100.0	1,000.0	1,000.0

Table 4.1 *Sample of 56 perch caught in a research trawl on Längelmävesi*

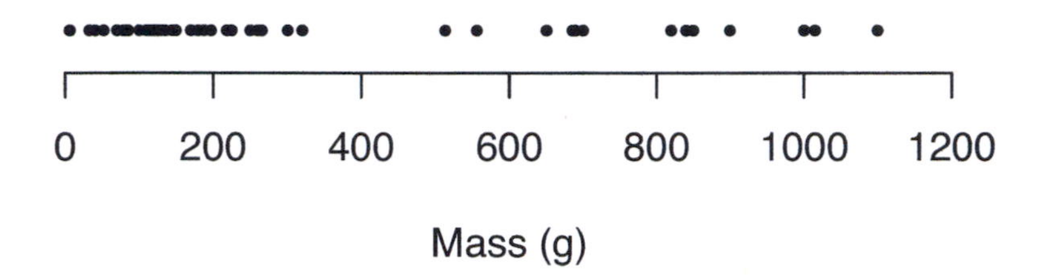

Figure 4.1 *Dotplot of mass of 56 perch caught in a research trawl on Längelmävesi*

4.3 The Dotplot

4.3.1 Definition

Consider observations for the mass of each of 56 perch (*perca fluviatilis*) caught in a research trawl on Längelmävesi (a freshwater lake in Finland). The data were published in 1917 in an article by Brofeldt [14]. The observations vary from 5.9 to 1100.0 grams and are listed in Table 4.1. These observations are depicted in the *dotplot* of Figure 4.1 that was produced by the following code using the `graphics` package in R.

```
stripchart(mass,method="overplot",xlab="Mass (g)",
pch=19,cex=0.5,xlim=c(0.,1200.))
```

Although one observation in Table 4.1 is recorded to the nearest 100 milligrams, it is apparent that some of the fisheries biologists rounded the data to the nearest 5 g while others rounded the data to the nearest 10 g. The result is that some of the observations occur more than once in the data set. Replicated observations occur as a single dot in Figure 4.1. The passed parameter of `method="overplot"` in the code for Figure 4.1 is the default and causes the overprinting of symbols in the dotplot. If observations occur sufficiently close together, the separation between dots is not apparent at a given scale and they can also appear to be a single dot.

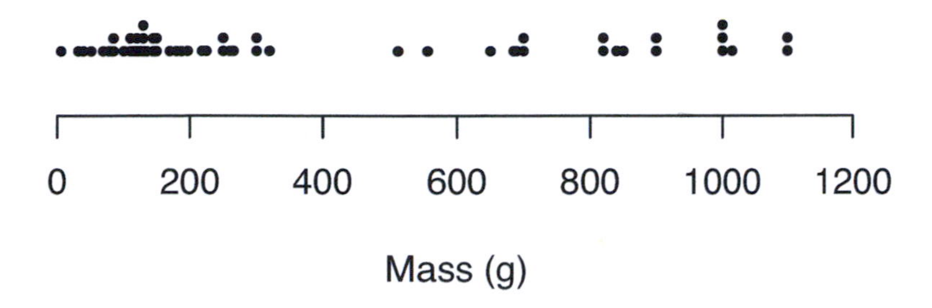

Figure 4.2 *Stacked dotplot of mass of 56 perch caught in a research trawl on Längelmävesi*

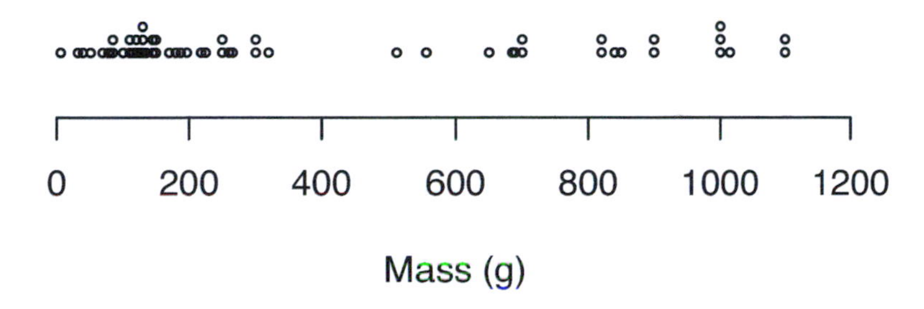

Figure 4.3 *Stacked dotplot of mass of 56 perch, caught in a research trawl on Längelmävesi, with open circles replacing dots*

4.3.2 Variations on the Dotplot

The overprinting of nearby or identical observations can be ameliorated by stacking replicates on top of each other as done in the *stacked dotplot* of Figure 4.2. This does not solve the problem of nearby points being indistinguishable. A series of nearby observations can result in smear over a range of values. This effect is visible in a few places in Figure 4.2. In an attempt to eliminate this effect, the dots of Figure 4.2 have been replaced by open circles in Figure 4.3.

The call to the R function `stripchart` for creating Figure 4.3 is as follows.

```
stripchart(mass,method="stack",offset=0.6,xlab="Mass (g)",
pch=21,cex=0.5,xlim=c(0.,1200.))
```

The data have been stored in the vector variable `mass` and the argument

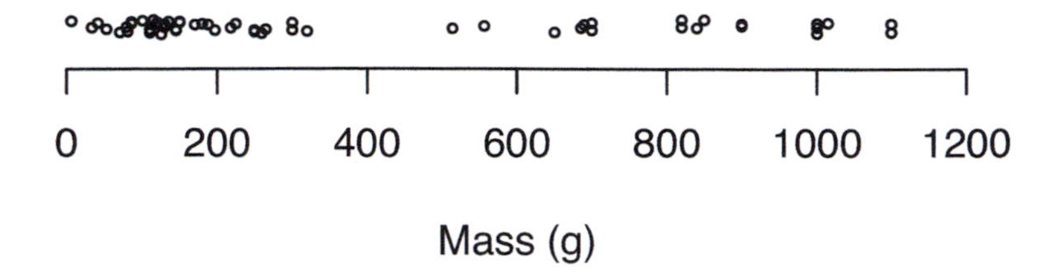

Figure 4.4 *Jittered dotplot of mass of 56 perch caught in a research trawl on Längelmävesi*

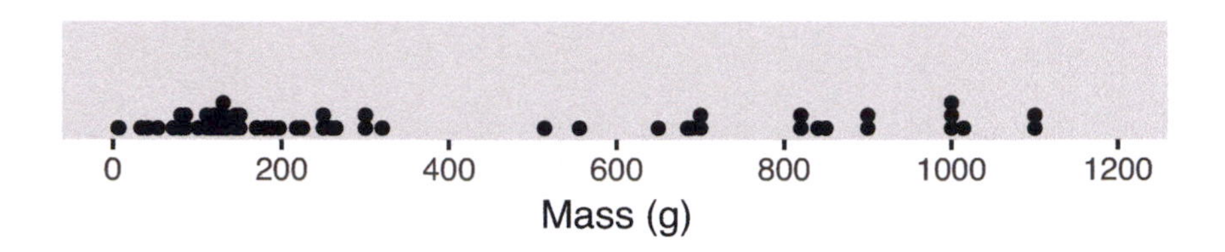

Figure 4.5 *Stacked dotplot of mass of 56 perch caught in a research trawl on Längelmävesi with* `ggplot2`

`method="stack"` creates the vertical stacking of dots. However, this does not fully solve the problem of distinguishing all data points.

Another approach to displaying replicates or nearby points in a dotplot is to randomly distribute these points in the vertical. This process is known as *jittering* and is depicted in the dotplot of Figure 4.4. This effect is achieved by the following R script.

```
stripchart(mass,method="jitter",offset=0.6,xlab="Mass (g)",
pch=21,cex=0.5,xlim=c(0.,1200.))
```

Arguably, the *jittered dotplot* with open circles offers an improvement over the basic dotplot of Figure 4.1. But the result again is not entirely satisfactory.

Another issue with the use of jitter in any type of graphical display is that it attempts to introduce *Clarity* at the cost of *Truthfulness.* Because the jitter introduced in Figure 4.4 is random, the jittered appearance changes with each execution in R. So rather than the Lie Factor being a fixed value for jittered dotplots, the Lie Factor is a random variable.

The package `ggplot2` was used to produce a version of Figure 4.1 in Figure 4.5. The `ggplot2` code to produce Figure 4.5 is as follows.

```
figure<-ggplot(perch, aes(x=mass)) +
```

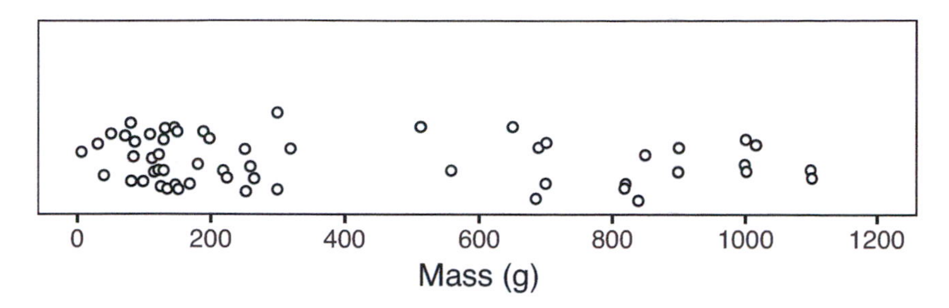

Figure 4.6 *Jittered dotplot of mass of 56 perch caught in a research trawl on Längelmävesi with* ggplot2

```
geom_dotplot(binwidth=5,dotsize=3) +
theme(axis.title.y=element_blank(),
axis.text.y=element_blank(),
axis.ticks.y=element_blank(),
panel.grid=element_blank()) +
scale_x_continuous(breaks=200*(0:6),limits=c(0,1200)) +
xlab("Mass (g)")

print(figure)
```

The `ggplot2` code is a fair bit longer than the `graphics` package code but then the `ggplot2` code is doing a few things. The vertical axis title, tick marks and their labels are shut off. Specific tick marks have been selected for the horizontal axis. The default `ggplot2` theme has been used in Figure 4.5, which results in a gray-colored background for the dotplot. The argument `panel.grid = element_blank()` passed to the function `theme` removes unnecessary vertical grid lines that would otherwise be present.

Figure 4.6 is a `ggplot2` counterpart to the jittered dotplot of Figure 4.4 that was produced by the `graphics` package. The code for Figure 4.6 is as follows.

```
figure<-ggplot(perch, aes(x=mass)) +
geom_dotplot(binwidth=3,dotsize=5,fill="white",
position="jitter") +
theme_bw() + theme(axis.title.y=element_blank(),
axis.text.y=element_blank(),
axis.ticks.y=element_blank(),
panel.grid=element_blank()) +
scale_x_continuous(breaks=200*(0:6),limits=c(0,1200)) +
xlab("Mass (g)")

print(figure)
```

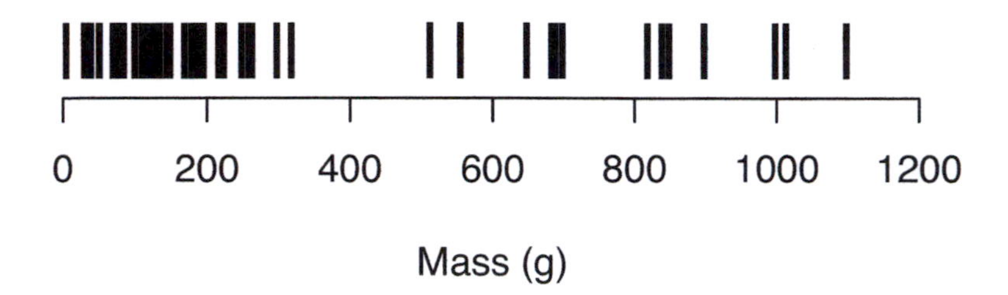

Figure 4.7 *Barcode plot of mass of 56 perch caught in a research trawl on Längelmävesi*

For this figure the black-and-white theme in `ggthemes` has been used via the function call `theme_bw()`. This provides for a white background, a bounding box for the figure, and gray labels for axis tick marks as defaults. With a bit of experimentation, a bin width of 3 and a dot size of 5 was found to produce a result similar to but still a little different from the jittered dotplot of Figure 4.4. Both figures use black circles with white fill as the plotting symbol. As with Figure 4.5, additional `ggplot2` code is needed to suppress the vertical axis in its entirety.

Yet another version of the dotplot is given in Figure 4.7 with the dot icon replaced by a vertical line segment. The image of line segments in Figure 4.7 resembles the barcode label seen on consumer items. So although some authors refer to Figure 4.7 as a *barcode plot*, it is merely a variation of the dotplot for a single quantitative variable.

Dotplots are sometimes referred to as *one-dimensional scatterplots*. A scatterplot is conventionally depicted in two dimensions and will be defined in the next chapter. Because a dotplot can be produced in a linear fashion from stripchart printers, the dotplot is sometimes referred to as a *stripchart*. Indeed, the graphical function in R used to produce the graphical displays of this section is called `stripchart`. Figure 4.7 is an example of an image that could be produced by a stripchart printer.

All the dotplot variations of this section are plagued with the problem of revealing the data in accordance with the axiom of *Clarity*. The next section presents a simple graphic that can avoid this issue.

4.4 The Stemplot

4.4.1 Definition

The starting point for depicting distributions of a numerical variable ought to be the *stemplot*, or the *stem-and-leaf display* as it was originally named by its

```
The decimal point is 2 digit(s) to the right of the |

   0 | 134578899
   1 | 01122233334555578 9
   2 | 0235567
   3 | 002
   4 |
   5 | 16
   6 | 599
   7 | 00
   8 | 2245
   9 | 00
  10 | 0002
  11 | 00
```

Figure 4.8 *Stemplot of the mass of 56 perch caught in a research trawl on Längelmävesi*

creator John Tukey. The first stem-and-leaf display appeared in print in 1972 in a paper written by Tukey [126] entitled "Some Graphic and Semigraphic Displays." This paper appeared in the monograph *Statistical Papers in Honor of George W. Snedecor*, edited by T. A. Bancroft. The stemplot is one of several *Exploratory Data Analysis (EDA)* plots presented by Tukey [127] in his book of the same name in 1977.

The stemplot is also known as the *stem-and-leaf plot*. The stemplot was introduced much more recently than the histogram—which will be encountered in the next chapter.

The stemplot has three distinct features as listed below.

1. For data sets of reasonable size, each observed value of the quantitative variable can be listed thereby giving a complete picture of all the individual data points as well as a picture of the distribution.
2. The stem can be prepared quickly and easily by hand for small data sets.
3. It is useful for hand calculation of quantiles and thus is useful in preparing another well-known EDA plot—the boxplot (to be presented in the next section)—and even a histogram.

See Figure 4.8 for an example of a stemplot drawn for the mass of 56 perch tabulated in the previous section. This figure was created by the R script after the prompt on the following line.

```
stem(mass,scale=2)
```

The parameter `scale` is used to expand the scale of the plot. Setting `scale=2` caused the plot to be roughly twice as long as the default.

The first order of business in drafting a stem-and-leaf plot is selecting the *stems* and *leaves*. In Figure 4.8, the stems are the digits representing hundreds of grams and the leaves are the digits from the observations representing tens of grams. Notice that the values of stems increase down the figure. This is conventional but also a matter of personal preference for the computer programmer creating the stemplot routine. Of course, 56 observations are not too onerous a number to be prepared manually. If done with word processing software, care must be taken to use fonts with characters of the same width that lack proportional spacing.

Note that leaves are not necessarily unique. Leaves might appear to represent a repeated observation if rounding or truncation has been performed. In this case additional information on lower-order digits is lost. For example, for stem 0 in Figure 4.8, the 2 leaves denoted by 8 refer to the observations of 78 g and 80 g, respectively. Also note that the leaf of 1 for stem 0 represents the observation of 5.9 g.

For each stem in Figure 4.8, leaves can be any one of the single digits 0 through 9. Other choices are possible. Figure 4.9 takes the stems of Figure 4.8 and splits them into two with leaf digits 0 through 4 for the first stem in the pair and leaf digits 5 through 9 for the second stem in the pair.

Figure 4.10 is another rendering of Figure 4.8, but instead according to the plotting convention of Tukey [127]. At first glance, an apparent difference between these two figures is that Figure 4.10 lacks a legend in comparison with Figure 4.8.

Tukey's [127] plotting convention for the stem-and-leaf plot lacks a legend so one would need to refer to the accompanying discussion in the case of a printed document in order to interpret the values for the stems and leaves. This is not efficient. An oral explanation for a presentation would be insufficient and a written legend would be helpful as in Figure 4.8.

A more important difference between Figures 4.8 and 4.10 is that some of the leaves have different values. This is due to rounding of the leaf digit in the stemplot of Figure 4.8 compared to truncation for the classical stem-and-leaf plot of Figure 4.10. Tukey [127] (p. 4) refers to truncation as "cutting" and states that either rounding or cutting is acceptable for exploratory data graphics.

Figure 4.11 adds a legend to Figure 4.10 and a further embellishment known as *depth*. The depth of the stem containing the median is its leaf count enclosed in parentheses. The depth of any other stem is the count of leaves at the stem plus the number of leaves at the other stems toward either the high or low

```
The decimal point is 2 digit(s) to the right of the |

   0 | 134
   0 | 578899
   1 | 01122233334
   1 | 5555789
   2 | 023
   2 | 5567
   3 | 002
   3 |
   4 |
   4 |
   5 | 1
   5 | 6
   6 |
   6 | 599
   7 | 00
   7 |
   8 | 224
   8 | 5
   9 | 00
   9 |
  10 | 0002
  10 |
  11 | 00
```

Figure 4.9 *Alternative stemplot of the mass of 56 perch, caught in a research trawl on Längelmävesi, with the number of stems doubled*

data extreme. Notice in Figure 4.11 that no depth is recorded for any stem without leaves.

Depth was first illustrated in a stem-and-leaf display by Tukey [126] in 1972.

Note that the legend of Figure 4.8 notes where to find the decimal point, whereas that of Figure 4.11 shows what an example stem-and-leaf represents: the value of the leaf unit and the number of observations depicted. The legend of the latter stemplot requires less effort on the part of a viewer.

A different definition of depth for stems was used by Tukey [127] in 1977. He then defined depth as the count of leaves per stem and places depth to the right of the stem-and-leaf plot in Figure 4.12. In comparison, Figure 4.11 has depth in columns to the left of the stem-and-leaf plot, as in the original stem-and-leaf display of 1972 given by Tukey [126].

```
0 | 034577888
1 | 0111222333344557889
2 | 125566
3 | 002
4 |
5 | 15
6 | 589
7 | 00
8 | 2245
9 | 00
```

Figure 4.10 *Stem-and-leaf plot in the style of Tukey [127] of the mass of 56 perch caught in a research trawl on Längelmävesi*

```
1 | 2: represents 120
 leaf unit: 10
            n: 56
   9     0 | 034577888
  28     1 | 0111222333344557889
  (6)    2 | 125566
  22     3 | 002
         4 |
  19     5 | 15
  17     6 | 589
  14     7 | 00
  12     8 | 2245
   8     9 | 00
   6    10 | 0001
   2    11 | 00
```

Figure 4.11 *Stemplot, with depth, of mass of 56 perch caught in a research trawl on Längelmävesi*

```
                                          (#)
 0 | 034577888                            (9)
 1 | 0111222333344557889                 (19)
 2 | 125566                               (6)
 3 | 002                                  (3)
 4 |
 5 | 15                                   (2)
 6 | 589                                  (3)
 7 | 00                                   (2)
 8 | 2245                                 (4)
 9 | 00                                   (2)
10 | 0001                                 (4)
11 | 00                                   (2)
                                        (56,√)
```

Figure 4.12 *Classical stem-and-leaf plot in the style of Tukey [127], with the count of leaves for each stem, of the mass of 56 perch caught in a research trawl on Längelmävesi*

Tukey [127] did not use the term depth. In 1977, Tukey [127] envisioned the stem-and-leaf plot as being drafted by hand for the most part. So the column to the right of the stemplot in Figure 4.12 has a sum given in the last line with a check mark to confirm that the number of points plotted in the stem-and-leaf plot is the same as the number of observations. As in Tukey [127], the check mark symbol is actually the character for a square root. Also added to Figure 4.12 is a legend at the top left, which would not be seen in Tukey [127].

Figure 4.13 is a version of Figure 4.9 with an additional capability to identify high and low observations. Observations 100 and lower have been reported on the "LO" stem while values higher than 1,000 are identified on the "HI" stem. The R function `stem.leaf` in the Rcmdr package was used to produce Figure 4.13. This function permits the specification of high and low cut-off values and can be used to produce the classical stem-and-leaf plots in the style of Tukey [127] for repeated stems. But the R function `stem.leaf` does not offer the option of printing the count of leaves for each stem as in the classical stem-and-leaf plot of Figure 4.12. The R function `stem` in the basic R Graphics library was used to produce Figure 4.9 and does not offer either the option to produce the classical stem-and-leaf plots in the style of Tukey [127] for repeated stems or the count of leaves to the right of the leaves.

Figure 4.13 is a version of the stemplot of Figure 4.9 with two branches for each stem digit. In addition to depth, Figure 4.13 adds the asterisk "*" and

```
1 | 2: represents 120
 leaf unit: 10
              n: 56
LO: 5.9 32 40 51.5 70 78 80 85 85
  22      1* | 0111222333344
  28      1. | 557889
  (2)     2* | 12
  26      2. | 5566
  22      3* | 002
          3. |
          4* |
          4. |
  19      5* | 1
  18      5. | 5
          6* |
  17      6. | 589
  14      7* | 00
          7. |
  12      8* | 224
   9      8. | 5
   8      9* | 00
          9. |
   6     10* | 0001
HI: 1100 1100
```

Figure 4.13 *Stem-and-leaf plot, with depth added, of mass of 56 perch, caught in a research trawl on Längelmävesi, with the number of stems doubled and high and low values noted*

the period "." to each stem in the manner of Tukey [127]. So doing calls further attention to the splitting of leaves from 0 to 9 to one group of $\{0, 1, 2, 3, 4\}$ and another of $\{5, 6, 7, 8, 9\}$.

In the case of five splits for each stem digit, Tukey [127] used the symbol "*" for the numbers 0 and 1, "t" for 2 and 3, "f" for 4 and 5, "s" for 6 and 7, and finally "." for 8 and 9. An example of a stem-and-leaf plot with 5 splits is given in Figure 4.14 for the ratio of maximal width to length from the mouth to tip of the tail for each of the 56 perch. A little reflection on the initial English letter for the pairs of numbers in the five splits will reveal the rationale behind the choices of the letters "t", "f", and "s."

The stemplots of Figures 4.8 through 4.14 ought to appear to be reminiscent of the stacked dotplots of Figures 4.2 and 4.3 in the previous section. The stacked dotplots and the stemplots are also quite similar in many respects to the pictograph of Figure 3.1 for depicting the distribution of a single qualitative variable. Both the dotplot and the stemplot are forms of the pictograph.

```
  1 | 2: represents 1.2
   leaf unit: 0.1
              n: 56
     1      t | 2
            f |
     2      s | 6
     3    13. | 9
          14* |
     4      t | 3
     6      f | 55
     9      s | 667
    12    14. | 889
    23    15* | 00000001111
    25      t | 23
    28      f | 445
   (2)      s | 67
    26    15. | 88999
    21    16* | 011
    18      t | 2333
            f |
            s |
    14    16. | 8
    13    17* | 00
    11      t | 3
    10      f | 5
     9      s | 66677
     4    17. | 89
     2    18* | 1
            t |
            f |
            s |
          18. |
          19* |
            t |
            f |
            s |
          19. |
          20* |
            t |
            f |
            s |
     1    20. | 9
```

Figure 4.14 *Stem-and-leaf plot of width ratio (percentage) for 56 perch caught in a research trawl on Längelmävesi*

Icons in pictographs ought to be nearly the same size so that distortion is minimized and adverse consequences of breaking the axiom of *Truthfulness* are avoided. This fact reinforces the need for the numbers in a stemplot to be in a nonproportional font because the numbers are inherently icons. So much the better if the numbers are also the same size and width. Doing so would be faithful to the original presentation of the stemplot by John Wilder Tukey [127].

The stemplot was created by Tukey [126, 127] at a time corresponding to the early widespread introduction of computers. At this time output was produced by computers on duo-fold paper with line printers that used metal type for alphanumeric symbols affixed to chains that struck an inked-ribbon before hitting the paper. The hammer that struck the metal type was a fixed size and so the characters were equally wide. The so-called mono-spaced Courier font is a relic of this bygone era and was used to draft the stemplots in this chapter. When the stemplot was first programmed, it permitted the rapid depiction of multiple data sets for comparison of distributions of quantitative variables.

To the discerning eye of either a graphics professional or a member of the public, the stemplot may be considered a crude, unappealing, and unsophisticated graphical display. The stemplot is not much used outside the statistics profession. This is a pity because it is easily explained to a public audience and administrators alike. Conveniently, the stemplot is usually available in most, if not all, statistical software packages.

4.5 The Boxplot

4.5.1 Definition

Another EDA graphical display created by John Tukey is the *box-and-whisker plot.* This appeared in Tukey's [127] book published in 1977. Some have shortened this name to *box-whisker plot.* Others have shortened this further to *boxplot.* The latter of the three names is adopted here.

Two different boxplot conventions are presented in this chapter. These are the conventions most frequently encountered. Both forms of the boxplot require the concept of a *quantile.*

To define the *quantile*, consider a random variable X and its probability distribution. Let c be a proportion between 0 and 1, inclusive. The c-th *quantile* of the random variable X is the value ξ_c such that a proportion of at least c of its probability distribution is less than or equal to ξ_c.

If the quantitative random variable X is continuous, then the value of the quantile ξ_c is given uniquely by solving the probability equation

$$P(X \leq \xi_c) = c. \tag{4.1}$$

This is the definition often given in introductory textbooks. But it represents an oversimplification that is permissible at the introductory level. If the quantitative variable can take discrete values, then the definition of the parameter ξ_c is not unique.

There are further complications regarding quantiles. Because ξ_c is a characteristic of the probability distribution of X, it is a parameter. When dealing with a random sample, the parameter ξ_c is estimated from the sample. The corresponding sample quantile is denoted x_c. This is the easy part. The difficult part is defining how to calculate x_c from the sample.

There is no consensus among statisticians regarding the calculation of the statistic x_c. This is usually glossed over in introductory statistical textbooks. A discussion of this situation is required as it pertains directly to the plotting convention for the boxplot. This discussion will be deferred until the more substantive features of the boxplot have been discussed and an example given. Before this can be done, there are a few less contentious points to be noted about quantiles.

The quantile can be encountered in one or more of the following forms. If the proportion c is a member of the set $\{1\%, 2\%, 3\%, \ldots, 99\%\}$, then the quantile ξ_c is called a *percentile*. If the proportion c is a member of the set $\{0.1, 0.2, 0.3, \ldots, 0.9\}$, then the quantile ξ_c is called a *decile*. If the proportion c is a member of the set $\{0.2, 0.4, 0.6, 0.8\}$ then the quantile ξ_c is called a *quintile*. And if the proportion c is a member of the set $\{0.25, 0.5, 0.75\}$, then the quantile ξ_c is called a *quartile*.

Some further comments regarding nomenclature with respect to the quartiles are as follows. The value $\xi_{0.25}$ is referred to as either the *lower quartile* or the *first quartile* of the distribution of the random variable X. Correspondingly, the value $\xi_{0.75}$ is referred to as either the *upper quartile* or the *third quartile*. There could be a middle or second quartile but it is never referred to as such. Instead, $\xi_{0.5}$ is called the *median*.

The quantile ξ_0 is referred to as the *minimum* and ξ_1 is referred to as the *maximum*. In our discussion of boxplots, we will be making use of the quartiles, the minimum, and the maximum. Corresponding to the five values $(\xi_0, \xi_{0.25}, \xi_{0.5}, \xi_{0.75}, \xi_1)$ of the probability distribution of X are those of the sample. Conventionally, the symbol Q_1 denotes the first quartile of the sample, the symbol M denotes the median of the sample, and the symbol Q_3 denotes the third quartile of the sample. Common to the plotting conventions for the two boxplots presented in this chapter is the depiction of these three quartiles in a box along a linear axis.

Note that the sample median M indicates the center of location for a distribution. Together the sample lower quartile Q_1 and the sample upper quartile Q_3 convey information regarding the spread of the distribution.

The sample quartiles (Q_1, M, Q_3) convey no information about the lower and

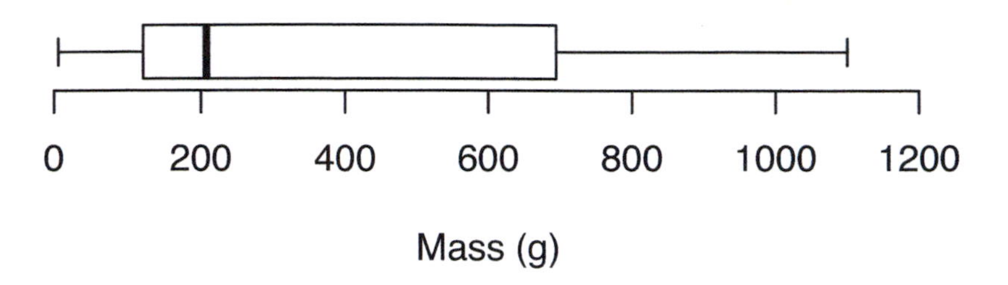

Figure 4.15 *Quantile boxplot of mass (g) for 56 perch caught in a research trawl on Längelmävesi*

upper tails of the distribution. The maximum and minimum values from the sample provide some information about the tails of the distribution of a random variable.

Putting the pieces together, information about the center, spread, and tails of the distribution according to a sample is given by the *five-number summary*:

$$\text{minimum}, Q_1, M, Q_3, \text{maximum}. \tag{4.2}$$

A visual representation of the five-number summary for the mass of 56 perch caught in a research trawl on Längelmävesi is given in the *quantile boxplot* of Figure 4.15.

The boxplot in Figure 4.15 was produced by the following call to the R function `boxplot`.

```
boxplot(mass,range=1.5,horizontal=TRUE,xlab="Mass (g)",
pars=list(boxwex=1.5),lty=1,ylim=c(0.,1200.),
yaxp=c(0.,1200.,6),outline=FALSE)
```

The orientation of the boxplot can be either horizontal or vertical (by setting `horizontal=FALSE`). Setting `boxwex=1.5` widens the box. Setting `lty=1` forces the whiskers to be plotted as solid lines. The parameter `ylim` sets the lower and upper limits for the axis, which is horizontal in Figure 4.15. Setting `yaxp=c(0.,1200.,6)` plots tick marks and labels every 200 g beginning at zero. Discussion of the definitions of the parameters `range` and `outline` is deferred to the following subsection.

As previously noted, horizontal orientations are preferred because it is thought that we are better judges of position and length when the orientation is horizontal compared to the vertical.

Before desktop computers, the required numerical estimates for the boxplot could be obtained by drafting a stemplot by hand. Nowadays, a stemplot or boxplot, or both, can easily be obtained in a blink of an eye from a statistical software package after the data have been entered. So the choice of whether to execute a stemplot or boxplot is one of personal preference.

The plotting convention for the quantile boxplot consists of:

- a rectangular box with edges determined by the lower and upper quartiles;
- the median denoted as a line segment splitting the rectangular box into two adjoining boxes;
- a *whisker* (that is, a line segment) from Q_1 to the minimum; and
- a whisker from Q_3 to the maximum.

Boxplots can be found in the technical journals of various disciplines in addition to statistics. The boxplot is not likely to be used at present or in the future in the popular press or media because it assumes a fair degree of background knowledge in statistics (for example, familiarity with the definition of the quartiles of a random sample).

4.5.2 Variations on the Boxplot

There have evolved a number of boxplot plotting conventions and so it is necessary to consult boxplot captions, or accompanying text. If the viewer does not recall the particular boxplot convention from long-term memory, a new memory will need to be created, or an old one re-created, to understand the image.

The quantile boxplot based on the five-number summary is one of two forms of the box-and-whisker plot described by John Tukey in 1977. Rather than referring to quartiles to construct the box, Tukey [127] used the term *hinges*. Others have extended Tukey's definition of hinges to quantiles other than quartiles so that it is possible to construct a boxplot with quantiles other than the quartiles. But this extension was not proposed by Tukey.

Tukey [127] describes the construction of the boxplot using the terms *lower hinge*, *middle hinge*, and *upper hinge*. The symbols H_1, H_2, and H_3 refer to the lower, middle, and upper hinges, respectively. Typically, the middle hinge is the median but the lower hinge could be a decile $x_{0.1}$, a quintile $x_{0.2}$, or anything else, with a matching symmetric selection for the upper hinge. Most commonly in a boxplot, the hinges are chosen to be the quartiles.

Tukey [127] also introduced the term *fences* in reference to distances from the lower and upper hinges. There are two sets of fences: the *inner fences* and the *outer fences*. Calculation of the fences requires calculating the *InterHinge Range*

$$IHR = H_3 - H_1 \tag{4.3}$$

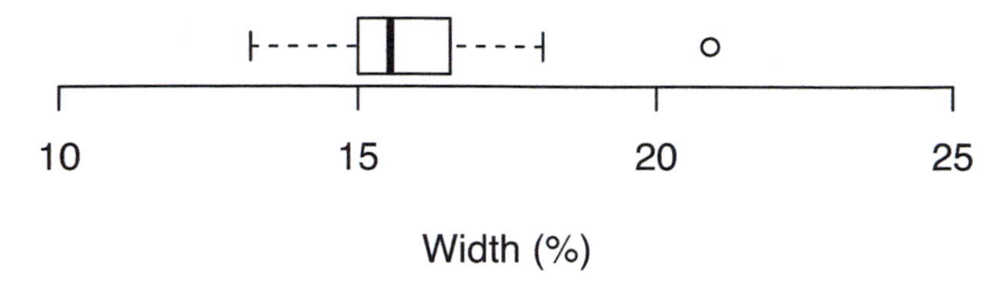

Figure 4.16 *Outlier boxplot of width ratio (percentage) for 56 perch caught in a research trawl on Längelmävesi*

and selecting the *inner fence factor* a and the *outer fence factor* b with a and b positive nonzero numbers representing multiples of the IHR with $a < b$.

Tukey [127] did not use the term InterHinge Range but instead coined the term *H-spread*. He also defined the term *step* as 1.5 times H-spread and set the inner fences at 1 step outside the hinges and the outer fences at 2 steps outside the hinges. These choices correspond to inner and outer fence factors of 1.5 and 3, respectively. He also chose the hinges to be quartiles. When the hinges are quartiles, we usually refer to the distance between the first and third quartiles as the *InterQuartile Range*:

$$IQR = Q_3 - Q_1. \tag{4.4}$$

The inner fences are used to draw the whiskers. Each whisker terminates at the observed value at each end closest to but inside the inner fence. The values where the whiskers terminate are called *adjacent*. Tukey [127] referred to data values between the inner and outer fences as *outside*. He referred to data values at the outer fences or beyond as *far out*. The outer fences are not drawn nor are their positions identified in any way.

More commonly, data points beyond the inner and outer fences are referred to as *outliers* and *extreme outliers*, respectively. In retrospect and to be faithful to Tukey, perhaps the term outsider should have been adopted instead of outlier. The main issue is that these outsiders or outliers need to be examined as to whether they are recording errors or contaminants from another population that somehow made it into the sample.

An example of an outlier boxplot is given in Figure 4.16. The data are again from the catch of 56 perch in a research trawl on Längelmävesi. The variable plotted in Figure 4.16 is maximal width as a percentage of the length measured from the end of the tail to the tip of the mouth. There is one outlier at 20.9% width.

Figure 4.16 was plotted by the following R script.

```
boxplot(width,range=1.5,horizontal=TRUE,
pars=list(boxwex=1.5),xlab="Width (%)", cex=1.0,
ylim=c(10.,25.),yaxp=c(10.,25.,3),outline=TRUE)
```

The hinges for the R function `boxplot` are the quartiles. The inner fence factor a is set to 1.5 by the syntax `range=1.5`. Setting `outline=TRUE` causes outliers, if any, to be plotted. The function `boxplot` does not distinguish between outliers and extreme outliers so there is only one plotting symbol for any type of outlier.

Justification for the choice of $a = 1.5$ in determining the inner fences and for flagging outliers is in order. These values were not chosen on a whim. The general principle behind the choice of hinges and fence multipliers is one of illustrating agreement or departure from a model of a probability distribution in any given simple random sample.

A reasonable distributional model to choose would be the normal distribution function. For any normal distribution the ratio of the interquartile range to the standard deviation is approximately 1.3490:1.

For $a = 1.5$, the area under any normal curve between the lower inner fence and the upper inner fence is thus 0.9930 or approximately 99%. If the random sample truly is normally distributed, then an outlier will be falsely detected approximately once in one hundred uses of the outlier boxplot.

The conventional choice of $b = 3.0$ for the outer fence factor can also be supported from a heuristic argument. However, for this choice, the area under any normal curve between the lower outer fence and the upper outer fence is thus approximately 99.999766%. If the random sample truly is normally distributed, then an extreme outlier will be falsely detected approximately once in 425,532 uses of the outlier boxplot. So there is justification for the modifier *extreme* in *extreme outlier* or the descriptor *far out* in Tukey's terminology. But the lingering question is why one in 425,532?

The values of $a = 1.5$ and $b = 3.0$ have been adopted as conventions by practitioners of statistics with broad acceptance but these choices are not carved in stone. Other values for a and b have been considered in the past in the context of reference to the normal distribution. If the reference distribution changes, it would certainly be desirable to alter a and b to take this into account. Moreover, if the reference distribution is skewed, consideration ought to be made regarding the selection of different values of a and b depending on whether the fences are lower or upper. Additionally, the hinges can also be chosen asymmetrically. Arguably, it is reasonable to stick with the median as the middle hinge. In summary, there appears to be ample opportunity for creative re-design of the boxplot.

The plotting convention for the *outlier boxplot* consists of:

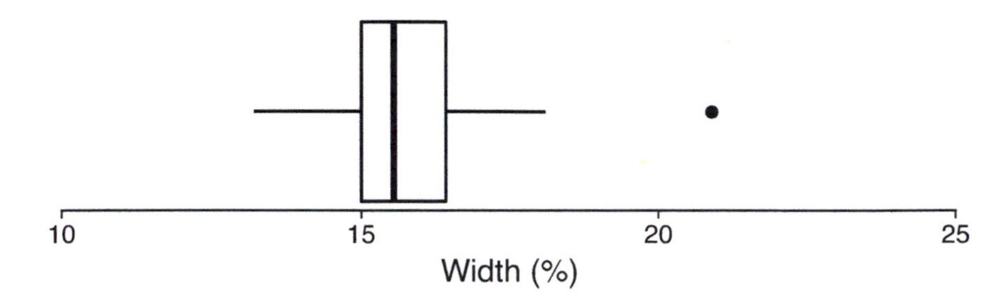

Figure 4.17 *Outlier boxplot of width ratio (percentage) for 56 perch caught in a research trawl on Längelmävesi with* `ggplot2`

- a rectangular box with edges determined by the lower and upper quartiles;
- the median denoted as a line segment splitting the rectangular box into two adjoining boxes;
- whiskers ending at the adjacent values; and
- open circles to represent points lying beyond the inner fences.

Notice that the whiskers in Figure 4.16 are dashed rather than solid. This is in accordance with the original plotting convention of Tukey [127] that also required that the crossbars be dashed. Tukey's name for the outlier boxplot is *schematic plot.* Tukey required that the whiskers be solid for his box-and-whisker plots so that these two different types of plots using boxes and whiskers could be more easily distinguished. But this sensible convention has not stood the test of time. McGill, Tukey, and Larsen [83] in 1978 used dashed whiskers in their illustrated convention for the quantile boxplot.

McGill, Tukey, and Larsen [83] referred to the box-and-whisker plot of Tukey [127] as the *box plot*—which is also notable. The first schematic plots appeared in print in 1972 in a paper written by Tukey [126] entitled "Some Graphic and Semigraphic Displays." This paper appeared in the monograph *Statistical Papers in Honor of George W. Snedecor*, edited by T. A. Bancroft.

Tukey [127] also insisted that the plotting symbols for the outside and far out values be distinct and impressive for the far out values. The outlier in Figure 4.16 is indeed far out beyond the outer fence but the R function `boxplot` used to produce Figure 4.16 supports only a common choice of symbol for both outliers and extreme outliers.

Figure 4.17 was drafted using the following `ggplot2` code.

```
figure<-ggplot(perch, aes(1,width)) +
geom_boxplot(color="black") + theme_linedraw() +
theme(panel.border = element_rect(fill=NULL,
color="white",linetype="solid"),
```

```
axis.title.y=element_blank(),
axis.text.y=element_blank(),
axis.ticks.y=element_blank(),
axis.line.x=element_line(),
panel.grid=element_blank()) +
scale_y_continuous(expand=c(0,0),
breaks=5+5*(1:4),limits=c(10,25)) +
ylab("Width (%)") + coord_flip()
> print(figure)
```

For this figure, the line-draw theme in **ggthemes** has been selected and it has been necessary to pass the parameter `panel.border` with the given arguments to suppress the background grid lines. Note that the parameter `color="black"` has been passed on calling function `geom_boxplot` so that the boxplot is drawn in black rather than the default gray shade. Code has also been added to suppress entirely the unnecessary vertical axis. The default orientation is vertical so `coord_flip` is called to convert to a horizontal orientation for the boxplot plot. Note that the default symbol for an outlier is a solid black circle. The whiskers are solid not dashed and do not terminate with a vertical line segment. The documentation for `geom_boxplot` clearly states that the lower and upper hinges are the first and third quartiles but these are not calculated using the hinges definition of Tukey [127] that equal the quartiles for odd sample sizes but differ for even sample sizes. Quartile (and percentile) estimation is discussed in greater detail in Section 5.5.

Tukey [127] required that the adjacent, outside, and far out values be labeled with full capital letters reserved for the far out labels. These are sensible but rarely used embellishments. An illustration is given in Figure 4.18 for the width of 56 perch. Notice that the orientation has been changed from horizontal to vertical to accommodate the label "FISH #143." Fortunately, the Finnish fisheries biologists were very thorough as the data were collected. Examination of fish number 143 revealed 6 roaches (*Leuciscus rutilus*) in its stomach and thus an explanation for this outlier.

To better illustrate the differences between the boxplot and the outlier boxplot, these two plots are depicted side by side in Figure 4.19 for the width ratio of perch collected by the Finnish fisheries biologists. Note that the box is identical in both of the two forms of the boxplot. But the outlier is only detected in the outlier boxplot.

Both forms of the boxplot are quick visual methods of depicting the distribution of a quantitative variable. Both forms have associated with them a degree of robustness. The presence of one or two outliers in a random sample from a normal population will cause a linear perturbation in the value of the sample mean but a much greater quadratic perturbation in the value of the sample variance. But typically very little alteration is noticed in the values of the quartiles.

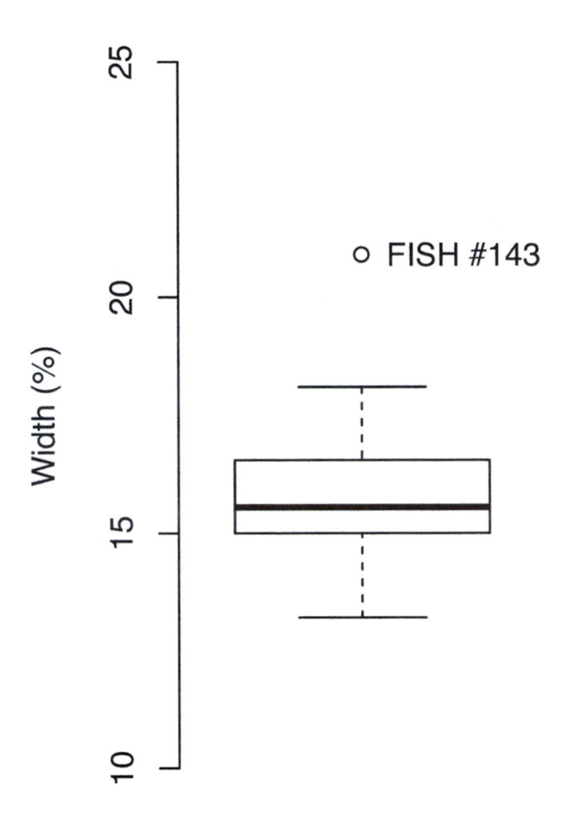

Figure 4.18 *Outlier boxplot (with labeled outlier) of width ratio (percentage) for 56 perch caught in a research trawl on Längelmävesi*

In the presence of outliers in a random sample from a normal population:

- the median is a more robust measure of location than the mean; and
- the interquartile range is a more robust measure of spread than the standard deviation.

So with both forms of the boxplot, one gets a robust view of the location and spread of the distribution. But the lengths of the whiskers of the quantile boxplot are strongly affected by the presence of outliers. In comparison, the whiskers of the outlier boxplot are not so affected. Strongly asymmetric whiskers in a quantile boxplot may not be an indicator of skewness in the population but rather a symptom of the presence of an outlier. This is a good reason to select the outlier boxplot over the quantile boxplot.

The important detail of an algorithm for estimating the quartile from data

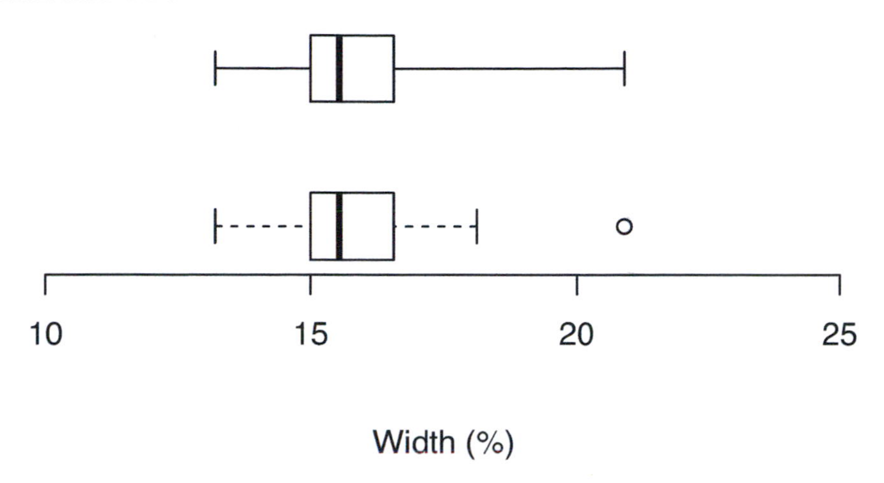

Figure 4.19 *Boxplot (top) and outlier boxplot (bottom) of width ratio (percentage) for 56 perch caught in a research trawl on Längelmävesi*

was glossed over so that we could get on with discussing the boxplot. But the choice of algorithm can strongly affect the appearance of the boxplot. We have reached the point where this discussion of the choice of algorithm ought to occur. The available options, however, are better appreciated in the context of the following plot. The discussion of algorithm is deferred to the next chapter.

4.6 The EDF Plot

4.6.1 Definition

Consider the random sample $\{x_i\}_{i=1}^n$. We can take this sample and arrange the outcomes in nondecreasing order:

$$x_{(1)} \leq x_{(2)} \leq x_{(3)} \leq \cdots \leq x_{(n)}. \tag{4.5}$$

The notation $\{x_{(i)}\}_{i=1}^n$ denotes a random sample in nondecreasing order. (The term nondecreasing is used to describe the ordering of the sample rather than increasing because in some samples there might be multiple observations of the same value.) Note that the random sample is a set of a finite number of discrete elements. A discrete approximation to the distribution of the random variable X is given by the *empirical distribution function (EDF)*:

$$S_n(x) = \begin{cases} 0 & \text{if } x < x_{(i)}, \\ \frac{i}{n} & \text{if } x_{(i)} \leq x < x_{(i+1)}, \\ 1 & \text{if } x \geq x_{(n)}. \end{cases} \tag{4.6}$$

Index i	Mass (g) $x_{(i)}$	EDF $S_n(x_{(i)})$	Index i	Mass (g) $x_{(i)}$	EDF $S_n(x_{(i)})$
1	5.9	1/56	29	218.0	29/56
2	32.0	2/56	30	225.0	30/56
3	40.0	3/56	32	250.0	32/56
4	51.5	4/56	33	260.0	33/56
5	70.0	5/56	34	265.0	34/56
6	78.0	6/56	36	360.0	36/56
7	80.0	7/56	37	320.0	37/56
8	85.0	9/56	38	514.0	38/56
10	100.0	10/56	39	556.0	39/56
12	110.0	12/56	40	650.0	40/56
13	115.0	13/56	41	685.0	41/56
15	120.0	15/56	42	690.0	42/56
16	125.0	16/56	44	700.0	44/56
19	130.0	19/56	46	820.0	46/56
20	135.0	20/56	47	840.0	47/56
22	145.0	22/56	48	850.0	48/56
24	150.0	24/56	50	900.0	50/56
25	170.0	25/56	53	1000.0	53/56
26	180.0	26/56	54	1015.0	54/56
27	188.0	27/56	56	1100.0	56/56
28	197.0	28/56			

Table 4.2 *Ordered sample (replicates omitted) of the mass of 56 perch caught in a research trawl on Längelmävesi with empirical distribution function* S_n

Note that the empirical distribution function is an approximation to the cumulative distribution function of the random variable X so it is also known as the *empirical cumulative distribution function (ecdf)*.

Table 4.2 presents the empirical distribution function for the mass of 56 perch given in Table 4.1. Notice that the replicate observations have been omitted in Table 4.2 so that the empirical function is well defined with exactly one value $S_n(x_{(i)})$ for a given $x_{(i)}$.

Table 4.2 is graphically represented in Figure 4.20. The actual observations are depicted by solid black circles. The amount of repetition of a specific value is indicated by the height of the vertical gap between line segments or solid black circles. The largest vertical gap in Figure 4.20 occurs for the mass observation of 1000.0 g, which is repeated 3 times in the data.

An EDF plot is useful for looking for groups of observations separated by gaps. The longest horizontal line segment in Figure 4.20 brings to our attention the gap in observations of perch mass between 320.0 g and 514.0 g. If not an artifact of the research trawl or an environmental event, this gap suggests a

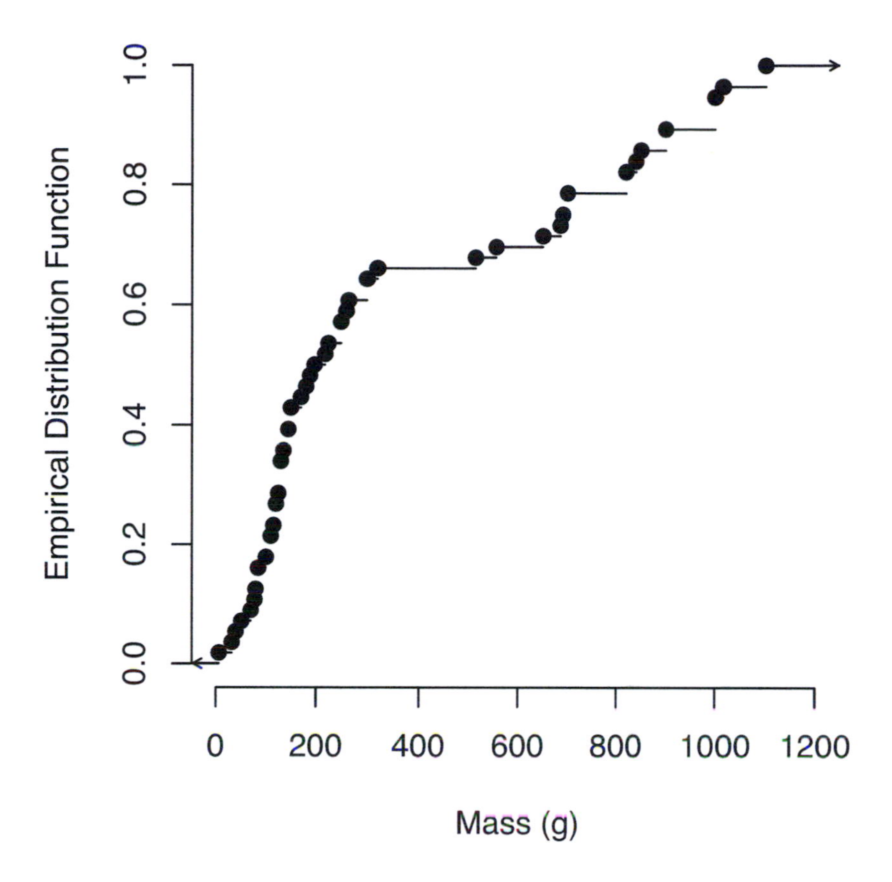

Figure 4.20 *EDF plot of mass for 56 perch caught in a research trawl on Längelmävesi*

biologically meaningful epoch in perch growth. This gap is also apparent in the dotplots and stemplots already presented for perch mass. These types of gaps generally are undetected in boxplots, as in Figure 4.15. This suggests the need to execute more than one plot for a quantitative variable, or any variable for that matter.

The version of the EDF plot presented in 1968 by Wilk and Gnanadesikan [135] interchanges the axes of Figure 4.20 so that the range of the empirical distribution function is plotted along the horizontal axis rather than the vertical axis. Consequently, their plot is actually the inverse of the empirical cumulative distribution function. Wilk and Gnanadesikan [135] in their article even acknowledge John Tukey as having had pointed this out to them. Tukey suggested that Wilk and Gnanadesikan [135] use the term *empirical representing function* for the inverse of the empirical cumulative distribution

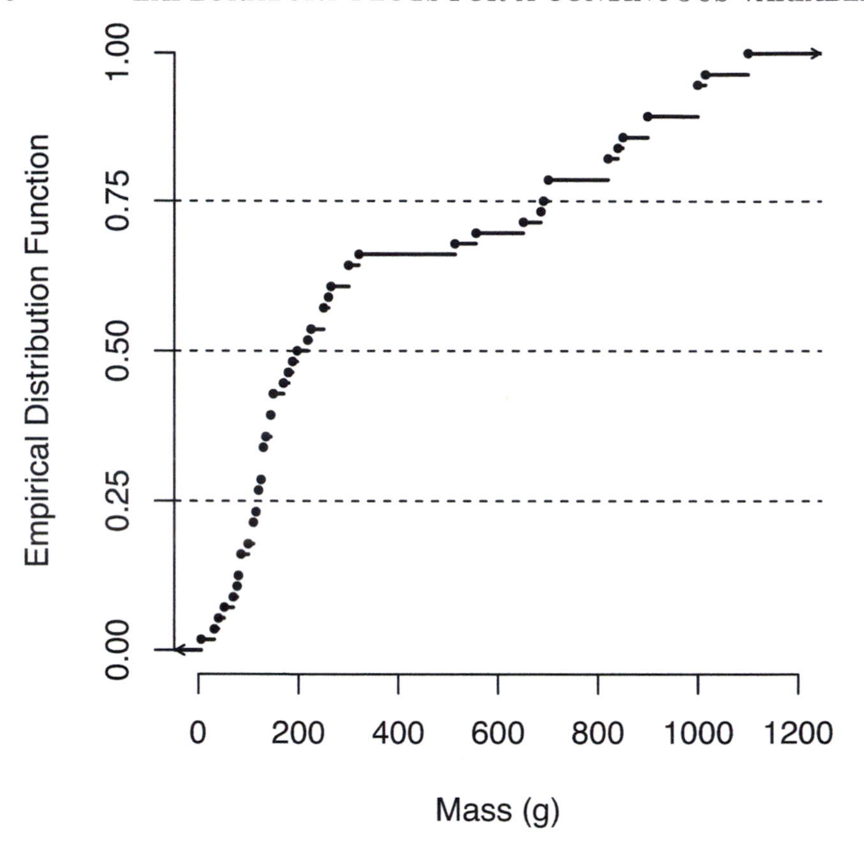

Figure 4.21 *EDF plot of mass for 56 perch, caught in a research trawl on Längelmävesi, with reference lines for estimation of quartiles*

function. However, neither the term "empirical representing function" nor the version of the EDF plot by Wilk and Gnanadesikan [135] have stood the test of time.

The EDF plot can be and, before computer graphics, was used in the visual estimation of quartiles. This is done in Figure 4.21. The R code for creating Figure 4.21 is as follows.

```
ecdfmass<-ecdf(mass)
plot.stepfun(ecdfmass,xlab="Mass (g)",
ylab="Empirical Distribution Function",
main=NULL,verticals=FALSE,do.points=TRUE,pch=19,
cex=0.5,xlim=c(0,1200),yaxp=c(0.,1.,4),lwd=1.75)

arrows(max(mass),1.0,1245,1.0,code=2,length=0.05)
```

```
arrows(-45.,0.0,min(mass),0.0,code=1,length=0.05)
abline(h=c(0.25,0.5,0.75),lty=2)
```

The R function `ecdf` can generate a plot of an empirical cumulative distribution. Instead, the EDF for the mass of 56 perch has been stored in the variable `ecdfmass` that is then plotted by the R function `plot.stepfun`. This latter function provides greater control over aspects of the plot. The quality of the plot produced with the default plotting standard of `ecdf` is simply not good enough.

By setting `verticals=FALSE` in the call to `plot.stepfun`, the vertical line segments, which link the horizontal line segments, are not plotted. The originating points for the horizontal line segments are plotted as a result of setting `do.points=TRUE`. By setting `pch=19`, the originating points are plotted as solid circles. By setting `cex=0.5`, the diameters of the solid circles have been reduced by a factor of one-half for Figure 4.21 compared to Figure 4.20. By setting `lwd=1.75`, the thickness of the horizontal lines has been increased for Figure 4.21 compared to Figure 4.20.

The two calls to the R function `arrow` add arrowheads to the two horizontal line segments at the left and right extremes of the data in Figure 4.21. The call to the R function `lines` adds the reference lines for the three quartiles. The plotting symbol size has been reduced and the line segments thickened in Figure 4.21 to aid in estimation of the quartiles.

The horizontal reference line corresponding to the lower quartile does not pass through any point on the EDF plot of Figure 4.21. One solution to this problem is given by the version of the EDF plot in Figure 4.22 that is known as the *step plot*. The vertical line segments in Figure 4.22 were obtained by setting `verticals=TRUE` in the call to the R function `plot.stepfun`.

A `ggplot2` version of Figure 4.22 is given in Figure 4.23 in the line-draw theme of `ggthemes`, which produced a bounding box for the plot as a default. Dispensing with saving the plot to an object in R, the code for producing Figure 4.23 is simply as follows.

```
ggplot(perch, aes(x=mass)) + stat_ecdf() +
theme_linedraw() + theme(panel.grid=element_blank()) +
scale_x_continuous(breaks=200*(0:6),limits=c(0,1200)) +
labs(x="Mass (g)",y="Empirical Distribution Function")
```

The `ggplot2` function for producing an EDF plot is `stat_ecdf`. The only added instructions are for labeling the axes, determining the limits and tick marks for the horizontal axis, and suppressing the background grid lines.

The embellishments of reference lines for estimating the sample quartiles are added to the step plot in Figure 4.24, and we find 120.0 g as our estimate of the lower quartile.

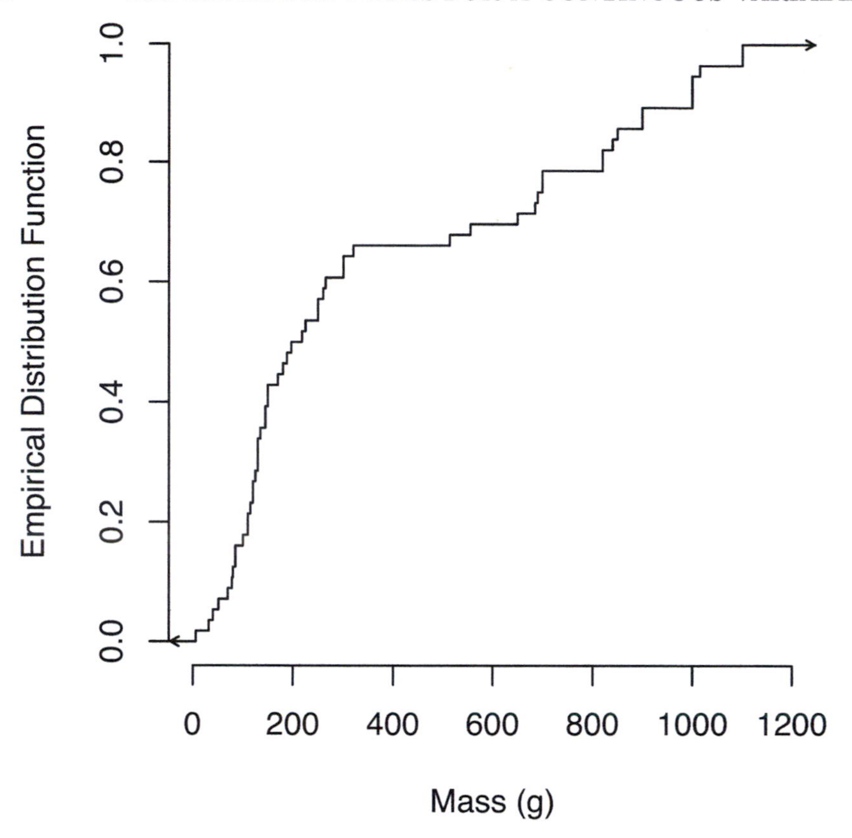

Figure 4.22 *Step plot version of the EDF plot of mass for 56 perch caught in a research trawl on Längelmävesi*

A closer examination of the EDF plot in Figure 4.21 at the reference line for the median reveals a second problem. The estimate of the median is, in fact, not a single data point but some value M such that 197.0 g $\leq M <$ 218.0 g. For perch mass, if a single point estimate of the median is required, it will be a point in this interval.

With respect to estimating the upper quartile, a closer examination of the step plot in Figure 4.24 at the horizontal reference at $y = 0.75$ reveals a third problem in the EDF plot of Figure 4.21 that is missed by all but a careful eye. There is no single point estimate for the upper quartile, but rather some value Q_3 such that 690.0 g $\leq M <$ 700.0 g.

If a single point estimate of the upper quartile is required, it will be a point in this interval. As with the estimate of the median for this data, the choice of estimate for the upper quartile is a matter of discussion if not contention.

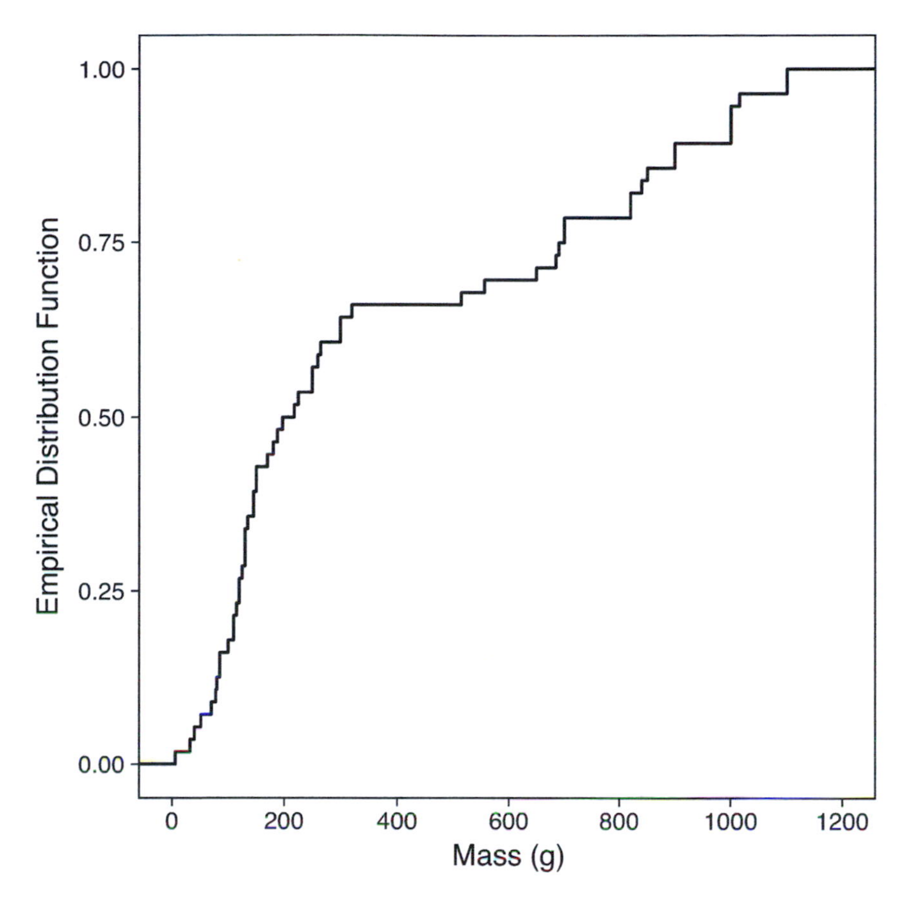

Figure 4.23 *Step plot version of the EDF plot of mass for 56 perch caught in a research trawl on Längelmävesi using* `ggplot2`

Approaches to finding a point estimate for each of the quartiles, or any other percentile, are considered in more detail in the next chapter.

From the viewpoint of mathematical rigor, the stepwise structure depicted in Figures 4.22, 4.23, and 4.24 is not a function. The rationale for this conclusion is that where there are vertical line segments, there is not a unique functional value defined on the vertical axis for the corresponding value on the horizontal axis.

Despite this fact, the step plot of Figures 4.22, 4.23, and 4.24 is available in most statistical software packages and thus is more frequently encountered in publications than the pedantically correct EDF plot of either Figure 4.20 or Figure 4.21. To some extent, Figures 4.22, 4.23, and 4.24 represent a legacy of computer plotters in the past that utilized pen cartridges.

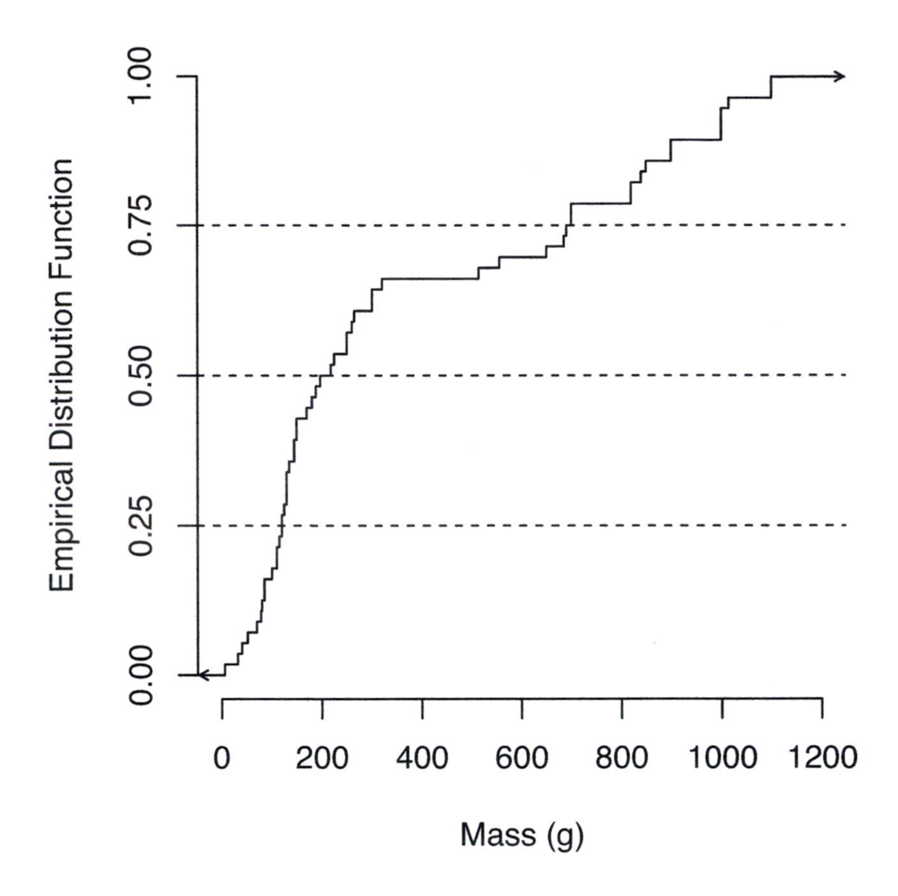

Figure 4.24 *Step plot version of the EDF plot of mass for 56 perch, caught in a research trawl on Längelmävesi, with reference lines for estimation of quartiles*

4.6.2 The EDF Plot as a Diagnostic Tool

A central theorem of statistics states that, as the size n of a random sample increases without bound, the empirical distribution function uniformly converges to the cumulative distribution function of the population. Visually, if one were to prepare a step plot like Figure 4.22, one would find with increasing sample size that the step sizes would become smaller and smaller until the EDF curve appears smooth. Such a smooth curve is superimposed on the step plot for perch mass in Figure 4.25. The smooth curve represents the cumulative normal distribution function with mean and standard deviation given by the corresponding sample estimates for perch mass.

The EDF plot or step plot can be used as a visual check for goodness-of-fit to a distribution by adding a theoretical curve of the distribution to the EDF

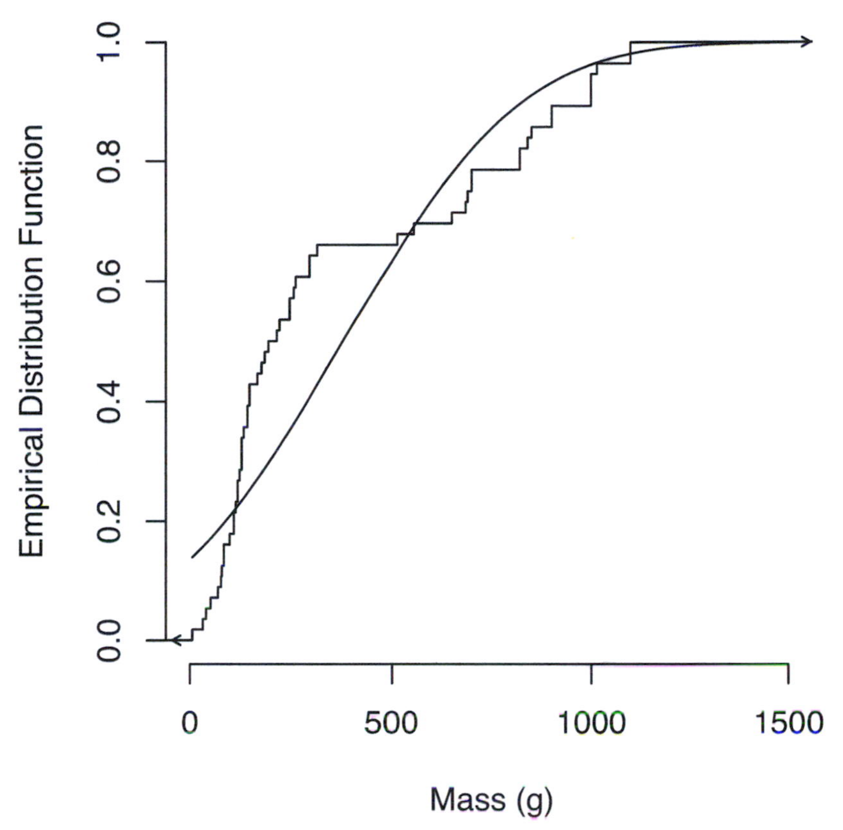

Figure 4.25 *Step plot version of the EDF plot of mass for 56 perch, caught in a research trawl on Längelmävesi, with cumulative normal distribution function added*

plot as done in Figure 4.25 for the perch mass data. In this figure, the fit is not good.

There are a number of different statistics available for testing the goodness-of-fit based on the empirical distribution function. See the article of Stephens [114] published in 1974. These statistics work by estimating the gap between the jagged empirical distribution function and the smooth cumulative distribution function. These statistics are known as *EDF statistics.* Despite the popularity of these statistics, the EDF plot, or step plot, is rarely displayed. This is a shame. Instead, the standard practice is to execute a stem-plot, or one of its alternatives, to depict density rather than the cumulative density while computing the appropriate EDF statistic.

There are variations on the EDF plot that are frequently used in a diagnostic setting. These will be explored in the next chapter.

4.7 Conclusion

The advocacy of simply executed but informative statistical graphics for exploratory data analysis can be traced to Tukey [127]. Dotplots, stemplots, and boxplots were proposed at a time when the widely available computer output devices were printers capable of only printing fixed-width type fonts.

The early boxplots were printed with the characters "|", "-", "o", and "*". So much for history. Yet even with statistical software, the boxplot and stemplot continue to be widely available because of the elegance and efficiency inherent in their designs.

Stemplots and boxplots have yet to make their way into the popular media. Given the technical details involved in their drafting, they are not likely to be seen anytime soon. On the other hand, the dotplot is more easily explained and comprehended. The stacked dotplot is probably the better option for the public given that the dotplot can obscure multiple or closely adjacent observations. The barcode plot can similarly obscure data.

The EDF plot is likely only to be encountered in technical and scientific settings. Unlike the dotplot, including the stacked dotplot, the EDF plot clearly illustrates repeated and closely adjacent observations. The EDF plot is not widely available in statistical software packages and outside of statistical journals it is a rare bird.

An important detail with both the boxplot and the EDF plot has been glossed over in this chapter. This detail concerns how quartiles and percentiles are estimated. This is the first topic of the next chapter.

4.8 Exercises

1. Consider the following data collected by Charles Darwin:

$$-67, -48, 6, 8, 14, 16, 23, 24, 28, 29, 41, 49, 56, 60, 75.$$

   These data represent the difference in height (in eighths of an inch) between the self-fertilized member and the cross-fertilized member of each of 15 pairs of plants.

   (a) Produce a stemplot and comment on any features you see.

   (b) Produce an EDF plot comparing features in this plot to the stemplot.

   (c) Comment on whether there are any outliers in the data.

   (d) Comment on whether the data appear to follow a skewed distribution. How would skewness be discernible in an EDF plot?

2. Consider the data for the total compensation in 2008 received by chief executive officers (CEOs) employed by industrial companies listed in Table 4.3.

   (a) Produce a dotplot for total compensation.

Company	Amount (US Dollars)	Company	Amount (US Dollars)
3M	11,730,000	Jabil Circuit	4,809,300
Aecom Technology	6,706,600	Jacobs Engineering	6,440,700
AGCO	7,561,800	L-3 Communications	12,318,900
Agilent Technologies	7,773,800	Lockheed Martin	17,845,400
Arrow Electronics	5,825,000	Manpower	10,326,000
Ball	7,121,000	MDU Resources	2,211,300
Boeing	15,606,200	MeadWestvaco	5,503,600
Burlington Northern	12,013,300	Navistar Int'l	6,535,500
Crown Holdings	10,767,700	Norfolk Southern	11,893,000
CSX	7,090,300	OshKosh	3,109,000
Danaher	4,367,300	Owens-Illinois	5,571,400
Deere	14,387,900	Paccar	10,826,200
Eaton	6,599,700	Rockwell Automation	5,995,800
Emerson Electric	6,380,000	Ryder System	4,291,200
Expeditors Int'l	6,028,600	Sanmina-SCI	2,029,400
Fluor	10,705,800	Shaw Group	8,077,100
General Dynamics	17,558,000	Sherwin-Williams	5,310,700
General Electric	4,935,700	SPX	9,252,000
Goodrich	9,601,900	Textron	9,182,300
Honeywell Int'l	14,309,000	Timken	5,204,300
Illinois Tool Works	9,345,800	Waste Management	6,045,600
ITT	10,101,500		

Table 4.3 *Total compensation in 2008 for chief executive officers of 43 industrial corporations*

(b) Produce a stacked dotplot for total compensation.

(c) Produce a jittered dotplot for total compensation.

(d) Which of the three dotplots of parts (a), (b), and (c) would you choose to use in a written presentation? Justify your answer.

(e) Do the plotting symbols in the dotplot of part (a) appear in clusters? Do these clusters appear to correspond to specific industrial sectors? If necessary, use an internet search to identify industrial sectors for companies that appear to cluster.

3. Consider the data for the total compensation in 2008 received by CEOs employed by industrial corporations listed in Table 4.3.

(a) Produce a series of stemplots with different stems and leaves. Choose the stemplot from among these that in your opinion best depicts total compensation.

(b) Comment on skewness in your chosen stemplot for part (a).

(c) Comment on whether any of the 43 industrial companies appear to be outliers with respect to compensating their CEOs.

4. Consider the data for the total compensation in 2008 received by CEOs employed by industrial corporations listed in Table 4.3.

   (a) Produce a quantile boxplot for total compensation.

   (b) Comment on skewness in the quantile boxplot of your answer to part (a).

   (c) Produce an outlier boxplot for total compensation.

   (d) Compare the quantile boxplot and the outlier boxplot of parts (a) and (c), respectively. What do you conclude regarding skewness and outliers?

   (e) Write a script for an oral presentation of the features in the outlier boxplot.

5. Consider the data in Table 4.3 for the total compensation in 2008 received by chief executive officers employed by 43 industrial corporations. If you haven't already done so, draft for these data: a dotplot, a stemplot, an outlier boxplot, and an EDF plot.

   (a) Select the most appropriate chart for an audience of the general public. Justify your choice.

   (b) Select what you think would be the most appropriate plot for an audience of economists. Justify your choice.

6. Consider the data in Table 4.3 for the total compensation in 2008 received by chief executive officers employed by 43 industrial corporations.

   (a) Using the package `ggthemes`, draft for these data a jitter dotplot with each of the following themes:

      i. `theme_gray`,
      ii. `theme_base`,
      iii. `theme_bw`,
      iv. `theme_linedraw`,
      v. `theme_light`,
      vi. `theme_dark`,
      vii. `theme_minimal`,
      viii. `theme_classic`,
      ix. `theme_void`.

   (b) Which of the preceding themes would you use for fast prototyping? Discuss.

   (c) Which of the preceding themes would you use for final production? Discuss.

7. In 1882, Simon Newcomb measured the time required for light to travel from his laboratory on the Potomac River to a mirror at the base of the Washington Monument and back, a total distance of about 7400 meters. These measurements were used to estimate the speed of light. Table 4.4 contains the estimated speed of light for 66 trials. These data were reported

Time (given as $\times 10^{-3} + 24.8$ in microseconds)							
28	22	36	26	28	28	26	24
32	30	27	24	33	21	36	32
31	25	24	25	28	36	27	32
34	30	25	26	26	25	-44	23
21	30	33	29	27	29	28	22
26	27	16	31	29	36	32	28
40	19	37	23	32	29	-2	24
25	27	24	16	29	20	28	27
39	23						

Table 4.4 *Newcomb's third series of measurements of the passage time of light (made July 24, 1882 to September 5, 1882)*

in an article by Steven Stigler [115] in 1977 in *The Annals of Statistics*. This data set is considered a classic.

(a) Produce a stacked dotplot for the data in Table 4.4.

(b) Do there appear to be any outliers in the stacked dotplot?

(c) Is it possible to determine whether the data is skewed by examining the stacked dotplot for your answer of part (a)?

8. Consider the time passage data for light in Table 4.4 that was obtained by Simon Newcomb.

(a) Produce a stemplot for the data in Table 4.4 using a statistical software package with the leaf unit set equal to 10.

(b) How did the statistical software package order the negative leaves?

(c) Produce a stemplot for the data in Table 4.4 using a statistical software package with the leaf unit set equal to 1. Experiment with 1, 2, and 5 leaves per stem.

(d) Of the three stemplots produced in part (c), which do you prefer? Justify your answer.

9. Consider the time passage data for light in Table 4.4 that was obtained by Simon Newcomb.

(a) Produce an outlier boxplot for the data in Table 4.4 using a statistical software package.

(b) Which observations were noted as outliers in the boxplot produced in part (a)?

(c) Are any of the outliers plotted in the boxplot extreme enough to be called "far out" in Tukey's [127] terminology? Does the statistical software package use different plotting symbols to differentiate between those outliers that are far out and those that are not?

10. Consider the time passage data for light in Table 4.4 that was obtained by Simon Newcomb and the boxplot produced by the previous question.

(a) Produce a revised data set with the outliers removed.

(b) Produce an outlier boxplot with the revised data set from part (a).

(c) Do any outliers remain?

(d) Does the revised data set appear to be skewed?

11. Consider the time passage data for light in Table 4.4 that was obtained by Simon Newcomb.

    (a) Produce an EDF plot with the complete data set.

    (b) Comment on any apparent outliers or skewness in the EDF plot produced in part (a).

    (c) Estimate the median, and the upper and lower quartiles from your answer to part (a).

12. Consider the time passage data for light in Table 4.4 that was obtained by Simon Newcomb.

    (a) Using the package `ggthemes`, draft for these data an EDF plot with each of the following themes:

        i. `theme_gray`,
        ii. `theme_base`,
        iii. `theme_bw`,
        iv. `theme_linedraw`,
        v. `theme_light`,
        vi. `theme_dark`,
        vii. `theme_minimal`,
        viii. `theme_classic`,
        ix. `theme_void`.

    (b) Which of the preceding themes would you use for fast prototyping? Discuss.

    (c) Which of the preceding themes would you use for final production? Discuss.

Chapter 5

Diagnostic Plots for the Distribution of a Continuous Variable

5.1 Introduction

Two diagnostic plots related to the EDF plot will be presented in this chapter. For purposes of checking whether the data follow an assumed distribution, these plots are more efficient than the EDF because the reference distribution is presented on either plot by a straight line. It is for this reason that sometimes a choice of both of these two diagnostic plots is available in a given statistical software package.

There is unfinished business left from the previous chapter. Estimation of quartiles for the boxplots and percentiles for the EDF plot was glossed over. This chapter concludes with a discussion of algorithms used for estimating quartiles and percentiles from simple random samples.

5.2 Learning Outcomes

When you complete this chapter, you will be able to do the following.

- Know what a quantile-quantile plot is and be able to create it with either the `graphics` package or the `ggplot2` package in R.
- Know what a probability plot is and be able to create it with either the `graphics` package or the `ggplot2` package in R.
- Optionally, understand that while the concept of a quartile or percentile is simple enough, there is not just one algorithm for estimating either from data. Know that not all software uses Tukey's [127] algorithm for estimating quartiles when producing a box-and-whisker plot. Know the default algorithm in R for estimating percentiles with the function `quantile` and know that this algorithm produces estimates that are mode unbiased. Know that

this is different from the default algorithm in SAS and SPSS that is approximately median unbiased only for normally distributed data. Know that the default algorithm in SAS and SPSS is the only method in MINITAB for estimating quantiles.

- Optionally, know what an ogive is and how to use either the `graphics` package or the `ggplot2` package in R to draft one.

5.3 The Quantile-Quantile Plot

The *quantile-quantile plot* is defined as the plot of two inverse distribution (or quantile) functions, $Q_1(p)$ and $Q_2(p)$, for $0 < p < 1$. The points $\{(Q_1(p_k), Q_2(p_k))\}$ are plotted in the Cartesian coordinate plane corresponding to selected values of $\{p_k\}$ determined from an ordered random sample. Potential values for p_k are given in formula (5.5). The quantile-quantile plot is also known as the *QQ plot* and the *quantile plot.* The quantile-quantile plot was first proposed by Wilk and Gnanadesikan [135] in 1968.

If the distribution corresponding to Q_2 is the uniform distribution function given by

$$P_2(x) = \begin{cases} 1 & \text{if } x > 1, \\ x & \text{if } 0 \leq x \leq 1, \text{ and }, \\ 0 & \text{if } x < 0 \end{cases}$$

then the order statistics for the sample are plotted along the vertical axis. This is the usual choice. The most common choices of values for Q_1 are the corresponding quantiles from the standard normal distribution $z_k = \Phi(p_k)$, respectively. In this situation, the resulting plot is commonly called a *normal quantile plot.* Wilk and Gnanadesikan [135] gave the first presentation of a normal quantile plot.

A *normal quantile-quantile plot* of the mass of 56 perch caught in the Finnish lake known as Längelmävesi is presented in Figure 5.1. Note that mass is plotted along the vertical axis. The quantiles corresponding to a standard normal distribution with mean 0 and standard deviation 1 are plotted along the horizontal axis in Figure 5.1. This has emerged as the conventional format for the *normal quantile-quantile plot*, which is usually referred to by the shorter title of *normal quantile plot.* But one does see from time to time the alternative presentation with the axes interchanged as in Figure 5.2.

Figure 5.1 was drafted by the following call to the R function `qqnorm`.

```
qqnorm(mass,main=NULL,xlim=c(-3,3),ylim=c(0,1200),
xlab="Standard Normal Quantiles",ylab = "Mass (g)")
```

The limits for the horizontal and vertical axes are set by the parameters `xlim` and `ylim`, respectively. Setting `main=NULL` in a call to `qqnorm` suppresses any main title centered above the quantile-quantile plot.

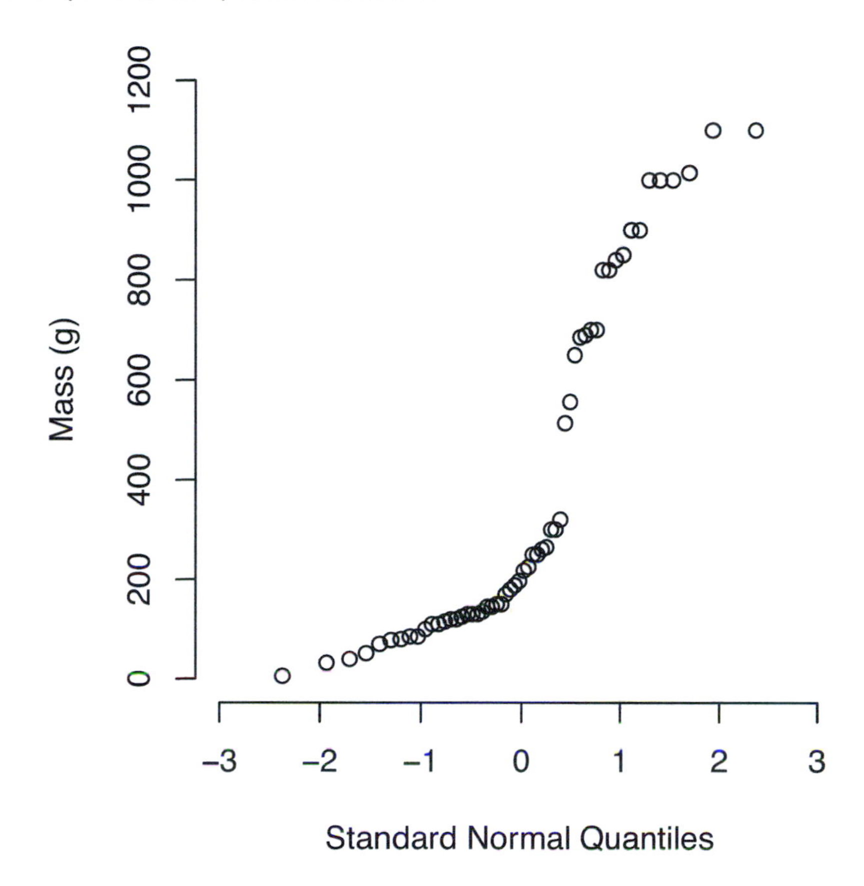

Figure 5.1 *Normal quantile-quantile plot of the mass of 56 perch*

To obtain the alternative presentation with the axes interchanged as in Figure 5.2, the following R code is used.

```
qqnorm(mass,main=NULL,datax=TRUE,ylim=c(0,1200),
xlab="Standard Normal Quantiles",ylab = "Mass (g)")
```

Setting `datax=TRUE` forces the data to be plotted along the horizontal axis. The default value for `datax` is `FALSE` in `qqnorm`.

Figures 5.3 and 5.4 are `ggplot2` versions of Figures 5.1 and 5.2. The code for generating Figure 5.3 is as follows.

```
ggplot(perch, aes(sample=mass)) + geom_qq()
```

The syntax for storing the plot as an object and then printing the object has been omitted to facilitate comparison with the code given previously for Figure 5.1. Less coding was required for using the function `geom_gg` in `ggplot2`

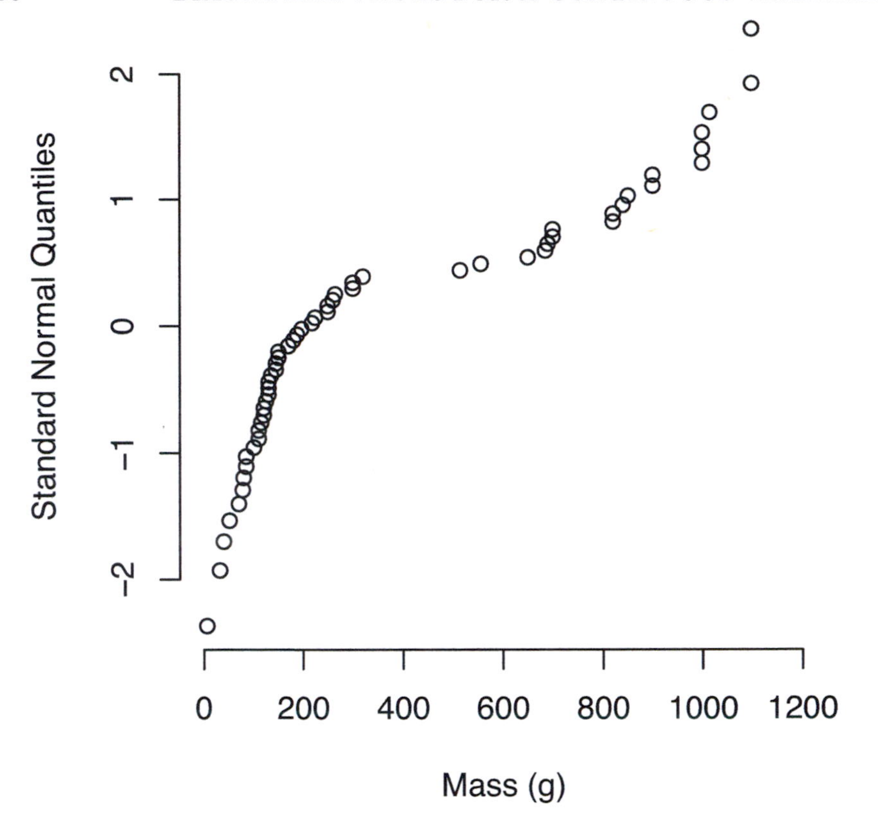

Figure 5.2 *Normal quantile-quantile plot of the mass of 56 perch with standard normal quantiles on the vertical axis*

than for qqnorm in the stats package in R. But this comes at a cost. The default axis labels are "theoretical" and "sample" for the horizontal and vertical axes, respectively, in Figure 5.3. Instead of the open circles in Figure 5.1, which allow the data to be more clearly perceived, the default plotting symbol in ggplot2 is the completely filled black circle. But these deficiencies are easily corrected in the following code for Figure 5.4.

```
ggplot(perch, aes(sample=mass)) +
geom_qq(shape=1) +
labs(x="Standard Normal Quantiles",y="Mass (g)") +
theme_linedraw() +
theme(panel.grid = element_blank()) +
coord_flip()
```

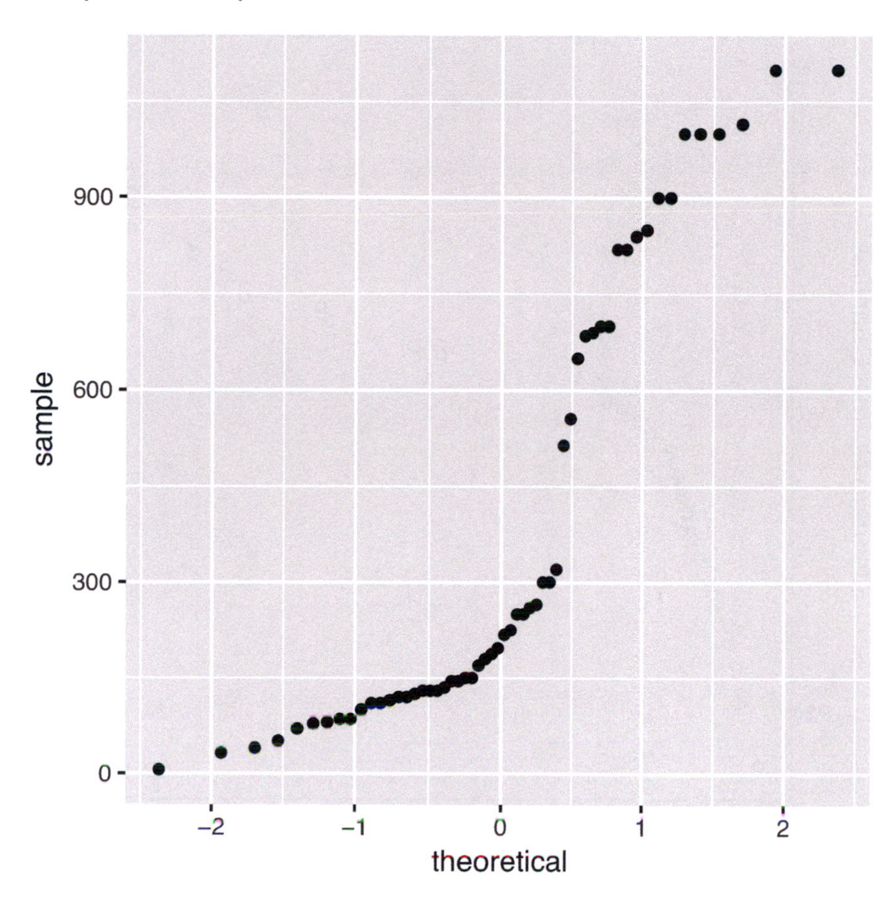

Figure 5.3 *Normal quantile-quantile plot of the mass of 56 perch with* `ggplot2`

The parameter `shape=1` passed on the call to `geom_qq` selects the same symbol in Figure 5.4 as used in Figures 5.1 and 5.2. The call to the function `labs` changes the axis titles to those of Figure 5.2. Rather than using the default theme of `ggplot2`, Figure 5.4 is drafted with the `linedraw` theme in the package `ggthemes`. Still, the call `theme(panel.grid = element_blank())` is needed to eliminate the major and minor grid lines that would otherwise appear as chartjunk. Finally, a call is made to the generic `ggplot2` function `coord_flip` to flip the axes. Note: instead of calling the function `geom_qq`, Figures 5.3 and 5.4 could have been drafted with a call to the synonym function `stat_qq` in `ggplot2`.

The great utility of the normal quantile plot is that if the data do follow a normal distribution, then the points in the plot should lie nearly along a straight line. Our eyes can play tricks when examining Figures 5.1 through 5.4. So the standard procedure is to add a reference representing a normal distribution.

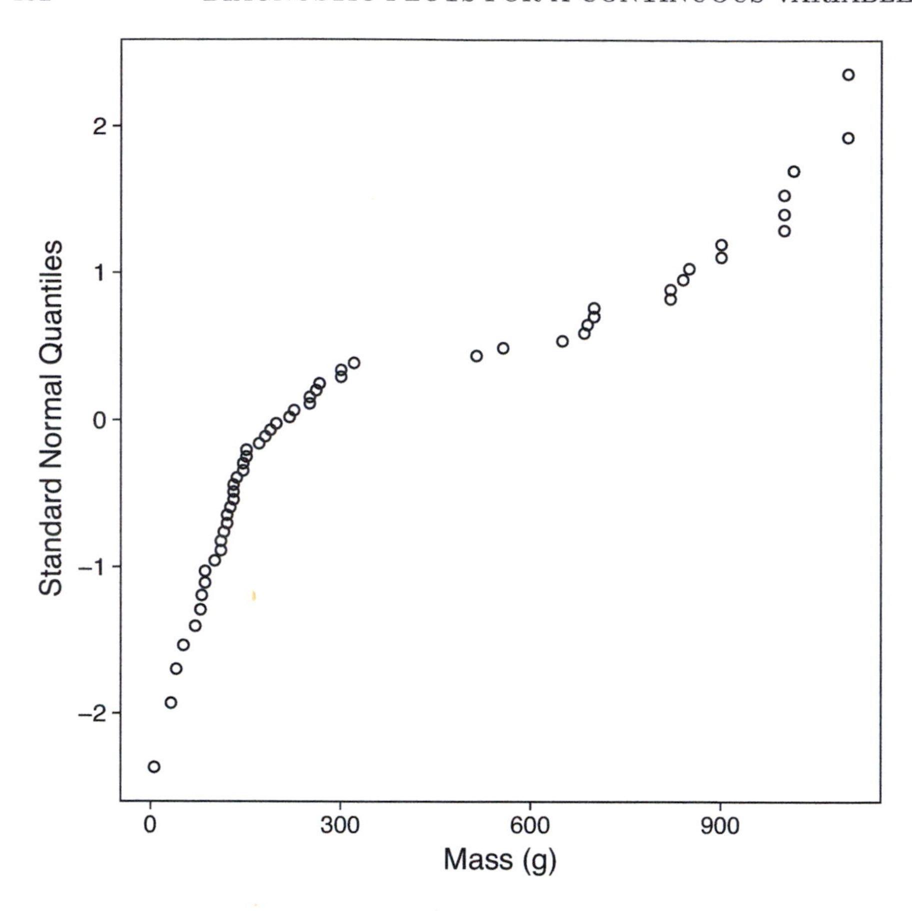

Figure 5.4 *Normal quantile-quantile plot of the mass of 56 perch with standard normal quantiles on the vertical axis with* `ggplot2`

If there is a hypothetical normal population to which the data are to be compared with known mean μ and standard deviation σ, then the appropriate reference line in Figure 5.1 ought to have a slope of σ and an intercept of μ. Typically, this is not the case, and the hypothesis is that the distribution is normal with mean and standard deviation unknown. The population mean μ and standard deviation σ are then replaced by the sample mean $\bar{x}$ and standard deviation s, respectively. Figure 5.5 is Figure 5.1 with the sample normal reference line added.

Normal quantile plots can be used to characterize data beyond simply checking normality. Note the following

- If all but a few points fall on the normal reference line, then these few points may be outliers.

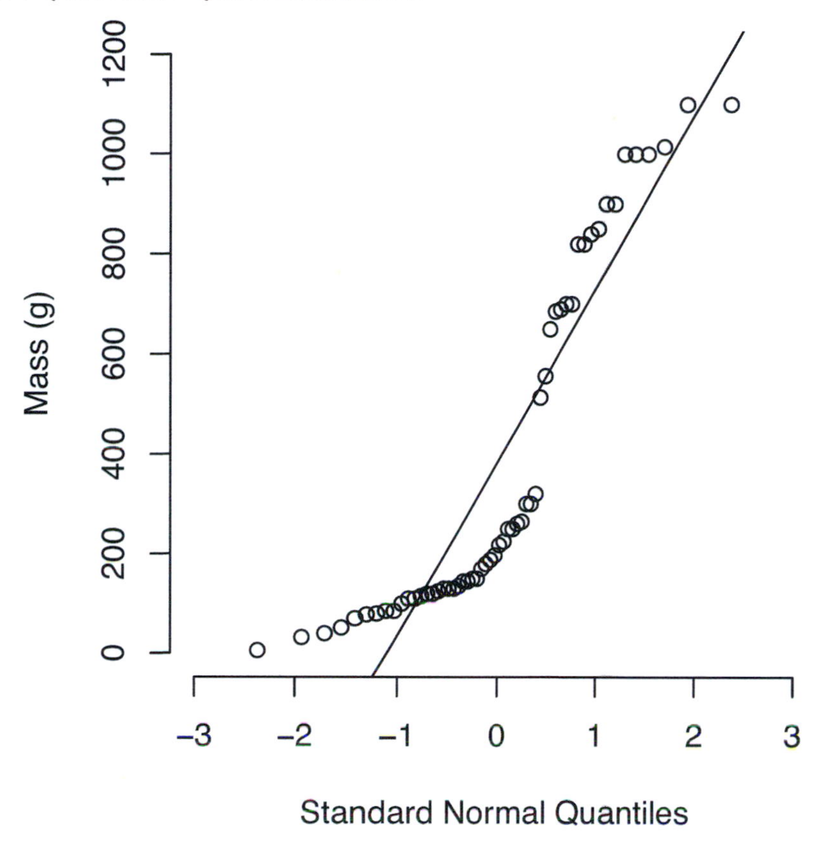

Figure 5.5 *Normal quantile-quantile plot of the mass of 56 perch with a normal reference line*

- If the left end of the data is above the line and the right end of the pattern is below the line, then the distribution may have short tails at both ends.
- If the left end of the data is below the line and the right end of the pattern is above the line, then the distribution may have long tails at both ends.
- If there is a curved pattern with the slope increasing from left to right, then the data are skewed to the right.
- If there is a curved pattern with the slope decreasing from left to right, then the data are skewed to the left.
- If there is a step-like pattern with plateaus and gaps, then this is an indication that the data have been rounded (or truncated) or are discrete.

With the aforementioned list in mind, a careful examination of the normal quantile plot in Figure 5.5 suggests that the data are not normal but skewed to the right. The rightward skewness was detected in our first visual examination of this data in the stemplot of Figure 4.8.

Because of the amount of statistical knowledge required, quantile-quantile plots are generally not suitable for the public, administrators, nor all disciplines. For the public or administrators the quantile-quantile plot may present them with more information than they really need for decision-making purposes. The quantile-quantile plot is rarely used by anyone other than statistical practitioners or those in other disciplines who specialize in the application of statistical methods.

There is a concern that those who use quantile-quantile plots without formal training may not understand all the implications. If one decides to present quantile-quantile plots to an audience that does not have the necessary training, the key point that ought to be communicated to anyone is whether the points are close to a straight line.

In this section, a general definition of quantile-quantile was used at the outset. The focus was placed on the normal quantile-quantile plot. But quantile-quantile plots can be used to judge whether observations follow a variety of distributions such as: the gamma distribution, the beta distribution, the Chi-square distribution, and the lognormal.

Typically one does not find these forms of QQ plots routinely available in statistical software routines and one must create them. The burden to produce, say, a gamma quantile-quantile plot using R is light. This is done in Figure 5.6 for the mass of 56 perch caught during a research trawl in Längelmävesi in Finland with the following code that uses the `plot` function in the `graphics` package.

```
sn<-ppoints(length(mass))
sshape<-mean(mass)/sd(mass)
sshape<-sshape*sshape
sscale<-var(mass)/mean(mass)
smass<-sort(mass)
sq<-qgamma(sn,shape=sshape,scale=sscale)

plot(sq,smass,ylab="Mass (g)",
xlab="Gamma Quantile",ylim=c(0,1200),
xlim=c(0,2000))
```

The following code produces a `ggplot2` version of Figure 5.6 in Figure 5.7.

```
sshape<-mean(mass)/sd(mass)
sshape<-sshape*sshape
sscale<-var(mass)/mean(mass)

figure<-ggplot(perch, aes(sample=mass)) +
```

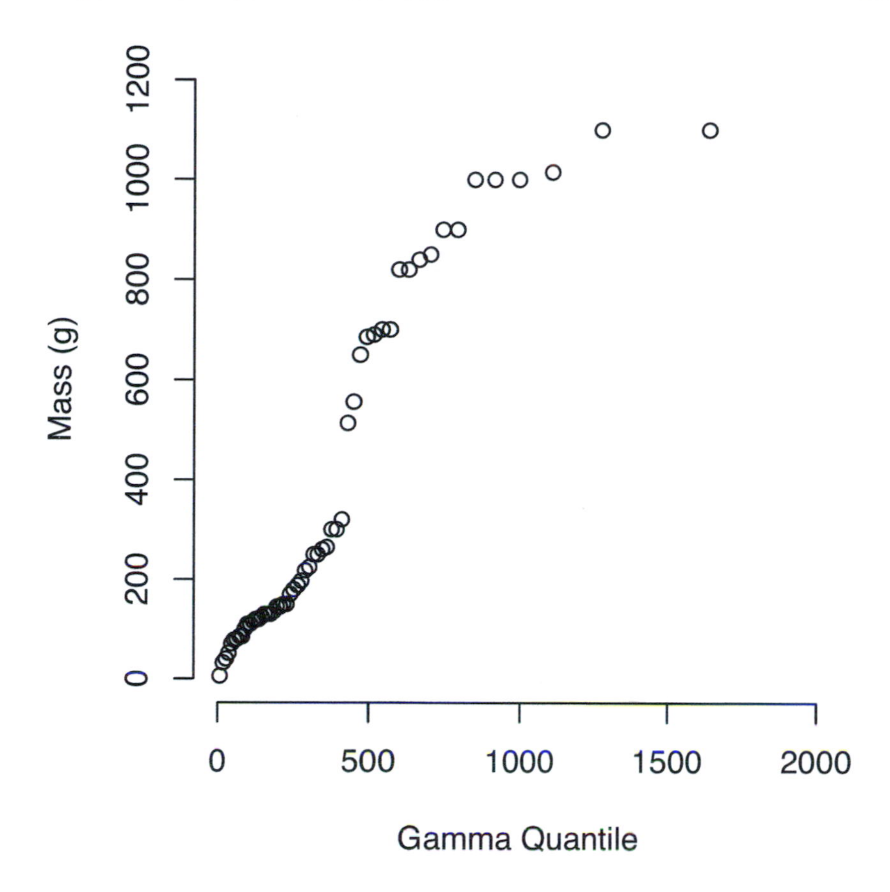

Figure 5.6 *Gamma quantile-quantile plot of the mass of 56 perch*

```
theme_linedraw() +
theme(panel.grid=element_blank(),
axis.title=element_text(size=12),
axis.text=element_text(size=12)) +
geom_qq(shape=1,distribution=qgamma,
dparams=list(shape=sshape,scale=sscale)) +
labs(x="Gamma Quantiles",y="Mass (g)") +
scale_x_continuous(limits=c(0,2000),
breaks = (0:4)*500) +
scale_y_continuous(limits=c(0,1200),
breaks = (0:6)*200)
print(figure)
```

In the ggplot2 version, while the parameters of the gamma still require code to be estimated, it is not necessary to write code to compute the gamma

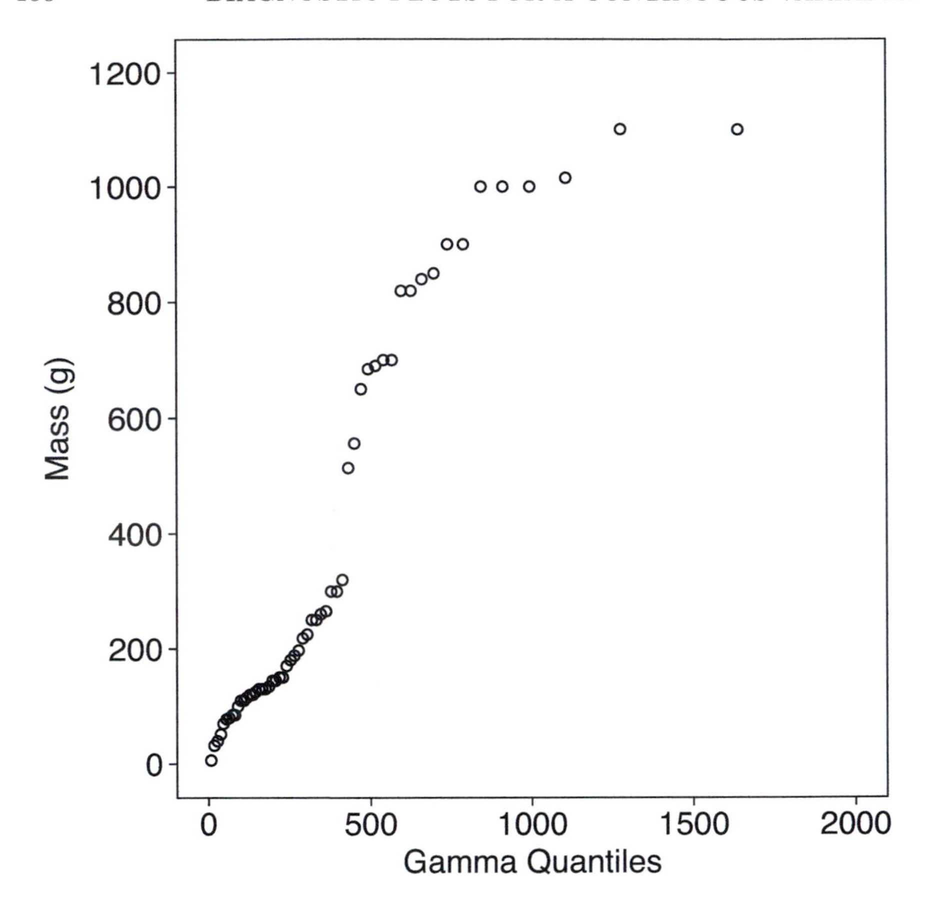

Figure 5.7 *Gamma quantile-quantile plot of the mass of 56 perch with* `ggplot2`

quantiles because `geom_qq` will do this provided the `distribution` parameter is changed from the default value for the normal distribution to the gamma distribution. It is still necessary, however, to estimate the gamma parameters and pass these as a `list` to the distribution parameters argument `dparams`. The default axis-ticks and axis-labels in `ggplot2` are not the same as those in the `graphics` package, so there is housekeeping code to do this for Figure 5.7 in addition to that required to suppress the grid lines in the line drawn theme of the `ggthemes` package.

The gamma distribution is a good guess for the mass of perch as it is a distribution with a heavy right tail. But the perch mass data are more heavily skewed to the right than even the gamma distribution as seen in Figures 5.6 and 5.7. The most frequently encountered quantile-quantile plot is by far the normal distribution. Other distributions are used but much less frequently. The topic of the next section is a variation on the quantile-quantile plot.

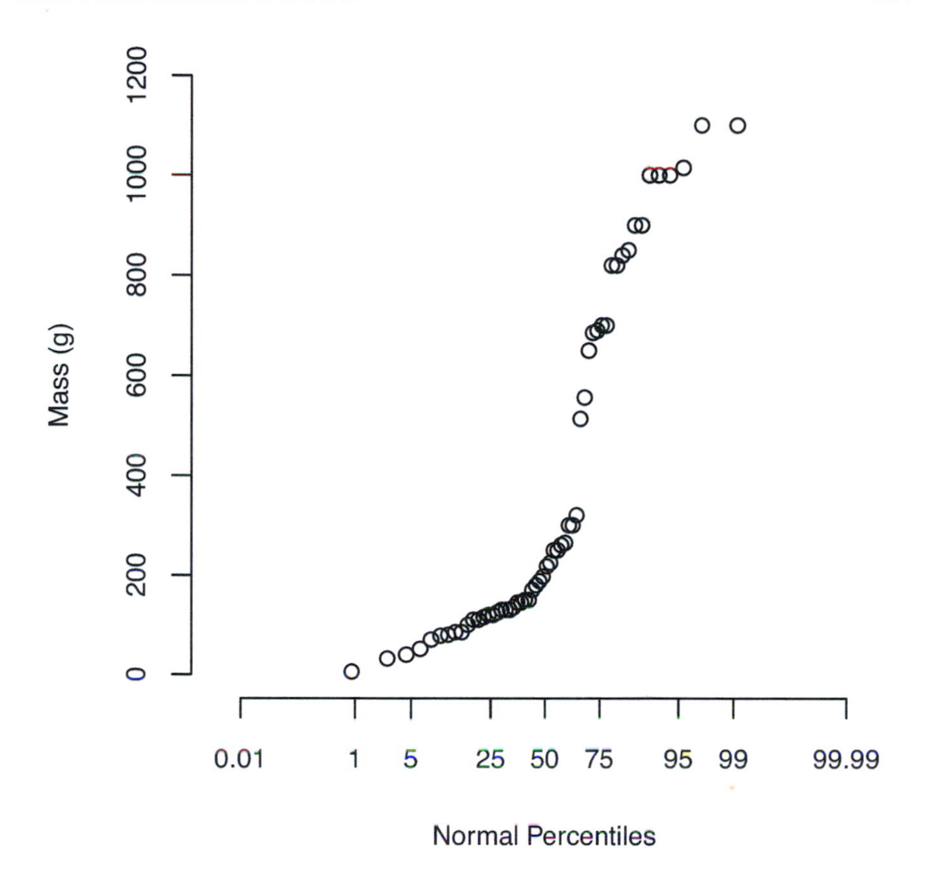

Figure 5.8 *Normal probability plot of the mass of 56 perch*

5.4 The Probability Plot

The *probability plot* is a variation of the quantile-quantile plot. The points plotted are $\{(Q_1(p_k), Q_2(p_k))\}$, just as in the case of the quantile-quantile plot. But the choice of scale for the reference distribution is chosen to be cumulative probability instead of quantile values. If the reference distribution is normal, then a *normal probability plot*, the most common, is obtained.

An example of a normal probability plot is given in Figure 5.8 for the mass of 56 perch caught in a research trawl on Längelmävesi in Finland. This graphic was produced by the R function `qqnorm`, just as the normal quantile plot of Figure 5.1, but the R function `axis` has been used to relabel and relocate ticks on the horizontal axis as done in the following R script.

```
qqnorm(mass,main=NULL,xaxt="n",xlim=c(-4,4),ylim=c(0,1200),
ylab = "Mass (g)",xlab="Normal Percentiles")
```

```
lprob<-c("0.01","1","5","25","50","75","95","99","99.99")
probly<-as.numeric(lprob)/100.
zprob<-qnorm(probly)
axis(1,zprob,lprob)
```

The only difference between Figures 5.1 and 5.8 is the horizontal axis and its labeling. So the rules described for detecting outliers and characterizing skewness and kurtosis (heaviness of tails) that apply to the normal quantile plot also apply to the normal probability plot, as do the caveats for presentation of probability plots to a nontechnical audience.

When Wilk and Gnanadesikan [135] proposed the quantile plot in 1968, they also proposed the probability plot but with a linear scale for both axes. This original version is now rarely seen. If the quantiles are plotted and labeled with probabilities according to the chosen reference distribution, as in Figure 5.8 for the standard normal distribution, the reference distribution is a good candidate for describing the data if the data fall nearly in a straight line. This is not the case for the probability plot as originally proposed by Wilk and Gnanadesikan [135].

The axes for the quantile plot can also be interchanged. The version of the normal probability plot in Figure 5.9 has interchanged the axes of Figure 5.8. In so doing, the opportunity has been taken to increase the font size of the axes' labels by using labels for ticks on the vertical normal percentile axis that run horizontally. Also tick marks have been added for 10% and 90%. The overall effect in Figure 5.9 is an increase in clarity compared to Figure 5.8.

Figure 5.10 provides an example of a gamma probability plot to demonstrate that other distributions can be used other than just normal. The gamma distribution is not providing a distributional model noticeably better than the normal for the mass of perch.

The search for a better model of the distribution of perch mass is continued in the next two chapters. A classical plot for a quantitative variable is studied in some detail and an associated computationally intensive way of modeling distributions is discussed in Chapter 6. A few older approaches for modeling the probability distribution of a continuous variable round out the discussion in Chapter 7.

5.5 Estimation of Quartiles and Percentiles*

The estimation of quartiles and percentiles is complicated by the fact that they are not necessarily uniquely defined for a simple random sample. This

*This section can be omitted without loss of continuity.

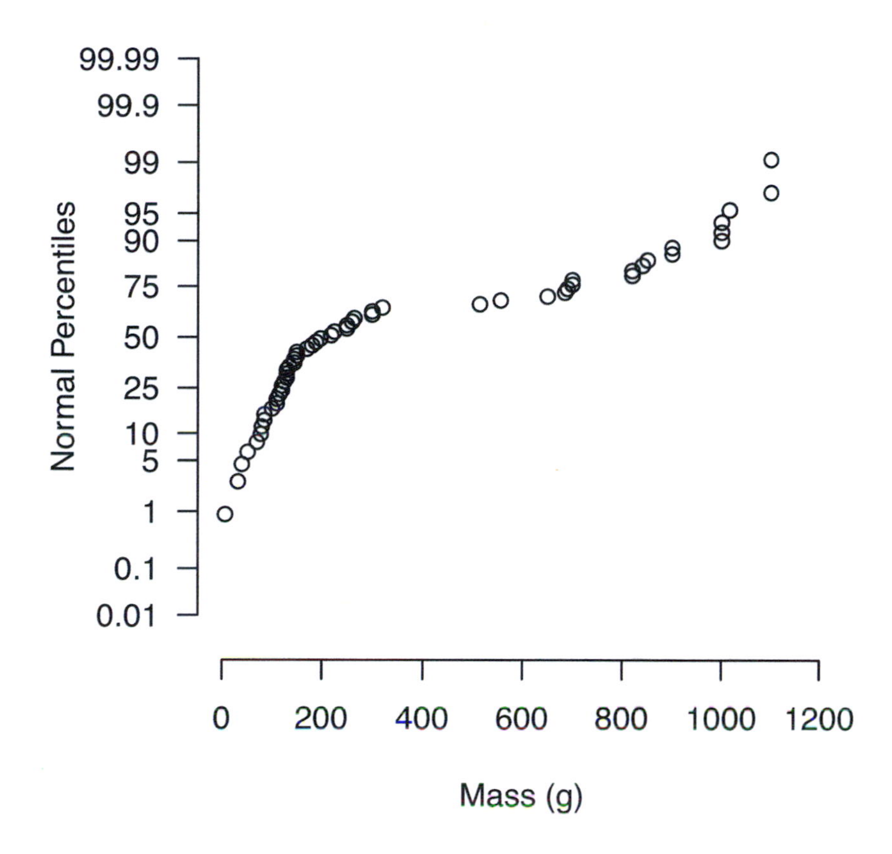

Figure 5.9 *Normal probability plot of the mass of 56 perch with normal percentiles on the vertical axis*

is compounded by the fact that there is no consensus among statisticians regarding which method is best.

In the case of boxplots, the choice of the method of estimation does have an impact on the value of the quartiles and, therefore, on the values for the fences. It shall be seen in this section that this can have a considerable impact on the appearance of the boxplot.

5.5.1 *Estimation of Quartiles*

Discussion of the methods of quartile estimation begins with respect to the lower quartile Q_1. The remaining quartiles are estimated by suitable modifications to the algorithms for the lower quartile. By way of notation, let $\lfloor x \rfloor$ denote the largest integer that does not exceed x. As in the last chapter, let

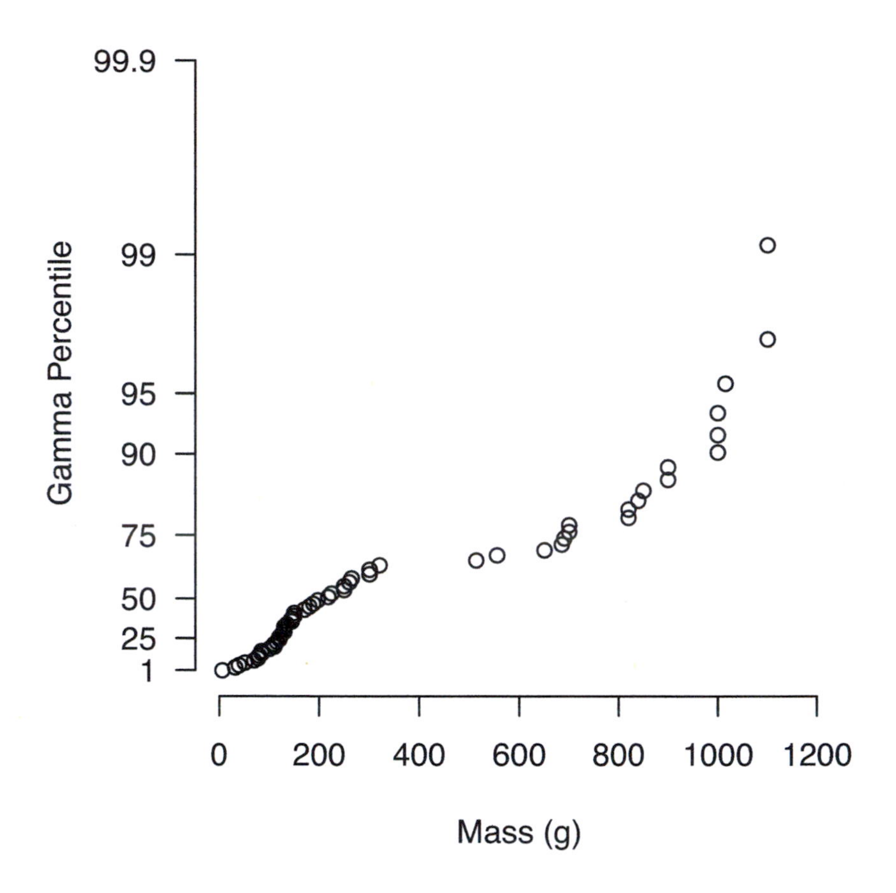

Figure 5.10 *Gamma probability plot of the mass of 56 perch with gamma percentiles on the vertical axis*

$x_{(1)} \leq x_{(2)} \leq \cdots \leq x_{(n)}$ denote an ordered sample of size n. Let j be an integer and g a real number such that $0 \leq g < 1$. Define $\delta(g) = 0$ if $g = 0$; $\delta(g) = 1$ if $g > 0$. Ten different methods for estimation of the first quartile are as follows.

1. **Aimed at $x_{(\frac{n}{4}+\frac{3}{4})}$ with averaging:**

$$Q_1 = (1-\gamma)x_{(j)} + \gamma x_{(j+1)}$$

where $\lfloor \frac{n+3}{2} \rfloor /2 = j + \gamma$. Note that either $\gamma = 0$ or $\gamma = \frac{1}{2}$. This is the method used by Tukey [127] when he created the box-and-whisker and schematic plots. One of its features is quick hand calculation.

2. **Weighted average aimed at $x_{(\frac{n}{4}+\frac{3}{4})}$:**

$$Q_1 = (1-\gamma)x_{(j)} + \gamma x_{(j+1)}$$

where $\frac{n}{4} + \frac{3}{4} = j + \gamma$. Tukey's [127] method is an approximation to this method.

3. Weighted average aimed at $x_{(\frac{n}{4}+\frac{1}{2})}$:

$$Q_1 = (1 - \gamma)x_{(j)} + \gamma x_{(j+1)}$$

where $\frac{n}{4} + \frac{1}{2} = j + \gamma$.

4. Weighted average aimed at $x_{(\frac{n}{4}+\frac{5}{12})}$:

$$Q_1 = (1 - \gamma)x_{(j)} + \gamma x_{(j+1)}$$

where $\frac{n}{4} + \frac{5}{12} = j + \gamma$.

5. Weighted average aimed at $x_{(\frac{n}{4}+\frac{1}{4})}$:

$$Q_1 = (1 - \gamma)x_{(j)} + \gamma x_{(j+1)}$$

where $\frac{n}{4} + \frac{1}{4} = j + \gamma$.

6. Aimed at $x_{(\frac{n}{4})}$ **with averaging to the right**:

$$Q_1 = (1 - \gamma)x_{(j)} + \gamma x_{(j+1)}$$

where $\frac{\lfloor \frac{n}{2} \rfloor + 1}{2} = j + \gamma$. Note that either $\gamma = 0$ or $\gamma = \frac{1}{2}$. This is one of the more popular methods taught in introductory courses for statistics.

7. Aimed at $x_{(\frac{n}{4})}$ **in the EDF with averaging**:

$$Q_1 = \frac{1}{2}\left\{\left[1 - \delta(\gamma)\right]x_{(j)} + \left[1 + \delta(\gamma)\right]x_{(j+1)}\right\}$$

where $\frac{n}{4} = j + \gamma$. This method is equivalent to using the empirical distribution function with averaging of the length of the horizontal line segments.

8. Observation numbered closest to $x_{(\frac{n}{4})}$:

$$Q_1 = (1 - \gamma)x_{(j)} + \gamma x_{(j+1)}$$

where $\frac{n}{4} + \frac{1}{2} = j + g$. If $g \neq \frac{1}{2}$, then $\gamma = 1$. But if $g = \frac{1}{2}$, then $\gamma = 0$ if j is even and $\gamma = 1$ if j is odd.

9. Aimed at $x_{(\frac{n}{4})}$ **in the EDF**:

$$Q_1 = x_{(j+\delta(\gamma))}$$

where $\frac{n}{4} = j + \gamma$. This method is a graphical method whereby one plots a horizontal line corresponding to $y = 0.25$ on a stepwise EDF plot, as in Figure 4.24, and notes the first point of intersection. It is a method formerly taught in introductory statistics courses before statistical software became ubiquitous.

10. Weighted average aimed at $x_{(\frac{n}{4})}$:

$$Q_1 = (1 - \gamma)x_{(j)} + \gamma x_{(j+1)}$$

where $\frac{n}{4} = j + \gamma$.

Method	Q_1	M	Q_3	IQR	Lower Inner Fence
1,2	78	85	92	14	75
3	$76\frac{1}{2}$	85	$93\frac{1}{2}$	17	53
4	76	85	94	18	53
5, 6, 7, 8, 9	75	85	95	20	53
10	$70\frac{1}{4}$	83	96	$25\frac{3}{4}$	53

Table 5.1 *Effects of different quartile estimation methods*

It is reasonable to assume that the algorithms given above for the lower quartile will not always produce the same estimate for a given data set nor, implicitly, the same estimate of the interquartile range, which is used in construction of fences for boxplots.

Eight of the ten methods enumerated above can be found in Frigge, Hoaglin, and Iglewicz [45]. Liberties have been taken in altering the numbering and presentation of the eight methods as described by them to accommodate the two additional methods presented. The differences between the methods will be small for large samples for most continuously distributed random variables. Hoaglin and Iglewicz [66] discuss fine tuning of the process of quartile estimation that could be important in small samples.

To make the definition of the empirical distribution function more concrete, consider the following set of 11 observations arranged in ascending order: {53, 56, 75, 81, 81, 85, 87, 89 95, 99, 100}. This is one of the rare instances in this book in which a fake set of data is being used. It is felt necessary to do so in this instance to illustrate the significance of the choice of the method of percentile estimation on the appearance of a boxplot. A step plot of the empirical distribution function for this data is given in Figure 5.11.

The methods have been listed in descending order of the interquartile range estimate that they generate for this example. Note that only methods 1 through 7, inclusive, treat the ends of the ordered sample symmetrically with respect to the estimation of the lower and upper quartiles.

The estimates of the quartiles, interquartile range, and lower inner fences are given comparatively for each of the ten methods with the illustrative data set in Table 5.1. For all but methods 1 and 2 the lower outer fence is 53, which is the minimum of the sample. The upper outer and inner fences are all equal to 100, which is the maximum in the sample.

Notice that methods 5 through 9, inclusive, produce the same quartile estimates for the illustrative data. For large sample sizes for which rounding or truncation is not a problem, we would expect all the methods to yield nearly similar results. The boxplot is typically used with small samples so any claim

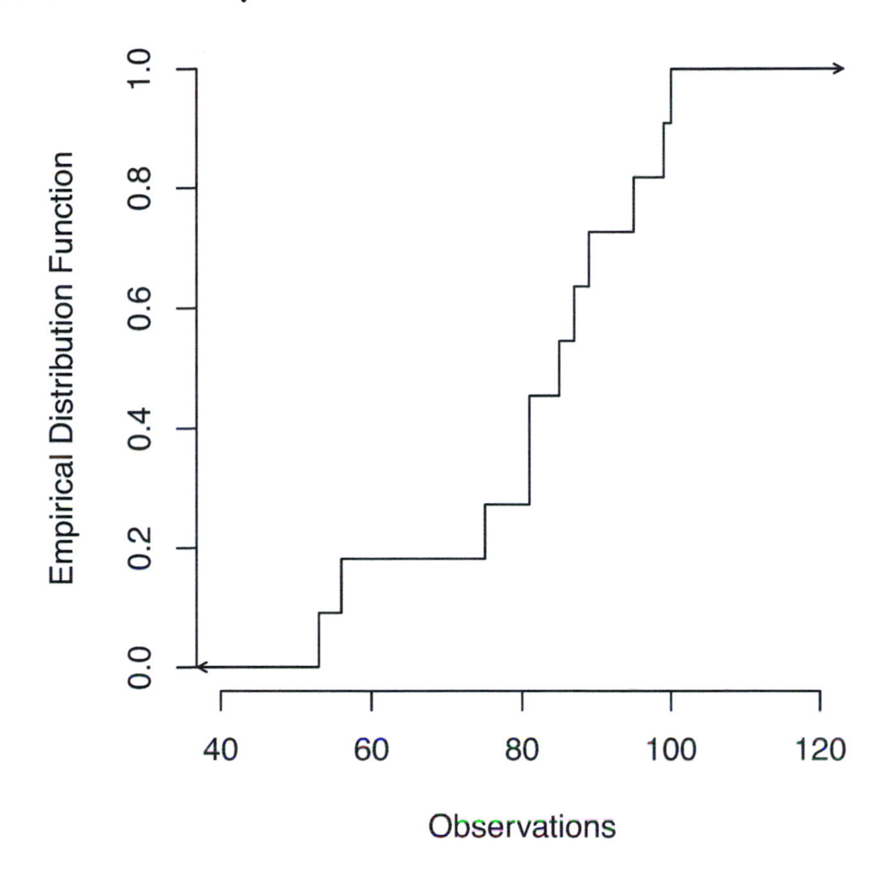

Figure 5.11 *Step plot of the EDF of the illustrative example*

that the choice of method is not a concern is suspect when all ten methods are considered.

There is also another specific concern regarding the choice of methods. Given that the sample size is odd, it is reasonable to expect that all methods would provide $x_{(6)} = 85$ as the estimate of the median. This is the case for all methods except method 10, which instead calculates $(x_{(5)} + x_{(6)})/2 = (81 + 85)/2 = 83$.

As to whether the differences among the ten methods as represented in Table 5.1 are concerns, the best approach is to look for the differences in the boxplots themselves generated by these methods as given in Figure 5.12.

It would appear, based on the considerable differences from top to bottom in Figure 5.2, that there is good reason for concern on two accounts. There is considerable difference in box dimensions between method 1 and method 10.

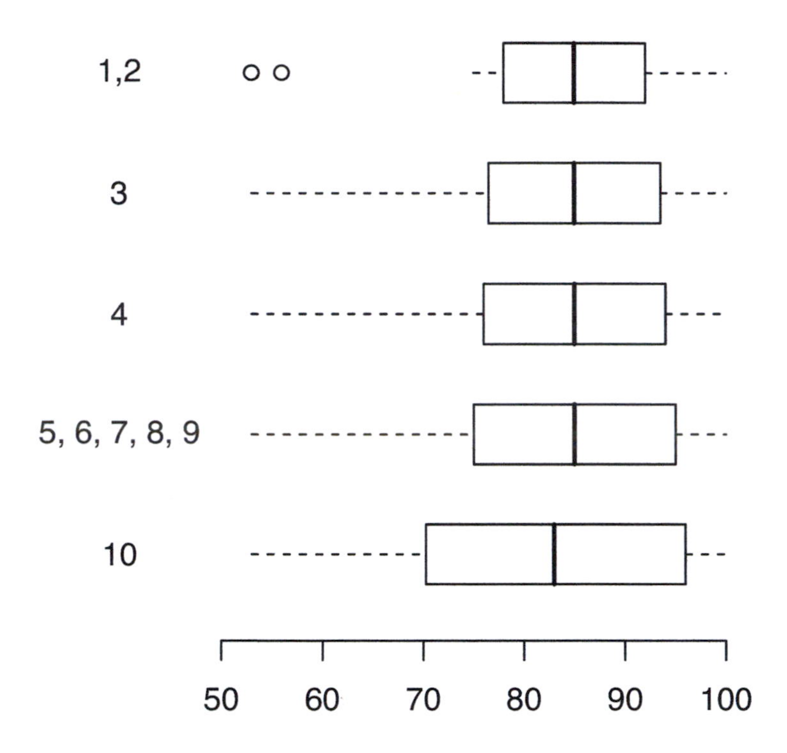

Figure 5.12 *Comparison of boxplots by the ten quartile estimation methods*

There is also the fact that methods 1 and 2 indicate outliers corresponding to the observations of 53 and 56, whereas the other methods do not.

The only axiom with respect to the **ACCENT** rule that bears scrutiny on comparison of the five outlier boxplots in Figure 5.12 would appear to be that of *Truthfulness*. If the two lowest observations truly are outliers with an explanation, then a boxplot constructed with the quartiles of methods 1 and 2 would be the more truthful. On the other hand, if the observations of 53 and 56 are not outliers, then is the boxplot with the quartiles of methods 1 and 2 a lie?

Arguably, an outlier can actually be obtained from the true distribution with approximately 0.1% probability. The flagging of outliers in Figure 5.12 for methods 1 and 2 only brings these observations to our attention. If further examination of these observations reveals nothing untoward, then they can be accepted. However, if they truly are outliers, due to contamination or error,

by formally defining the quantile of a distribution function F as

$$Q(p) = F^{-1}(p) = \inf\{x : F(x) \geq p\} \tag{5.1}$$

where $0 < p < 1$. Here inf is an abbreviation for infimum, which is the smallest real value x that satisfies the inequality $F(x) \geq p$.

A sample quantile $\hat{Q}(p)$ based upon the order statistics $\{x_{(i)}\}$ provides a non-parametric estimator of its population counterpart. This is not uniquely defined. One general format, used by Hyndman and Fan [69], is

$$\hat{Q}_i(p) = (1 - \gamma)x_{(j)} + \gamma x_{(j+1)} \tag{5.2}$$

where

$$\frac{j - m}{n} \leq p < \frac{j - m + 1}{n} \tag{5.3}$$

for some real number m and $0 \leq \gamma \leq 1$. The value of γ is a function of $j = \lfloor pn + m \rfloor$ and $g = pn + m - j$. Because more than one sample quantile will be considered in this discussion, the subscript i in the definition of $\hat{Q}_i(p)$ will be used to index the different sample quantiles.

For the previous discussion involving ten different possible point estimates of the lower quartile, in the context of formula (5.2) for $\hat{Q}_i(p)$: set $p = 1/4$.

A further issue in finding sample quantiles is that the empirical distribution function is not the only function used to approximate cumulative distribution functions. A difficulty with the empirical distribution function is that it is discontinuous. If only the coordinates of the EDF are plotted and line segments joining the points are drawn, then the *ogive* of Figure 5.5 is obtained. For purposes of comparison, the step plot of the EDF has been added to Figure 5.13.

When doing statistical analysis in bygone years without resort to computer graphics, drawing the step plot of the EDF with its many short horizontal and vertical line segments was considered tedious. Often to save time, the points were plotted and then joined with straight line segments. For the purposes of nostalgia, the ogive for the perch data is presented by itself in Figure 5.14. The ogive is a piecewise linear continuous function that approximates the stepwise discontinuous empirical distribution function. Among the ten methods presented for estimating the first quartile, the ogive was implicitly used in methods 2 through 5, inclusive, and method 10. The code for producing the ogive in Figure 5.14 is as follows.

```
knots<-sort(unique(mass))
n<-length(mass)
y<-cumsum(tabulate(match(mass,knots)))/n

plot(knots,y,col=gray(0.5),type="l",
lwd=1.5,xlab="Mass (g)",
```

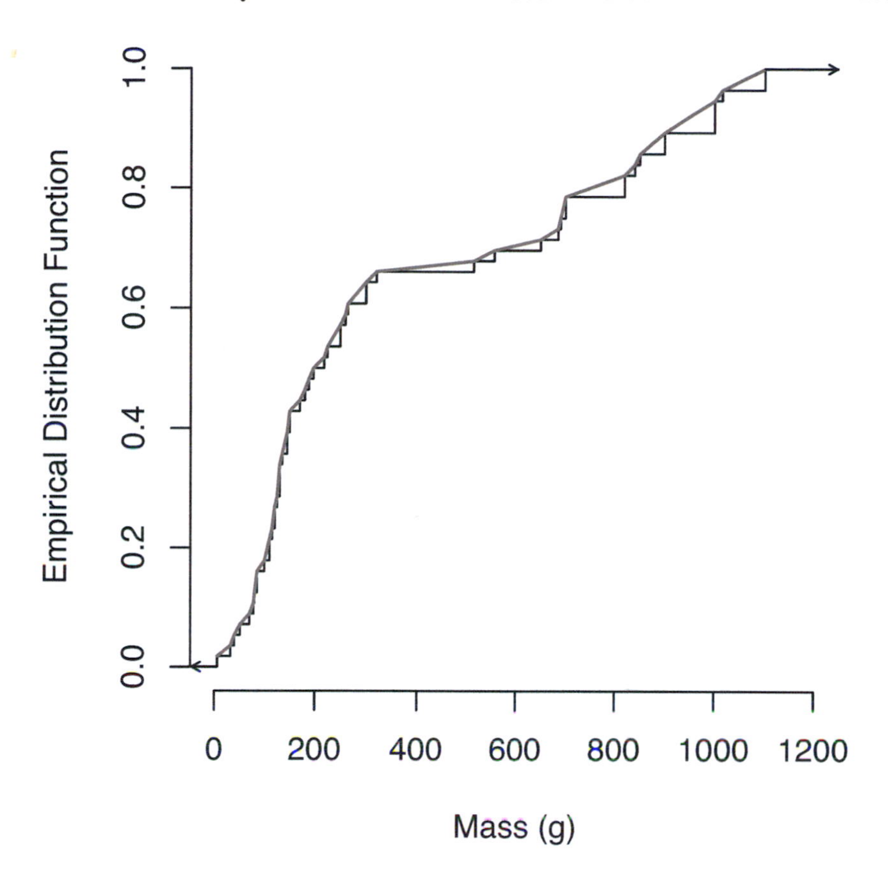

Figure 5.13 *Ogive and step plot of the empirical distribution function for perch mass (the ogive is given by the thick gray line-segments)*

```
ylab="Empirical Distribution Function",
main=NULL,xlim=c(0,1200),ylim=c(0,1))
```

It is not necessary to write code to compute the knot points for linear interpolation when using `ggplot2` to produce this same ogive. The following `ggplot2` code for Figure 5.15 merely requires overriding the default `geom="step"` with `geom="line"` when calling the function `stat_ecdf`.

```
ggplot(perch, aes(x=mass)) + stat_ecdf(geom="line") +
theme_linedraw() + theme(panel.grid = element_blank(),
axis.title=element_text(size=12),
axis.text=element_text(size=12)) +
scale_x_continuous(breaks=200*(0:6),limits=c(0,1200)) +
labs(x="Mass (g)",y="Empirical Distribution Function")
```

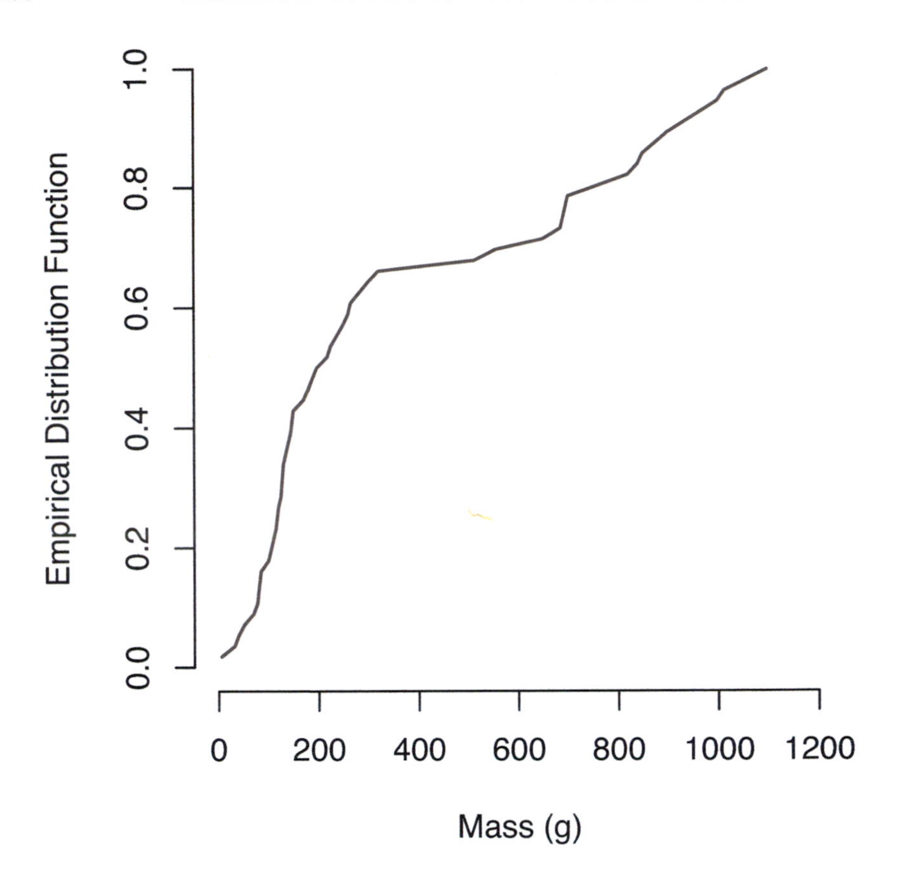

Figure 5.14 *Ogive of the empirical distribution function for perch mass*

The choice to use the EDF itself rather than its ogive to estimate quantiles leads to even more choices that need to be addressed. In the situation for which the empirical distribution function suggests an interval estimate, the simplest options are to choose the left endpoint, the right endpoint, or the simple average of the two endpoints. More complicated options include choosing the endpoint of the two that is the nearest, or the nearest even or nearest odd observation in terms of order of the sorted unique observations among the data. There can also be algorithms to select quantiles based upon combinations of these approaches.

Method 1 due to Tukey [127] is one such combination method that also lends itself to easy hand calculation or computer determination. To find the first quartile, the lower half of the data is considered. With odd sample sizes, the median is included in the lower half of the data. If there are an odd number of observations in the lower half of the data, the middle observation in the

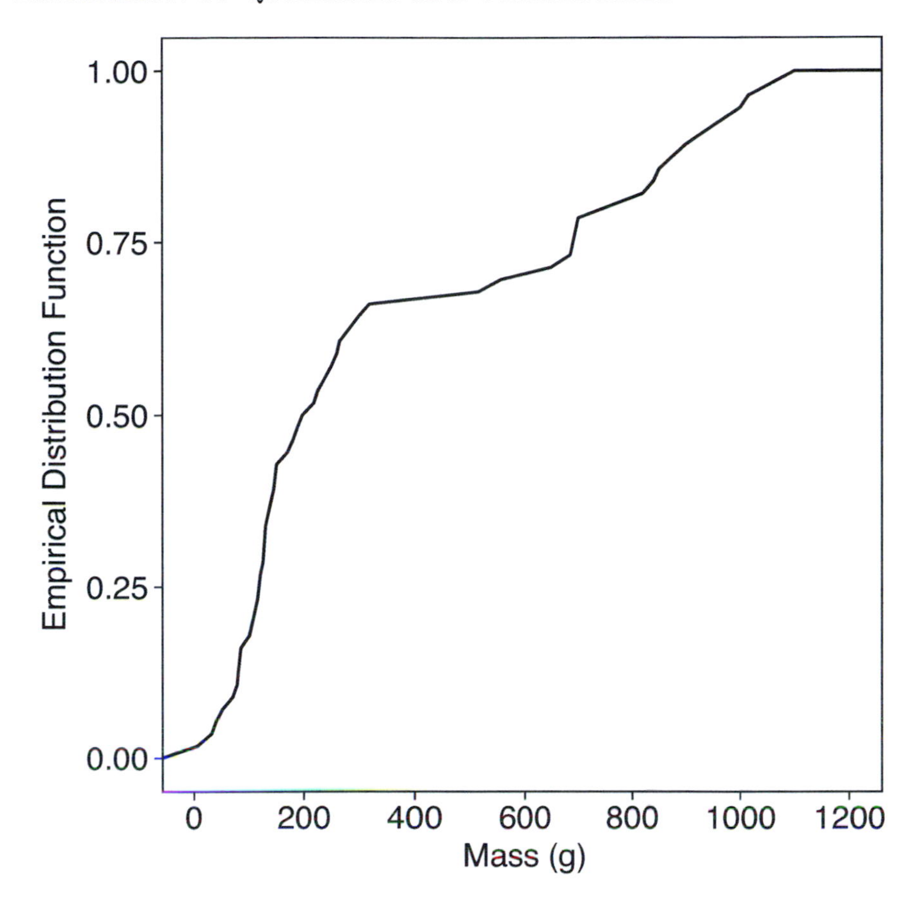

Figure 5.15 *Ogive of the empirical distribution function for perch mass with* `ggplot2`

lower half of the data is selected; otherwise the two middle observations are averaged. The two middle observations may or may not be endpoints of an interval estimate for the lower quartile.

Method 6 for finding the lower quartile is another EDF method that selects either the left endpoint or averages the two observations bracketing $x_{(\lfloor \frac{n}{4} \rfloor)}$.

Method 7 for finding the lower quartile averages the length of the horizontal line segment in the EDF function. Quite similar is method 9, which dispenses with the averaging.

Implicit in these discussions has been the definition of the empirical distribution function. The definition in equation (4.6) is used by the vast majority of statisticians and just about everyone else. But probabilists, and also engineers, have been known to use the following variation:

$$T_n(x) = \begin{cases} 0 & \text{if } x \leq x_{(i)}, \\ \frac{i}{n} & \text{if } x_{(i)} < x \leq x_{(i+1)}, \\ 1 & \text{if } x > x_{(n)}. \end{cases} \tag{5.4}$$

The issue at hand is whether the cumulative distribution function is defined as $F(x) = P(X < x)$ or $F(x) = P(X \leq x)$. The bigger issue is whether the empirical distribution function should be used as is, or modified. A generalized alternative to $S_n\left(x_{(k)}\right)$ is

$$p_k = \frac{k - \alpha}{n - \alpha - \beta + 1}. \tag{5.5}$$

This provides an even richer environment for the estimation of quantiles by adding two tuning parameters α and β. To simplify the situation, one can choose to set $m = \alpha + p(1 - \alpha - \beta)$ and $\gamma(\lfloor pn + m \rfloor, g) = g$ and interpolate between the points $\left\{\left(p_k, x_{(k)}\right)\right\}$.

Hyndman and Fan [69] defined nine different types of sample quantile estimation algorithms using this approach. These are listed in Table 5.3, plus a tenth type, with values for the tuning parameters of m, γ, α, and β.

Types 1 through 3, inclusive, of Hyndman and Fan [69] rely on the EDF. Types 6 through 9, inclusive, rely on linear interpolation with the ogive. The type 4 sample quantile interpolates the EDF and in doing so uses the ogive and is an alternative to method 1, which uses the step function of EDF and does not interpolate.

Of the various software packages listed in Table 5.3, only R offers all nine types of sample quantiles given in Hyndman and Fan [69]. The R function to do this is called `quantile`.

SAS is the only other statistical software package to offer more than one type. The SAS procedure `STDIZE` offers five of the methods in Table 5.3. Percentile definition 1 of SAS procedure `STDIZE` is type 4 in Table 5.3. Percentile definition 2 is type 3. Percentile definition 3 is type 1. SAS percentile definition 5 is type 2. SAS percentile definition 4 is type 10. The only difference between type 6 of Hyndman and Fan [69] and SAS percentile definition 4 is that SAS percentile definition 4 sets β equal to 1 instead of 0.

As to why ten different algorithms exist for estimating quantiles from a random sample, the reason is that not one of these has met the ironclad property of statistical optimality. That is, no unbiased minimum-variance estimator of a quantile has been created.

Sample quantile type 9 is only approximately median unbiased for normally distributed data. Of all the nine candidate algorithms considered by Hyndman and Fan [69], only sample quantile type 8 is approximately median unbiased for

Type	m	γ	α	β	Software
1	0	$\gamma = \begin{cases} 0 & \text{if } g = 0 \\ 1 & \text{if } g > 0 \end{cases}$	0	1	R, SAS
2	0	$\gamma = \begin{cases} \frac{1}{2} & \text{if } g = 0 \\ 1 & \text{if } g > 0 \end{cases}$	0	1	R, SAS
3	$-\frac{1}{2}$	$\gamma = \begin{cases} 0 & \text{if } g = 0 \text{ and } j \text{ even} \\ 1 & \text{if } g = 0 \text{ and } j \text{ odd} \\ 1 & \text{if } g > 0 \end{cases}$	0	1	R, SAS
4	0	$\gamma = g$	0	1	R, SAS
5	$\frac{1}{2}$	$\gamma = g$	$\frac{1}{2}$	$\frac{1}{2}$	R
6	p	$\gamma = g$	0	0	R MINITAB SPSS
7	$1 - p$	$\gamma = g$	1	1	R
8	$\frac{1}{3}(p+1)$	$\gamma = g$	$\frac{1}{3}$	$\frac{1}{3}$	R
9	$\frac{1}{4}p + \frac{3}{8}$	$\gamma = g$	$\frac{3}{8}$	$\frac{3}{8}$	R SAS
10	p	$\gamma = g$	0	1	

Table 5.3 *Types of quantile estimators with associated tuning parameters and availability by statistical software package*

all distributions and receives their recommendation. Yet the type 7 algorithm is mode unbiased and this is the default for the function `quantile` available in R since release 2.0.0. Ironically, this code is authored by Ivan Frohne and Rob J. Hyndman, the lead author of Hyndman and Fan [69].

For sensible sample sizes, say at least 30 or more observations, there are no practical differences among any of the quantiles generated by the ten different algorithms for estimating quantiles. But for smaller sample sizes, the differences are sufficiently important for these algorithms to have been discussed.

Regarding the optimal choice of the parameters α and β in formula (5.5) for p_k, type 9 in Table 5.3 produces an approximately median-unbiased quantile for the normal distribution. SAS has adopted type 9 as the default when producing quantile-quantile plots with its `UNIVARIATE` procedure.

The SAS products INSIGHT and ANALYST also make calls to this procedure when a probability plot is requested by the user. The SAS procedure `UNIVARIATE` also permits the user to override the default values of $\alpha = \beta = 3/8$ and substitute any values of their own choosing for α and β when producing a quantile-quantile plot.

Type	α	β	SPSS method	Other Software
6	0	0	Van der Waerden	SAS
5	$\frac{1}{2}$	$\frac{1}{2}$	Rankit	R (Sample size $>$ 10)
8	$\frac{1}{3}$	$\frac{1}{3}$	Tukey	SAS
9	$\frac{3}{8}$	$\frac{3}{8}$	Blom	MINITAB, R (Sample Size $\leq$ 10), SAS (default)

Table 5.4 *Types of quantile estimators with associated tuning parameters and availability by statistical software package for producing normal quantile-quantile plots*

The R function for producing a normal quantile plot is `qqnorm`. This function relies on a call to another R function, called `ppoints`, to estimate the empirical distribution function. The function `ppoints` uses the type 9 approximation of Table 5.3 with $\alpha = \beta = 3/8$ when the sample size $n \leq 10$ and the type 5 approximation with $\alpha = \beta = 1/2$ otherwise. This rather unusual choice is documented by Becker, Chambers, and Wilks [8] for the precursor to R known as S. The rationale for this is due to Blom [11].

In his thesis published in 1958 by John Wiley and Sons, Blom [11] considered sample quantile types 5, 6, 8, and 9. Blom [11] only gave a passing reference to type 8. He provided a table comparing expected quantiles for types 5, 6, and 9. This table was determined by calculation for sample sizes of 5, 10, and 15 drawn at random from a normal population. Based upon Table 2 on page 162 of Blom [11], his conclusion to recommend type 9 from among the three is justified.

Table 5.4 lists type 9 quantile estimation as being offered by the four software packages MINITAB, R, SAS, and SPSS. Type 9 is the default for SAS and SPSS regardless of sample size. Type 9 is the only option for MINITAB.

SPSS offers type 9 quantile estimates as the default and offers three other methods when producing quantile-quantile plots.

According to SAS [71] documentation, the SAS procedure `RANK` also offers the Van der Waerden, Tukey, and Blom methods described in Table 5.4 when computing normal scores. The Blom method is the default algorithm in the `RANK` procedure.

One pairing of values of tuning parameters that has not been selected by any of the four statistical software packages considered here is $\alpha = 0$ and

$\beta = 0$. The reason for this in the case of the normal quantile-quantile plot is that it would be necessary to calculate and plot a point with a coordinate of $\Phi(p_n) = \Phi(1) = \infty$. This would require one heck of a lot of graph paper.

For sensible sample sizes, say at least 30 or more observations, there are no practical differences among any of the quantiles generated by the ten different algorithms for estimating quantiles. But for smaller sample sizes, the differences are sufficiently important for these algorithms to have been discussed.

5.6 Conclusion

The quantile-quantile plot or probability plot, and sometimes both, can be found in virtually any software package that does statistical analysis. In response to some software commands, these plots are often produced automatically by default. No doubt this has led to curious researchers investigating these plots and their properties—leading to wider adoption of these plots.

A concern shared with the boxplots and the EDF plot of the previous chapter is the choice of method of percentile estimation. The concept of hinges is due to Tukey [127] (pp. 32–33): "by counting half-way from each extreme to the median." Nine other methods for estimating the quartiles when executing the boxplot have been presented. Ten different methods have been presented for estimating quantiles. A variety of these are available in different statistical software packages but only the package R, at the time of writing, offers as many as nine of the ten algorithms for quantiles.

Curiously, only a fraction of these methods are available for normal quantile-quantile plots. SPSS offers four of the ten quantile estimation methods. SAS offers three of the SPSS methods and an option for the user to supply their own values for the tuning parameters α and β in the quantile estimation algorithm given in formula (5.5). The function `qqnorm` in the R statistical software package offers only one algorithm that was chosen on the basis of optimality considerations and Monte Carlo simulations.

If variety is the spice of life, then there is quite a selection of graphical displays available for depicting the distribution of a single continuous variable. In the next chapter, graphical displays for depicting the density of a single continuous variable will be considered.

5.7 Exercises

1. Consider the following data collected by Charles Darwin:

$$-67, -48, 6, 8, 14, 16, 23, 24, 28, 29, 41, 49, 56, 60, 75.$$

   These data represent the difference in height (in eighths of an inch) between the self-fertilized member and the cross-fertilized member of each of 15 pairs of plants.

(a) Produce a normal quantile plot or a normal probability plot. Add a normal distribution reference line to the plot.
(b) Comment on any features you see in your answer to part (a).
(c) Execute an outlier boxplot for the data.
(d) Compare the outlier boxplot from part (c) with your answer to part (a). Discuss the two plots with respect to skewness and outliers.

2. Consider the data for the total compensation in 2008 received by chief executive officers employed by industrial companies listed in Table 4.3.
   (a) Produce a normal quantile plot or a normal probability plot. Add a normal distribution reference line to the plot.
   (b) Do the data in the plot of part (a) appear to be normally distributed? Discuss.
   (c) Would you choose to use your plot for part (a) in a written media presentation? If not, then which type of plot would you use instead? Justify your choice.
3. Consider the data for the total compensation in 2008 received by chief executive officers employed by industrial companies listed in Table 4.3.
   (a) Produce a normal quantile plot or a normal probability plot.
   (b) Do there appear to be outliers in the plot of part (a)? Discuss.
   (c) Draft an outlier boxplot to confirm your conclusion in part (c).
   (d) Is it reasonable to expect that there would be outliers in CEOs' total compensation? What forces would be at play in the marketplace to discourage outliers in total compensation?
4. Consider the data for the total compensation in 2008 received by chief executive officers employed by industrial companies listed in Table 4.3.
   (a) Produce an EDF plot for the total compensation data. Add a normal curve to the EDF plot.
   (b) Produce a normal quantile plot or a normal probability plot for the total compensation data. Add a normal distribution reference line to the plot.
   (c) Which of the two graphics from parts (a) and (b) would you prefer to represent to a class of students registered in a Master of Business Administration program? Discuss.
   (d) Is it reasonable to present either graphic from parts (a) or (b) to a public audience? Discuss.
5. In 1882, Simon Newcomb measured the time required for light to travel from his laboratory on the Potomac River to a mirror at the base of the Washington Monument and back, a total distance of about 7400 meters. These measurements were used to estimate the speed of light. Table 4.4 contains the estimated speed of light for 66 trials. These data were reported in an article by Steven Stigler [115] in 1977 in *The Annals of Statistics*.
   (a) Produce a normal quantile plot or a normal probability plot for the data in Table 4.4.

Density of the Earth Relative to Water					
5.5	5.61	4.88	5.07	5.26	5.55
5.36	5.29	5.58	5.65	5.57	5.53
5.62	5.29	5.44	5.34	5.79	5.1
5.27	5.39	5.42	5.47	5.63	5.34
5.46	5.3	5.75	5.68	5.85	

Table 5.5 *Cavendish's 1798 determination of the density of the earth relative to that of water*

(b) Do there appear to be any outliers in your plot for part (a)? Discuss.

(c) Does there appear to be any skewness in your plot for part (a)? Discuss.

(d) Does the data appear to follow the normal distribution in your plot for part (a)? Discuss.

6. Consider the time passage data for light in Table 4.4 that was obtained by Simon Newcomb.

   (a) Produce an outlier boxplot for the data in Table 4.4.

   (b) Produce a revised data set with the outliers removed.

   (c) Produce a normal quantile plot, or a normal probability plot, for the data from your answer to part (b). Add a normal distribution reference line to the plot.

   (d) Do the data in your plot for part (c) appear to be normally distributed? Discuss.

7. On June 21st in 1798, Henry Cavendish [17] read before The Royal Society of London a paper concerning his experiments to determine the density of the earth. He modified an apparatus originally conceived by the Rev. John Michell. Inside a mahogany case was suspended a six-foot long slender rod from which two-inch diameter lead balls were suspended at either end. To estimate the density of the earth, Cavendish measured the displacement of the rod by the gravitational pull of a pair of eight-inch diameter lead balls positioned nearby and suspended by copper rods at the end of a wooden bar. Cavendish's experiments resulted in the 29 estimates for the density of the earth as given in Table 5.5.

   (a) Produce a dotplot for Cavendish's estimates of the density of the earth relative to water.

   (b) Produce a stemplot for Cavendish's estimates of the density of the earth relative to water.

   (c) Produce an EDF plot for Cavendish's estimates of the density of the earth relative to water.

   (d) From your answers to parts (a) through (c), inclusive, do the data appear to be skewed? Discuss.

8. Consider Cavendish's estimates for the density of the earth in Table 5.5.

(a) Produce a quantile boxplot for the estimates in Table 5.5.

(b) Produce an outlier boxplot for the estimates in Table 5.5.

(c) Produce a normal quantile plot or a normal probability plot for the estimates in Table 5.5.

(d) Do there appear to be any outliers among Cavendish's 29 estimates? Discuss.

(e) Do Cavendish's estimates for the density of the earth appear to be normally distributed? Discuss.

9. Consider Cavendish's estimates for the density of the earth in Table 5.5.

   (a) Produce a normal quantile plot or a normal probability plot for the estimates in Table 5.5.

   (b) Add a normal distribution reference line to the plot you produced for part (a).

   (c) Do Cavendish's estimates for the density of the earth appear to be normally distributed? Discuss.

10. The estimates of the earth's density by Cavendish [17] are reported in sequence in Table 5.5 as he recorded them. Table 5.5 is to be read row-wise from left to right beginning with the first row. The first six estimates in Table 5.5 were obtained as pairs from three experiments before Cavendish changed the wire suspending each of the two-inch lead balls to a stiffer one. Cavendish [17] was concerned about stretching affecting the results of his experiment. In 1977, Stigler [115] (see page 1073) also appears to consider the change in wire to be potentially important. Do you? Discuss using appropriate plots.

11. This exercise pertains to material in Section 5.5.

    (a) What approximation method is the default in the function `quantile` in the package `stats`?

    (b) Which approximation does the function `ppoints` in the package `stats` use?

    (c) What calculation does the function `geom_qq` in the package `ggplot2` do for estimating the percentiles used to find the theoretical quantiles?

    (d) Compare your answers to parts (a), (b), and (c). Discuss.

12. This exercise pertains to material in Section 5.5. Consider the data for the total compensation in 2008 received by chief executive officers employed by industrial companies listed in Table 4.3. Produce an ogive plot for the total compensation data. Add a normal curve to the ogive plot. Do the data appear to be normally distributed? Discuss.

Chapter 6

Nonparametric Density Estimation for a Single Continuous Variable

6.1 Introduction

The most widely recognized graphical display of a density of a single continuous variable is the histogram. A histogram is constructed without assuming a statistical model and estimating its parameters from data. Hence, it is a nonparametric procedure.

There are other nonparametric alternatives to the histogram. The kernel density estimate has come in to its own. It is the most widely used alternative. Typically both the histogram and kernel density estimate are displayed together. But this is not necessary.

An older alternative to the kernel density estimate is also presented in this chapter. The spline density estimate is not used as often as the kernel density estimate. The spline density estimate is typically displayed with a histogram as well.

Discussion in this chapter begins with the histogram.

6.2 Learning Outcomes

When you complete this chapter, you will be able to do the following.

- Be able to draft a histogram in either the `graphics` package or the `ggplot2` package in R.
- Be familiar with the various rules of thumb for determining either the number or width of classes when constructing a histogram.
- Be familiar with the circular variation on the histogram: the rose diagram of Florence Nightingale.

- Optionally, be familiar with the kernel density estimate as a nonparametric alternative to the histogram, know the mathematical forms for six different kernels, and various options available for estimating bandwidth when performing kernel estimation.
- Optionally, be able to use the `density` function in the `stats` package together with the `graphics` package to display a kernel density estimate alone or overlaid with a histogram or use the `ggplot2` package to accomplish either task.
- Optionally, be able to draft a violin plot with either the `lattice` package or the `ggplot2` package in R.
- Optionally, be knowledgeable about the limitations of spline methods for estimating the probability density of a continuous random variable.
- Optionally, be aware of how the Kullback-Leibler distance in information theory can guide consideration of selection of a method to graphically portray the probability density of a continuous random variable.

6.3 The Histogram

6.3.1 *Definition*

A *histogram* is constructed by noting the full range of values observed for a quantitative variable in a random sample, and subdividing this range into intervals, usually of equal length. After deciding whether each interval includes or excludes its right or left endpoint, the number of observations in each interval is tallied and plotted as a rectangle in a figure. The intervals are commonly referred to as *classes*.

An example of a histogram is given in Figure 6.1 for perch mass. As the vertical axis corresponds to the counts, or frequency, for each interval, Figure 6.1 is often characterized as a *frequency histogram*. Notice that a convenient value has been chosen for the number of classes, so that the length of each class is 100 g. Figure 6.1 has been produced with the R function `hist` with its default to include observations at the right endpoint but not the left in the count for a given interval.

Figure 6.1 was created by the following R instruction.

```
hist(mass,freq=TRUE,xlim=c(0,1200),breaks=12,ylim=c(0,20),
main=NULL,xlab="Mass (g)")
```

In the call to `hist`, the counts (or frequency) in each class were plotted by setting `freq=TRUE`. The limits of the horizontal axis were set at the left to 0 g and at the right at 1,200 g by the statement `xlim=c(0,1200)`. The number of classes for this range was set to 12 by the statement `breaks=12`. Note that each class width in Figure 6.1 is set evenly to 100 g. Setting `ylim=c(0,20)` in

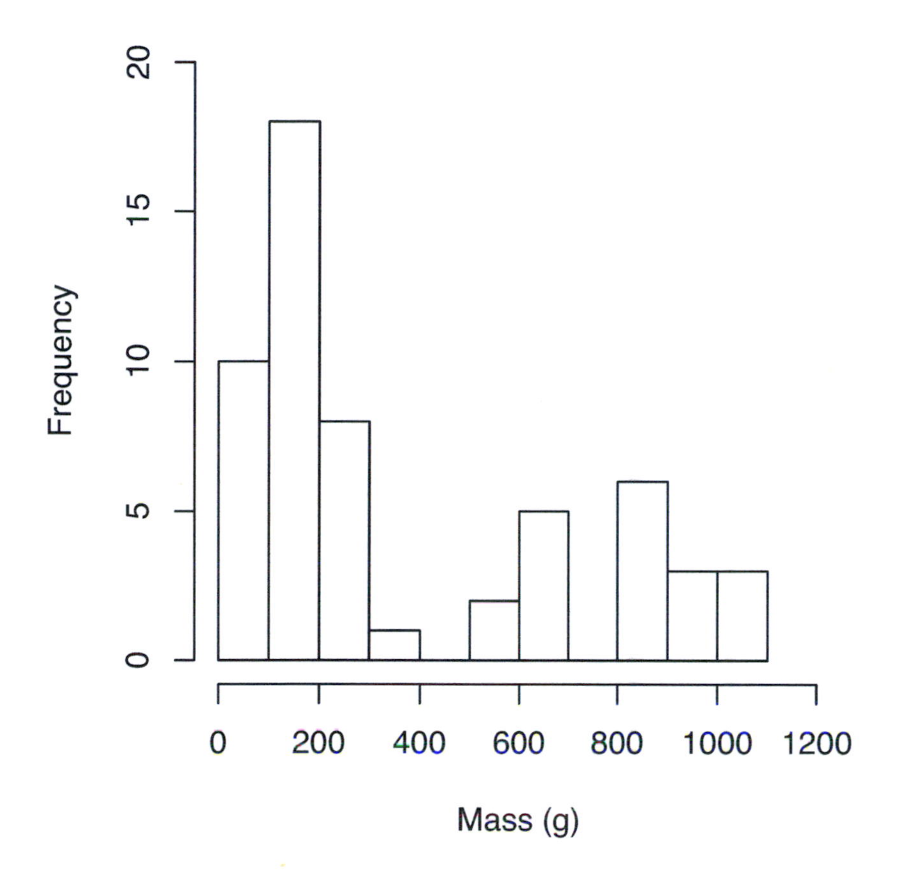

Figure 6.1 *Frequency histogram of the mass of 56 perch with twelve classes*

the call to `hist` was done to produce a range of 0 and 20 for the vertical axis of counts with ticks evenly spaced and labeled at intervals of 5.

Figure 6.2 is a ggplot2 counterpart to Figure 6.1 created by the following code.

```
figure<-ggplot(perch, aes(mass)) +
geom_histogram(breaks=(0:11)*100,closed="right",
color="black",fill="white") +
theme_base() + theme(plot.background
element_rect(fill = NULL, color = "white",
linetype = "solid"),
panel.border = element_rect(fill=NULL,
color="white",linetype="solid"),
axis.line=element_blank(),
```

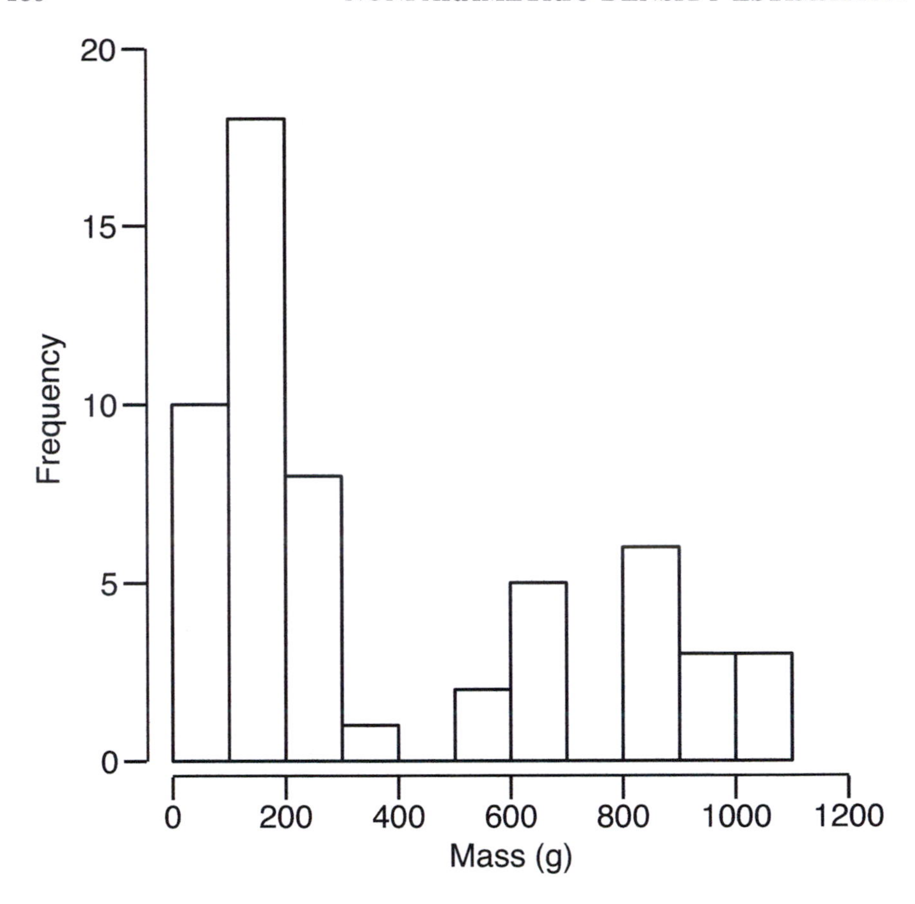

Figure 6.2 *Frequency histogram of the mass of 56 perch with twelve classes using* ggplot2

```
axis.ticks=element_line(lineend="square"),
axis.title=element_text(size=12),
axis.text=element_text(size=12)) +
scale_y_continuous(expand=c(.0,0),limits=c(-0.5,20.25)) +
scale_x_continuous(expand=c(0.0,0),breaks=200*(0:6),
limits=c(-50,1250)) +
annotate("segment",x=0,xend=1200,y=-.425,yend=-.425,
size=.5) +
annotate("segment",x=-46,xend=-46,y=0,yend=20,size=.5) +
labs(x="Mass (g)",y="Frequency")

print(figure)
```

There is considerably more ggplot2 code for Figure 6.2 than that needed to use the graphics function hist to produce Figure 6.1. But the comparison is not fair without noting that the ggplot2 was written using the ggthemes function theme_base and additional code to produce a facsimile of Figure 6.1 in the base style of the graphics package in R. Extra ggplot2 coding was needed to turn off the bounding boxes framing the axes and the figure as a whole. More coding was needed to terminate the axes lines in the graphics package style of both figures. A bit of experimentation was done to position the axes of the required length using the annotate function in ggplot2 to place the lines adjacent to their respective axis ticks after first eliminating the axes lines. The graphics and ggplot2 coding both required setting the breaks for the histogram, the horizontal axis title, and the limits of the horizontal and vertical axes. The default in both the graphics function hist and the ggplot2 function geom_histogram is to include observations at the right endpoint but not the left in the count for a given interval but a clear instruction was given in the call of geom_histogram. The default fill color for the histogram bars in ggplot2 function geom_histogram is dark gray but this was overridden to be white in the function call and the parameter color="black" was passed on calling geom_histogram to get the black lines for the bars.

If the counts are divided by the sample size and used to indicate the height of the rectangles with the vertical axis changed accordingly, the result is the *relative frequency histogram* as depicted in Figure 6.3 for the perch data.

The relative frequencies in Figure 6.3 were plotted by setting freq=FALSE in the call to the R function hist. Alternatively, the relative frequencies could have been obtained by stating probability instead of freq=FALSE. Note that freq defaults to TRUE if and only if breaks are equidistant (and probability is not specified).

The histogram is one of the most frequently encountered graphical displays for a single quantitative variable. At first glance, a histogram appears to be a simple plot that is easily executed. This is deceptive. The histogram is also one of the most poorly executed graphical displays and is easily manipulated by researchers to conceal rather than reveal information from a sample. In fact, the construction of a histogram is at least as complex as creating a quantile-quantile plot, if not more so. The advantage of a properly constructed histogram is that its format is familiar to members of the public.

In the histogram, the areas of the rectangles ought to be proportional to the number of observations in a given range of values (so-called classes). This precise definition can be ignored if all of the classes are of the same width as is typically done in an introductory course in statistics.

Common problems associated with histograms are:

- bars are separated by a gap as done for a bar chart for a qualitative variable;
- outliers are omitted,

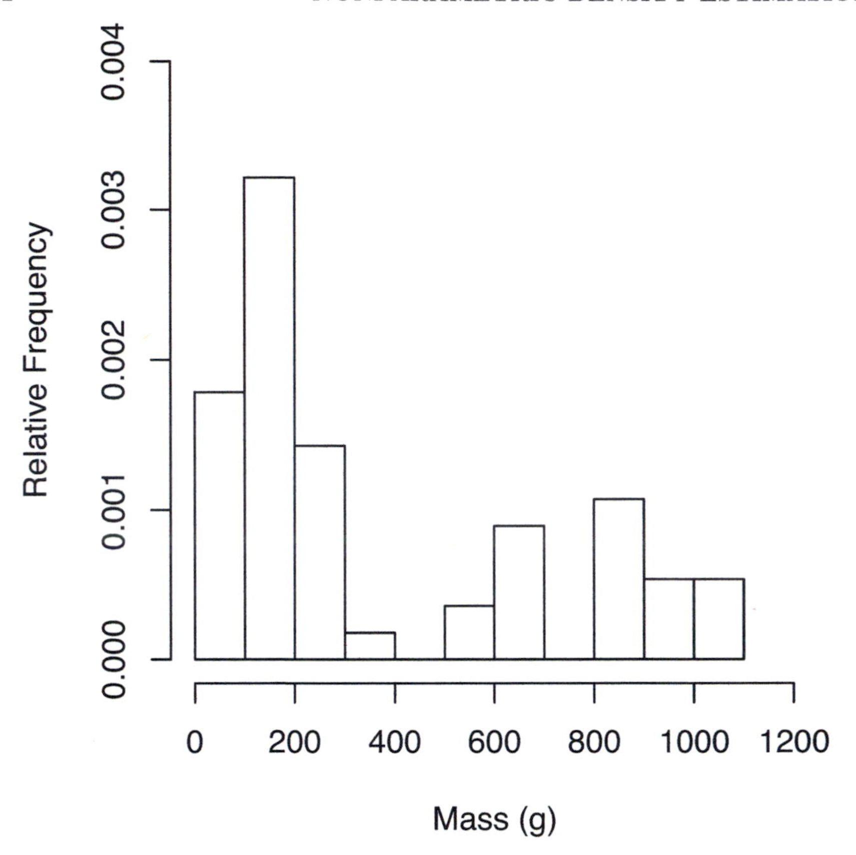

Figure 6.3 *Relative frequency histogram of the mass of 56 perch with twelve classes*

- uncertainty as to whether each class includes the left endpoint or the right endpoint,
- improper choice of the number of classes, and
- unequal class widths not properly accommodated.

Despite these difficulties, histograms remain popular with the public and administrators. Many disciplines still require histograms and will not readily accept any of the better alternatives previously discussed. The convenience of a well-used and familiar graphical display in these disciplines is considered paramount.

Guidance for the selection of the number of classes, or equivalently a class width, when all are of equal width is in the form of the following five rules. Let the sample size be denoted by n.

Rule of Twelve: Use approximately 12 classes to span the range of the random sample. This is a popular *ad hoc* rule.

Robust Rule of Twelve: Use approximately 12 classes to span 4.45 IQR of the sample. Note that other robust alternatives to the range could be used. The choice here is theoretically equivalent to a span of six standard deviations, or 6σ, for a normal distribution. This accounts for 99.7% of the probability mass under a normal curve.

Sturges' Rule: The number of classes M_S is given by

$$M_S = 1 + \frac{\log(n)}{\log(2)}. \tag{6.1}$$

Sturges [118] created this rule so that the class frequencies will comprise a binomial series for a random sample drawn from a normal population.

Doane's Rule: Set the class width to be

$$M_D = 1 + \frac{\log(n) + \log(1 + c_1)}{\log(2)} \tag{6.2}$$

where the standardized measure of skewness is

$$c_1 = \frac{m_3}{{m_2}^{3/2}} \left[\frac{(n+1)(n+3)}{6(n-2)} \right]^{1/2} \tag{6.3}$$

with the biased central moments given by

$$m_j = \sum_{i=1}^{n} (X_i - \overline{X})^j / n \tag{6.4}$$

with $j = 2, 3$. Doane [32] commented that sample data are seldom symmetric, let alone normally distributed. He felt that Sturges' Rule does not provide enough classes to reveal the shape of a severely skewed distribution and so proposed his own modification of Sturges' Rule.

Scott's Rule: The class width w_C is given by

$$w_C = 3.49 s n^{-1/3} \tag{6.5}$$

where s is the standard deviation of the sample. Scott [105] assumed a normal distribution in deriving this rule, which minimizes the integrated mean square error (IMSE) between the relative frequency histogram and the probability density. Scott did theoretically consider his proposed rule for examples of distributions that are skewed, heavy-tailed, or bimodal and concluded that his proposed rule leads to class widths that were too large in each of these three cases.

Freedman-Diaconis Rule: The class width w_{FD} is given by

$$w_{FD} = 2\ IQR\ n^{-1/3} \tag{6.6}$$

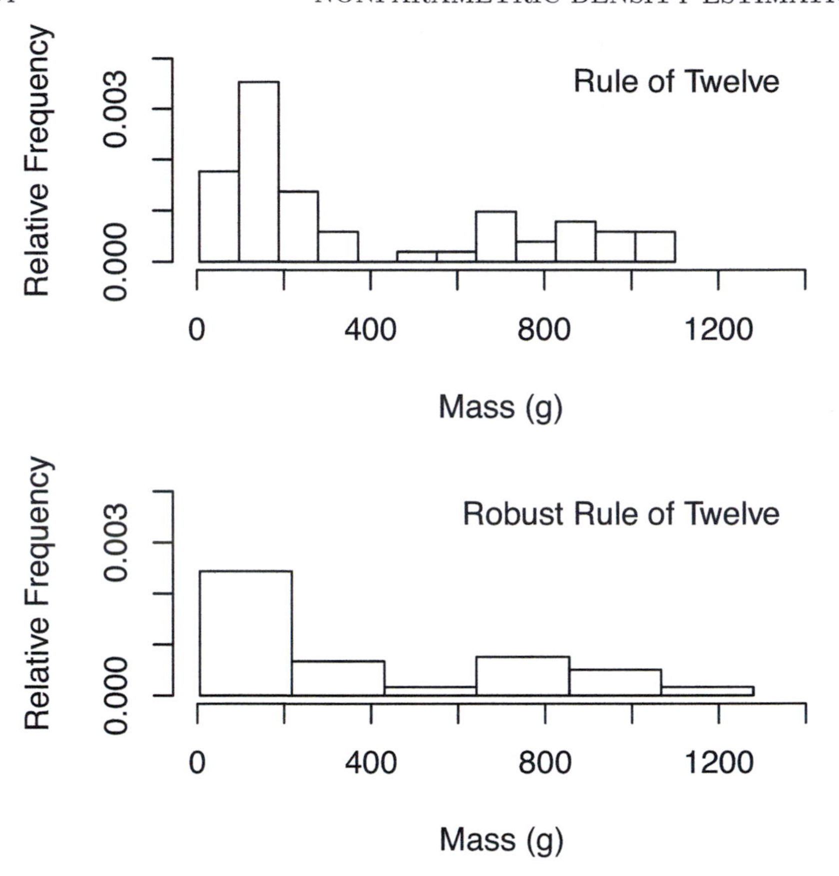

Figure 6.4 *Relative frequency histograms of the mass of 56 perch depicting the Rule of Twelve and its robust counterpart*

where IQR is the interquartile range of the random sample. Freedman and Diaconis [44] developed this rule as an approximation to the solution they found for minimizing the mean of the integrated square error (MISE) between the histogram and the probability density. Because the IQR is a robust alternative to the sample standard deviation s, the Freedman-Diaconis rule of 1981 can be considered to be a robust version of Scott's rule of 1979. But this *post hoc* view ignores the fact that Freedman and Diaconis [44] sought to minimize the MISE in contrast to Scott [105] who sought to minimize the IMSE.

It is to be understood in the definitions above for Sturges' and Doane's rules that M_S and M_D are to be taken to be the smallest integer larger than the value on the right side of their respective equations.

For 56 perch caught in the research trawl on Längelmävesi, the range is 1094.1

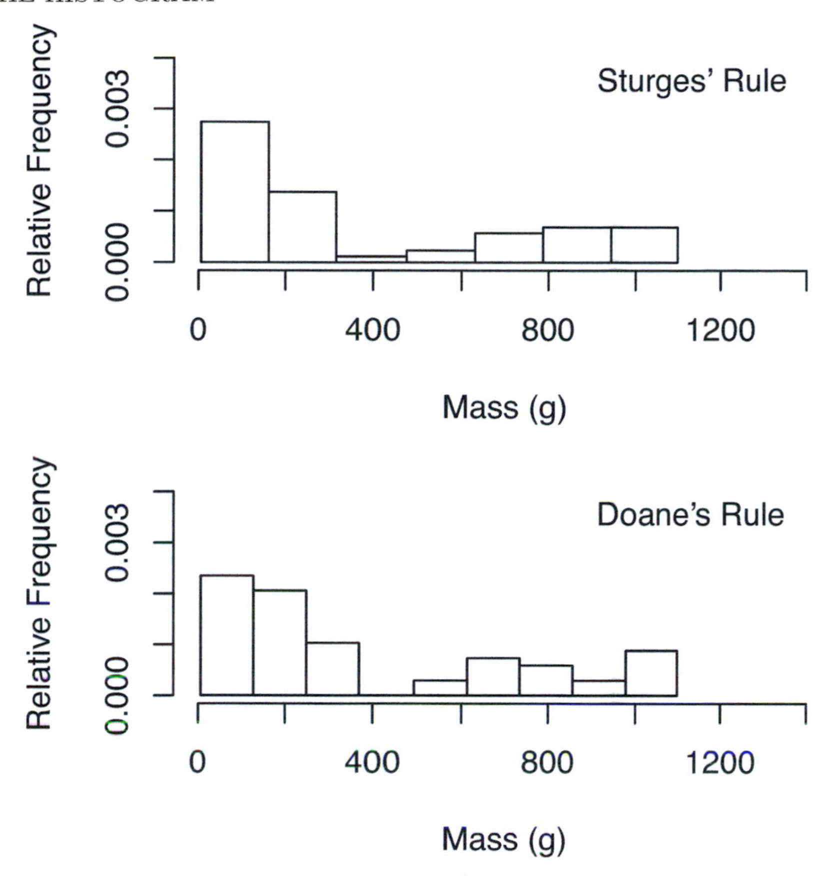

Figure 6.5 *Relative frequency histograms of the mass of 56 perch depicting Sturges' and Doane's Rules*

g. So by the Rule of Twelve, the class width is 91.175 g. The IQR from the `quantile` function in R with the default percentile estimation method (type 7) is 572.5 g. So by the Robust Rule of Twelve, the class width for constructing a histogram is 212.302 g. The histograms by the Rule of Twelve and its robust counterpart are presented in Figure 6.4.

Because the sample size for this data is $n = 56$, the number of equal size classes by Sturges' rule is

$$M_S = 1 + \frac{\log(n)}{\log(2)} = 1 + \frac{\log(56)}{\log(2)} = 6.81. \tag{6.7}$$

Rounding this figure upward to 7 and dividing 7 into the range, the class width is 156.300 g.

There is the impression that the distribution of the mass of perch is not

	Method									
Software	1	2	3	4	5	6	7	8	9	10
Minitab	•									
R	•									
SAS					4		5	2	3	1
SAS/Analyst							•			
SAS/INSIGHT							•			
SPSS							•			

Table 5.2 *Availability of methods for quartile estimation for boxplots in various statistical software packages (the numbers in the row corresponding to SAS are the method numbers in the documentation for this package, and the default method for the SAS* `BOXPLOT` *procedure is SAS method 5)*

then they would go completely undetected by the other eight methods. If one is averse to this risk, then it is important to know which method of quartile estimation is being used. When drafting boxplots, it is also important to consider whether the potential viewers are risk averse.

It is sound practice when relying on statistical software to know which method of quartile estimation is in use and communicate this information to viewers. To do so assumes that the viewers are technically competent with respect to these details. This would not be a concern if all software packages were to use the same method for quartile estimation. Alas, there is considerable variability among statistical software packages concerning the selection of quartile-estimation method. An indication of the availability of these methods among the various statistical software packages is given in Table 5.2. It would be desirable if all packages offered all methods but this is not the case.

Note that methods 3, 4, and 6 are not available in any of the statistical software packages cited in Table 5.2. Methods 3 and 4 have features that their creators hoped would lead to their adoption but this has not been the case. Methods 1 and 6 lend themselves to quick hand calculation and are encountered most often in introductory statistics courses. However, method 6 is really only intended as a quick and easy substitute for either of methods 5 or 7.

Ideally, all software packages ought to provide method 1, due to Tukey [127], in addition to any other method or methods. According to Table 5.2, this is not the case.

5.5.2 Estimation of Percentiles

On the topic of estimation of any percentile from a random sample, the EDF function defined in the previous section is one of many approaches. We begin symmetric. Doane's rule is used with skewed data instead of Sturges' rule because it is believed that for skewed data Sturges' rule yields too few classes. Doane's rule requires calculation of the skewness measure:

$$c_1 = \frac{33,590,596}{118,680.25^{3/2}} \left[\frac{(57)(59)}{(6)(54)} \right]^{1/2} = 2.66492. \tag{6.8}$$

Thus the number of classes by Doane's rule is

$$M_D = 1 + \frac{\log(56) + \log(1 + 2.65)}{\log(2)} = 8.67403, \tag{6.9}$$

and after rounding upward to 9, the class width is determined to be 121.567 g. The histograms by Sturges' and Doane's rules are given in Figure 6.5. Comparing the two rules reveals some merit in the argument that the greater number of classes by Doane's rule works better for data that might be skewed to the right. The histogram by the Rule of Twelve in Figure 6.4 appears to be just as good as the one by Doane's rule.

Scott's rule requires calculation of the standard deviation of the sample. This value is 347.618 g for the mass of 56 perch. This results in a class width of 317.101 g. The Freedman-Diaconis rule uses the IQR as a measure of spread instead and this rule produces a class width of 299.279 g. Histograms by both these rules are depicted in Figure 6.6.

It is generally believed that Scott's rule will yield too few classes for skewed data. This belief apparently also holds for the Freedman-Diaconis rule that substitutes the more robust measure of spread given by the IQR in place of the sample standard deviation s. With respect to Doane's rule and robustness, the biased central second-order moment m_2 is not much different from the sample variance s^2 and is similarly not robust. Higher-order moments, such as the third-order central moment m_3, are even worse off. So while Doane's rule is designed for skewed data, outliers can cause havoc.

Note that the six rules for determining class widths for histograms do not yield aesthetically pleasing class widths. These class widths would not have been used in the days of hand drafting histograms on graph paper. The rule of Sturges [118] was proposed in 1926. The rule of Doane [32] was proposed in 1976. Even with a separation of 50 years, the same approach to the selection of class widths was advocated. Each rule was proposed as a rule of thumb to select a number W of the form $W = A \cdot 10^B$ where A is a member of the set $\{1, 2, 5\}$ and B is any integer. That is, B is a member of the set $\{\ldots, -3, -2, -1, 0, 1, 2, 3, \ldots\}$. Doane [32] actually went further than Sturges [118] in developing his algorithm to find the best combination of A and B. But this has not stood the test of time. Even Scott [105] for his rule advocates a convenient choice of class width either slightly larger or smaller.

The class widths for the six rules are listed in Table 6.1 together with their

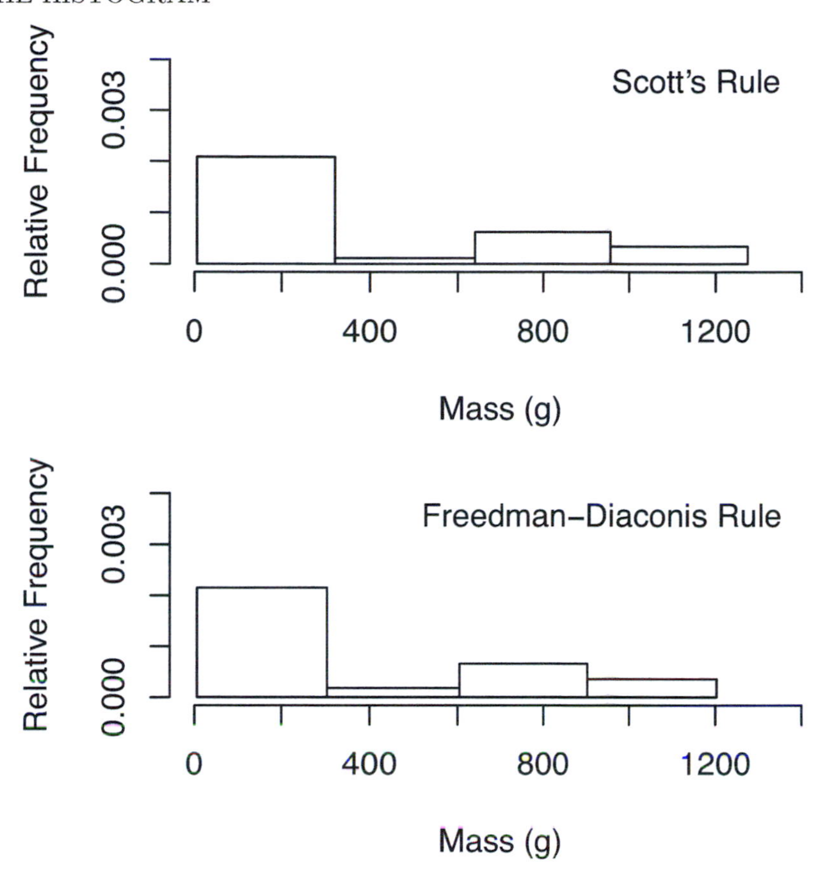

Figure 6.6 *Relative frequency histograms of the mass of 56 perch depicting Scott's and the Freedman-Diaconis Rules*

convenient counterparts. The Rule of Twelve and Doane's rule suggest a class width of 100 g while the other four rules agree on 200 g. The relative frequency histograms for these two class widths are depicted in Figure 6.7. Both do a good job of capturing what might be a long right tail or a bimodal distribution.

It must be noted that Scott's Rule and the Freedman-Diaconis Rule produce class widths that are optimal in terms of IMSE and MISE, respectively, but only if used without adjustment. If these widths are adjusted to convenient values, as they typically are, then the result is some degree of suboptimality.

In each of the histograms in Figures 6.1 through 6.7, inclusive, each class is of the same width. This uniformity is not a universal plotting standard for the histogram. But most statistical software packages are not capable of any other type. An exception is the function `hist` in R that allows the user to supply the interval endpoints.

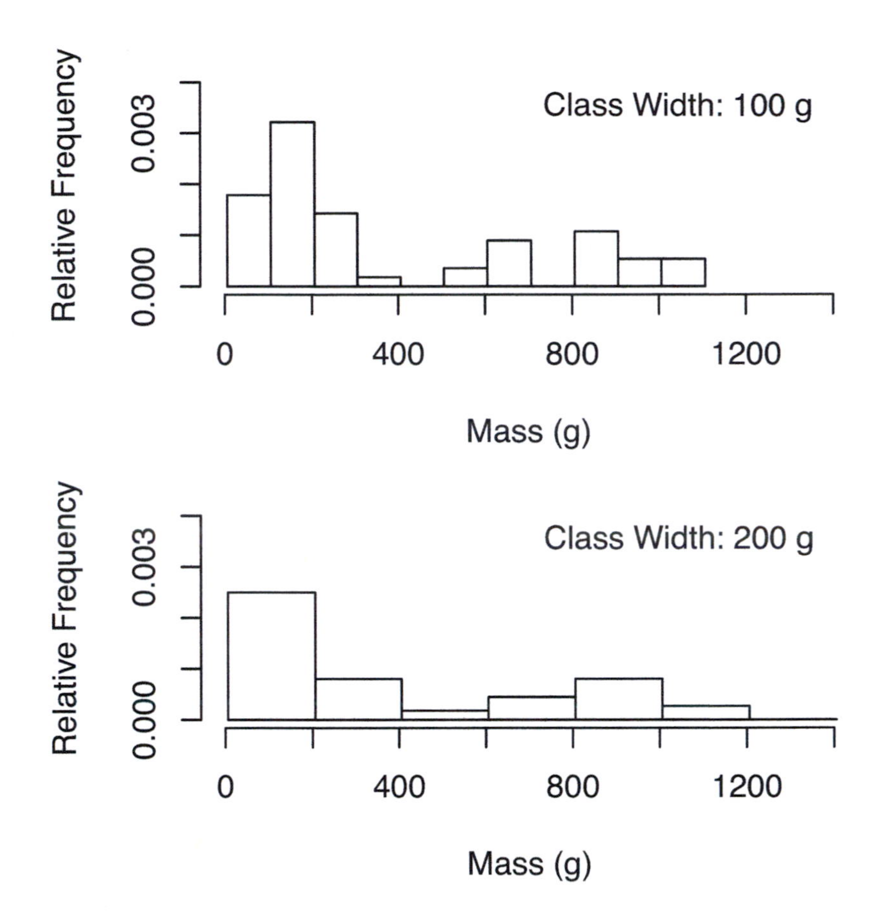

Figure 6.7 *Relative frequency histograms of the mass of 56 perch with class widths of 100 and 200 grams*

Rule	**Class Width**	**Closest Convenient Class Width**
Twelve	91.175	100
Robust Twelve	212.302	200
Sturges'	156.300	200
Doane's	121.567	100
Scott's	317.101	200
Freedman-Diaconis	299.279	200

Table 6.1 *Class widths and convenient alternatives for the six rules for histograms*

A histogram is an approximation to the probability density function of a continuous random variable. The probability of a random variable lying in any one particular interval is by definition equal to the area under the probability density function for the particular interval. Consequently, the total area underneath the probability density function must be equal to 1. So a crucial requirement for a histogram to appear to be a model of the probability density function based on a random sample is that the total area enclosed by the bars be equal to one. Suppose a histogram consists of M classes for a sample of size n. Let the width of the ith class be denoted by w_i and allowed to vary among the classes. Suppose there are n_i observations in the ith class. Clearly,

$$n = \sum_{i=1}^{M} n_i. \tag{6.10}$$

Let the height of the rectangle for the ith class be defined as f_i. The estimate of the probability density function given by the histogram must satisfy the constraint

$$\sum_{i=1}^{M} f_i w_i = 1. \tag{6.11}$$

One possible solution that satisfies this equation is

$$f_i = \frac{n_i}{n w_i}. \tag{6.12}$$

Table 6.2 presents values of $\{f_i\}$ for the 56 observations of mass for perch caught from Längelmävesi in 1917 using the formula above for classes of different widths. The histogram corresponding to the information in Table 6.2 is given in Figure 6.8. In part, the R script for generating this figure is as follows.

```
brks<-c(0,100,200,300,600,900,1200)

hist(mass,freq=FALSE,xlim=c(0,1400),breaks=brks,xaxt="n",
ylim=c(0,0.004),main=NULL,xlab="Mass (g)",
ylab="Relative Frequency")

axis(1,at=(0:7)*200,labels=c("0","","400","","800","",
"1200",""))
```

The desired breaks for the classes have been stored in the vector variable `brks`, which are then passed in the argument `breaks=brks` in the call to `hist`. Note that the range for the horizontal axis has been set to run from 0 g to 1,400 g by setting `xlim=c(0,1400)`. By setting `xaxt="n"` the placement of tick marks and labels by the function `hist` has been suppressed. Instead, the R function `axis` on the last line of the preceding code has been used to place the tick marks and labels in a more aesthetically pleasing manner.

Class	Interval	Count	Width	Relative Frequency
i		n_i	w_i	f_i
1	$[0, 100)$	10	100	0.001786
2	$[100, 200)$	18	100	0.003214
3	$[200, 300)$	8	100	0.001429
4	$[300, 600)$	3	300	0.000179
5	$[600, 900)$	11	300	0.000655
6	$[900, 1200)$	6	300	0.000357

Table 6.2 *Frequency table for histogram of perch mass with variable class width*

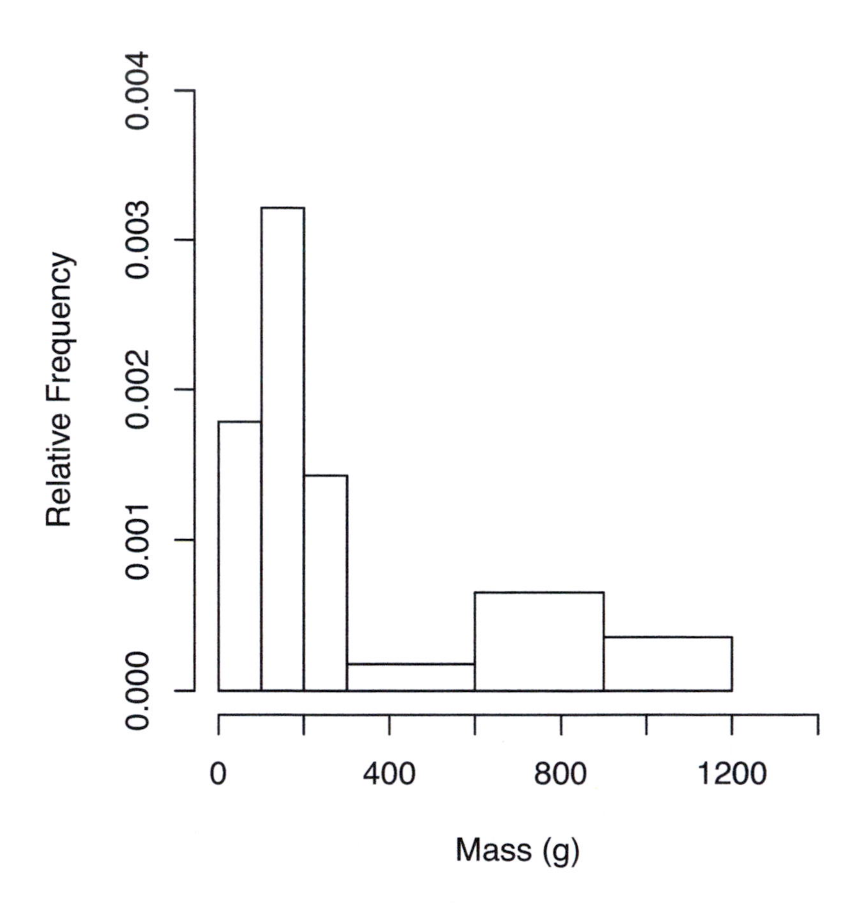

Figure 6.8 *Relative frequency histogram of the mass of 56 perch with varying class widths*

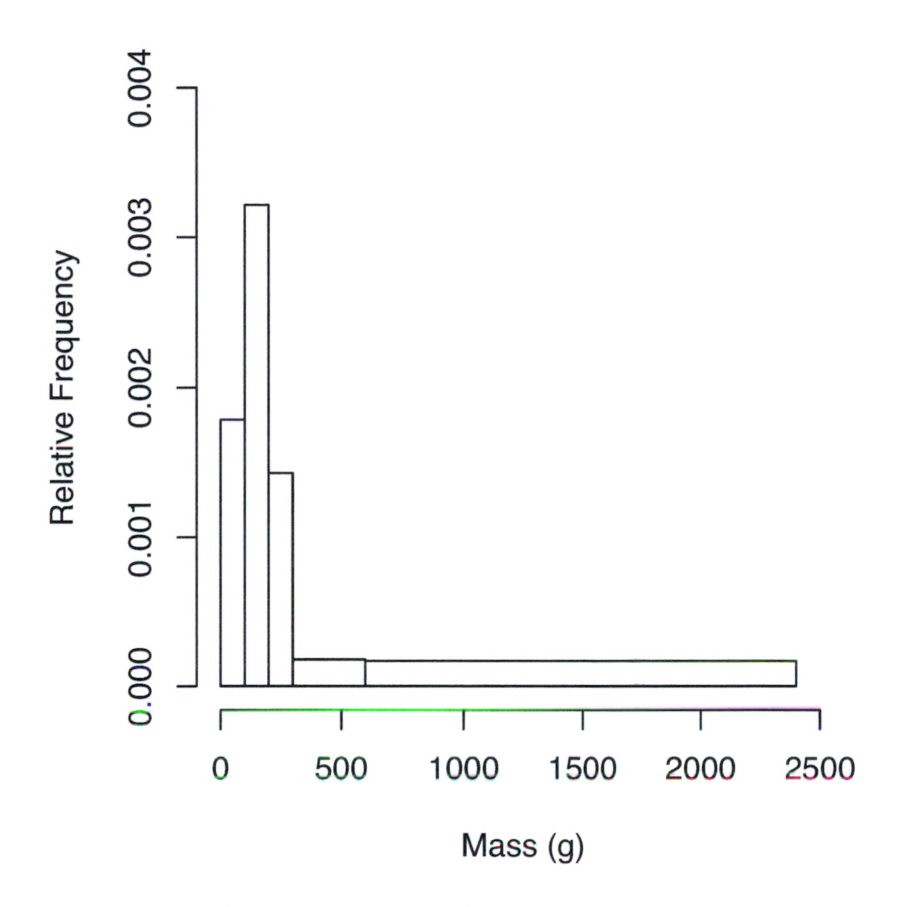

Figure 6.9 *Relative frequency histogram of the mass of 56 perch with varying class widths selected to conceal*

In Figure 6.8 there is a suggestion that instead of the data being a random sample from a distribution with a single peak, the data might be a random sample from a mixture of two different distributions with different locations and spreads. But a careful selection of class widths can change this picture, as in Figure 6.9.

The class widths have been deliberately manipulated to conceal bimodality in the distribution of mass in perch. The histogram of Figure 6.9 is intended to portray the data as from a distribution with a single peak that is skewed to the right. But if one is careful, one is not so easily deceived. In relative frequency histograms, such as Figure 6.9, the areas of the rectangles are to be compared not their heights. Given a previous discussion concerning the order of difficulty of certain graphical tasks, judgments concerning area are more difficult than judgments concerning length. A careful examination of the areas of the bars of Figure 6.9 would suggest that there is something more happening in the right tail of this histogram as the area of the rightmost bar is considerably greater than its nearest neighbor.

In this example of the mass of perch caught in a Finnish lake around 1917, we are exploring real data. If we knew the true underlying distribution of fish mass for this species, then we would know which of the class widths were to be preferred for the purposes of estimating the probability density function. But we do not. We can conjecture that since the perch spawn only once per year in the spring and, if the research trawl occurred in the summer, then the population of perch may consist of a large number of juveniles compared to a small number of fish from previous seasons. The reasoning being that the mortality risks for juveniles are greater as they tend to get eaten by larger fish of the same and other species. Hence, it would be reasonable from this argument to conclude that the mass of the perch population in Längelmävesi could be characterized by the mixture of two distributions: one distribution for juveniles and another distribution for adult fish. This would result in the appearance of the two modes in the lower histogram of Figure 6.7 and the classical stem-and-leaf plot of Figure 4.10.

Histograms are good graphical devices for detecting bimodality or multimodality, and a mixture of distributions with different locations and scales. This is the key advantage of the histogram over the boxplot. Given the close similarity among the histogram, the stacked dotplot, and the stemplot, it follows that the stacked dotplot and the stemplot are likewise superior to the boxplot in this regard as well.

A prudent caution in light of the perch data is not to rely on default settings in statistical software packages for determining histogram class widths. Explore the function for producing a histogram and the available options. If using variable class widths is an option, verify that the relative frequencies are correctly calculated. It is not unknown to find that for the ith class that $f_i^* = n_i/n$ has been computed and plotted instead of $f_i = n_i/(nw_i)$. This mistake in programming goes undetected if the class widths are all the same.

A remaining question is which of the rules for determining class widths for histograms represents the best estimate of the probability density function? This is a difficult question to answer. And this is why many statisticians have strong reservations about the use of the histogram. The histogram can conceal more than it reveals. This can cause problems if one uses the histogram as an exploratory tool and relies on software defaults. One recommendation is to plot and consider histograms produced by all the rules. Better yet, examine the data with an EDF plot or a normal quantile-quantile plot. The advantage of these two types of plots that deal with the empirical distribution is that any two researchers will likely produce similar plots as a result of the fact that the plotting conventions for these charts leave little room for artistic interpretation.

Note that the cumulative distribution function *(cdf)* represents the area underneath the probability density function *(pdf)*. The EDF plot is a graphical estimate of the cumulative density function, whereas the histogram is a graphical estimate of the probability density function.

Note that the EDF plot and the histogram are both produced without the estimation of any parameters in the population such as the mean and standard deviation in the case of a normal population. Thus, we say the EDF plot is a nonparametric estimate of a cumulative distribution function and the histogram is a nonparametric estimate of a probability density function. The EDF plot and histogram have discrete jumps associated with observations while the curves they estimate are smooth. We shall encounter in the next section another possible nonparametric density estimate of the probability density function which yields a smooth alternative.

6.3.2 A Circular Variation on the Histogram: The Rose Diagram

The number of births in Sweden in each month of 2004 is presented in a histogram in Figure 6.10. The three months in Figure 6.10 with the highest birth rates are March, April, and May. These months are consecutive. The four months in Figure 6.10 with the lowest birth rates are January, February, November, and December which are in the winter.

Month, hour, or direction are not measurements along a linear axis but rather along a circular axis. Because time is continuous, the months represent a convenient binning of the calendar year. It would be more truthful to display the birth data for Sweden in a manner that gives this impression. The *rose diagram* of Figure 6.11 does just this. Note that radial distance in Figure 6.11 is proportional to the square root of the number of births.

The first published use of the rose diagram was by Florence Nightingale [87] in 1858 on matters relating to the efficiency of the hospital administration in the British Army. Her rose diagrams were quite effective in demonstrating that the British soldier during the Crimean War had more to fear from disease than the enemy insofar as mortality rates were concerned.

It is sometimes erroneously claimed that Florence Nightingale referred to the rose diagram as a *coxcomb*. In fact, she used the term coxcomb to refer to her collection of rose diagrams. She did not use the term *polar area diagram* to refer to a rose diagram but this is also used. The term polar area diagram is quite accurate. Initially, Nightingale had her rose diagrams constructed with radial length proportional to frequency. But she soon realized her error. Her corrected rose diagrams were altered so that proportionality for sectors were based upon area. This is done by making the square root of radial distance proportional to frequency.

The rose diagram can perform well for data with striking differences in frequency as a function of angle. This is despite the absence of a common axis for

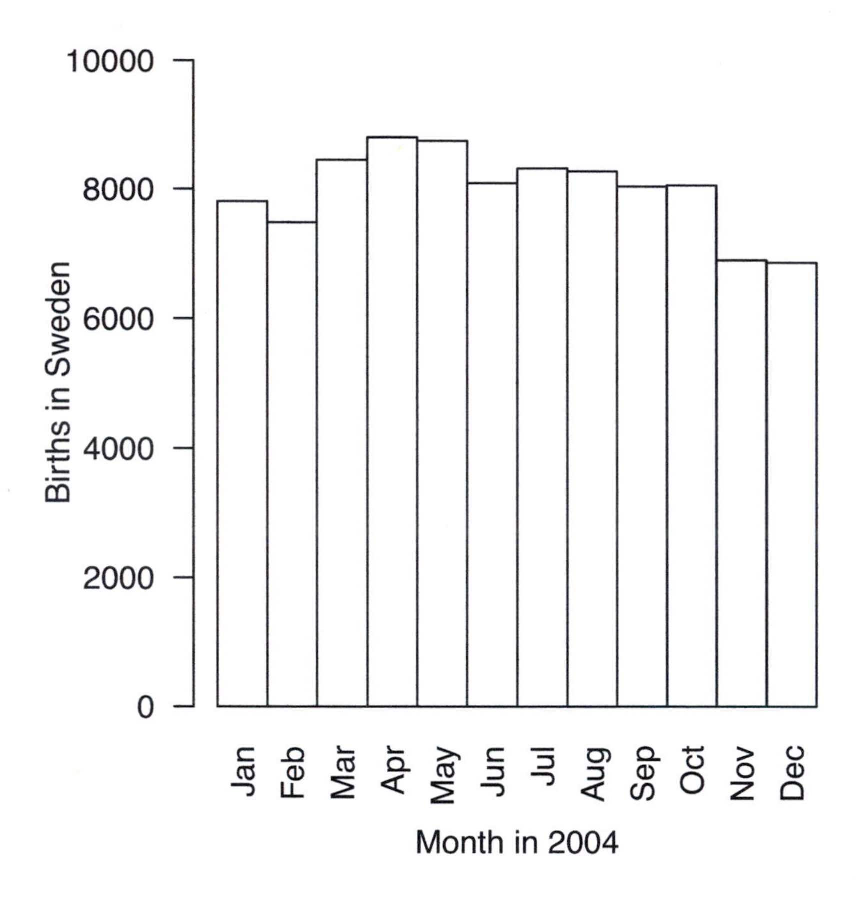

Figure 6.10 *Frequency histogram of births in Sweden in 2004*

comparison. It does take more effort to compare the areas of the segments in the rose diagram of Figure 6.11 than to compare the lengths of the bars in the histogram of Figure 6.10. It is possible to do quite well in comparing adjacent segments of a rose diagram but beyond that, comparisons are tougher to do.

This brings us to the comparison of the rose diagram in Figure 6.11 with the pie chart of Figure 6.12 for Swedish births. With pie charts there are two problems, as previously discussed: areal comparisons are more difficult than comparisons of length (or length of arc), and the presenter can never be sure which viewer is using which comparison method. There is also the possibility that a viewer will rely on comparison of sector angles to judge differences among sectors.

Regardless of whether arc length, sector angle, or area is used to gauge differences among the sectors, judging these differences in searching for a pattern is hard going for the monthly birth rates depicted in Figure 6.12. On the

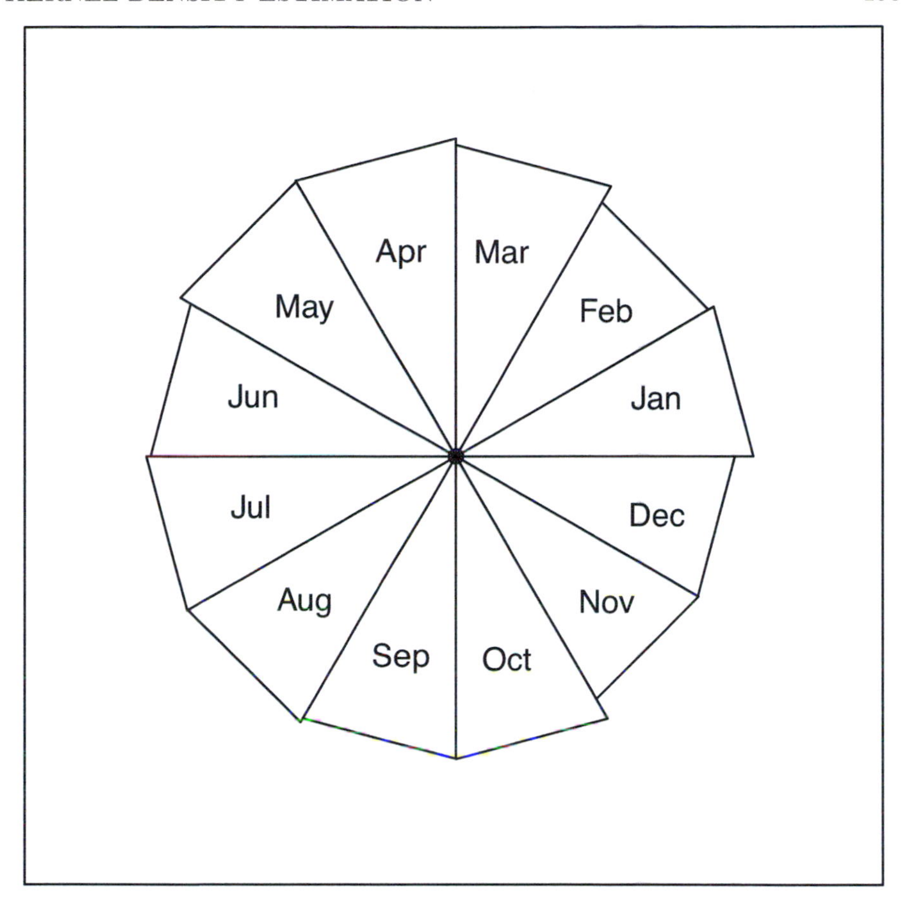

Figure 6.11 *Rose diagram of births in Sweden in 2004 (radial distance is proportional to the square root of the number of births)*

other hand, the pattern of low birth rates in the winter months of November through February, peaking with high birth rates in the spring months of March through May, is discernible in the rose diagram of Figure 6.11.

6.4 Kernel Density Estimation*

The histogram, the stemplot, and the stacked dotplot are examples of graphical displays that are produced by density estimation. Recall that a distribution of a continuous quantitative random variable can be represented by a probability density function. The probability that a random variable takes a value within some interval is given by the area underneath its probability density

*This section can be omitted without loss of continuity.

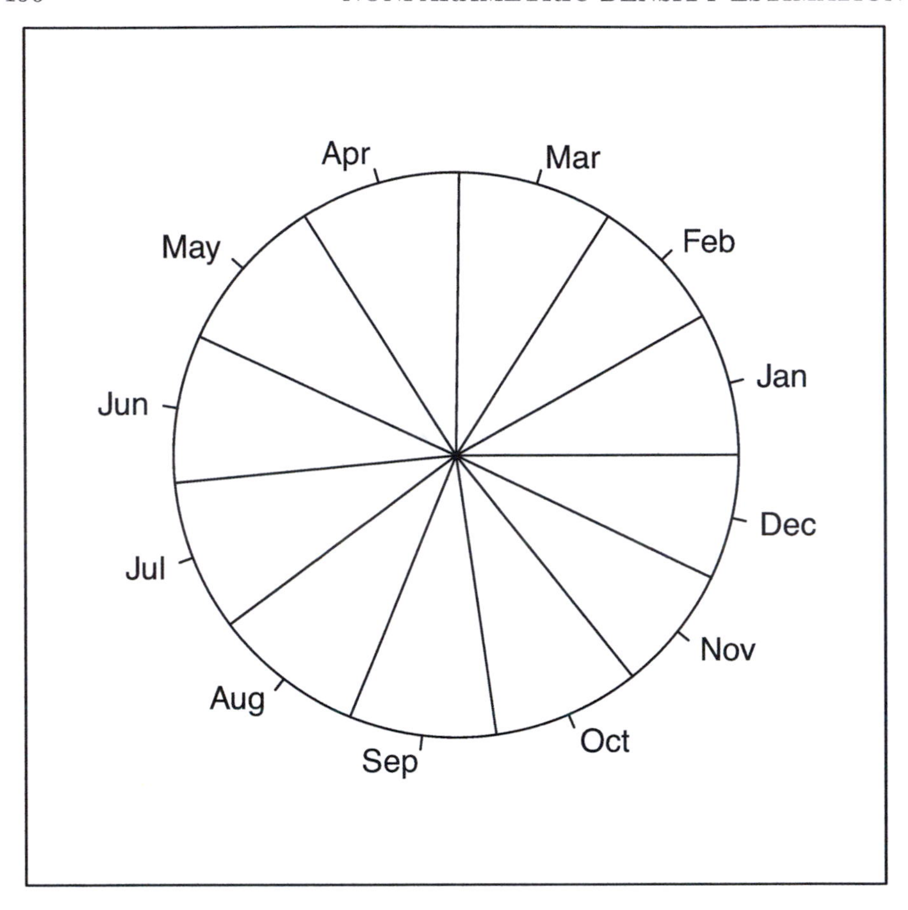

Figure 6.12 *Pie chart of births in Sweden in 2004*

function on the interval. The goal of density estimation is to generate an estimate of the true but unknown probability density function from a sample.

In essence, from a finite sample, the goal is to provide a point estimate for each of an infinite number of points—a very difficult task. The histogram represents one such density estimator. The histogram is not a particularly good estimator, however, because it does not vary smoothly as it is composed of horizontal line segments typically with jumps at regular intervals if the class widths are of equal length.

The kernel method is an example of nonparametric density estimation that is continuous and does not have discrete jumps. The kernel method essentially produces a smoothed version of a histogram from a function K by

$$\hat{f}_\lambda(x) = \frac{1}{n\lambda} \sum_{i=1}^{n} K\left(\frac{x - X_i}{\lambda}\right) \tag{6.13}$$

for observations $X_1, X_2, \ldots, X_n$ and a parameter λ known as the bandwidth. The function K is called the *kernel*, hence the name *kernel density estimation* for this form of density estimation.

The bandwidth λ is not a true parameter because its value can be chosen by the researcher for aesthetic reasons. In essence, kernel estimation produces a smoothed estimate of the probability density function by way of a sliding window that moves across the graph from left to right. The bandwidth λ represents the effective width of that window. Generally, with increasing sample size n, one reduces the bandwidth λ toward zero.

Commonly used kernels include the following.

Epanechnikov:

$$K_E(z) = \begin{cases} \frac{3}{4\sqrt{5}}\left(1 - \frac{1}{5}z^2\right) & \text{for } |z| \leq \sqrt{5}; \\ 0 & \text{otherwise.} \end{cases} \tag{6.14}$$

Biweight (or Quartic):

$$K_B(z) = \begin{cases} \frac{15}{16}(1 - z^2)^2 & \text{for } |z| \leq 1; \\ 0 & \text{otherwise.} \end{cases} \tag{6.15}$$

Triangular:

$$K_T(z) = \begin{cases} 1 - |z| & \text{for } |z| \leq 1; \\ 0 & \text{otherwise.} \end{cases} \tag{6.16}$$

Gaussian:

$$K_G(z) = \frac{1}{\sqrt{2\pi}}{}^{-z^2/2} \tag{6.17}$$

Rectangular (or Uniform):

$$K_R(z) = \begin{cases} \frac{1}{2} & \text{for } |z| \leq 1; \\ 0 & \text{otherwise.} \end{cases} \tag{6.18}$$

Cosine:

$$K_C(z) = \begin{cases} \frac{\pi}{4}\cos\left(\frac{\pi}{2}z\right) & \text{for } |z| \leq 1; \\ 0 & \text{otherwise.} \end{cases} \tag{6.19}$$

General properties that all kernel functions must satisfy are that they are unimodal, symmetric about zero, and contain a unit area above the z-axis. The functions listed above satisfy all these requirements and one other not necessarily required of kernel functions in general: nonnegativity. This last restriction is added because the definition of the probability density function stipulates nonnegativity.

Figure 6.13 depicts the Epanechnikov kernel density estimate, with each of its 56 constituent kernels, for perch mass. Figure 6.14 overlays the Epanechnikov kernel density estimate of Figure 6.13 on a histogram for perch mass for comparison. Note that for illustrative purposes the vertical scale of each

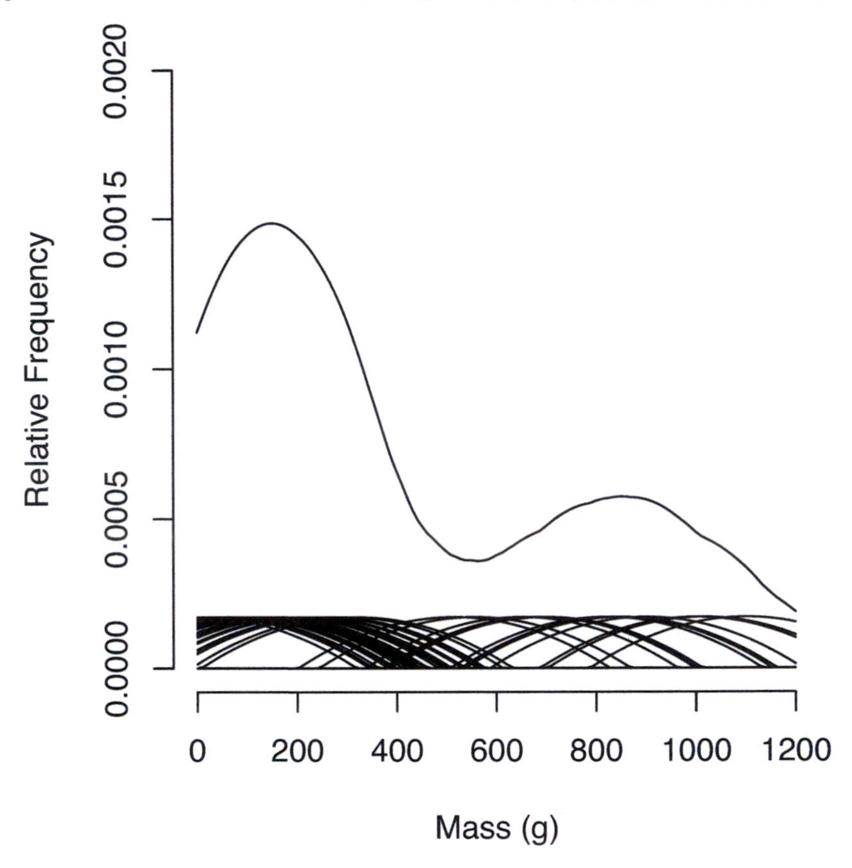

Figure 6.13 *Epanechnikov kernel density estimate for perch mass with constituent kernels (note that the vertical scale of each kernel has been exaggerated by a factor of four for the purposes of illustration)*

constituent kernel in Figure 6.13 has been exaggerated by a factor of four so that the kernels are more easily distinguished.

The R code for producing Figure 6.14 is as follows.

```
brks<-(0:12)*100

hist(mass,breaks=brks,freq=FALSE,xlim=c(0,1200),xaxt="n",
ylim=c(0,0.004),main=NULL,xlab="Mass (g)",
ylab="Relative Frequency")

axis(1,at=(0:6)*200,
labels=c("0","200","400","600","800","1000","1200"))
```

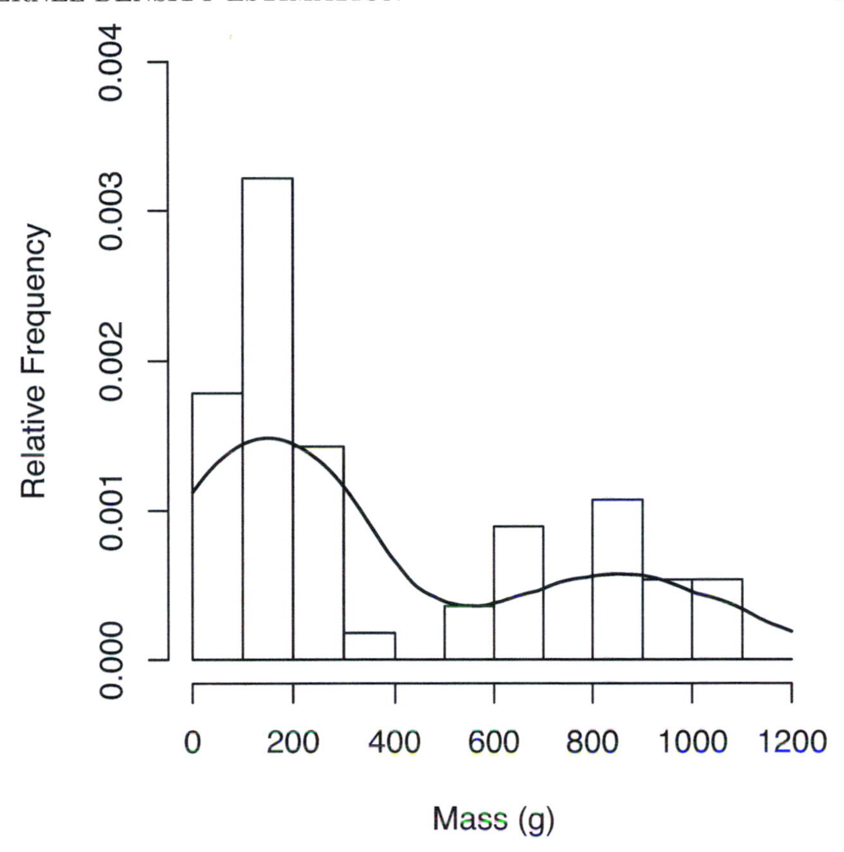

Figure 6.14 *Frequency histogram of perch mass with Epanechnikov kernel density estimate*

```
denmass<-density(mass,kernel="epanechnikov",from=0,to=1200)
lines(denmass$x,denmass$y,lwd=1.5)
```

The function `hist` produced the histogram in Figure 6.14 with class boundaries every 100 g. Setting `xaxt="n"` suppresses the plotting of the horizontal axis by `hist`. The R function `axis` was then used to add the horizontal axis with tick marks and labels every 200 g.

After the histogram is plotted, the R function `density` is executed with the Epanechnikov kernel from 0 g to 1,200 g and the result is stored in the R list `denmass`. The component `x` in this list stores the gridded mass and the component `y` stores the corresponding kernel estimate. The length of each of `denmass$x` and `denmass$y` as a result of the preceding R code is 512 elements. This is a consequence of the default in `density` that computes the density estimate at 512 equally spaced points. The argument `n=` can be used to set the

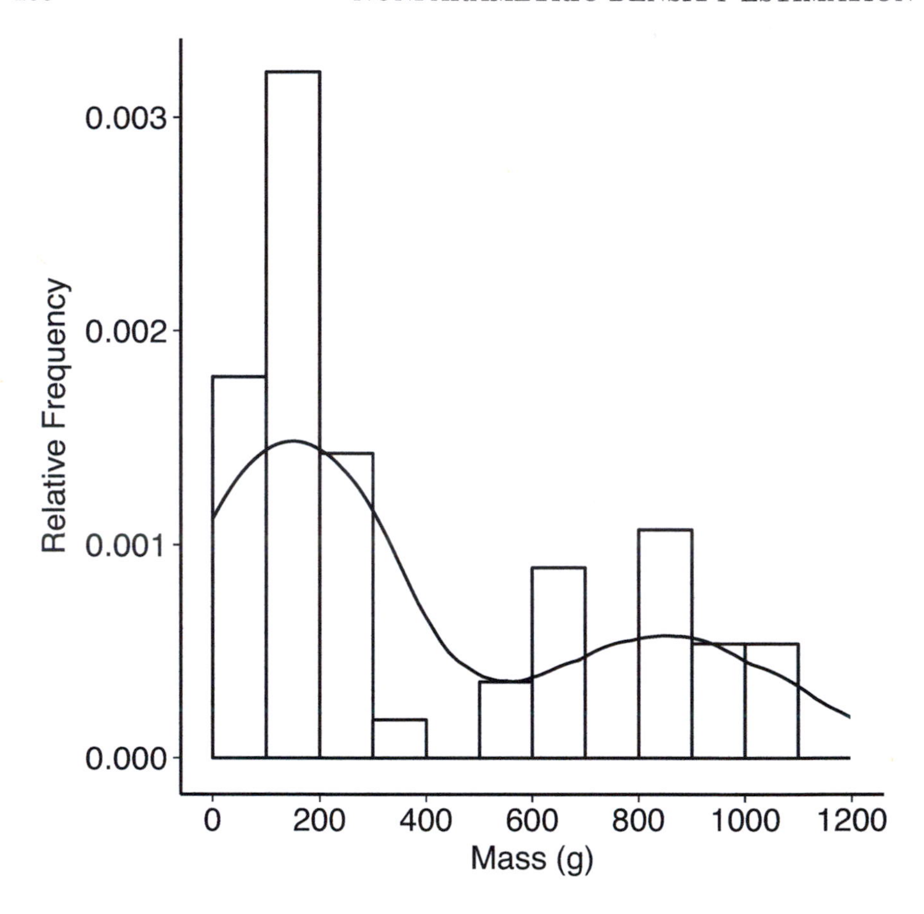

Figure 6.15 *Frequency histogram of perch mass with Epanechnikov kernel density estimate using* `ggplot2`

number of equally spaced points to a greater number. The function `density` rounds the number up to the next power of 2 for efficiency reasons.

The last function call in the preceding R script is to the function `lines`. This plots the kernel density estimate. Note that the argument `lwd=1.5` increases the width of the curve of the kernel density estimate by 50 percent compared to the width of the line segments in the histogram. This is done for the purpose of clarity.

Figure 6.15 is a `ggplot2` counterpart to Figure 6.14. The code for Figure 6.15 is as follows.

```
figure<-ggplot(perch, aes(mass,y=..density..)) +
geom_histogram(breaks=(0:11)*100,closed="right",
color="black",fill="white") +
```

```
geom_density(kernel="epanechnikov") +
theme_linedraw() + theme(panel.grid=element_blank(),
panel.border=element_rect(fill=NULL,
color="white",linetype="solid"),
axis.line=element_line(),
axis.title=element_text(size=12),
axis.text=element_text(size=12)) +
scale_x_continuous(breaks=200*(0:6),
limits=c(0,1200)) +
annotate("segment",x=1200,xend=1200,y=-0.00001,
yend=0.0003,color="white") +
labs(x="Mass (g)",y="Relative Frequency")

print(figure)
```

The ggplot2 function for plotting a kernel density estimate is geom_density, which internally calls the stats package function density to do the actual computations. Note that a request for the Epanechnikov kernel is passed by the parameter kernel in the call of geom_density. Note that the parameter y=..density.. is passed in the call to aes so that a relative frequency histogram is produced lest the kernel density estimate be indistinguishable from the horizontal axis in this example. To keep the ggplot2 code to a minimum yet still produce a crisp and simple visual, the line-draw theme in ggthemes is used to produce Figure 6.15. The line-draw theme is a reasonable compromise for producing a figure similar to the base graphics theme in Figure 6.14. Unfortunately, geom_density places bounding vertical line segments at the left and right edges. This is not a problem on the left where there is an edge of a histogram bar at 0 g. But the call to annotate prints a white line that covers completely the right black bounding line segment at a mass of 1200 g as a simple fix to take care of the problem.

A comparison of the Gaussian, Epanechnikov, and rectangular kernels is given in Figure 6.16 for perch mass. The other three kernels are not depicted because they are very similar to each other and the Gaussian kernel.

The terms Gaussian and normal are both used to describe the same distribution function:

$$f(x) = \frac{1}{\sigma\sqrt{2\pi}} e^{-\frac{(x-\mu)^2}{2\sigma^2}} \tag{6.20}$$

where the parameter μ is the mean and σ the standard deviation. The equation for the Gaussian kernel is that of a normal distribution with a mean of zero and a standard deviation of one. An alternative name for the normal distribution is the Gaussian distribution, in honor of the great mathematician Carl Gauss.

Silverman [110], on page 45 of his monograph on density estimation, recommended using the Gaussian window for normal distributions. The issue of choosing the bandwidth still remains in these situations.

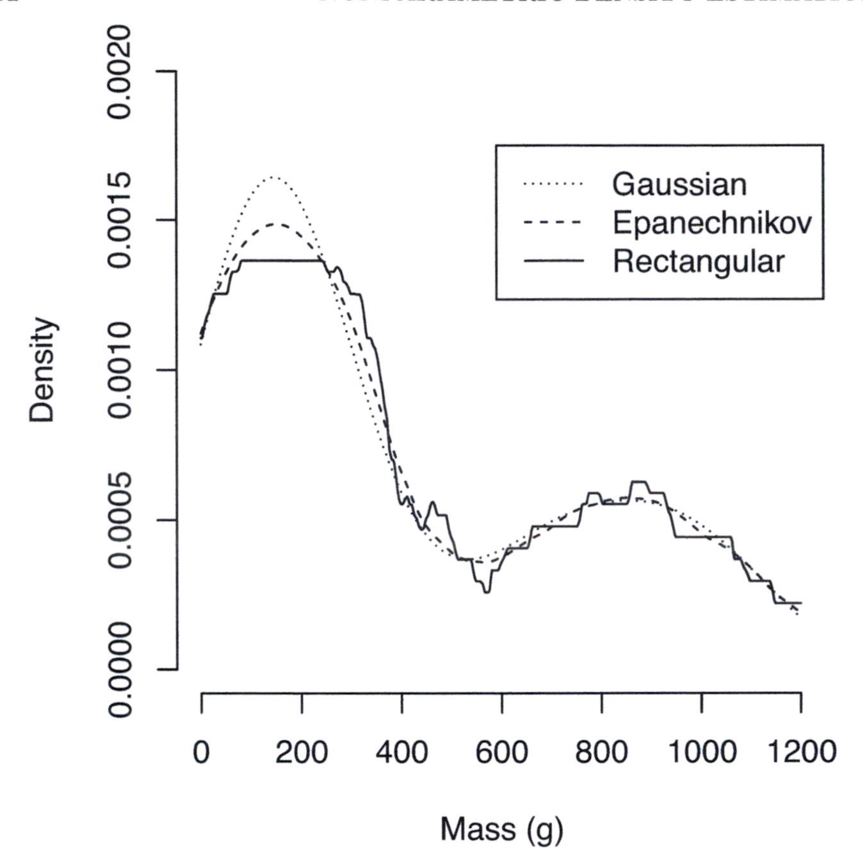

Figure 6.16 *Comparison of Gaussian, Epanechnikov, and rectangular kernel density estimates for perch mass*

Tukey [127] (page 623) advocated using the normal distribution as a reference standard for the shapes of distributions to which comparisons are to be made. Tukey also recommended using the normal distribution as a crude approximation to actual behavior. Tukey did not refer to the Gaussian distribution as the normal distribution because he felt the adjective normal to be misleading. But he found himself in a minority group in this regard among statisticians.

The problem of choosing a bandwidth for a random sample can be reduced to the optimization problem of finding the value of λ for a probability density f and an arbitrary kernel K that minimizes the mean integrated square error approximated by

$$AMISE = \frac{1}{n\lambda}\int [K(x)]^2\,dx + \frac{1}{4}\lambda^4\kappa_2^2\int [f''(x)]^2\,dx \qquad (6.21)$$

for a random sample of size n where

$$\kappa_2 = \int x^2 K(x) dx. \tag{6.22}$$

The optimal value of the bandwidth is

$$\lambda_{AMISE} = \left\{ \int \left[K(x)\right]^2 dx \right\}^{1/5} \left\{ n\kappa_2^2 \int \left[f''(x)\right]^2 dx \right\}^{-1/5}. \tag{6.23}$$

An unfortunate problem with this solution is that it depends on the probability density function f that is generally unknown. After all, the purpose of kernel density estimation is to estimate unknown f.

If the distribution is normal with standard deviation σ, then the optimal bandwidth for the Gaussian kernel is

$$\lambda_G = \left(\frac{4}{3}\right)^{1/5} \sigma n^{-1/5} \approx 1.06\sigma n^{-1/5}. \tag{6.24}$$

Because the population standard deviation σ is unknown, it is replaced by its estimate from the sample so that the estimate of optimal bandwidth becomes

$$\hat{\lambda}_{SNR} = 1.06 s n^{-1/5} \tag{6.25}$$

with the subscript letters SNR signifying a simple normal reference distribution.

A concern with the bandwidth estimate $\hat{\lambda}_G$ is that it is optimal for the normal distribution only. On the basis of extensive simulations with several different distributions, Silverman [110] recommended the following modification

$$\check{\lambda}_{SNR} = 0.9 s n^{-1/5}, \tag{6.26}$$

which has a small reduction in the multiplicative coefficient. Silverman [110] reported that this modification would be satisfactory for heavy-tailed symmetric distributions, skewed distributions, and bimodal distributions (such as the perch mass example).

A problem does remain with both bandwidth estimators $\hat{\lambda}_{SNR}$ and $\check{\lambda}_{SNR}$. Both estimators depend on the sample standard deviation s, which is known not to be robust. A robust alternative to using s is to note that the InterQuartile Range for a normal distribution is equivalent to 1.349 standard deviations and use $IQR/1.349$ instead of s in either formula for $\hat{\lambda}_{SNR}$ or $\check{\lambda}_{SNR}$. Silverman's [110] opinion is that this makes things worse for bimodal distributions by oversmoothing.

An example of an oversmoothed kernel density estimate is given in Figure 6.17 for the perch data. In this figure, the Epanechnikov kernel has been used with

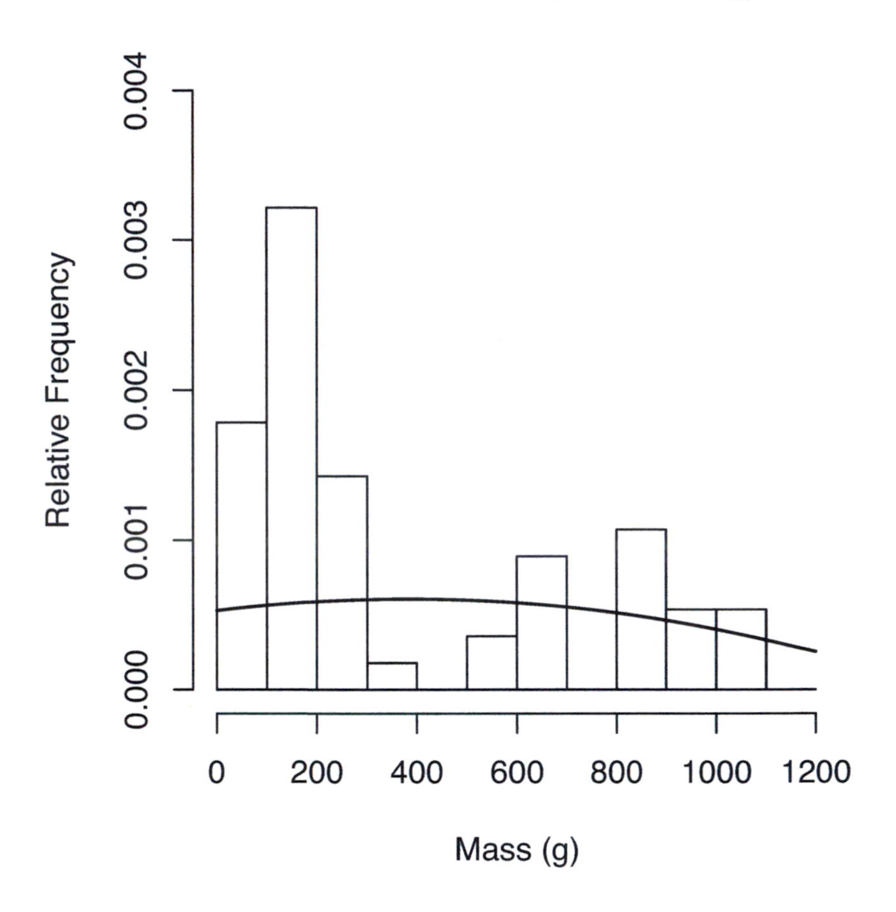

Figure 6.17 *Relative frequency histogram of perch mass with oversmoothed Epanechnikov kernel density estimate ($\lambda = 500$)*

the bandwidth λ arbitrarily set to 500. The result is a kernel density estimate with a single mode.

On the other hand, if the bandwidth is set too narrow, undersmoothing results. See Figure 6.18 for an example of an undersmoothed kernel density estimate with the bandwidth λ arbitrarily set to 40. The resulting kernel density estimate is not bimodal but instead is multimodal with no fewer than six modes.

There are a couple of technical considerations with respect to adjusting bandwidth in addition to artistic license. Although the details are omitted, narrow bandwidths produce nonparametric density estimates with low bias and high variance while broad bandwidths produce nonparametric density estimates with high bias and low variance. A heuristic justification for these considerations can be obtained from the examination of Figures 6.17 and 6.18.

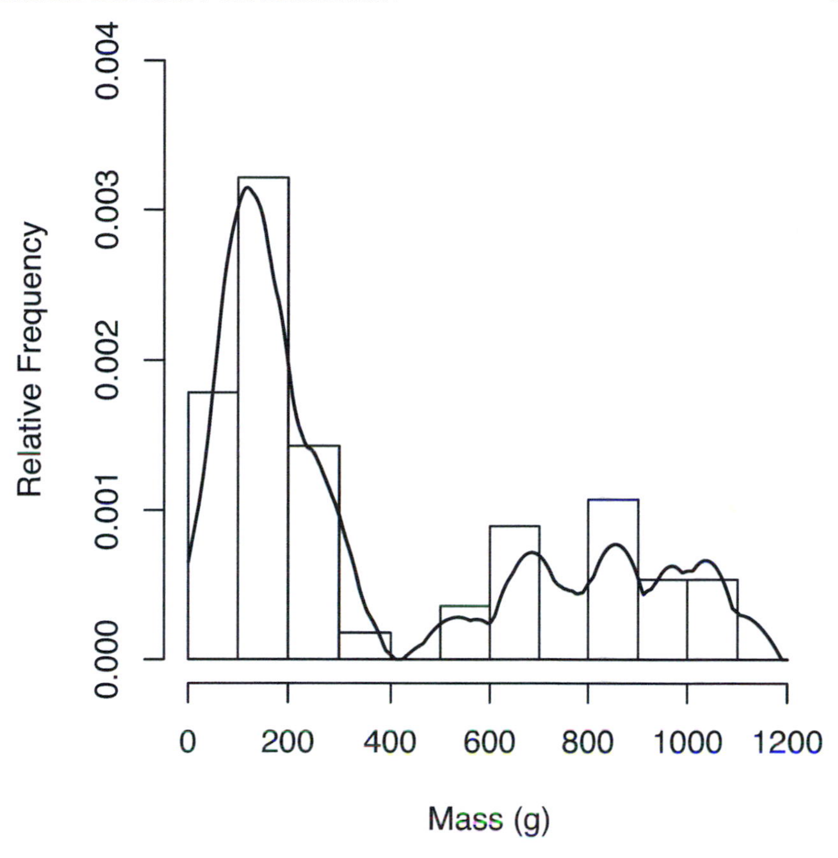

Figure 6.18 *Relative frequency histogram of perch mass with undersmoothed Epanechnikov kernel density estimate* ($\lambda = 40$)

Notice that the smooth curve for the broad bandwidth in Figure 6.17 can depart from the histogram according to the Rule of Twelve by a considerable distance for certain classes, whereas the jagged curve for the narrow bandwidth in Figure 6.18 fits the tops of the bars for the classes more closely but jumps about a fair bit. The jumpiness in Figure 6.18 is referred to as *noise*, a term borrowed from signal processing and time series analysis where kernel estimation is also employed in spectral frequency estimation. In an acoustic setting, noise implies unwanted variation.

Because of the trade-off between bias and variance, the usual criterion for choosing bandwidth is finding λ that minimizes the mean integrated square error. The bandwidth that approximately minimizes MISE for the observations of perch mass results in the smooth curve in Figure 6.14. The bandwidth estimate used to produce Figure 6.14 is the following rule of thumb due to

Silverman [110]:

$$\hat{\lambda}_{SROT} = 0.9An^{-1/5} \tag{6.27}$$

where

$$A = \min(s, IQR/1.349). \tag{6.28}$$

The subscript letters $SROT$ signify *Silverman's Rule of Thumb.* Note that Silverman [110] did not use the divisor of 1.349 in the preceding formula but instead chose to round this number to 1.34. Other authors instead round 1.349 to 1.35 so all three versions of the divisor are found in the literature.

Terrell [121] deliberately produced an oversmoothed density estimate:

$$\hat{\lambda}_{OS} = 1.144sn^{-1/5}. \tag{6.29}$$

This was developed from the maximum smoothing principle of Terrell and Scott [122]. The basic idea is to choose the largest degree of smoothing compatible with estimated scale of the density. Taking the variance σ^2 as the scale parameter, Terrell [121] found that the beta(4,4) family of distributions with variance σ^2 minimizes

$$\int [f''(x)]^2 \, dx. \tag{6.30}$$

For the Gaussian kernel, this leads to the oversmooth bandwidth estimator $\hat{\lambda}_{OS}$.

Another method for determining optimal band width is based on the integrated square error of the kernel density estimator $\hat{f}_\lambda$ given by

$$ISE = \int \left[\hat{f}_\lambda(x) - f(x)\right]^2 dx. \tag{6.31}$$

Expansion of the integrand in the above formula leads to

$$\begin{aligned} ISE \quad = \quad & \int \left[\hat{f}_\lambda(x)\right]^2 dx - 2\int \hat{f}_\lambda(x) f(x) dx \\ & + \int [f(x)]^2 \, dx. \end{aligned} \tag{6.32}$$

Observing that the last term on the right of the above formula is not a function of the bandwidth λ suggests that it is necessary only to optimize

$$LSCV = \int \left[\hat{f}_\lambda(x)\right]^2 dx - 2\int \hat{f}_\lambda(x) f(x) dx, \tag{6.33}$$

which is referred to as the *least-squares cross-validation* ($LSCV$) criterion.

Hall [57] derived the following approximation

$$LSCV \approx \int \left[\hat{f}_\lambda(x)\right]^2 dx - \frac{2}{n}\sum_{i=1}^{n} \hat{f}_{-i}(x_i) \tag{6.34}$$

where $\hat{f}_{-i}$ is the leave-one-out density estimate.

Least-squares cross-validation is also referred to as *unbiased cross-validation* (UCV) because

$$E\left\{LSCV + \int [f(x)]^2\,dx\right\} = MISE. \tag{6.35}$$

Consequently, the bandwidth λ that minimizes the $LSCV$ approximation of Hall [57] in 1983 is denoted by $\hat{\lambda}_{UCV}$. There is no closed formula for this estimate of bandwidth, which is found by numerical optimization. Alternatively, one can plot Hall's [57] $LSCV$ function and visually search for the bandwidth that minimizes $LSCV$.

Figure 6.19 depicts the $LSCV$ function versus bandwidth for perch mass. By visual inspection of Figure 6.19, the optimal value $\hat{\lambda}_{UCV}$ is nearly 30 (g). Sheather [108] recommended that in practice it is prudent to plot $LSCV(\lambda)$ and not just rely on the result of a minimization routine. This is sage advice.

The R function `bw.ucv` in turn relies on a call to the R function `optimize` to find the value $\hat{\lambda}_{UCV}$ that minimizes the $LSCV$. The R function `optimize` is not pushed to its limits in this example of perch mass as the $LSCV$ function in Figure 6.19 is quite smooth with a single minimum value. The R function `bw.ucv` yields $\hat{\lambda}_{UCV} = 34.33$ g.

The $LSCV$ function can have more than one local maximum. In this situation, Wand and Jones [131] repeat the suggestion of Marron [82] to select as $\hat{\lambda}_{UCV}$ the value that corresponds to the largest local minimum. Alternatively, one could plot the kernel density estimates for all the local minima and then make a choice regarding which value to select as $\hat{\lambda}_{UCV}$.

Five years after unbiased cross-validation was proposed, Scott and Terrell [107] introduced a method called *biased cross-validation* (BCV) that chooses a bandwidth $\hat{\lambda}_{BCV}$ that minimizes an approximation to the mean integrated square error (formula 6.35) in which

$$\int [f''(x)]^2\,dx \approx \int \left[\hat{f}''_{\lambda}(x)\right]^2 dx - \frac{1}{n\lambda^5}\int [K''(x)]^2\,dx \tag{6.36}$$

where $\hat{f}''_{\lambda}$ is the second derivative of the kernel density estimate. The quantity being subtracted in the preceding approximation is a bias term.

There is no closed formula for biased cross-validation bandwidth estimate $\hat{\lambda}_{BCV}$. This is a consequence of using cross-validation to approximate both terms in formula (6.35). In this instance *leave-one-out cross-validation* is performed for a random sample of size n whereby the estimate is calculated n times for each subset of the sample with one observation deleted. The n estimates are then averaged.

Leave-one-out cross-validation is also known as *jackknifing*, which falls under

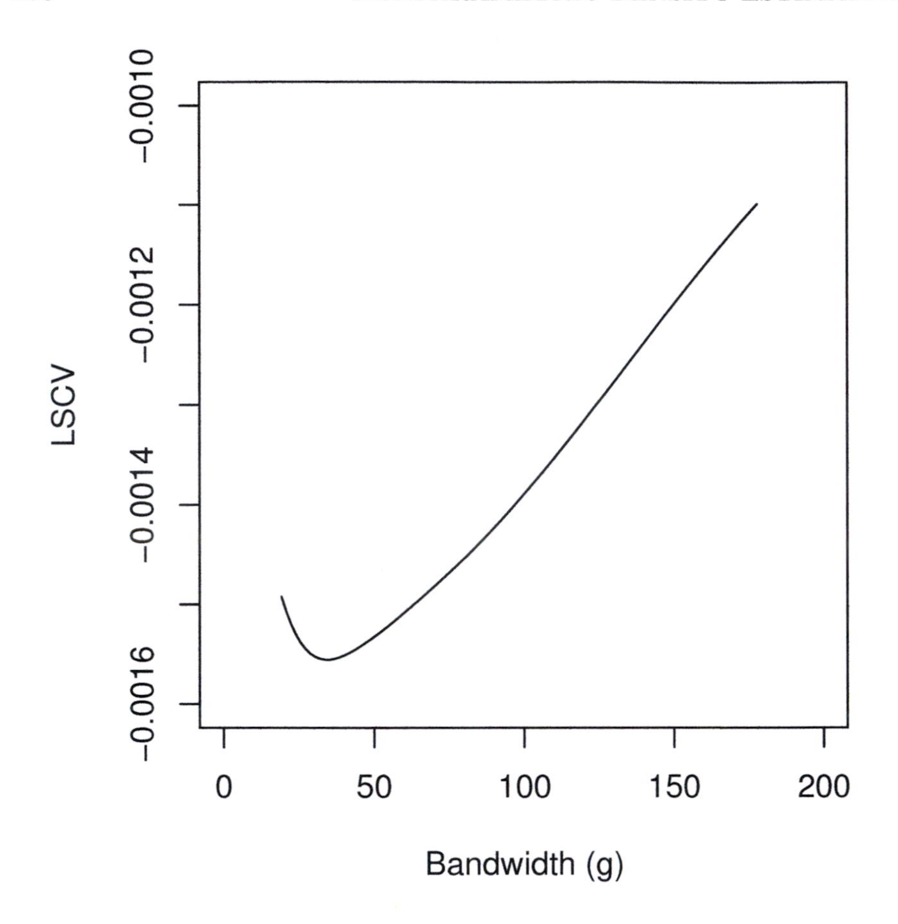

Figure 6.19 *Plot of LSCV versus bandwidth for perch mass*

the general rubric of *bootstrap methods* although it is a much older technique than any of these.

A numerical algorithm can be used to find the value $\hat{\lambda}_{BCV}$ that minimizes the cross-validated version of the right side of

$$
\begin{aligned}
AMISE \approx & \frac{1}{n\lambda}\int [K(x)]^2\, dx \\
& + \frac{1}{4}\lambda^4\kappa_2^2 \left\{ \int \left[\hat{f}_\lambda''(x)\right]^2 dx - \frac{1}{n\lambda^5}\int [K''(x)]^2\, dx \right\}.
\end{aligned}
\tag{6.37}
$$

Examining a plot of $AMISE$ versus bandwidth to determine $\hat{\lambda}_{BCV}$ is also

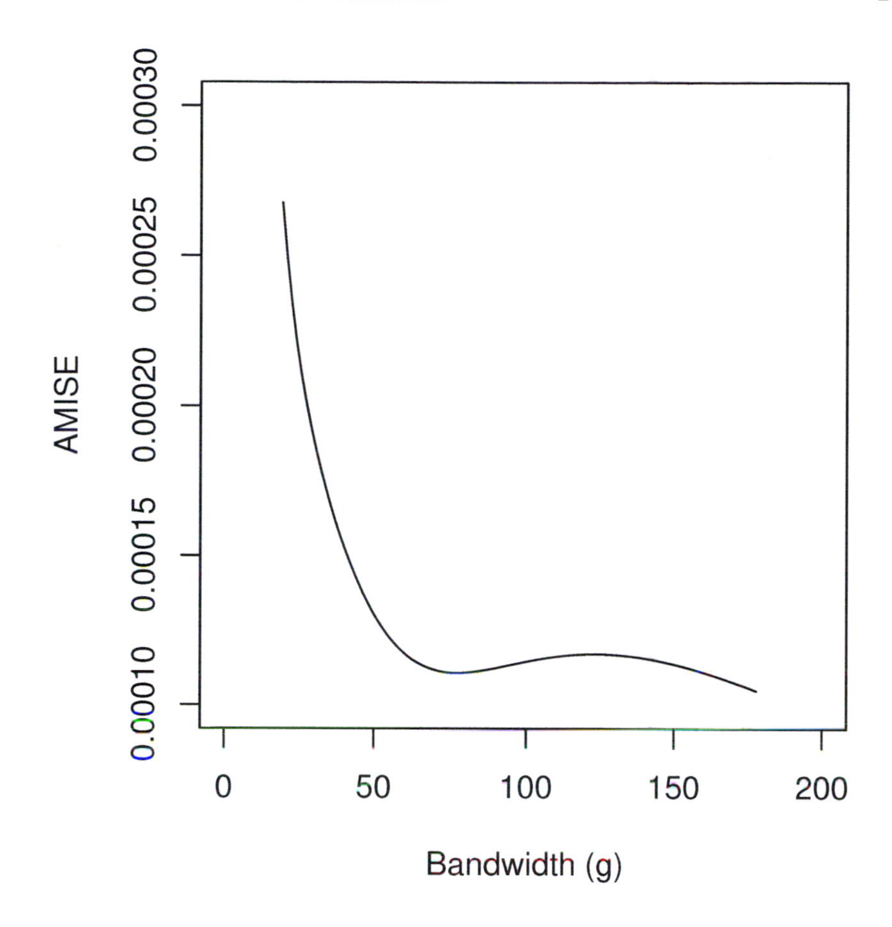

Figure 6.20 *Plot of AMISE versus bandwidth for perch mass*

worthwhile. Figure 6.20 provides one such plot for the mass of perch caught in Finland.

In 1992, Scott [106] recommended that $\hat{\lambda}_{BCV}$ be the largest minimizer less than or equal to the oversmoothed bandwidth $\hat{\lambda}_{OS}$ of Terrell [121]. On the other hand, Marron [82] in 1993 cited empirical experience in recommending the smallest local minimizer. Based upon empirical performance, Jones, Marron and Sheather [73] in 1996 recommended selecting as $\hat{\lambda}_{BCV}$ the smallest local minimizer rather than the global minimizer. An examination of Figure 6.20 will reveal that the value $\hat{\lambda}_{BCV} = 77.29$ g satisfies the criterion of being the largest local minimizer. The global minimum at the extreme right end of the curve corresponds to $\hat{\lambda}_{OS} = 177.78$ g.

The R function `bw.bcv` was used to find the value $\hat{\lambda}_{BCV} = 77.29$ g. This R function also calls the R function `optimize` to produce $\hat{\lambda}_{BCV}$. In seeking the minimum value, both R functions `bw.ucv` and `bw.bcv` restrict the search to bandwidths on the interval $[\hat{\lambda}_{OS}/10, \hat{\lambda}_{OS}]$. For the Finnish perch data, $\hat{\lambda}_{OS} = 177.78$ g, hence the range for the horizontal axis from 0 g to 200 g in each of Figures 6.19 and 6.20.

It is worthwhile to theoretically consider how the bandwidth estimates from unbiased and biased cross-validation compare. Scott and Terrell [107] showed that the ratio of the two asymptotic variances for the Gaussian kernel is

$$\frac{\text{var}\left(\hat{\lambda}_{UCV}\right)}{\text{var}\left(\hat{\lambda}_{BCV}\right)} \approx 15.7. \tag{6.38}$$

So at the cost of adding a little bias, a large reduction in variability of the bandwidth estimate is obtained when biased cross-validation is selected over unbiased cross-validation.

A problem with both the biased and unbiased cross-validation approaches to determining bandwidth is their slow rate of convergence. Scott and Terrell [107] in 1987 showed that

$$n^{1/10}\left(\frac{\hat{\lambda}_{UCV}}{\lambda_G} - 1\right) \tag{6.39}$$

has an asymptotic $N\left(0, \sigma^2_{UCV}\right)$ distribution assuming the data and kernel are both Gaussian. This fact was also established in 1987 by Hall and Marron [58]. Scott and Terrell [107] also showed under the same assumptions that

$$n^{1/10}\left(\frac{\hat{\lambda}_{UCV}}{\lambda_G} - 1\right) \tag{6.40}$$

has an asymptotic $N\left(0, \sigma^2_{UCV}\right)$ distribution.

In 1991, Sheather and Jones [109] proposed a new estimator of bandwidth that minimizes MISE with a better rate of convergence. Their approach can be characterized as *smoothed cross-validation* (SCV). Let ϕ denote the standard normal density

$$\phi(x) = \frac{1}{\sqrt{2\pi}} e^{-\frac{x^2}{2}} \tag{6.41}$$

and $\phi^{(n)}(x)$ its n-th order derivative with respect to x. The Sheather and Jones [109] algorithm begins with two estimates of bandwidth

$$\hat{\alpha} = 0.920\,[IQR]\,n^{-1/7} \tag{6.42}$$

and

$$\hat{\beta} = 0.912\,[IQR]\,n^{-1/9} \tag{6.43}$$

that are used in kernel-based density estimators

$$\hat{S}(\hat{\alpha}) = \frac{1}{n(n-1)\hat{\alpha}^5} \sum_{i=1}^{n} \sum_{j=1}^{n} \phi^{(4)} \left(\frac{x_i - x_j}{\hat{\alpha}} \right) \tag{6.44}$$

of

$$\int \left[f''(x) \right]^2 dx \tag{6.45}$$

and

$$\hat{T}(\hat{\beta}) = -\frac{1}{n(n-1)\hat{\beta}^7} \sum_{i=1}^{n} \sum_{j=1}^{n} \phi^{(6)} \left(\frac{x_i - x_j}{\hat{\beta}} \right) \tag{6.46}$$

of

$$\int \left[f'''(x) \right]^2 dx, \tag{6.47}$$

respectively. The Sheather-Jones *solve-the-equation* bandwidth $\hat{\lambda}_{SJ}$ is the numerical solution of equation (6.23), that is,

$$\left\{ \int \left[K(x) \right]^2 dx \right\}^{1/5} \left\{ n\kappa_2^2 \hat{S} \left[\hat{\alpha}_1(\lambda) \right] \right\}^{-1/5} - \lambda = 0 \tag{6.48}$$

where

$$\hat{\alpha}_1(\lambda) = 1.357 \left[\frac{\lambda^5 \hat{S}(\hat{\alpha})}{\hat{T}(\hat{\beta})} \right]^{1/7}. \tag{6.49}$$

For a suitably smooth probability density function,

$$n^{5/14} \left(\frac{\hat{\lambda}_{SJ}}{\lambda_G} - 1 \right) \tag{6.50}$$

has an asymptotic $N\left(0, \sigma_{SJ}^2\right)$ distribution. Thus the Sheather-Jones bandwidth has a much higher rate of convergence than either of the two cross-validation methods but at the cost of more smoothness of probability density function.

The Sheather-Jones bandwidth estimator $\hat{\lambda}_{SJ}$ is implemented in the R statistical software package by the function `bw.SJ` for the Gaussian kernel only. There are a couple of minor differences in how R computes this bandwidth estimator. Instead of the pilot bandwidth estimates $\hat{\alpha}$ and $\hat{\alpha}$, the function `bw.SJ` uses without justification the following albeit reasonable alternatives

$$a = 1.24An^{-1/7} \tag{6.51}$$

and

$$b = 1.23An^{-1/9}, \tag{6.52}$$

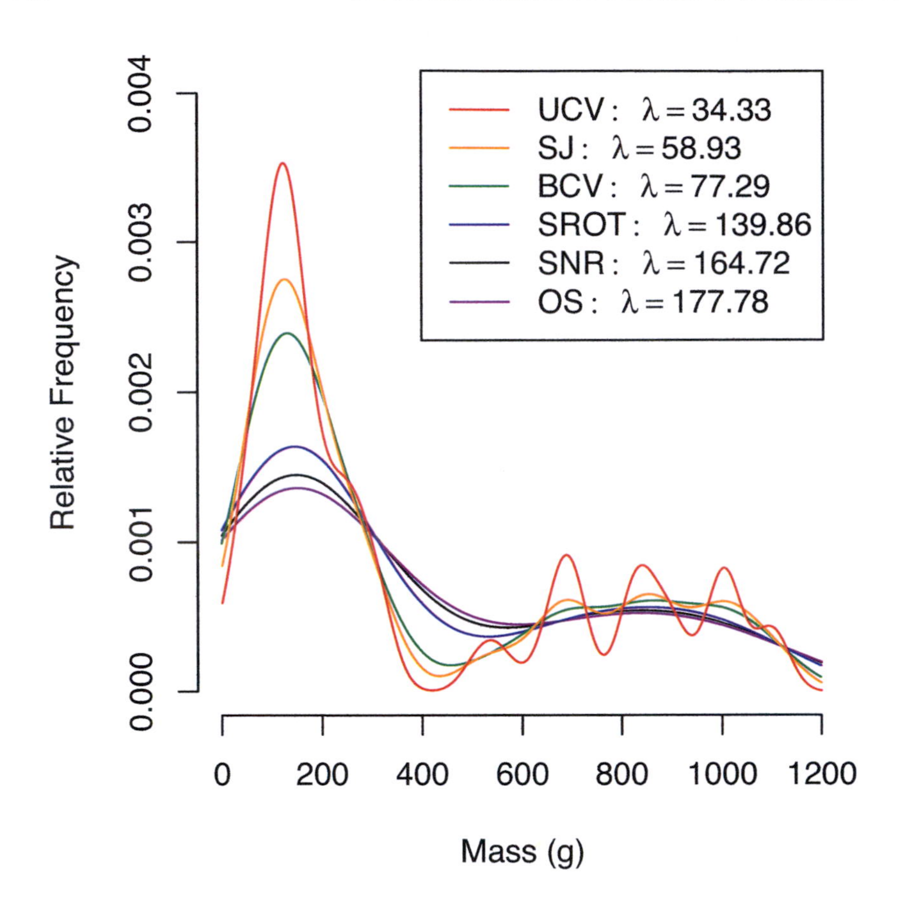

Figure 6.21 *Gaussian kernel density estimates of perch mass with different bandwidths*

respectively. The function also provides the following *direct plug-in* alternative to $\hat{\lambda}_{SJ}$:

$$\tilde{\lambda}_{SJ} = \left[\frac{1}{2\sqrt{\pi}\hat{S}(\hat{\alpha}_2)}\right]^{1/5} \tag{6.53}$$

where

$$\hat{\alpha}_2 = \left[\frac{2.394}{n\hat{T}(b)}\right]^{1/7}. \tag{6.54}$$

For perch mass, the Sheather-Jones direct plug-in estimate $\tilde{\lambda}_{SJ}$ is $= 76.67$ g and the Sheather-Jones direct solve-the-equation estimate $\hat{\lambda}_{SJ}$ is $= 58.93$ g.

As noted by Sheather [108] in 2004, numerous authors recommend that more than one density estimate be calculated. See Table 6.3 for a comparison of the various bandwidth estimates for the mass of Finnish perch. These bandwidths

Method	Symbol	Estimate
Simple Normal Reference	$\check{\lambda}_{SNR}$	164.72
Silverman's Rule of Thumb	$\hat{\lambda}_{SROT}$	139.86
Oversmoothed	$\hat{\lambda}_{OS}$	177.78
Unbiased Cross-Validation	$\hat{\lambda}_{UCV}$	34.33
Biased Cross-Validation	$\hat{\lambda}_{BCV}$	77.29
Sheather-Jones direct-plug-in	$\tilde{\lambda}_{SJ}$	76.67
Sheather-Jones solve-the-equation	$\hat{\lambda}_{SJ}$	58.93

Table 6.3 *Bandwidth estimates for kernel density estimation of perch mass*

are used with the Gaussian kernel and applied to the perch mass data in Figure 6.21. The choice of kernel as Gaussian was a consequence of using the R function `bw.SJ` to calculate the bandwidth estimates by the approach of Sheather and Jones [109]. The kernel density estimate with the Sheather-Jones direct plug-in bandwidth estimate of 76.67 g does not appear in Figure 6.21 because of its close proximity to the kernel density estimate with the biased cross-validation bandwidth estimate of 77.29 g.

The two shortest bandwidths in Table 6.3 lead to multimodality in the kernel density estimates for perch mass. This is not desirable if the data truly do follow a bimodal distribution.

The bandwidth by the biased cross-validation method results in a slight ripple in the kernel density estimate near 700 g, but the overall impression from the Gaussian kernel density estimate in Figure 6.21 with this bandwidth of 77.29 g is one of bimodality.

The largest bandwidth estimates of 139.86 g, 164.72 g, and 177.78 g are by the older first-generation methods: Silverman's rule of thumb (SROT), the standard normal reference (SNR) rule of thumb, and the oversmoothing (OS) rule of thumb, respectively. It needs to be noted that the unbiased cross-validation method that selects the smallest local minimizer also selects the oversmoothed estimate of bandwidth. The corresponding Gaussian kernel density estimates for these bandwidth estimates in Figure 6.21 are bimodal and quite similar.

It would be imprudent to make a recommendation regarding the choice of bandwidth based upon a single data set, in this case a collection of measurements on 56 perch captured in a research trawl on the Finnish lake Längelmävesi around 1917. Sheather's [108] review article of 2004 recommends plotting a family of kernel density estimates for several values of bandwidth, as done in Figure 6.21.

Jones, Marron, and Sheather [73] in their review article of 1996 summarize a major simulation study involving 15 normal mixture densities with sample sizes of 100 and 1,000. They found that the distribution for $\hat{\lambda}_{SROT}$ "had a mean that was usually unacceptably large." But they commented that "its variance was usually much smaller than for the other methods."

Jones, Marron, and Sheather [73] felt that the distribution for $\hat{\lambda}_{UCV}$ "was centered correctly but was unacceptably spread out." On the other hand, $\hat{\lambda}_{BCV}$ was considered erratic because its mean was too large and it was too variable, although $\hat{\lambda}_{UCV}$ was less variable than $\hat{\lambda}_{UCV}$ in their simulations.

Of the bandwidth estimators here, Jones, Marron, and Sheather [73] felt that $\hat{\lambda}_{SJ}$ was a useful compromise between $\hat{\lambda}_{UCV}$ and $\hat{\lambda}_{SROT}$.

Wand and Jones [131] in 1995 recommended versions of the direct plug-in and solve-the-equation bandwidth estimators by smoothed cross-validation rather than unbiased or biased cross-validation.

Sheather [108] recommended the Sheather-Jones plug-in bandwidth estimator $\hat{\lambda}_{SJ}$ and the unbiased cross-validation bandwidth $\hat{\lambda}_{UCV}$. His reasons were that the Sheather-Jones estimator tends to oversmooth hard-to-estimate densities for which $|f''_{\lambda}(x)|$ varies widely as a function of x while the unbiased cross-validation estimator tends to undersmooth in this situation.

Sheather [108] also recommended plotting the $LSCV$ of equation (6.34) and not relying on the result of the optimized estimate from computer software. This function is plotted for the perch mass in Figure 6.19. The approximation for the $AMISE$ given in equation (6.37), which is needed for determining the biased cross-validation estimate, is plotted for perch mass in Figure 6.20.

In Figure 6.21, the Sheather-Jones estimate leads to a ripple of three modes near the minor mode. On the assumption that perch mass is bimodal, the unbiased cross-validation bandwidth based on the largest local minimizer leads to a more acceptable result with two modes for the Gaussian kernel density estimate. But the result in this example is not as smooth as the density estimates produced in this figure by oversmoothing (which in this example produces a bandwidth estimate that coincides with that of unbiased cross-validation estimate based on the smallest local minimizer), the simple normal reference, and Silverman's [110] rule of thumb.

Wand and Jones [131] note that in many situations choosing the bandwidth by eye after looking at several density estimates over a range of bandwidths is satisfactory. One strategy they discuss is to begin with a large bandwidth and decrease the amount of smoothing until fluctuations that are more random than structural start to appear. In Figure 6.21, it would seem that the nonoptimal methods of oversmoothing, simple normal reference, and Silverman's [110] rule of thumb produce the most pleasing bandwidth estimates for perch mass. It just so happens, however, for the example of perch mass that the nonoptimal oversmoothing estimate is equivalent to the optimal estimate of bandwidth by the unbiased cross-validation method that selects the smallest local minimizer.

In arguing in favor of an objective method for selecting bandwidth, Wand and Jones [131] note two reasons. One reason being that subjective selection of bandwidth for a kernel density estimator can be time consuming. The other reason being that in many instances there is no prior knowledge about the shape of the distribution of the observations. Thus, they argue that there is justification for automatic bandwidth selection. But on the basis of the solitary example of perch mass, it is hard to support this viewpoint.

The decision to use a kernel density estimator is not without concerns. To summarize, disadvantages associated with a kernel density estimator include:

- choice of kernel;
- choice of bandwidth;
- depending on the choice of kernel, the smoothed histogram might have negative values;
- the area underneath the smoothed histogram might not equal 1.

The kernel estimator does offer advantages. If variable histogram class widths are not an option for the statistical software package of your choice, it is likely that a kernel density estimator is available. If shielded from complexity of the kernel density estimation process, members of the public have no difficulty in examining histograms with smooth curves, such as in Figure 6.22. This figure has been drafted with a Gaussian kernel and Silverman's rule of thumb for the bandwidth of 139.86 g, which are the default settings in the R function `density`. Figure 6.23 provides a `ggplot2` counterpart to Figure 6.22 for comparison. The kernel density estimate in both of these figures is put to good use in artistically bringing to the attention of viewers an opinion that the distribution of mass is bimodal. An older approach to adding a curve to a histogram, but based upon a tool for drafting smooth curves, is given in the next section.

Before proceeding to the next section, an alternative to the nonparametric density estimates of the histogram and Gaussian kernel density estimate of Figure 6.22 is given in Figure 6.24. The *violin plot* in Figure 6.24 is produced by placing an outlier boxplot between two kernel density estimates.

The violin plot was created in 1998 by Jerry Hintze and Ray Nelson [65]. Their original plotting convention considers only the use of the rectangular kernel. Hintze and Nelson [65] refer to the rectangular kernel density estimate not by that name, but the term *density trace*. The finite extent of the rectangular kernel is responsible for the jagged appearance in the upper half of the violin in Figure 6.24 compared to the smooth appearance of the Gaussian density estimate of Figure 6.22.

The R code, in part responsible for the violin plot of Figure 6.24, is as follows.

```
massf<-data.frame(mass=mass,type=rep(" ",length(mass)))
```

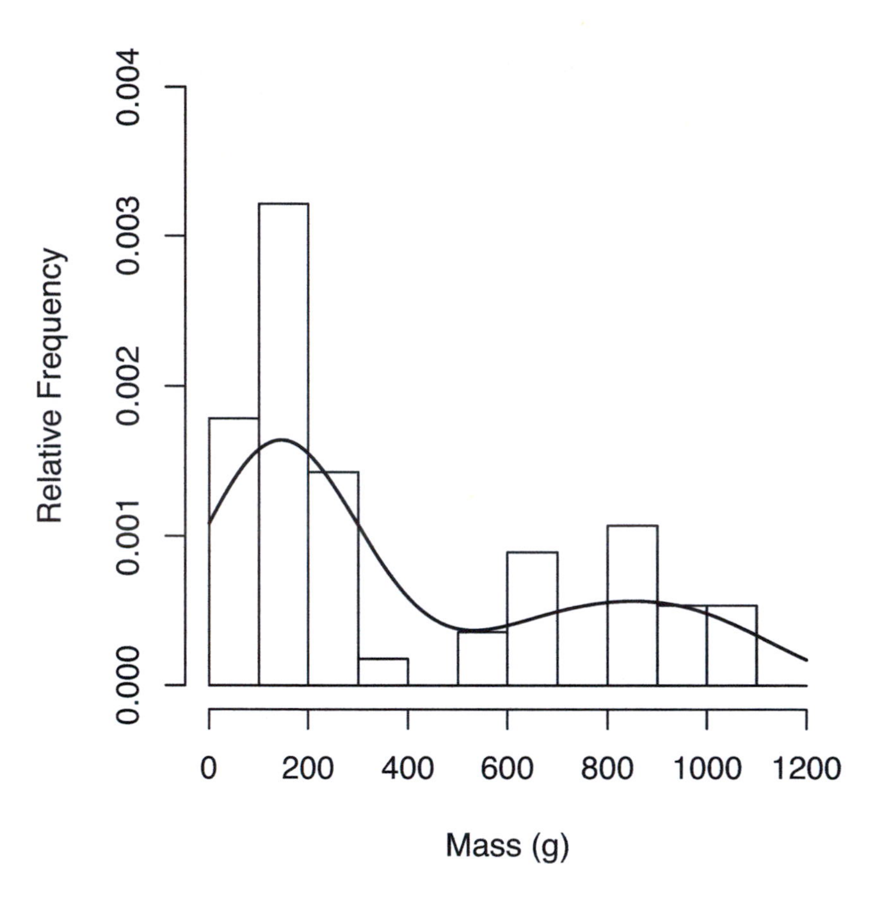

Figure 6.22 *Histogram and Gaussian kernel density estimates of perch mass with bandwidth 139.86 g by Silverman's rule of thumb*

```
trellis.par.set(box.rectangle=list(col="black"),
box.umbrella=list(col="black",lty=1))

figure<-bwplot(mass~type,massf,ylab="Mass (kg)",
xlab=" ",horizontal=FALSE,
scales=list(y=list(limits=c(0,1200),at=(0:6)*200,axs="i")),
panel=function(...,box.ratio) {
panel.violin(...,col="transparent",varwidth=FALSE,
box.ratio=box.ratio,kernel="rectangular")
panel.bwplot(...,fill="black",box.ratio=.025,col="white")})

print(figure)
```

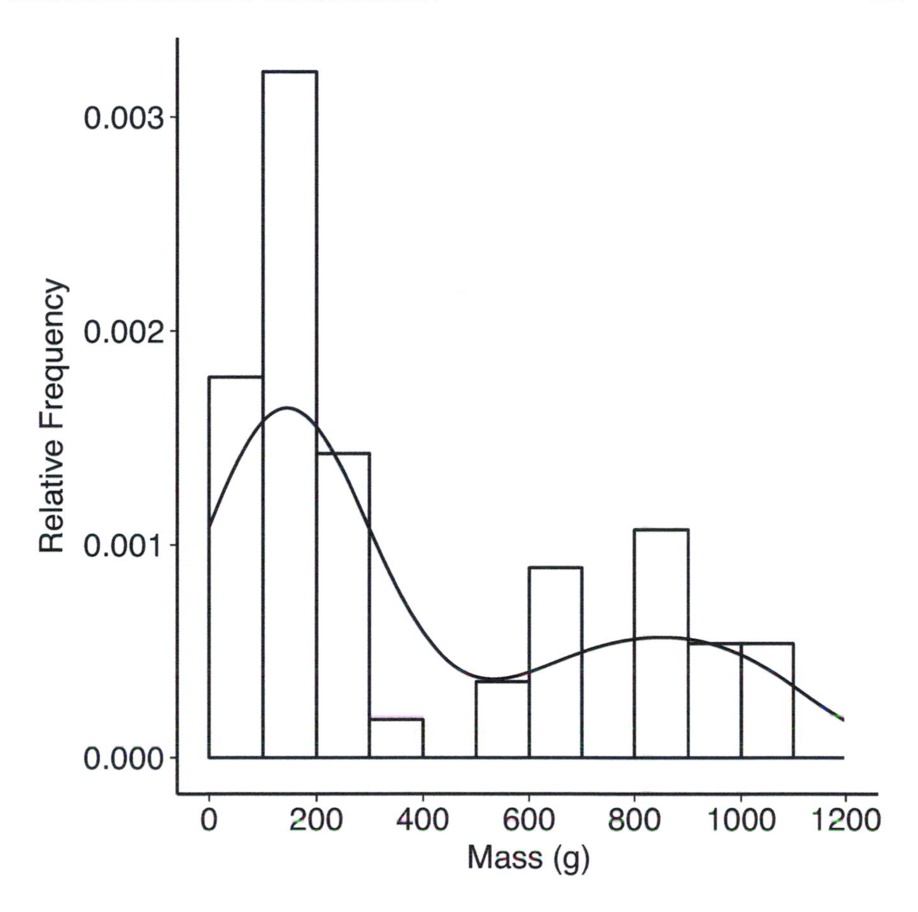

Figure 6.23 *Histogram and Gaussian kernel density estimates of perch mass with bandwidth 139.86 g by Silverman's rule of thumb using* `ggplot2`

To work, the preceding code requires loading the `lattice` graphics package in R. The `lattice` package is an implementation in R of the Trellis graphics package developed at Bell Labs by Richard Becker, William Cleveland, and Ming-Jen Shyu [9, 10]. Documentation for this package was initially published in 1996.

Notice that assigning the plot object to `figure` and then issuing the command `print(figure)` is required when using `lattice` within an R function, otherwise the figure will not appear. This code fragment is from within an R function to produce Figure 6.24. It is not required otherwise. In this way the packages `lattice` and `ggplot2` are similar. Note that this process is not needed with plotting functions from the `graphics` package when listed inside an R function.

The call to the `lattice` function `bwplot` produces the violin plot. Prior to

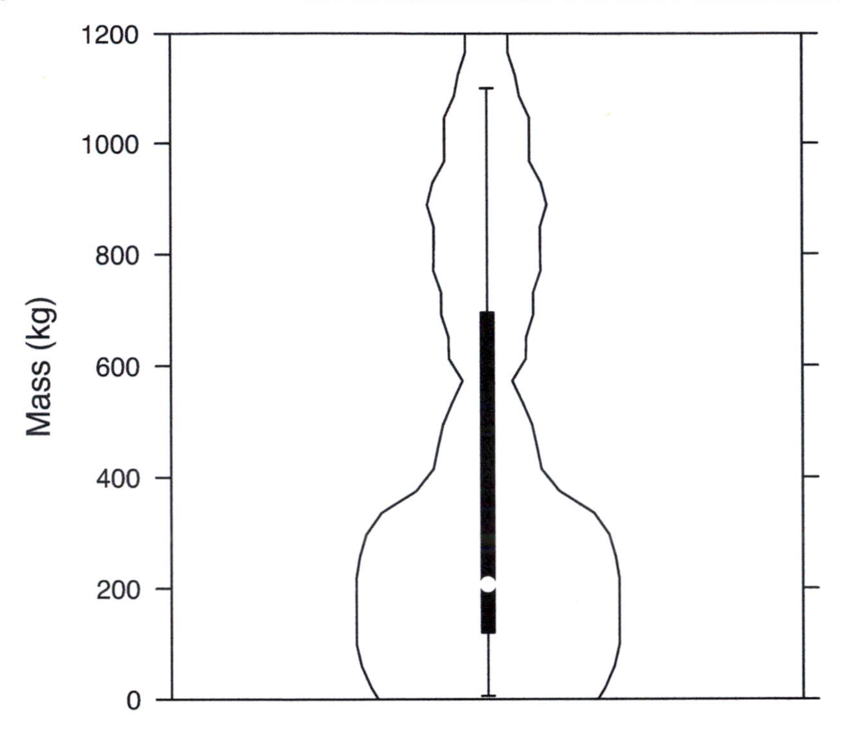

Figure 6.24 *Violin plot of the mass of 56 perch (outlier boxplot between two back-to-back rectangular kernel density estimates with bandwidth 139.86 g by Silverman's rule of thumb)*

this call, the vector variable `mass` is processed into a data frame `massf` as required by `bwplot`. The argument `panel=function(...,box.ratio)` creates a function that produces the single panel that is Figure 6.24. In this function, the `lattice` function `panel.violin` is called, which in turn calls the `stats` package function `density`. The argument `kernel="rectangular"` requests that the rectangular kernel in the function `density` be used.

The `lattice` function `panel.violin` graphs only the back-to-back kernel density estimates. A further call to the function `panel.bwplot` in the `lattice` graphics package produces the boxplot between the kernel density estimates.

The rendering of the kernel density estimate in the violin plot of Figure 6.25 uses the Gaussian kernel and bandwidth of 139.86 by Silverman's rule of thumb as used in Figure 6.22.

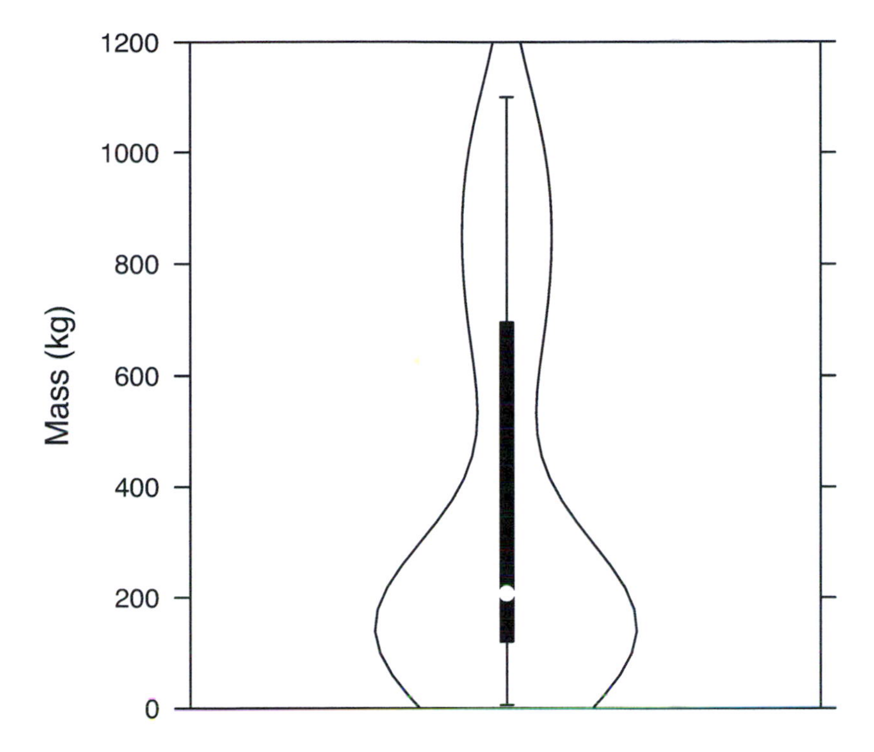

Figure 6.25 *Violin plot of the mass of 56 perch (outlier boxplot between two back-to-back Gaussian kernel density estimates with bandwidth 139.86 g by Silverman's rule of thumb)*

As noted by Hintze and Nelson [65], "The name *violin plot* originated because one of the first analyses that used the envisioned procedure resulted in the graphic with the appearance of a violin." The image of Figure 6.25 is close to a violin, but perhaps a bass fiddle is a closer match.

There is inherently a redundancy in choosing to illustrate two nonparametric density estimates in Figure 6.24. A single nonparametric density estimate ought to be sufficient. Although a single nonparametric density estimate is duplicated in Figure 6.25, the boxplot adds information.

The original plotting convention of Hintze and Nelson [65] includes a modification of the outlier boxplot that stipulates that the outliers are not to be plotted. No outliers were found for perch mass and so none are actually missing from Figures 6.24 and 6.25. So it is essential that viewers must be

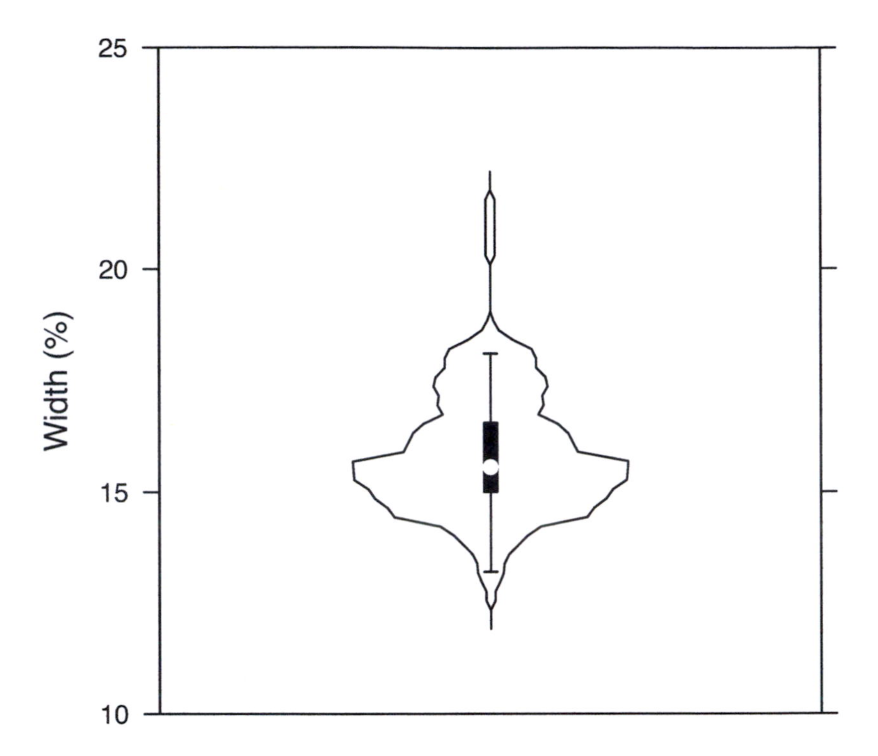

Figure 6.26 *Violin plot of the width of 56 perch (outlier boxplot, without outliers, between two back-to-back rectangular kernel density estimates with bandwidth by Silverman's rule of thumb)*

informed, when viewing a violin plot, as to whether the plotting convention for the outlier boxplot actually plots outliers.

To consider the potential impact of a violin plot using an outlier boxplot that suppresses the illustration of one or more outliers, consider a second example previously encountered. A violin plot of the width of perch as a percentage of length is given in Figure 6.26. The outlier of fish #143 is absent from Figure 6.26.

Figure 6.26 presents a density estimate that is jagged as a consequence of choosing the rectangular kernel. The impact of the outlier is visible in the upper tail of the violin but, with the original plotting convention of Hintze and Nelson [65], it is not possible from Figure 6.26 alone to conclude that the bulge in the upper tail is due to a single outlier. Figure 6.27 presents a

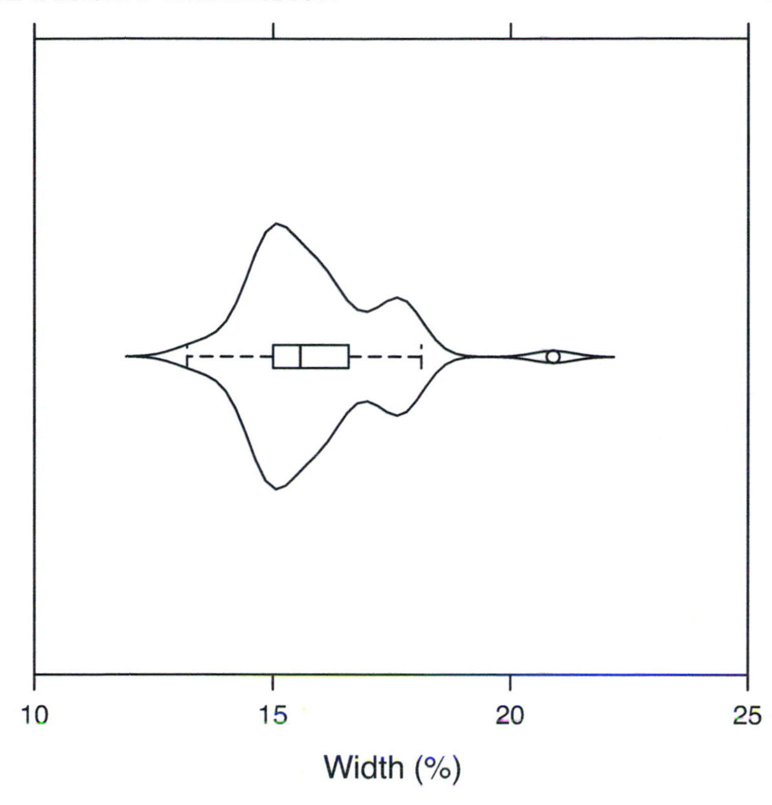

Figure 6.27 *Horizontal violin plot of the width of 56 perch (outlier boxplot, including outliers, in between two back-to-back Gaussian kernel density estimates with bandwidth by Silverman's rule of thumb)*

modification of the original plotting convention of Hintze and Nelson [65], as depicted in Figure 6.26 for perch width, to address this and other issues.

The information added by the outlier boxplot in Figure 6.27 is the presence of an outlier. This identifies the third of three modes depicted in Figure 6.27 as an artifact of a single outlier. An examination of Figure 6.27 reveals that the plotting convention for the outlier boxplot is the one given in a previous chapter. Gone is the black fill for the boxes in Figures 6.24, 6.25, and 6.26. As a result of the black fill, Hintze and Nelson [65] decided to use a large white dot to plot the location of the median. This is not as efficient as the vertical line segment in Figure 6.27 for indicating the location of the median.

The `lattice` package code for producing Figure 6.27 is as follows.

```
widthf<-data.frame(width=width,type=rep(" ",length(width)))

trellis.par.set(box.rectangle=list(col="black"),
box.umbrella=list(col="black",lty=5),
plot.symbol=list(pch=1,col="black",cex=0.8))

figure<-bwplot(type~width,widthf,xlab="Width (%)",
ylab=" ",horizontal=TRUE,
scales=list(x=list(limits=c(10,25),at=(2:5)*5,axs="i")),
panel=function(...,box.ratio) {
panel.violin(...,col="transparent",
varwidth=FALSE,box.ratio=box.ratio)
panel.bwplot(...,fill=NULL,box.ratio=.045,col="black",
pch="|")})

print(figure)
```

Note that the orientation of the violin plot and boxplot is set to horizontal with the parameter `horizontal=TRUE` passed to the `lattice` function `bwplot` in the code for Figure 6.27.

A comparison of outlier boxplots in Figures 6.26 and 6.27 with respect to plotting convention reveals a difference with respect to plotting the whiskers. The plotting convention published by Hintze and Nelson [65] has solid rather than dashed whiskers. This is not in keeping with the original proposal of Tukey [127].

The outlier boxplot in Figure 6.27 has been drafted with the dashed whiskers as recommended by Tukey [127] for quick visual identification of outlier boxplots. According to the conventions of Tukey [127], a viewer would conclude that the whiskers in Figure 6.26 extend to the extremes of the data when they, in fact, do not.

When drafting Figure 6.27, the opportunity was taken to use the Gaussian kernel instead of the rectangular kernel, with a smoother appearance of the kernel density estimate compared to Figure 6.26 as the result.

With respect to the vertical orientations of Figures 6.24, 6.25, and 6.26 versus the horizontal orientation of Figure 6.27, there is empirical evidence that we are better judges of length with respect to a horizontal orientation rather than vertical. Hence, the decision to render the violin plot of Figure 6.27 in a horizontal orientation. The added advantage of the horizontal orientation of Figure 6.27 is that the upper kernel density estimate appears in a more familiar orientation with the value of the variable plotted as increasing to the right along the horizontal axis.

Figure 6.28 provides a `ggplot2` version of a boxplot within a violin plot. The code for producing Figure 6.28 is as follows.

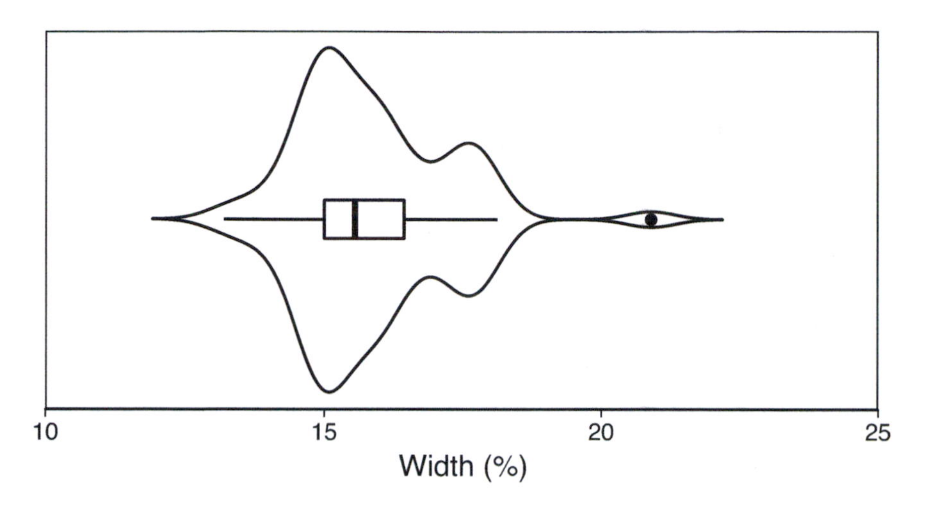

Figure 6.28 *Horizontal violin plot of the width of 56 perch (outlier boxplot, including outliers, in between two back-to-back Gaussian kernel density estimates with bandwidth by Silverman's rule of thumb) with* ggplot2

```
>perchw<-data.frame(width=width)

figure<-ggplot(perchw, aes(1,width)) +
geom_violin(trim=FALSE) +
geom_boxplot(width=0.1) +
theme_linedraw() +
theme(axis.title.y = element_blank(),
axis.text.y = element_blank(),
axis.ticks.y = element_blank(),
axis.line.x = element_line(),
panel.grid = element_blank()) +
scale_y_continuous(expand=c(0,0),
breaks=5+5*(1:4),limits=c(10,25)) +
ylab("Width (%)") + coord_flip()

print(figure)
```

The default kernel for both the `ggplot2` function `geom_violin` and the `lattice` function `panel.violin` is the Gaussian kernel as both functions rely on the `stats` package function `density` for kernel density estimation and its default bandwidth that is determined by Silverman's Rule of Thumb.

With the four changes made to the original plotting convention of Hintze and Nelson [65] for the violin plot, it is argued that a more efficient graphic is the result.

The next section considers an older solution to the issue of density estimation.

6.5 Spline Density Estimation*

Spline functions can be used to produce a nonparametric density estimate. Spline functions are discussed in detail in Chapter 12 with regard to fitting a smooth curve through bivariate data. The history of these functions can be traced back to the time when a steady hand and ink pen were used to draw curves on histograms. Thin strips of wood veneer were used to create a smooth curve to be traced. Later, French curves in the form of thin clear plastic templates were used. Spline functions are mathematical analogies to these thin strips of wood and are constructed out of polynomials.

A spline density estimate is given in Figure 6.29 on the histogram for perch mass. The spline density estimate was produced by the R function `spline`. The user does not have control over the rigidity of the splines with this function. But the R statistical software package offers a second function `smooth.spline`, which allows the user to specify a smoothing parameter. Figure 6.30 overlays two spline estimates on the perch mass histogram with the smoothing parameter set to 0.15 and 0.50.

In Figure 6.29, the spline density estimate unrealistically takes on negative values between 400 and 500 g of mass. The smoother of the two spline density estimates in Figure 6.30 does not have this problem. Also, the smoother of the two spline density estimates of Figure 6.30 portrays a picture of bimodal distribution but lacks the smoothness of the kernel density estimates in Figures 6.14 and 6.22.

Spline density estimation has not been used much since kernel density estimation became more widely available. But spline density estimates are still seen from time to time, typically executed by scientists not exposed to the topic of kernel density estimation in their formal education. In the context of density estimation, spline estimators suffer from the same advantages and disadvantages noted previously for kernel estimators.

6.6 Choosing a Plot for a Continuous Variable*

A number of different plots for presenting the distribution of a single continuous variable have been presented in this and two previous chapters. These choices include: the dotplot, the stemplot, variations on the boxplot, the EDF plot, the quantile-quantile plot, the probability plot, the histogram, and smoothed representations of a histogram such as a kernel density estimate or a spline estimate. There is limited guidance available for choosing among these.

*These sections can be omitted without loss of continuity.

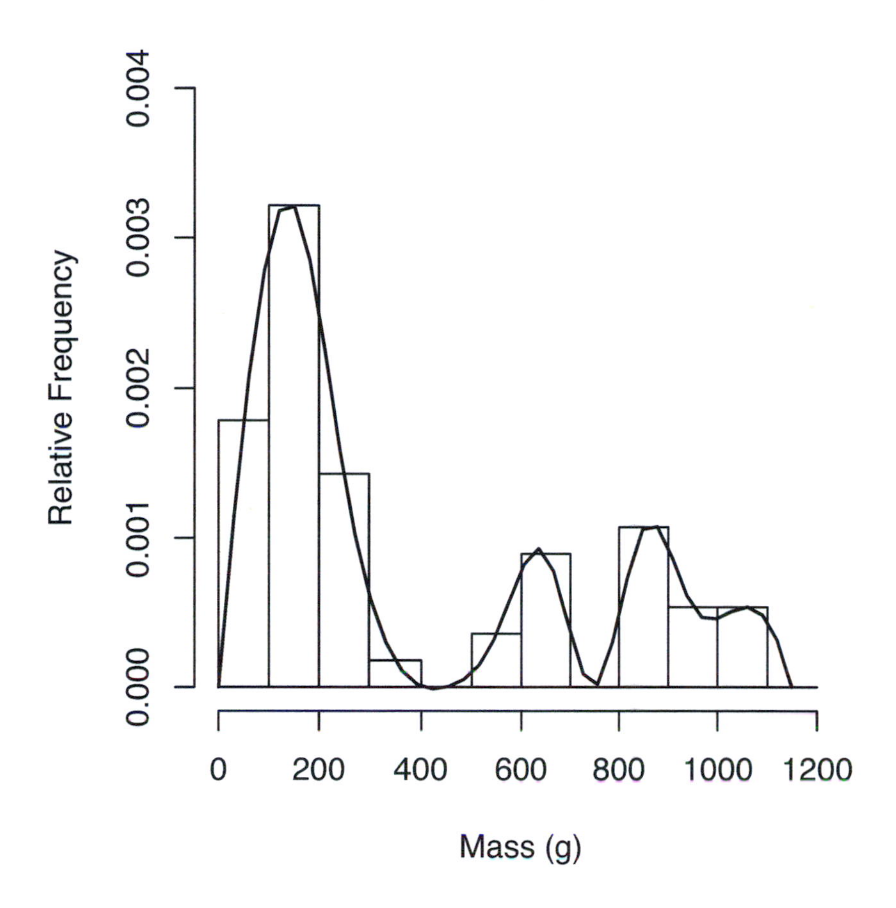

Figure 6.29 *Relative frequency histogram of perch mass with a spline density estimate*

Many researchers opt to draft several of these options during the analysis phase and then choose one or two for presentation live or by print.

There is guidance available for choosing among the quantile boxplot, the quantile-quantile plot, or the histogram. Lise Manchester considered this problem in her doctoral thesis defended in 1985 at the University of Toronto. Her results were summarized in a refereed article by Manchester [81] published in 1991. The statistic she chose for this task was the Kullback-Leibler divergence.

The Kullback-Leibler divergence, also known as the Kullback-Leibler distance, comes from information theory. The first of two fundamental goals in information theory is the formulation of theoretical limits on achievable performance when communicating a set piece of information over a communication channel using coding schemes. The second goal is the development of coding schemes that provide performance that are reasonable in comparison to the theoretical optimum.

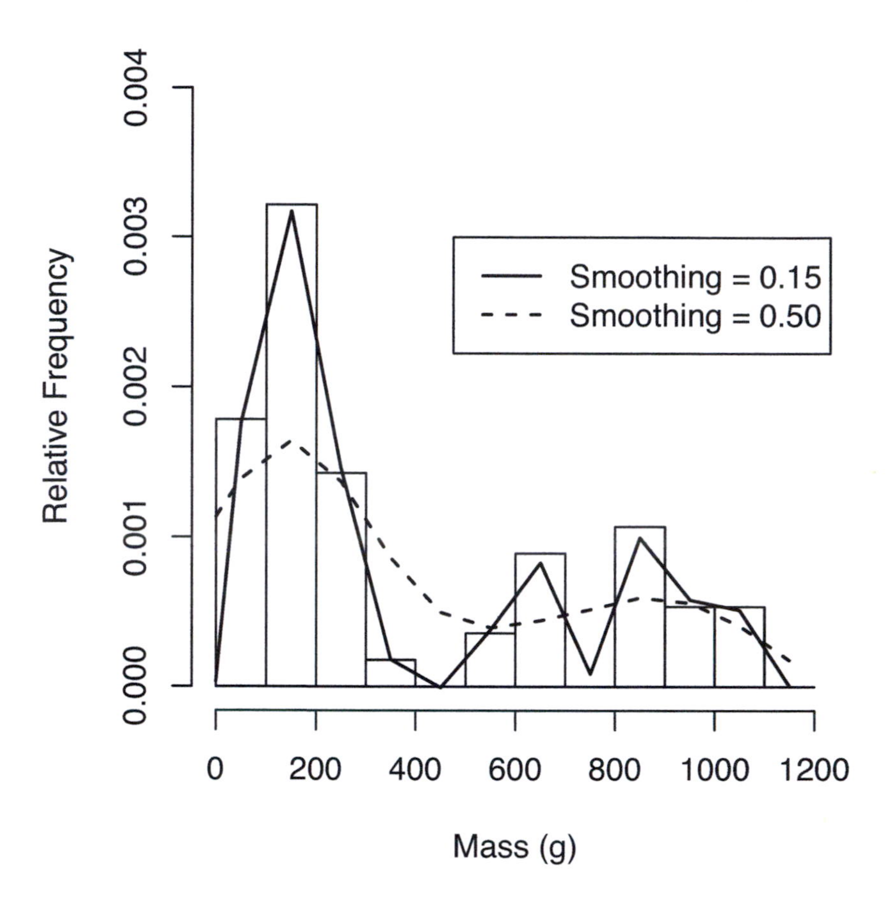

Figure 6.30 *Relative frequency histogram of perch mass with two spline density estimates*

Traditionally, information theory has been concerned with the transmission of textual information. Other areas of application of information theory include: voice communication via landlines or wireless; transmission of high-definition images for television; and the internet transmission of photographic and graphical images by any one of a number of visual encoding schemes.

Often in the development of scientific concepts in one discipline, the concept of another discipline is borrowed and used to serve as an analogy. The thermodynamic concept of entropy has been borrowed by the discipline of information theory. Entropy is the thermodynamic variable that measures the disorder of a given system. Entropy is defined as

$$S = k \ln p \tag{6.55}$$

where p is the probability that the system will be in the state it is in relative to all possible states it could be in and k is some constant (actually, Boltzmann's constant).

Let the probability of finding a single molecule in an initial volume V_i be

$$p = c\, V_i \tag{6.56}$$

where c is a constant. Assuming independence, the probability of finding N molecules simultaneously in an initial volume V_i is

$$p_i = (c\, V_i)^N. \tag{6.57}$$

Substituting equation (6.57) into equation (6.55), the entropy of the system is given by

$$S_i = kN(\ln c + \ln V_i). \tag{6.58}$$

The change in entropy for an ideal gas undergoing isothermal expansion from an initial volume V_i to a final volume V_f is

$$\begin{aligned} \Delta S &= S_f - S_i \\ &= kN \ln \frac{V_f}{V_i} \\ &= k \ln \frac{p_f}{p_i} \\ &\propto (p_f - p_i) \ln \frac{p_f}{p_i}. \end{aligned} \tag{6.59}$$

To apply this analogy in information theory, consider a set of codes $\{x_i\}$ and two associated probability mass functions such that $p_i = P(x_i)$ and $q_i = Q(x_i)$. Applying equation (6.59) by analogy yields the *Kullback-Leibler divergence* of P with respect to Q as

$$D(P, Q) = \sum_i (p_i - q_i) \ln \frac{p_i}{q_i}. \tag{6.60}$$

Notice that $D(P, Q) = D(Q, P)$ so D is symmetric. Another important feature is that $D(P, Q) = 0$ if and only if $p_i = q_i$ for all i.

If the codes $\{x_i\}$ represent features (say, points, lines, or curves) in two comparable graphs P and Q for which the values $\{p_i\}$ and $\{q_i\}$ are constrained to lie between 0 and 1, inclusive, then we have a statistic that can be used to assess differences in how two graphs depict the same data elements.

If functions F and G are used to approximate features that ought to be revealed in graphs P and Q, then it is also possible to calculate $D[F(P), G(Q)]$. If F and G are continuous functions, the resulting graphs $F(P)$ and $G(Q)$ are referred to as smooth versions of the original graphs. Typically, F and G are smoothed to the same degree and this is generally reflected by being dependent on a common smoothing parameter.

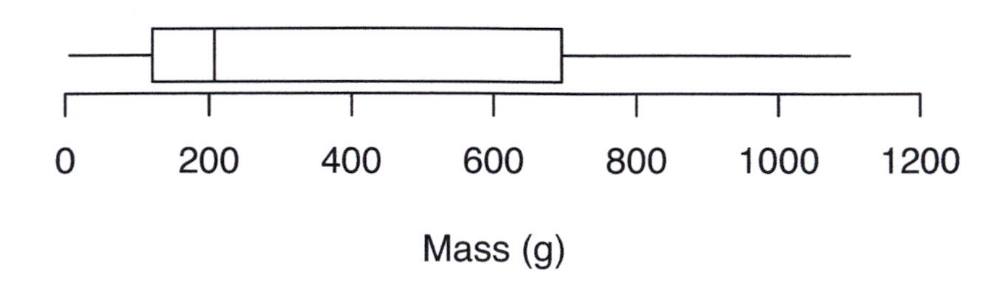

Figure 6.31 *Quantile boxplot of mass (g) for 56 perch caught in a research trawl on Längelmävesi*

A version of the quantile boxplot of mass for 56 perch caught in a research trawl on Längelmävesi is given in Figure 6.31. This is slightly different with respect to Figure 4.15 in that small perpendicular line segments are not drawn at the ends of the two whiskers. Another difference in the plotting conventions for these two figures is that the thickness of the line segment representing the median in Figure 6.31 is no different than any of the other line segments in Figure 6.31. The line segment for the median in Figure 4.15 is thicker for emphasis.

Figure 6.32 depicts a smoothed version of the quantile boxplot in Figure 6.31. If P denotes the quantile boxplot in Figure 6.31, then $F(P)$ denotes its smoothed version in Figure 6.32. Smoothing is achieved by an extension to two dimensions of the Gaussian kernel density estimator. For all (x, y) in the two-dimensional Cartesian coordinate plane

$$F(x, y) = \frac{1}{L\sqrt{2\pi\sigma^2}} \sum_{i=1}^{L} e^{-\frac{\delta_i^2}{2\sigma^2}} \tag{6.61}$$

where δ_i is the shortest distance between (x, y) and the line segment ℓ_i, L is the number of line segments $\{\ell_i\}$ in the graph P, and the standard deviation σ is the smoothing parameter.

In the quantile boxplot of Figure 6.31, there are 3 vertical line segments and 4 horizontal line segments so that $L = 7$. The smoothing parameter σ was set to 10 after some experimentation. It is evident from Figure 6.32 that the function F is maximized at corners in the quantile boxplot in Figure 6.31.

To select among the different plots for depicting the distribution of a single continuous variable, Manchester [81] executed Monte Carlo simulations for three different experiments. One experiment was with regard to detection of skewness, another for discrimination of the length of the tail of a distribution, and the third was with regard to detection of two modes. Two levels of smoothness, light and heavy, were considered. Class widths for the histogram

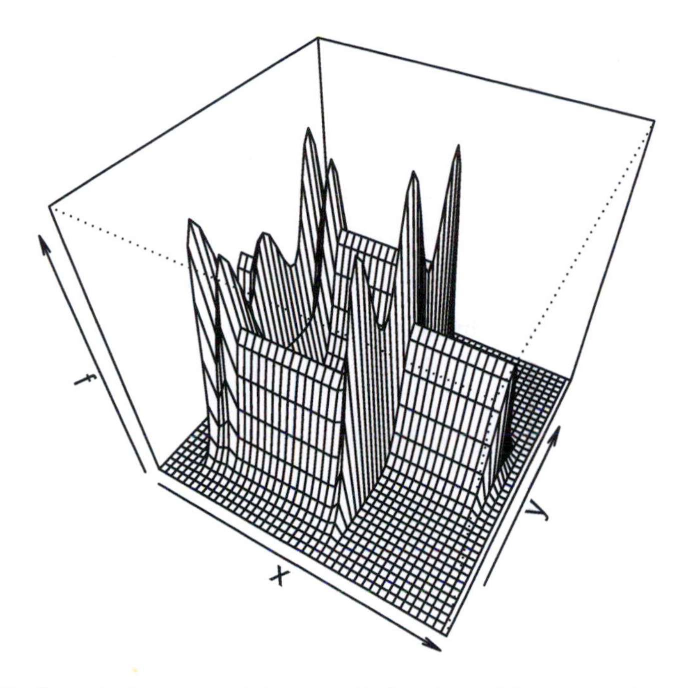

Figure 6.32 *Smoothed version of the quantile boxplot of Figure 6.31 for the mass (g) for 56 perch caught in a research trawl on Längelmävesi*

were determined by Doane's rule. The reference distribution for the quantile-quantile plot was the standard normal. The results of her experiments are summarized as follows.

In the skewness experiment, the population distribution was taken to be the single-parameter gamma with shape parameter equal to 2. The quantile boxplot and the quantile-quantile plot were more effective than the histogram on the basis of comparison of the sample median of the Kullback-Leibler divergence. This is the case for two levels of the smoothing parameter.

In the experiment concerning the length of the tail of the distribution, the population distribution was chosen to be the single-parameter gamma with the shape parameter equal to 10. The histogram was superior in comparison to the boxplot and the quantile-quantile plot. Again, this was true for both levels of smoothing. For each level of smoothing, the median divergence ranked from greatest to least was for the histogram, followed by the quantile-quantile plot, and then the boxplot.

In the experiment examining bimodality, the population distribution was a mixture of two normals. For a narrow separation between the two modes, the histogram had a greater median Kullback-Leibler divergence than either the boxplot or the quantile-quantile plot. The sample median divergence for the quantile-quantile plot was intermediate to the other two plots for light smoothing. But the sample median divergence for the boxplot was intermediate for heavy smoothing. For the widest separation between the two modes, the ranking of sample median divergence from greatest to least was the quantile-quantile plot followed by the boxplot and then the histogram. This was the case regardless of whether the smoothing was light or heavy.

Manchester [81] considered a fourth plot in the bimodality experiment. Tukey [126] proposed the *rootogram* as modification of the histogram in which the height of each bar is proportional to the square root of the frequency in each class. As noted by Manchester [81], some think the rootogram less likely to show spurious bimodality by virtue of plotting the square roots of the class frequencies, which serves to diminish the amount of variation compared to plotting the class frequencies in the histogram.

The median divergence was greater for the rootogram compared to the histogram except for the single case of heavy smoothing and the least separation between the two modes. In this case, the median divergences were nearly equal. In the case of the light smoothing, Manchester [81] judged the difference in median divergence between the rootogram and the histogram to be statistically insignificant. Her recommendation in case of bimodality was that either the histogram or rootogram is the graphical display of choice. The rootogram has not stood the test of time and is rarely seen.

A general observation among the experiments is that the histogram's performance tended to improve, relative to the other plots, as the smoothing parameter increased from light to heavy. In the case of the heavy-tail experiment, Manchester [81] herself stated "that a heavily smoothed histogram may be the best of the three graphical methods at displaying differences in tail length." The kernel density estimate can be considered to be the limiting case of maximal smoothing.

Perhaps not unexpectedly, a single plot with overall superiority did not emerge from Manchester's [81] extensive Monte Carlo simulations. But perhaps the case can be made for the quantile-quantile plot, which was never markedly inferior by itself in any of the experiments. A case can be made for the kernel density estimate as an alternative choice in all circumstances.

6.7 Conclusion

A discussion of the dotplot and its relationship to the histogram can be found on page 187 of Wilkinson [136]. The version of the stacked dotplot in Chapter 3 stacks dots only for identical values. In fact, it is possible to stack dots with one dot for each observation in a class interval. The histogram does the same thing but uses rectangles stacked without a gap and erasures of the overlapping lines. The figure on page 187 of Wilkinson [136] superimposes a histogram on this modified version of the dotplot.

If you have the time and inclination to see the rose diagrams of the Lady of the Lamp, then view Nightingale [87]. Roses in other forms have also been used. Familiar to mariners is the compass rose of navigation charts but aside from variation this provides no statistical information.

Familiar to sailors of yore and aircraft pilots of today are the wind roses of the meteorologists. Predating the work of Florence Nightingale is the wind rose of Lalanne [79] published in 1843. A wind rose by Lalanne in 1830 and more on roses can be found in Wainer [130] (pp. 103–110). For a complete discussion of plots on the circle, consult the standard reference for the statistical analysis of circular data by Fisher [36] (pp. 1–37).

Six named rules for determining class widths for histograms were presented in this chapter. Additionally, there is personal choice, which also includes the possibility of varying the widths among the classes. With respect to optimality considerations, Scott's rule intends to minimize the integrated mean square error (IMSE) between the relative frequency histogram and a normal density. Alternatively, the Freedman-Diaconis rule is an approximation to minimizing the mean of the integrated square error (MISE) between the relative frequency histogram and an arbitrary probability density.

Kernel density estimation has a counterpart to the issue of selecting a common width for all classes of a histogram. This is the issue of bandwidth. Kernel density estimation also introduces flexibility in the choice of kernel function. A reference used in preparation of the discussion on this topic is Jacoby's [72] book published in 1997 (see page 22). This chapter considers six different functions for kernel density estimation.

Seven different named rules for considering bandwidth have been considered in this chapter. An eighth rule would be personal choice. The unbiased cross-validation (UCV) bandwidth estimator $\hat{\lambda}_{UCV}$ minimizes an approximation to the IMSE. The UCV bandwidth estimator is a counterpart to Scott's rule for class width. The biased cross-validation (BCV) bandwidth estimator $\hat{\lambda}_{BCV}$ minimizes an approximation to the MISE. The UCV bandwidth estimator is a counterpart to the Freedman-Diaconis rule for class width.

Working with the Gaussian kernel, Sheather and Jones [109] sought to minimize the approximate MISE and proposed an approach called smooth cross-validation. This leads to two estimators. The Sheather-Jones solve-the-equation bandwidth estimator $\hat{\lambda}_{SJ}$ is found by numerical optimization. Sheather and Jones [109] also proposed a direct plug-in estimator $\tilde{\lambda}_{SJ}$ that does not require software for numerical optimization. Both Sheather-Jones estimators minimize the MISE with better rates of convergence than the biased cross-validation estimator.

Despite research considering square error, which led to improved estimators for histogram class width and kernel bandwidth, Sturges' [118] rule for class width and Silverman's [110] rule of thumb continue to be popular. Sturges' rule is the default for class width in the R function `hist`. Silverman's rule of thumb is the default for bandwidth in the R function `density`.

On the topic of choice of kernel function, this chapter presented six different functions for kernel density estimation. There is nothing to stop someone from creating a new kernel function of their own choosing. It is also possible to base the choice of kernel function on an objective basis. Consider the formula for the approximate mean integrated square error (AMISE) given in equation (6.21). With basic calculus, the optimal bandwidth for the AMISE is given by λ_{AMISE} in equation (6.23). Substituting this value for λ_{AMISE} back into equation (6.21) yields

$$AMISE = \frac{5}{4} C(K) \left\{ \int [f''(x)]\, dx \right\}^{1/5} n^{-4/5}$$

where the constant

$$C(K) = \kappa_2^{2/5} \left\{ \int [K(t)]^2\, dt \right\}^{4/5}.$$

Without loss of generality, the problem of minimizing $C(K)$ reduces to the problem of minimizing $\int [K(t)]^2\, dt$ subject to the constraints that $\int K(t)dt = 1$ and $\kappa_2 = \int t^2 K(t)dt = 1$.

In 1956, Hodges and Lehmann [67] showed that this problem is solved by setting $K(t)$ equal to the Epanechnikov kernel in formula (6.14). In 1969, Epanechnikov [34] was the first to suggest the use of this function in the context of nonparametric estimation of a multidimensional probability density. Hodges and Lehmann [67] were searching for a nonparametric alternative for the parametric t-test.

In addition to the six kernels for nonparametric density estimation given in this chapter, the R function `density` offers a seventh. It is a variation on the cosine kernel

$$K_C(z) = \begin{cases} \frac{\pi}{4} \cos\left(\frac{\pi}{2} z\right) & \text{for } |z| \le 1; \\ 0 & \text{otherwise.} \end{cases}$$

The alternative cosine kernel is given by

$$K_A(z) = \begin{cases} \frac{1}{2} [1 + \cos(\pi z)] & \text{for } |z| \le 1; \\ 0 & \text{otherwise.} \end{cases} \qquad (6.62)$$

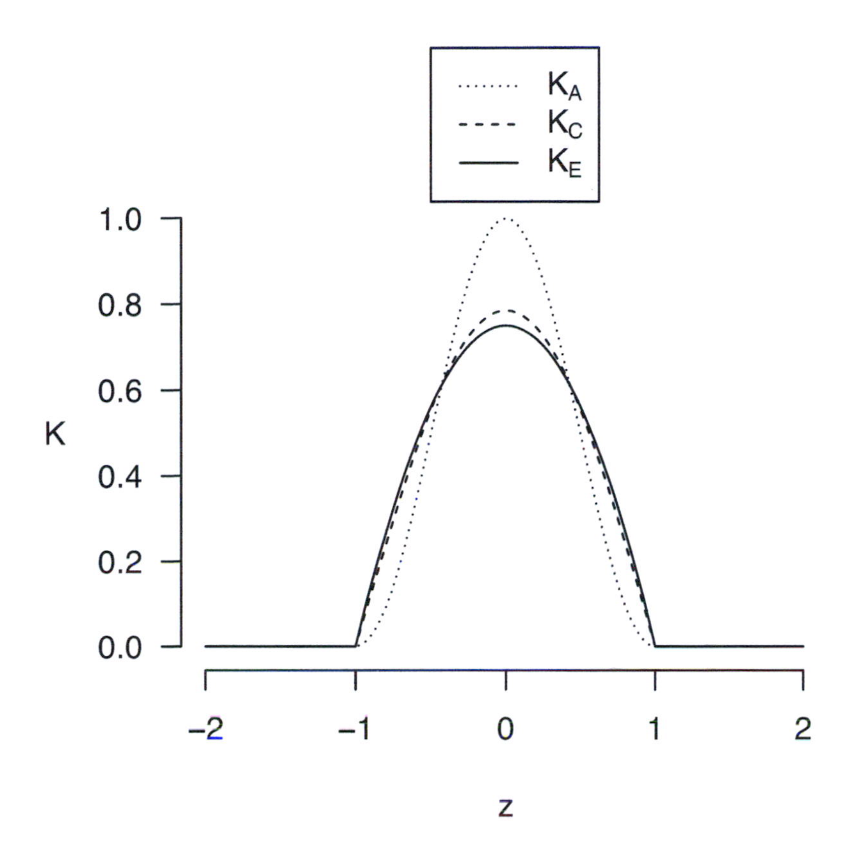

Figure 6.33 *Kernel functions (alternative cosine K_A, cosine K_C, and Epanechnikov K_E)*

Because the cosine function is differentiable everywhere, both cosine kernels are differentiable on the open interval $(-1, 1)$. But of the two kernels, only K_A is differentiable everywhere because its derivative $K\prime_A(z)$ vanishes as z approaches both -1 and 1 from either the left or right. So the alternative cosine kernel K_A is smoother than the cosine kernel K_C. However, the cosine kernel K_C is more similar to the Epanechnikov kernel, which also is nondifferentiable at -1 and 1.

Figure 6.33 depicts both cosine kernels and the Epanechnikov kernel. The latter has been rescaled to the interval $(-1, 1)$ from $(-\sqrt{5}, \sqrt{5})$ for comparison purposes. Note that the cosine kernel K_C more closely resembles the Epanechnikov kernel. Hence we expect the cosine kernel K_C to have optimality properties close to that of the Epanechnikov kernel.

Silverman [110] defined the efficiency of a kernel K relative to the Epanech-

Kernel	Exact	Approximate
Cosine	$\frac{48}{25}\frac{\sqrt{5}}{\pi\sqrt{\pi^2-8}}$	0.999451
Biweight (or Quartic)	$\frac{21}{125}\sqrt{35}$	0.993901
Alternative Cosine	$\frac{4}{25}\frac{\pi\sqrt{15}}{\sqrt{\pi^2-6}}$	0.989651
Triangular	$\frac{9}{50}\sqrt{30}$	0.985901
Gausssian	$\frac{6}{25}\sqrt{5\pi}$	0.951199
Rectangular	$\frac{6}{25}\sqrt{15}$	0.929516

Table 6.4 *Efficiency of kernels relative to the Epanechnikov kernel*

nikov kernel to be

$$\text{eff}(K) = \frac{C(K_E)}{C(K)}. \tag{6.63}$$

The efficiencies of the other six kernels relative to the Epanechnikov kernel are given in Table 6.4. The cosine kernel K_C is the nearest of the six to the Epanechnikov kernel with respect to efficiency. The cosine kernel K_C is also better than the alternative cosine kernel K_A. This is as expected based upon Figure 6.33.

Most of the research with respect to optimality of bandwidth has been done with respect to the Gaussian kernel. But, according to Table 6.4, the Gaussian kernel is the second worst with respect to minimizing the AMISE. To paraphrase Silverman [110], the message of Table 6.4 is that there is very little to choose between the various kernels on the basis of approximate mean integrated square error.

Spline density estimation has been included in this chapter largely for historical reasons, but also because Chapter 12 includes a discussion of the use of splines for obtaining a smooth curve through bivariate continuous data. The wealth of research on optimality in kernel density estimation that began in the mid 1980s, and then ran for nearly a decade, pretty much renders spline density estimation obsolete. As a consequence of this research, some advocate that the best approach to nonparametric density for a given data is to produce a kernel density estimate and not even bother with a histogram. The violin plot with its pairing together of a kernel density estimate and an outlier boxplot might be the best of both worlds. Given the technical nature of the outlier boxplot, it is probably not the best choice for a general audience. But a kernel density plot without a histogram could be readily apprehended by any audience.

6.8 Exercises

1. Consider the following data collected by Charles Darwin:

$$-67, -48, 6, 8, 14, 16, 23, 24, 28, 29, 41, 49, 56, 60, 75.$$

   These data represent the difference in height (in eighths of an inch) between the self-fertilized member and the cross-fertilized member of each of 15 pairs of plants.

   (a) The sample size for this data is probably at the lower range for non-parametric density estimation and certainly for the production of a histogram, but produce a frequency histogram for the data with uniform class widths. State and justify the rule you used for determining class width.

   (b) Use statistical software to produce a kernel density estimate for Darwin's data. Use the software's default choices for kernel function and bandwidth. Report these default settings.

   (c) Obtain an outlier boxplot for Darwin's data. Are there any outliers? Discuss.

   (d) The frequency histogram produced for part (a) lists the count of observations along the vertical axis. The smooth kernel density estimate produced for part (b) usually has relative frequency along the vertical axis and gives no indication of sample size. How easy would it be to mislead an audience with the kernel density estimate from part (b) and convince them that there is clear evidence of bimodality in Darwin's data? Discuss.

2. Consider the data for the total compensation in 2008 received by chief executive officers employed by industrial companies listed in Table 4.3.

   (a) Produce a relative frequency histogram for the data with the Rule of Twelve.

   (b) Produce a relative frequency histogram for the data with the Robust Rule of Twelve.

   (c) Produce a relative frequency histogram for the data with Sturges' Rule.

   (d) Produce a relative frequency histogram for the data with Doane's Rule, which is a robust version of Sturges' Rule.

   (e) Comment on the effect of using the robust rules compared to their non-robust counterparts when the data appear to be symmetric and without outliers.

Jan	Feb	Mar	Apr	May	Jun
7280	6957	7883	7884	7892	7609

Jul	Aug	Sep	Oct	Nov	Dec
7585	7393	7203	6903	6552	7132

Table 6.5 *Births in Sweden in 1935*

3. Consider the data for the total compensation in 2008 received by chief executive officers employed by industrial companies listed in Table 4.3.
   (a) Produce a relative frequency histogram for the data with Scott's Rule.
   (b) Produce a relative frequency histogram for the data with the Freeman-Diaconis Rule.
   (c) Compare the histograms of parts (a) and (b).
   (d) Discuss whether the Freedman-Diaconis Rule can be considered a robust version of Scott's Rule.
4. The data on the distribution of births in Sweden in 2004 depicted in the rose diagram of Figure 6.11 were motivated by a classic data set published by the statistician Harald Cramér of the University of Stockholm. Cramér's [28] textbook entitled *Mathematical Methods of Statistics* first appeared in Sweden in 1945. This was published in the United States in the following year by the Princeton University Press. This textbook was arguably the first textbook on mathematical statistics. The ordering of the chapters has been duplicated in many such textbooks by other authors. Given in Table 6.5 are the number of children born in Sweden in 1935 as reported on page 447 of Cramér's [28] textbook.
   (a) Depict the count of births in Sweden for the twelve months in 1935 in a rose diagram. If your statistical software package cannot generate a rose diagram, then draft the rose diagram for the data of Table 6.5 manually.
   (b) Compare your answer to part (a) with the rose diagram for the frequency of births in Sweden in 2004 given in Figure 6.11. Comment on whether there has been a change in pattern in 69 years.
5. Not all circular data comes in a form already summarized as counts for hours of the day or months of the year. Table 6.6 lists directions reported in degrees taken by 76 turtles after treatment with radiation. The data were collected to determine whether turtles have a preferred direction. The data were reported by Stephens [113] in 1969 in a technical report published at Stanford University. A version of Table 6.6 can also be found in Fisher's [36] book on the statistical analysis of circular data.
   (a) Create a stemplot for the directions. Use unit degrees for the leaves and tens of degrees for the stems.
   (b) From your answer to part (a), create a stacked dotplot on the circle.

8	9	13	13	14	18	22	27	30	34
38	38	40	44	45	47	48	48	48	48
50	53	56	57	58	58	61	63	64	64
64	65	65	68	70	73	78	78	78	83
83	88	88	88	90	92	92	93	95	96
98	100	103	106	113	118	138	153	153	155
204	215	223	226	237	238	243	244	250	251
257	268	285	319	343	350				

Table 6.6 *Directions (in degrees) taken by 76 turtles after radiation treatment*

Recall that the dotplots of Chapter 4 were all along a straight-line axis so you are being asked to do something new and not previously presented. Plot the circular dotplot manually if your statistical software package does not. The function `rose.diag` in the CircStats package for R can be used to produce circular dotplots.

(c) Create a rose diagram for the data in Table 6.6. This can be done with the function `rose.diag` in R.

(d) Based on the plots produced for parts (a) through (c), inclusive, what can be said about the distribution of directions taken by the turtles after radiation treatment?

6. For the data in Table 6.6 discussed in Exercise 5, consider the modifications needed to produce a kernel density estimate on the circle. With or without these modifications, produce a kernel density estimate for the directions taken by turtles and depict the result on the circle.

7. Consider the data for the total compensation in 2008 received by chief executive officers employed by industrial companies listed in Table 4.3.

(a) Produce a relative frequency histogram for the total compensation data. Justify your choice of rule for determining the width of each class interval.

(b) Add an EDF to the histogram of part (a).

(c) Produce a kernel density estimate for the total compensation data. Add this to your histogram of part (a).

(d) Which do you prefer: adding an EDF plot or a kernel density estimate to a histogram? Justify your choice.

8. In 1882, Simon Newcomb measured the time required for light to travel from his laboratory on the Potomac River to a mirror at the base of the Washington Monument and back, a total distance of about 7400 meters. These measurements were used to estimate the speed of light. Table 4.4 contains the estimated speed of light for 66 trials. These data were reported in an article by Steven Stigler [115] in 1977 in *The Annals of Statistics*.

(a) Produce a relative frequency histogram for the data in Table 4.4 with the default settings of your software package.

(b) Do there appear to be any outliers in your histogram for part (a)? Discuss.

(c) Produce a separate kernel density estimate for the data in Table 4.4 with the default software settings.

(d) Do there appear to be any outliers in your plot for part (c)? Discuss.

(e) Produce a violin plot for the data in Table 4.4.

(f) Based on your answers to parts (a), (c), and (e), which does a better job of depicting data with outliers: a histogram, a kernel density estimate, or a violin plot? Discuss.

9. Examine the `lattice` and `ggplot2` code used to produce the violin plots and boxplots of Figures 6.27 and 6.28, respectively. Which of these two R packages would you prefer to use on a routine basis? Discuss.

10. Consider the time passage data for light in Table 4.4 that was obtained by Simon Newcomb.

    (a) Produce a kernel density for the data in Table 4.4 using each of the seven different kernel functions presented in this chapter. In each instance use the bandwidth as determined by Silverman's Rule of Thumb.

    (b) Based on your kernel density estimates for part (a), do you have a preferred choice for kernel function or is one as good as another? Justify your answer.

11. Consider the time passage data for light in Table 4.4 that was obtained by Simon Newcomb.

    (a) Produce a kernel density estimate for the data in Table 4.4 using only the Gaussian kernel but with each of the seven different bandwidth estimators listed in Table 6.3.

    (b) Based on your kernel density estimates for part (a), do you have a preferred choice for bandwidth estimator for the data in Table 4.4 or is one as good as another? Justify your answer.

12. An interesting versatile plot for a single continuous random variable is the BLiP plot created by Lee and Tu [80]. Look up the reference and have fun with it.

Chapter 7

Parametric Density Estimation for a Single Continuous Variable

7.1 Introduction

At the heart of parametric density estimation lies a statistical model expressed in a mathematical formula with parameters to be estimated from the data. If the statistical model is a close match to the true state of affairs in the data, then parametric estimation or inference will outperform any nonparametric alternative. On the other hand, if the assumed statistical model is not a close fit, then the results of any statistical analysis can be misleading, even erroneous. In this case, the parametric analysis is not *robust*. On the other hand, nonparametric methods can be robust because they do not make precise assumptions regarding the shape of a distribution.

The nonparametric density estimation methods of the previous chapter are ideal when the nature of the underlying distribution is unknown. There is the additional advantage that these methods can still be used when researchers have a fairly good idea about the underlying statistical model for the data. The drawback for using a nonparametric method when a parametric method is available is that greater sample sizes are needed for the nonparametric method to achieve the same power in the case of an hypothesis test, or the same width in the case of a confidence interval.

In a situation when there are ample data for parametric density estimation but no clue initially as to the parametric model for the distribution, nonparametric density estimation can be used for exploration. Nonparametric density estimation can be done by either a histogram or kernel density estimate alone, or both in combination.

In this chapter, four methods for parametric density estimation are discussed. The first method consists of taking the data and estimating the mean and variance parameters of a normal distribution and then overlaying the density on a histogram. The Gaussian distribution is referred to as the normal distribution because this distribution is found to be a good fit to the data in most instances. It comes then as no surprise that the most frequently encountered density estimate overlaid on a histogram is the normal density curve rather than a nonparametric kernel density estimate.

If the data are not normally distributed, then there is a chance that the distribution can be described by a member of another family of distributions. If so, this family might be one of Pearson's curves. The method of Pearson's curves is inherently a graphical technique for finding a parametric model of a distribution for a random sample. Although this method has gone out of fashion with respect to instruction in data analysis, it still provides the quickest route to finding the probability density function from among the most commonly encountered distributions.

The last two methods of parametric density estimation discussed are based on the normal distribution but intended for nonnormal data. The Gram-Charlier series expansion involves taking a sum of Chebyshev-Hermite polynomials and then multiplying this sum by the standard normal density. In a sense, this is a parametric counterpart to Gaussian kernel density estimation. Gram-Charlier series expansions are an alternative to Pearson's curves and can produce a good result when the method of Pearson's curves does not.

Perhaps the most popular approach for dealing with nonnormal data is the second parametric density estimation method considered in this chapter. The idea behind this method is very simple and widely successful based upon statistical practice. If the data are not normal, then check to see whether a transformation of the data by a simple function will result in a new variable that is nearly normal. The survey in this chapter of parametric density estimation techniques, however, begins with the assumption that the data are normal.

7.2 Learning Outcomes

When you complete this chapter, you will be able to do the following.

- Be able to perform normal density estimation and add a normal density estimate to a histogram with either the `graphics` package or the `ggplot2` package.
- Understand the principal features of the Box-Cox transformation to normality from another non-normal distribution.
- Optionally, be knowledgeable about the Pearson's curves family of probability distribution functions and be able to use the nomogram for Pearson's curves to infer which member of a family of 8 distributions best represents the underlying distribution of a random sample.
- Optionally, be able to model a mixture of two normal distributions and overlay this probability density on an histogram with either the `graphics` package or the `ggplot2` package.

- Optionally, be able to model a probability density function with a Gram-Charlier expansion and overlay this density on a histogram with either the `graphics` package or the `ggplot2` package.

7.3 Normal Density Estimation

Suppose a random sample $\{x_i\}_{i=1}^{n}$ is normally distributed with unknown mean μ and unknown variance σ^2. Calculated from the random sample are the sample mean

$$\bar{x} = \sum_{i=1}^{n} \frac{x_i}{n} \tag{7.1}$$

and the sample variance

$$s^2 = \sum_{i=1}^{n} \frac{(x_i - \bar{x})^2}{n-1}. \tag{7.2}$$

A smooth estimator of the probability density function is

$$\hat{f}(x) = \frac{1}{\sqrt{2\pi}s} \exp\left\{-\frac{(x-\bar{x})^2}{2s^2}\right\}. \tag{7.3}$$

Depiction of this estimator is often available as an option for histograms in some statistical software packages.

An example of a histogram with a normal density curve superimposed is given in Figure 7.1. In a rare instance for this book, the data depicted in Figure 7.1 are not real but faked. A pseudo-random-number generator was used to generate 500 observations from a normal distribution with mean 65 and standard deviation 10. The R script for producing these random numbers and Figure 7.12 is as follows.

```
set.seed(seed=136073225)
z<-rnorm(500,65,10)

hist(z,breaks=(5:20)*5,freq=FALSE,xlim=c(20,100),xaxt="n",
ylim=c(0,0.05),main=NULL,xlab="Normal Deviate",
ylab="Relative Frequency")

axis(1,at=(1:5)*20,labels=c("20","40","60","80","100"))

curve(dnorm(x,mean=mean(z),sd=sd(z)),add=TRUE,lwd=1.5)
```

The R function for generating the sample of 500 observations is `rnorm`. It is preceded by a call to the function `set.seed` that sets the pseudo-random-number seed to 136073225. This is done so that whenever this R script is executed, the same random sample is produced.

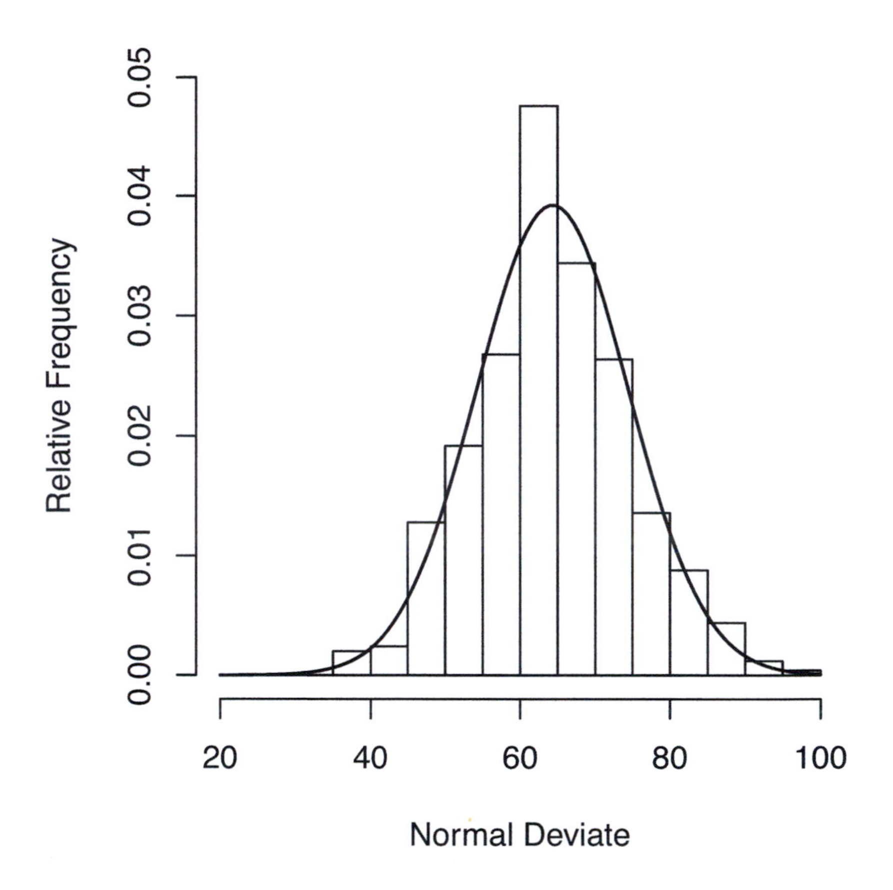

Figure 7.1 *Relative frequency histogram of 500 normal deviates with normal density estimate*

The function `hist` is called to plot the histogram. With the argument `xaxt="n"`, the horizontal axis is not plotted by `hist`. The following call to the R function `axis` produces a more eye-appealing horizontal axis with tick marks and labels every 20 units.

The function `dnorm` calculates the normal probability density function with the mean and standard deviation set equal to those of the sample. The `graphics` package function `curve` does the plotting of `dnorm`.

Figure 7.2 plots a version of Figure 7.1 using `ggplot2`. The code for Figure 7.2 using the line-draw theme in `ggthemes` is as follows.

```
set.seed(seed=136073225)
z<-rnorm(500,65,10)
normdev<-data.frame(z=z)
```

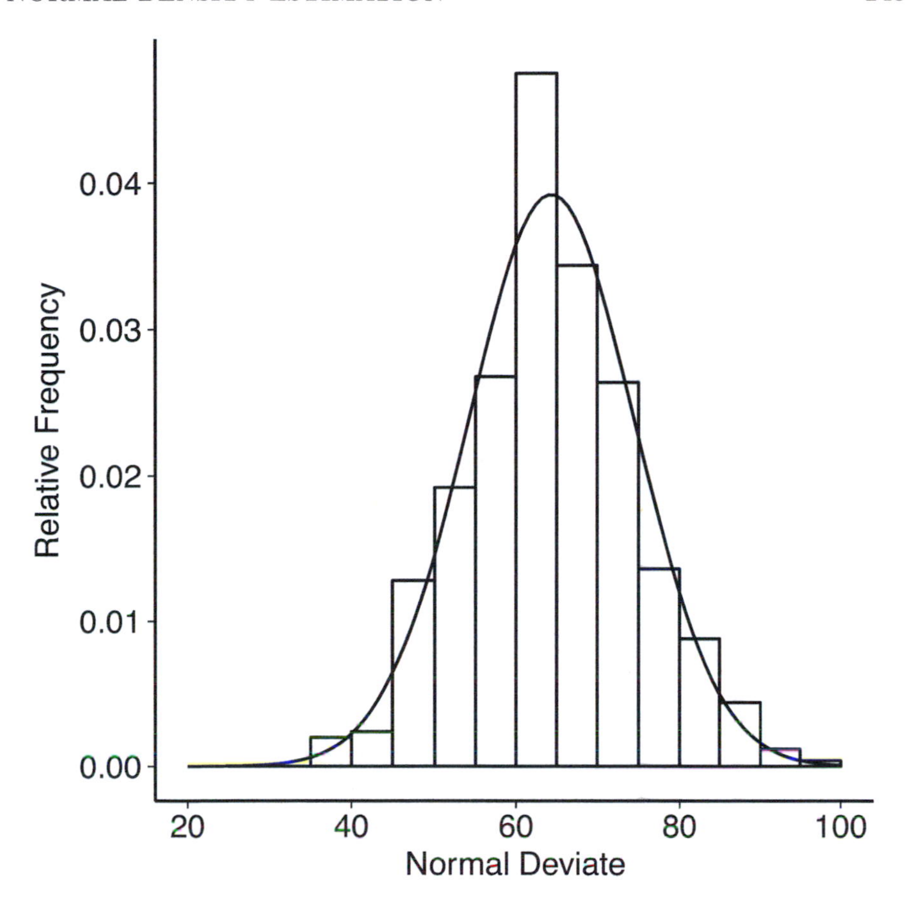

Figure 7.2 *Relative frequency histogram of 500 normal deviates with normal density estimate using* ggplot2

```
figure<-ggplot(normdev, aes(x=z)) +
geom_histogram(breaks=(5:20)*5,
closed="right",color="black",fill="white",
aes(x=z,y=..density..)) +
stat_function(fun=dnorm,
args=list(mean=mean(z),s =sd(z))) +
theme_linedraw() +
theme(panel.grid=element_blank(),
panel.border=element_rect(fill=NULL,
color="white",linetype="solid"),
axis.line=element_line(),
axis.title=element_text(size=12),
```

```
axis.text=element_text(size=12)) +
scale_x_continuous(breaks=20*(1:5),
limits=c(20,100)) +
labs(x="Normal Deviate",y="Relative Frequency")

print(figure)
```

The ggplot2 function stat_function adds the normal density curve. It too uses the function dnorm just as the curve function did for Figure 7.1. Notice the different manner in which stat_function and curve handle the parameter arguments for the stats function dnorm. Notice as well the grammatical difference in how a curve is added to a plot by the graphics and ggplot2 packages.

The rationale for using simulated rather than real data is that in order to assess the merits of parametric density estimation, it is important to see how this can be ideally accomplished. Figure 7.1 presents one such image of how well this can be done. The normal curve has been plotted with the estimate of 65.318 for the mean and 10.166 for the standard deviation based upon all 500 simulated values.

Figure 7.3 presents a histogram with a normal density curve for the same distribution but for only the first 30 of the 500 observations depicted in Figure 7.1. The fit of a normal density curve to a histogram can be worse than in Figure 7.3, but this figure is a good example of a poor fit even when the underlying distribution is truly normal. The normal curve has been plotted with the estimate of 62.698 for the mean and 10.127 for the standard deviation based on the 30 simulated values.

A problem with parametric estimation of the normal density is that it can very easily be incorrectly added to a histogram. It is imperative that it be added in such a way that the areas under the histogram and normal curve both sum to one and the full bell curve is visible. This has been done in Figures 7.1 and 7.3.

Another problem with parametric estimation of the normal density is that although the placement of the normal curve is accurate with regard to the sample mean $\bar{x}$, artistic license is used in choosing either the vertical scale or the variance so that the curve visually better fits the data with truthfulness being sacrificed.

The overlay of the normal density, or bell curve as it is popularly known, upon a histogram is a graphical display that is quite familiar to members of the public. For a technical audience, or a group of suitably prepared administrators, it could be argued that it is better to use a normal quantile plot to assess normality. The justification for this recommendation is that there is less opportunity for artistic license as in the case of a histogram with the overlay of a normal density curve. An alternative recommendation would be that researchers routinely use normal quantile-quantile plots in their own work and

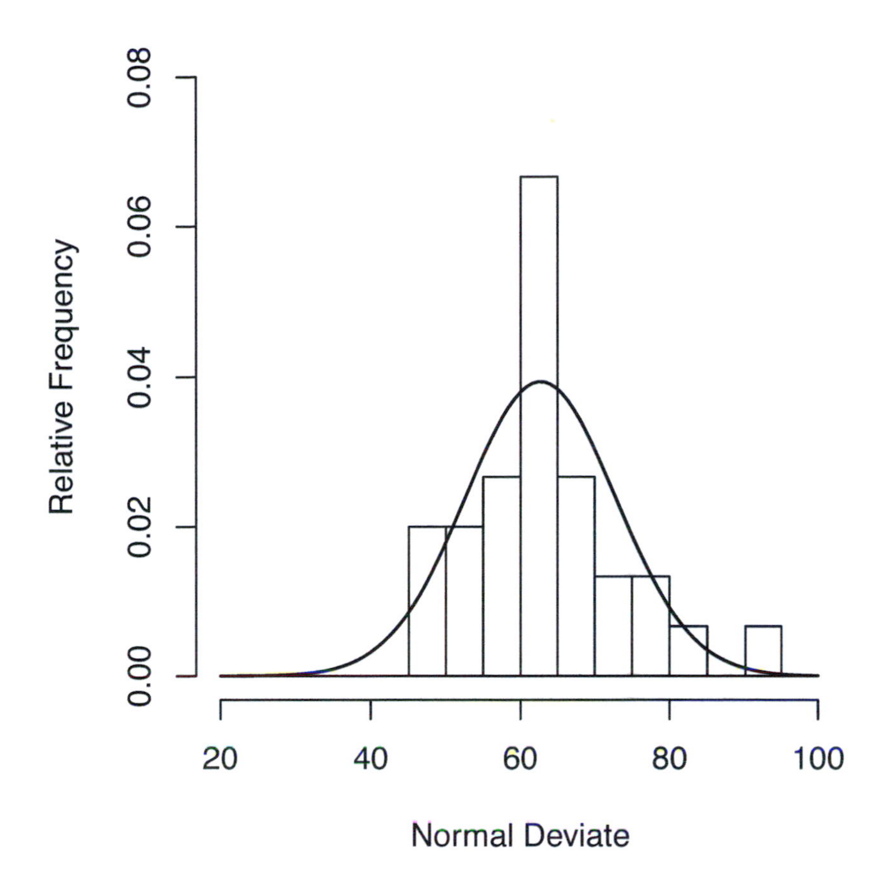

Figure 7.3 *Relative frequency histogram of 30 normal deviates with normal density estimate*

resort to using the histogram with the bell curve overlaid as necessary to communicate with those who might not be fully familiar with the quantile-quantile plots.

When the underlying distribution is not normal, the preceding discussion for estimating the normal density can also be modified by estimating the unknown parameters with any acceptable method including:

- maximizing the likelihood function;
- minimizing the square of the distance between the empirical distribution function and the hypothesized cumulative distribution function; or
- minimizing the square of the distance between a density estimate and the hypothesized probability density function.

The latter two methods have strong parallels with the kernel density estimation. And, of course, there is always the nonparametric density estimation technique of kernel estimation previously discussed as an alternative to parametric density estimation. There are also parametric techniques that work for a distribution specified from a broad class of distributions. These methods are the topics for the next two sections.

7.4 Transformations to Normality

In many instances a transformation of the original observations by a simple mathematical function can yield a new random variable that is very nearly normally distributed. The following family of transformations has been shown in many instances to produce a new variable that is nearly normal despite the fact that the transformation is really intended to reduce skewness:

$$y = \begin{cases} \dfrac{x^\lambda - 1}{\lambda} & \text{if } \lambda > 0; \\ \ln(x) & \text{if } \lambda = 0. \end{cases} \tag{7.4}$$

Transformations of this type are known as *Box-Cox transformations* after G. E. P. Box and Sir David Cox [12] who proposed them in 1964.

The trial-and-error approach to using Box-Cox transformations relies on examining formula (7.4) and noting special cases for λ of -1, 0 (that is, a logarithmic transformation), 0.5, and 2. One tries these special cases and then picks the value of λ that leads to a normal quantile-quantile plot that is the closest to a straight line.

With the use of the graphical approach of trial-and-error and enough practice comes experience. One becomes able in similar situations to accurately guess which transformation will be best. Experience has shown that the logarithm of skewed variables such as income, chemical concentration, or mass to be distributed reasonably close to the normal distribution. The Box-Cox transformation corresponding to $\lambda = 0$ is the natural logarithm. But the common logarithm to base 10 can be justifiably used instead. Figure 5.5 displays the normal quantile plot for the mass of 56 perch caught in Längelmävesi. There are glaring indications of a departure from normality in Figure 5.5. This is consistent with the bimodal mixture of normals that has been explored in a parametric model for this data.

Figure 7.4 displays the normal quantile plot for the common (base 10) logarithm of mass of 56 perch caught in Längelmävesi. There is an indication in Figure 7.4 that the issue of bimodality is not fully solved by a logarithmic transformation of the original data. This prompts the question as to whether a different Box-Cox transformation with $\lambda \neq 0$ could be approximately normalizing.

One solution to the problem of finding the best value of λ is obtained from the

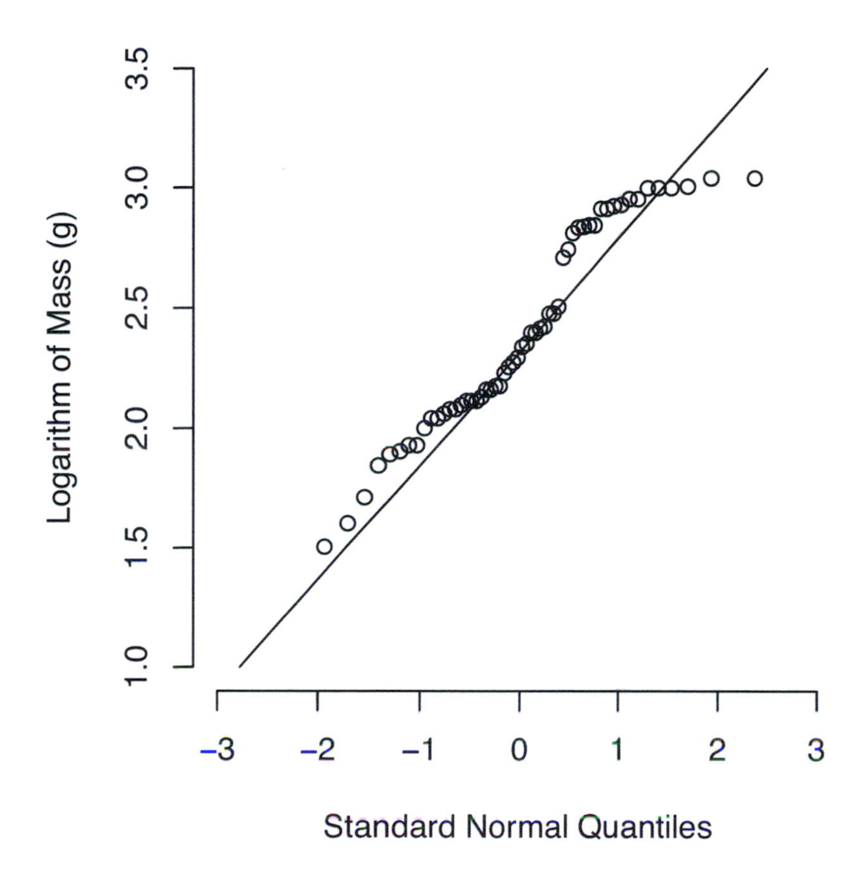

Figure 7.4 *Normal quantile-quantile plot of the common logarithm of mass of 56 perch*

`boxcox` function in the `MASS` package of the R statistical analysis system. The `boxcox` function can be used to calculate, and optionally plot, the profile log-likelihood for the parameter power λ of the Box-Cox transformation. One then chooses the value of λ by selecting the value that maximizes the log-likelihood function.

An illustration of the profile log-likelihood for the power parameter λ for perch mass is given in Figure 7.5. Note that a 95% confidence interval for λ is also depicted in Figure 7.5. The 95% confidence interval for the power parameter λ also includes the value zero that corresponds to the logarithmic transformation, which has already been tried with limited success.

With a little coaxing from the R function `boxcox`, the value of 0.16667 emerges as maximizing the profile log-likelihood for λ in Figure 7.5. It just so happens that $1/6 \approx 0.16667$ to five digits to the right of the decimal. As a convenience,

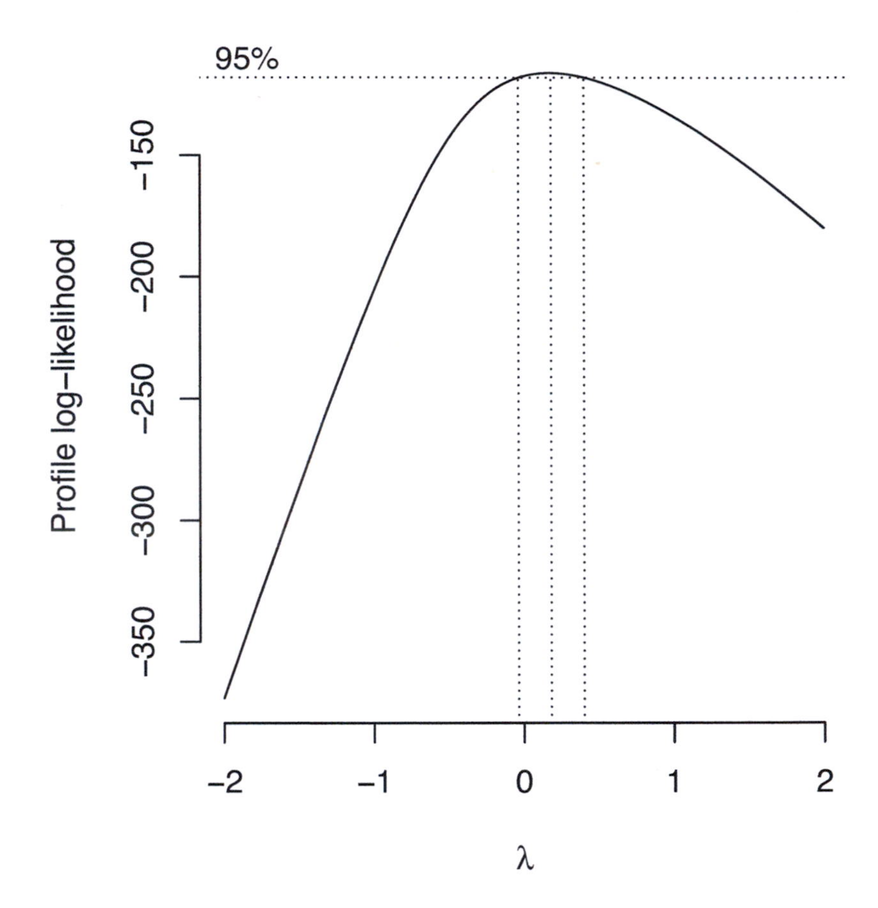

Figure 7.5 *Profile log-likelihood of the Box-Cox power parameter λ for the mass of 56 perch with 95% confidence interval*

setting $\lambda = 1/6$ leads to a transformation given by taking the sixth root of the mass of the perch caught in Längelmävesi. The normal quantile-quantile plot for this transformation is depicted in Figure 7.6 together with the normal quantile-quantile plots for the original untransformed data and the data after transformation by the common logarithm.

Both the logarithmic and sixth root transformations bring the data closer to normality. But it is debatable as to which of the two transformations is better. In Figure 7.6, it is also debatable whether either of the two transformations actually gets the job done.

As to how the log transform and sixth root transform of perch mass stack up against a fitted mixture of two normal distributions, Figure 7.7 reveals that the mixture of normals is the clear winner. Figure 7.7 gives the step plot of the

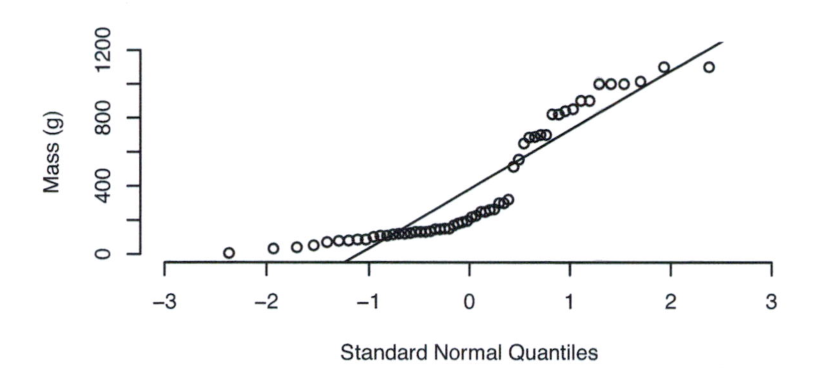

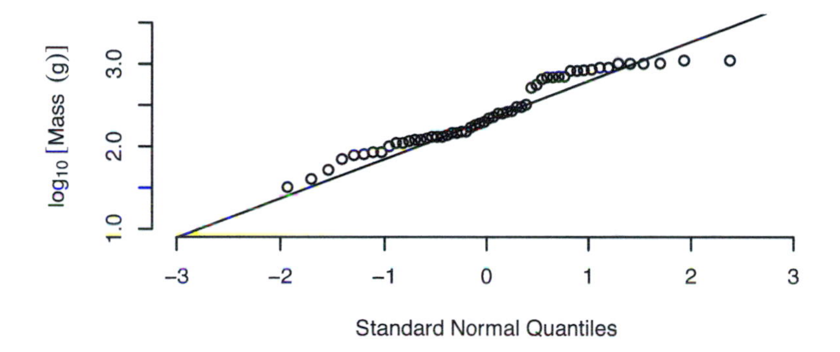

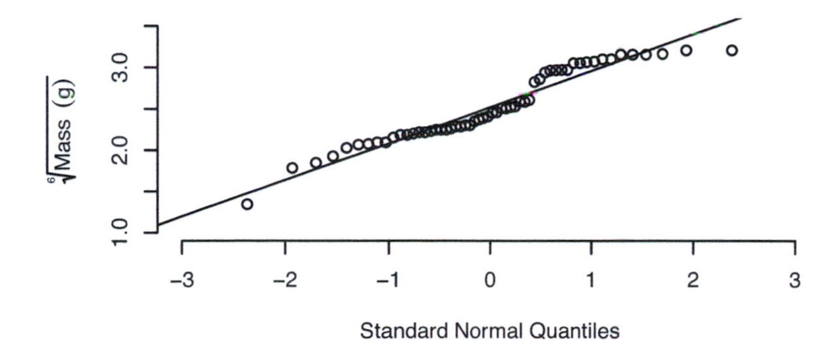

Figure 7.6 *Normal quantile-quantile plots of mass, the common logarithm of mass, and the sixth root of mass for 56 perch*

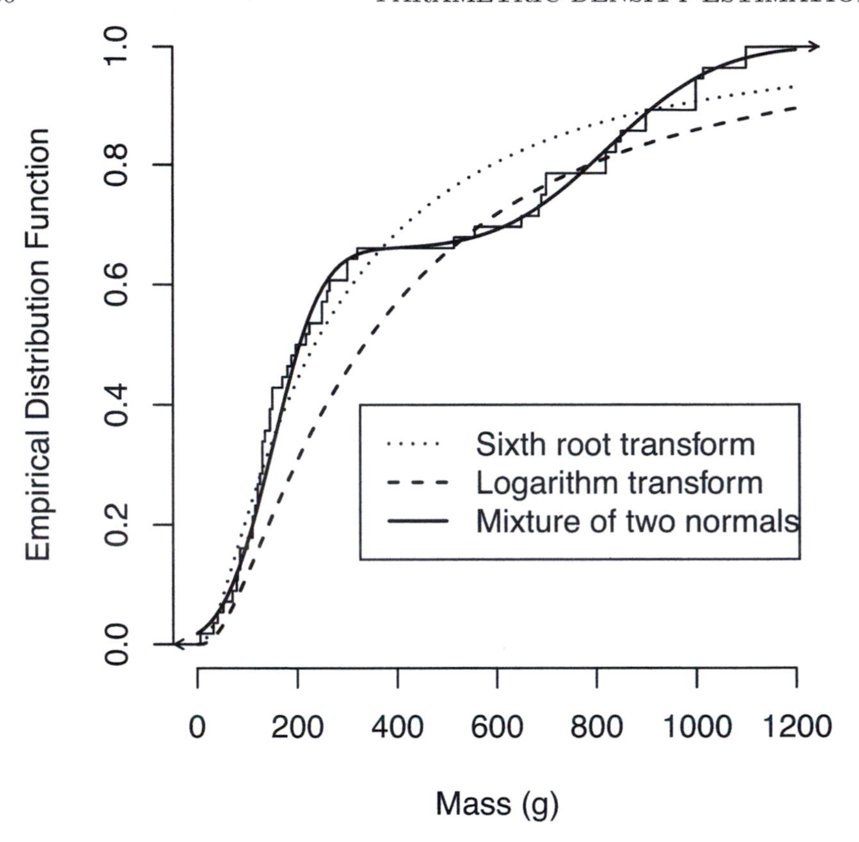

Figure 7.7 *Step plot of the empirical distribution function for perch mass with cumulative distribution function for a mixture of two normals, the common logarithm transformation, and the sixth root transformation*

empirical distribution of perch mass with the fitted cumulative distributions for the two transformations and the mixture of normals.

There are a couple of cautions with respect to the use of Box-Cox transformations. Sometimes it is difficult to use a Box-Cox transformation from a discipline-specific viewpoint. If the transformation does not make sense from a physical or biological science perspective, then there will be resistance to adopting it. In the case of the general public, or a less technically oriented audience, there can be barriers to overcome in explaining the transformation. Despite these cautions, the Box-Cox transforms have withstood the test of time and are widely used.

7.5 Pearson's Curves*

With a series of three papers published between 1895 and 1916, Karl Pearson [90, 93, 94] sought to create a family of distributions to satisfactorily represent observed data. This series of publications was actually the smaller part of a larger series of nineteen articles published in the *Philosophical Transactions of the Royal Society of London.* Pearson's [89] initial choice for the title of the larger series was *Contributions to the Mathematical Theory of Evolution.* His focus for this series was the presentation of statistical tools for human genetics (then called *eugenics*) with the first installment appearing as an abstract published in 1893.

Being an applied mathematician, Karl Pearson proposed his system of probability density functions based upon solutions to the ordinary differential equation:

$$\frac{df}{dx} = \frac{(x-a)f}{b_0 + b_1 x + b_2 x^2}, \tag{7.5}$$

the arbitrary coefficients b_0, b_1, and b_2 determined from sample moments.

Recall that the jth central sample moment m_j for a random sample $\{X_1, X_2, X_3, \ldots, X_n\}$ with sample mean $\overline{X}$ is given by

$$m_j = \sum_{i=1}^{n} (X_i - \overline{X})^j)/n \tag{7.6}$$

for $j = 2, 3, 4, \ldots$.

Keeping the mathematical details to a minimum, the use of Pearson's curves requires two more statistics in addition to $\overline{X}$ and m_2. One of these additional statistics estimates measures the lack of symmetry in the sample and it is called *skewness.* The other additional statistic is called *kurtosis* and it measures the heaviness of the tail or tails of the distribution. There are a number of different candidate measures of skewness and kurtosis. The measures used in the context of Pearson's curves calculating the sample skewness and kurtosis are defined, respectively, to be

$$\beta_1 = \frac{m_3^2}{m_2^3}; \tag{7.7}$$

$$\beta_2 = \frac{m_4}{m_2^2}. \tag{7.8}$$

Depending on the values of the normalized estimates of skewness β_1 and kurtosis β_2, the curve may be normal or one of several other types. Pearson [94] proposed twelve types in addition to the normal but only seven have withstood the test of time. These seven distributions, called types by Pearson, have the following probability density functions.

---

*This section can be omitted without loss of continuity.

Type I: (Beta Distribution)

$$f(x) = \frac{1}{B(\alpha, \beta)} x^{\alpha-1}(1-x)^{\beta-1}, \quad 0 \leq x \leq 1, \tag{7.9}$$

where $B(\alpha, \beta)$ is the beta function—a constant so that the area underneath the probability density function is equal to one over the interval $[0, 1]$.

Type II:

$$f(x) = \frac{1}{aB(\frac{1}{2}, \beta)} \left(1 - \frac{x^2}{a^2}\right)^{\beta-1}, \quad -a \leq x \leq a. \tag{7.10}$$

Type III: (Gamma Distribution)

$$f(x) = \frac{1}{\Gamma(\alpha)} x^{\alpha-1} e^{-x}, \quad x \geq 0, \alpha > 0, \tag{7.11}$$

where $\Gamma(\alpha)$ is the gamma function—a constant so that the area underneath the probability density function is equal to one.

Type IV:

$$f(x) = \kappa \left(1 + \frac{x^2}{\alpha^2}\right)^{-\beta} \exp\left[-\gamma \arctan\left(\frac{x}{\alpha}\right)\right], \quad \beta > \frac{1}{2} \tag{7.12}$$

where κ is a constant so that the area underneath the probability density function is equal to one. (This distribution is difficult to handle in practice owing to its complexity. The parameters α, β, γ, and κ used are obtained through numerical integration.)

Type V: (Inverse Gamma Distribution)

$$f(x) = \frac{\beta^\alpha}{\Gamma(\alpha)} x^{-(\alpha+1)} e^{-\beta/x}, \quad x \geq 0. \tag{7.13}$$

Type VI:

$$f(x) = \frac{1}{B(\alpha, \beta)} \frac{x^{\alpha-1}}{(1+x)^{\alpha+\beta}}, \quad x \geq 0, p > 0, q > 0. \tag{7.14}$$

This is sometimes called the *Beta distribution of the second kind*. The type I distribution being the *Beta distribution of the first kind*.

Type VII: (Student's t Distribution)

$$f(x) = \frac{1}{\alpha B(\frac{1}{2}, \beta)} \left(1 + \frac{x^2}{\alpha^2}\right)^{-(\beta+1)/2}. \tag{7.15}$$

This distribution is symmetric ($\beta_1 = 0$) with kurtosis $\beta_2 > 3$. It is a special case of the type IV distribution for which $\gamma = 0$. The usual Student's t distribution from sampling theory for the normal distribution is obtained

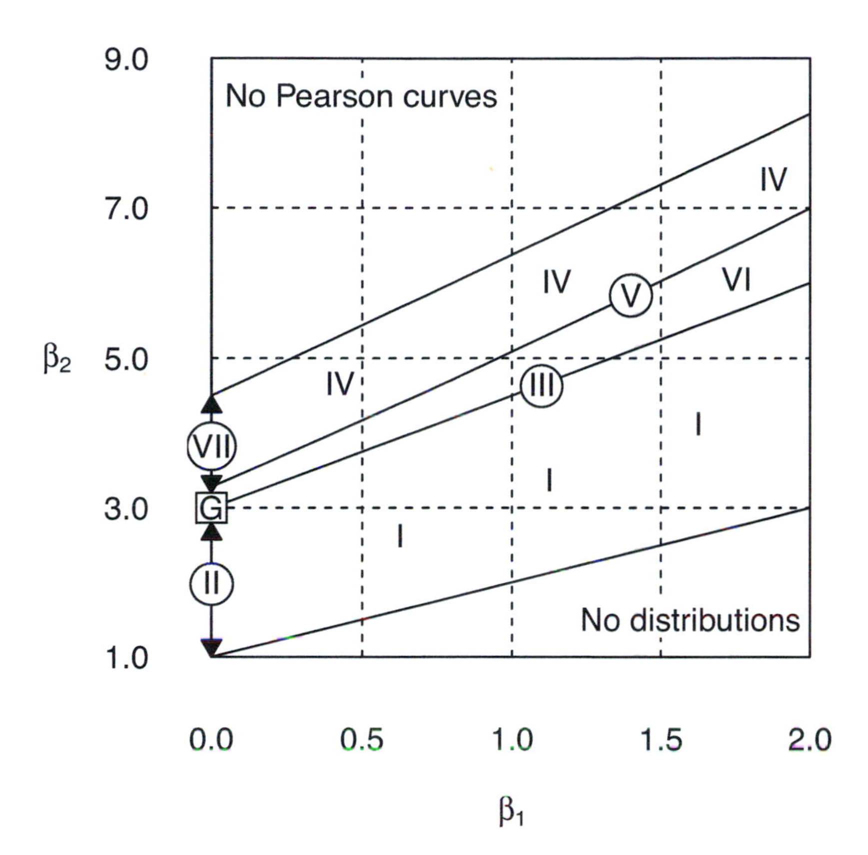

Figure 7.8 *Nomogram for Pearson's curves (note that any normal distribution corresponds to the point* $(0, 3)$*, and distributions of types II, III, V, and VII lie along lines or line segments as indicated)*

when $\alpha = \sqrt{\beta}$. In this case, β is referred to as the *degrees of freedom*. A parenthetical note: William S. Gosset (Student) deduced the form of the distribution that bears his name by obtaining a large number of random samples and using the method of Pearson's curves. (Sir Ronald A. Fisher was responsible for proving Student's result mathematically.)

Type G: (Gaussian Distribution)

$$f(x) = \frac{1}{\sigma\sqrt{2\pi}} e^{-\frac{(x-\mu)^2}{2\sigma^2}}. \tag{7.16}$$

This distribution is more frequently referred to as the normal distribution.

Pearson's type as a function of skewness β_1 and kurtosis β_2 is depicted in Figure 7.8. Note that the process of determining Pearson's type of a density curve is itself a graphical technique. The point corresponding to the coordinate pair (β_1, β_2) is plotted in the nomogram of Figure 7.8 and the type of distribution is determined by its location.

Note that every normal distribution is represented by the same single point in Figure 7.8, that is, where $\beta_1 = 0$ and $\beta_2 = 3$. On the other hand, the other types are represented by either lines, line segments, or regions in the Cartesian coordinate plane.

A version of Figure 7.8 first appeared in 1909 in an article by Rhind [98] who was working in Pearson's Biometric Laboratory in University College London. In fact, the article was published in the journal *Biometrika* edited by none other than Karl Pearson. A modified version of Rhind's [98] figure was published by Pearson [94] in the *Philosophical Transactions of the Royal Society of London, Series A*. Both versions are unusual and different from Figure 7.8 in having the direction of increase of β_2 being downward rather the usual upward for a vertical axis.

The distributions of types IV and VI are rarely used. The type V (inverse gamma distribution) is encountered in Bayesian theory as a conjugate prior for the sample variance when sampling from a normal distribution with known mean. It is remarkable that the normal, gamma, beta, and Student's t distributions are found as solutions of Pearson's ordinary differential equation.

To illustrate the process of the method of Pearson's curves, consider again the 56 observations of perch mass obtained by fisheries biologists in a research trawl on the Finnish freshwater lake known as Längelmävesi. For this sample, the estimate of the skewness statistic $\beta_1 = 0.674994$ and the kurtosis statistic $\beta_2 = 2.100796$. For the time being, ignore any estimates of variability for these two statistics. With the plot of the skewness and kurtosis statistics as a single point in the nomogram of Figure 7.9, the distribution of perch mass is apparently Pearson's type I.

The type I distribution is more commonly known as a beta distribution. The probability density function for this distribution is given in formula (7.9). Note that the beta distribution is nonzero only on the interval $[0, 1]$ but the maximum observed perch mass is 1100 g. To use formula (7.9) for the probability distribution of the beta distribution, the range of observations must be normalized. One could divide by 1100 g. On the other hand note the variable class-width histogram of Figure 6.8 that has utilized a range of 1200 g.

To estimate the parameters α and β in the beta distribution, the method of moments is used by noting that the mean μ and variance σ^2 of the distribution are as follows:

$$\mu = \frac{\alpha}{\alpha + \beta}; \tag{7.17}$$

$$\sigma^2 = \frac{\alpha\beta}{(\alpha + \beta)^2(\alpha + \beta + 1)}. \tag{7.18}$$

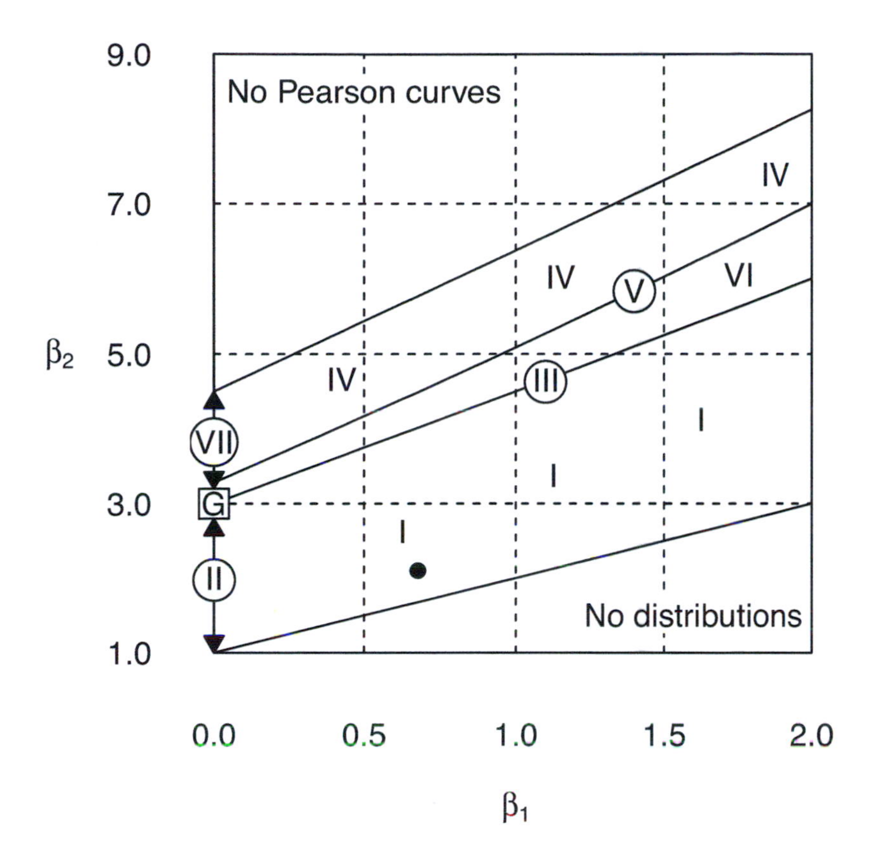

Figure 7.9 *Nomogram for Pearson's curves with data point for perch mass*

From the 56 observations, the sample mean $\bar{x} = 382.2$ g and the sample standard variance $s^2 = 120838.1$ g^2. By setting the sample mean equal to μ and the sample variance equal to σ^2 and substituting into equations (7.16) and (7.17), with an adjustment of 1200 g to standardize the data to an interval of 1, one obtains the following simultaneous nonlinear equations:

$$\frac{\bar{x}}{1200} = \frac{\alpha}{\alpha+\beta}; \tag{7.19}$$

$$\frac{s^2}{1200^2} = \frac{\alpha\beta}{(\alpha+\beta)^2(\alpha+\beta+1)}. \tag{7.20}$$

Solving these two equations leads to the estimates $\hat{\alpha} = 0.52$ and $\hat{\beta} = 1.11$ of the parameters in the beta distribution. Using these estimates and dividing all the observations by 1200 g when calculating the beta density, one obtains the beta density curve overlaid on the histogram in Figure 7.10.

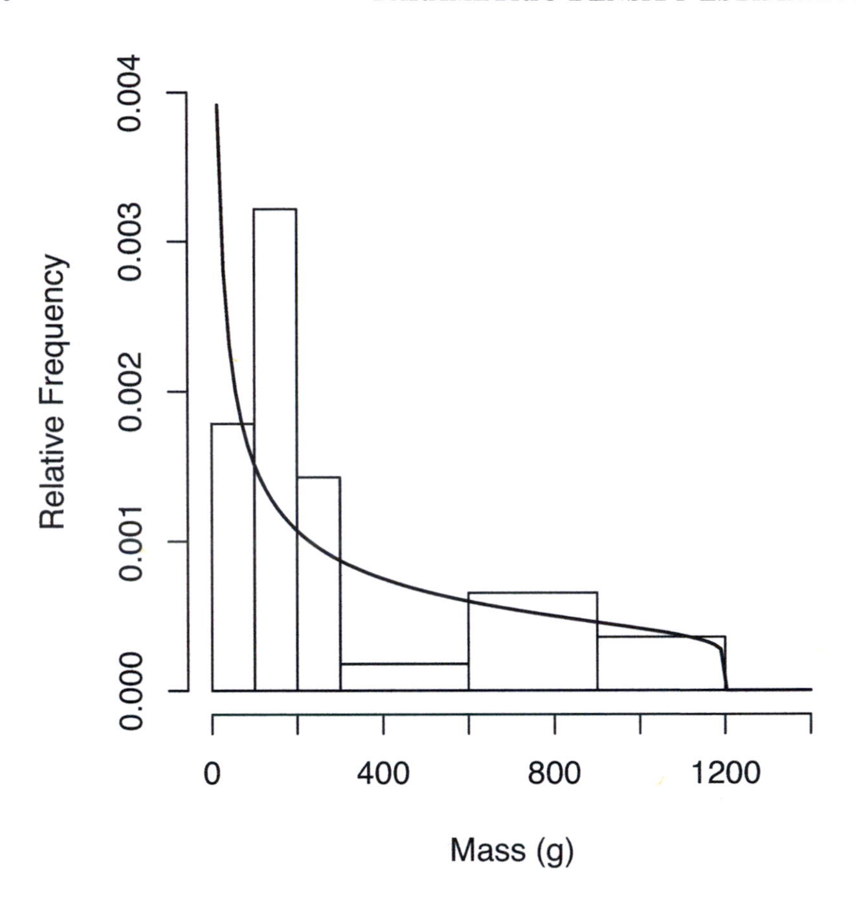

Figure 7.10 *Relative frequency histogram of the mass of 56 perch with varying class widths and a beta density function*

Note that Figure 7.10 also contains the variable class widths of Figure 6.8. The fit is not good. The presence of the smooth curve for the density in Figure 7.10 adds to the impression of the distribution being asymmetric with a heavy tail.

There is a caution to be made when considering the use of Pearson's curves. This was noted in 1909 by Rhind [98] who wrote: "... in practice these high moments are subject to very large percentage errors, rendering their use extremely undesirable ... ." It can be shown for normal samples that:

$$\operatorname{var}(\sqrt{\beta_1}) = \frac{6n(n-1)}{(n-2)(n+1)(n+3)} \approx \frac{6}{n}; \tag{7.21}$$

$$\operatorname{var}(\beta_2) = \frac{24n(n-1)^2}{(n-3)(n-2)(n+3)(n+5)} \approx \frac{24}{n}. \tag{7.22}$$

Even for the normal distribution, the variability associated with the estimates

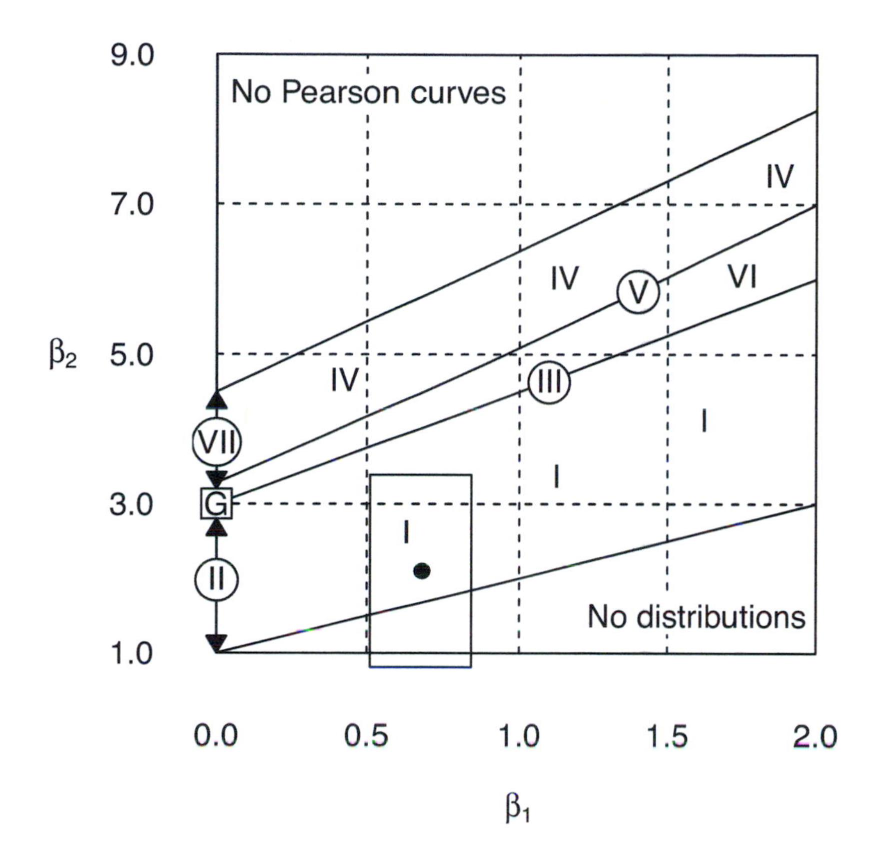

Figure 7.11 *Nomogram for Pearson's curves with data point and confidence rectangle for perch mass*

of the skewness and kurtosis can be quite high. This could result in a number of different suggestions as to Pearson's type of curves even in samples of moderate size drawn from a normal population. This can limit the applicability of this method.

For an illustration of the application of this theory, the Finnish perch mass data is used once again. A confidence rectangle with sides equal to two standard errors in either direction for both the skewness β_1 and kurtosis β_2 is added to the Pearson's nomogram of Figure 7.11. For any sample of size 56, the rectangle is quite large and could include all of Pearson's types of curves. In this particular example of perch mass, however, there are only two possible choices: Pearson's type I distribution or none at all.

In the example presented in this section, the beta distribution has been fitted to the data with the method of moments. There are more efficient ways to estimate model parameters than this, the maximum likelihood method being one in many situations.

In looking at the fit of the beta distribution to the perch mass data in Figure 7.10, one possible question is whether it is possible to get a better fit with a different parametric model than one of Pearson's curves. It was previously commented in the context of exploring variable class widths with perch mass, that there could be a biological explanation for the apparent bimodality in perch mass.

It was conjectured that the perch spawn only once per year in the spring. If the research trawl occurred in the summer, then the population of perch might consist of a large number of juveniles compared to a small number of fish from previous seasons. The reasoning is that the mortality risks for juveniles are greater as they tend to get eaten by larger fish of the same and other species. Hence, the mass of the perch population in Längelmävesi could be characterized by the mixture of two distributions: one distribution for juveniles and another distribution for adult fish.

Figure 7.12 depicts a histogram of perch mass with the probability density function corresponding to a mixture of two normal distributions. This mixture consists of one normal distribution with a mean of 150.2 g and a standard deviation of 78.1 g and another with a mean of 826.0 g and standard deviation of 171.9 g with a weighting of 65.94% for the first of these two normal distributions. These parameter estimates have been determined by the method of maximum likelihood.

The fit of the mixture of normals in Figure 7.12 to the variable class-width histogram of perch mass looks considerably better than that of the beta distribution in Figure 7.10. One of Pearson's curves does not seem to have fit the bill. But since the normal distribution is one of Pearson's types, it can be argued that a mixture of Pearson's curves can do quite well.

Figure 7.13 is the counterpart of Figure 7.12 rendered by `ggplot2`. Both plots rely on the general purpose optimization function `optim` in the package `stats` to find the maximum likelihood estimates of the means, standard deviations, and the proportion for the mixture of two normal distributions. The following simple R function for the probability density of the mixture of two normal distributions is used when plotting the density function in both Figure 7.12 and Figure 7.13.

```
dmixnorm<-function(x,p) {
p[5]*dnorm(x,mean=p[1],sd=p[3])+
(1.0-p[5])*dnorm(x,mean=p[2],sd=p[4])
}
```

In the previous two sections, nonparametric density estimates by kernels and splines were considered. To see how they stack up against a mixture of two normals for perch mass, see Figure 7.14.

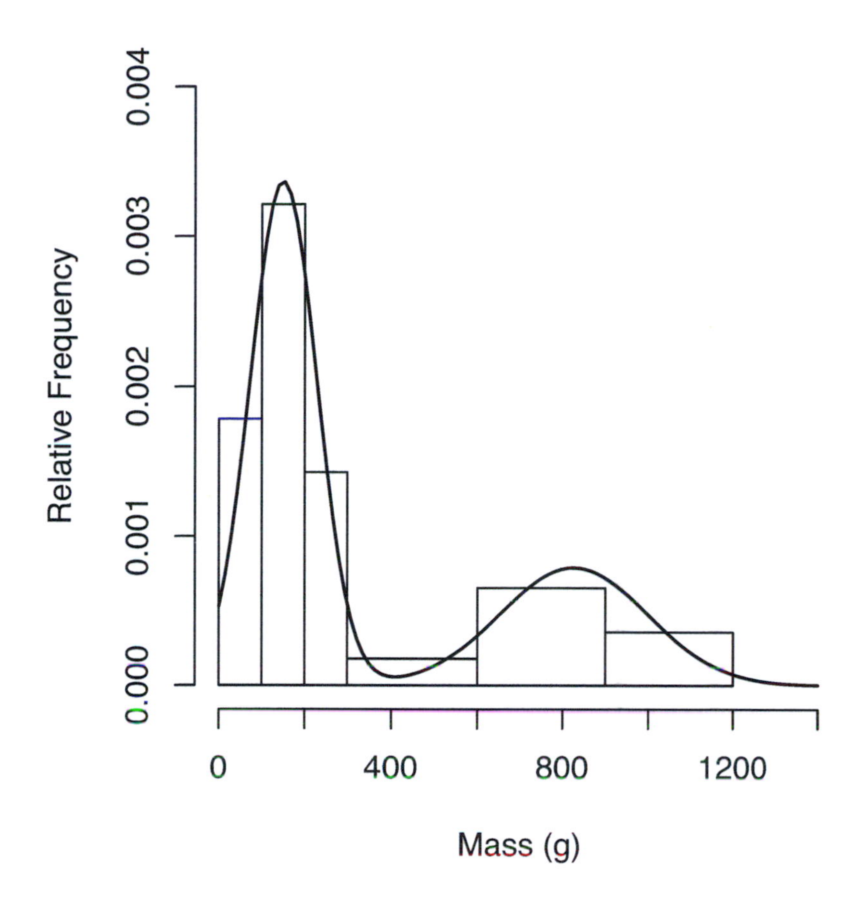

Figure 7.12 *Relative frequency histogram of the mass of 56 perch with varying class widths and a mixture of two normal distributions*

Undeniably, the better fitting density curve is the parameter one given by a mixture of two normals. Based upon smoothness, second place could be awarded to the Gaussian kernel density estimate with bandwidth estimated by oversmoothing with $\hat{\lambda}_{OS} = 177.78$ g.

Arguably, one purpose of nonparametric density estimation is to draw attention to features in a histogram such as skewness, heavy tails, or more than one mode. This would suggest possible paths for selecting a parametric model for a probability density function. In this sense, the kernel and spline density estimates were successful for perch mass.

The suggestion of a major mode near 200 and a second minor mode near 800 suggested that a parametric model formed from the sum of two normal densities might just do the trick.

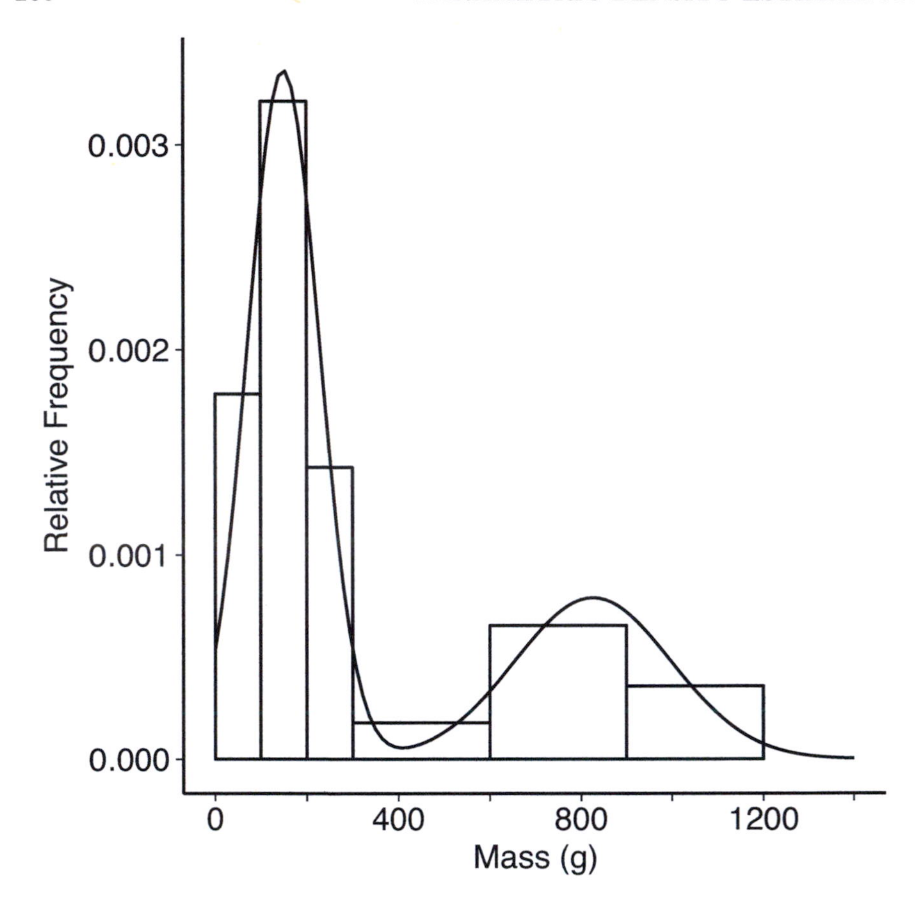

Figure 7.13 *Relative frequency histogram of the mass of 56 perch with varying class widths and a mixture of two normal distributions with* `ggplot2`

The parametric density estimate in Figure 7.14 is formed from the sum of two Gaussian density functions while the nonparametric kernel density estimate in Figure 7.14 is formed from the sum of 56 Gaussian density functions.

The next section considers a different way of summing Gaussian densities to produce a parametric density estimate.

7.6 Gram-Charlier Series Expansion*

Another approach for estimating densities is to approximate the density with a finite number of terms from an infinite series expansion. One method applies

*This section can be omitted without loss of continuity.

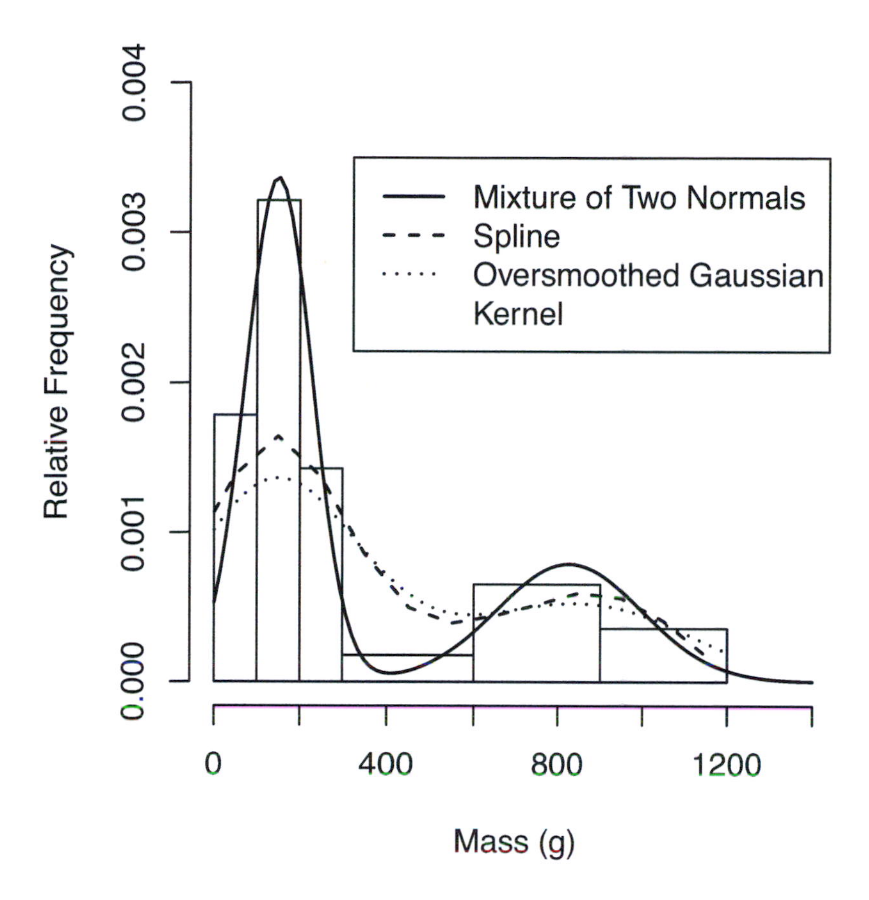

Figure 7.14 *Relative frequency histogram of the mass of 56 perch with different density estimates*

the Gram-Charlier series of type A:

$$f(x) = \phi(x)\left\{1 + c_1 H_1(x) + c_2 H_2(x) + c_3 H_3(x) + c_4 H_4(x) \cdots\right\} \tag{7.23}$$

where

$$\phi(x) = \frac{1}{\sqrt{2\pi}} \exp\left\{-\frac{x^2}{2}\right\} \tag{7.24}$$

is the standard normal density and the functions $\{H_i\}_{i=1}^{\infty}$ are Chebyshev-Hermite polynomials:

$$\begin{aligned} H_0(x) &= 1, \\ H_1(x) &= x \\ H_2(x) &= x^2 - 1, \\ H_3(x) &= x^3 - 3x, \end{aligned}$$

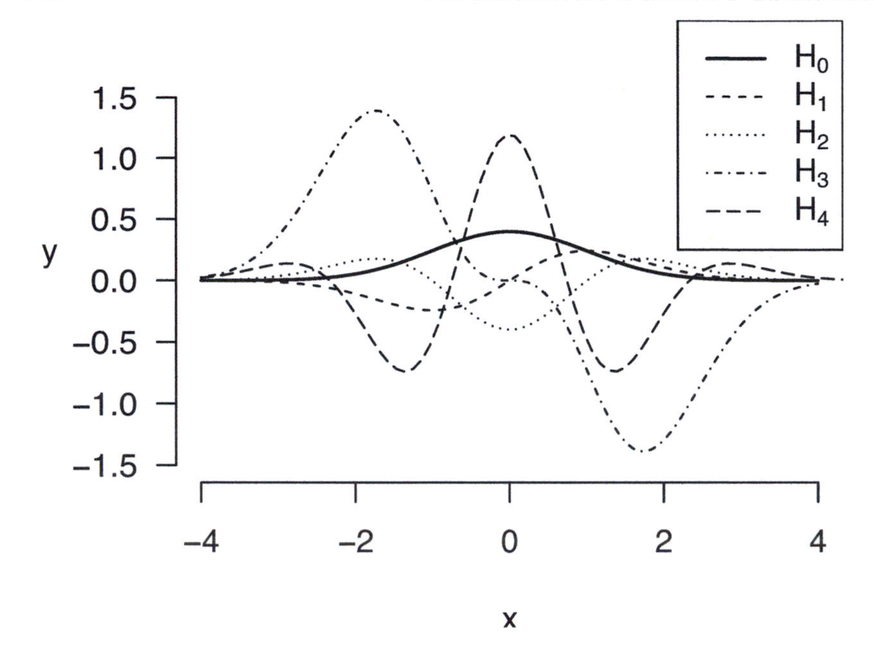

Figure 7.15 *First five functions in the Gram-Charlier series of Type A*

$$H_4(x) = x^4 - 6x^2 + 3,$$
etc.

The first five functions $\{H_i(x)\phi(x)\}_{i=1}^{2}$ of the Gram-Charlier series of type A are plotted in Figure 7.15.

Let the central moments about the mean μ be defined by $\mu_j = \mathrm{E}\left[(X-\mu)^j\right]$ for $j = 2, 3, \ldots$. Then coefficients $\{c_i\}$ in formula (7.22) are functions of the central moments:

$$\begin{aligned} c_1 &= 0, \\ c_2 &= \frac{1}{2}(\mu_2 - 1), \\ c_3 &= \frac{1}{6}\mu_3, \\ c_4 &= \frac{1}{24}(\mu_4 - 6\mu_2 + 3), \\ &\text{etc.} \end{aligned}$$

With only a random sample available, each population parameter μ_j is estimated by its sample counterpart m_j.

Application of the Gram-Charlier expansion is eased by standardization of the random sample before estimating the coefficients $\{c_i\}$ in the series expansion. A random variable x with known mean μ and standard deviation σ is standardized as follows:

$$z = \frac{x - \mu}{\sigma}. \tag{7.25}$$

Then the Gram-Charlier series expansion of type A simplifies to

$$f(z) = \alpha(z)\left[1 + \frac{1}{6}\mu_3^* H_3 + \frac{1}{24}(\mu_4^* - 3)H_4 + \ldots\right] \tag{7.26}$$

where μ_3^* and μ_3^* denote the third- and fourth-order central moments of the standardized variable z, respectively. With only a random sample available, these parameters are replaced by their corresponding sample estimates.

A parametric density estimate for the sample of the mass of 56 perch is given in Figure 7.16 using the Gram-Charlier series of type A. This figure also includes the constituent terms of H_0, H_3, and H_4 as perch mass was standardized according to equation (7.25) with the population parameters replaced by their statistical counterparts.

Note that the density estimate in Figure 7.16 has a single mode. The terms involving H_3 and H_4 have the effect of adding skewness to the H_0 term. Figure 7.17 presents the histogram for perch mass with variable class widths with just the density estimate by the first four terms included in the Gram-Charlier series of type A. Other than the mode in the series estimate and the mode in the histogram coinciding, the fit is not good. Improvement might be gained by adding additional terms to the infinite series expansion. But by using sample estimates of moments rather than the population moments, the outlook is not hopeful for a sample of size 56 as the variance of the moment estimates only increases with increasing order.

7.7 Conclusion

Brown and Huang [15] discuss how best to approximate a histogram by a normal density. Instead of using the usual unbiased estimators of the mean and variance of a normal population, they suggest a numerical method to estimate these parameters that minimizes the integral of the squared difference between the nonparametric density estimate and the normal probability density function. In their paper the nonparametric density estimate that they choose is that of the unsmoothed histogram of fixed class widths.

For an alternative review of Pearson's curves, see pages 159 to 166 in Volume I of *The Advanced Theory of Statistics* by Kendall and Stuart [76]. The approach in this chapter has been to fit curves to the probability density function. Kendall and Stuart [76] [pp. 185–186] briefly discussed and provide references for fitting curves to the cumulative distribution function using Burr's distributions.

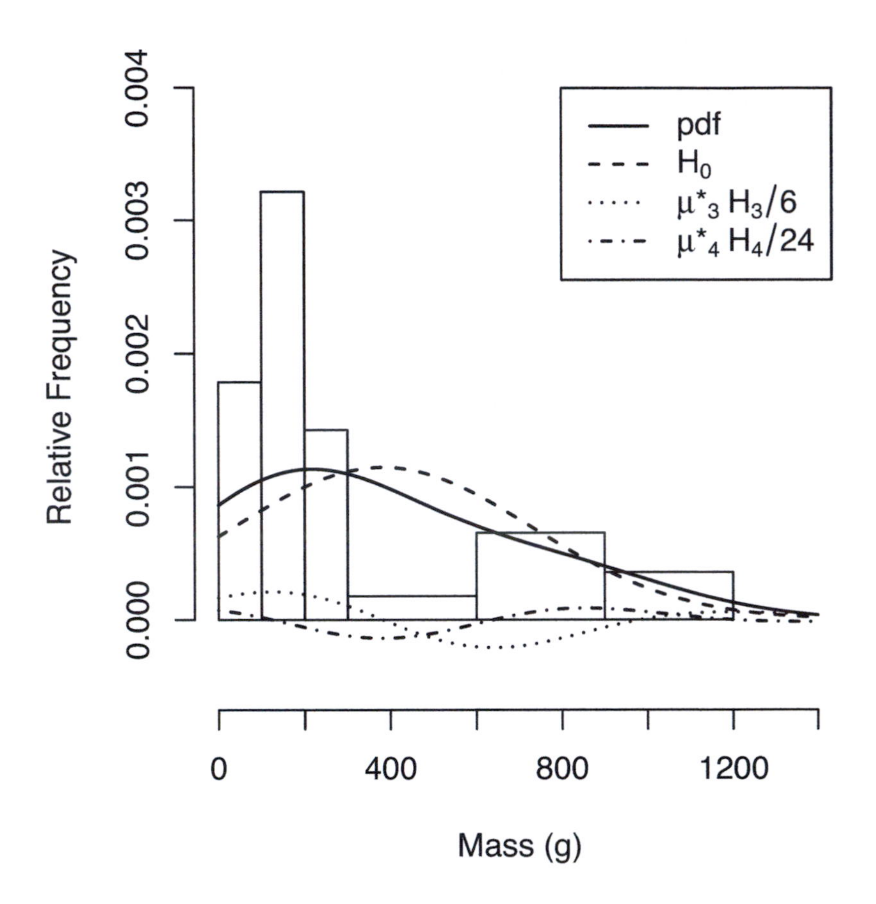

Figure 7.16 *Gram-Charlier type A density estimate for the mass of 56 perch (the components sum to the probability density estimate [pdf])*

For a discussion of Gram-Charlier series expansions refer to Kendall and Stuart [76] (pp. 166–172). The use of Chebyshev-Hermite polynomials has two different representations. In this chapter, we have discussed the Gram-Charlier series of type A, which requires parametric estimation of moments. Kendall and Stuart [76] (pp. 166–172) also discussed an equivalent representation known as Edgeworth's form of the type A series, which relies instead on estimates of the cumulants.

Although all infinite series expansions will converge to the function being approximated, the rate of convergence will vary depending on the series chosen. For example, for distributions similar to the normal distribution, the Chebyshev-Hermite polynomials tend to converge with fewer number of terms than other series. This results from the weight function ϕ being a normal density. It is reasonable to choose the infinite series expansion with a weight

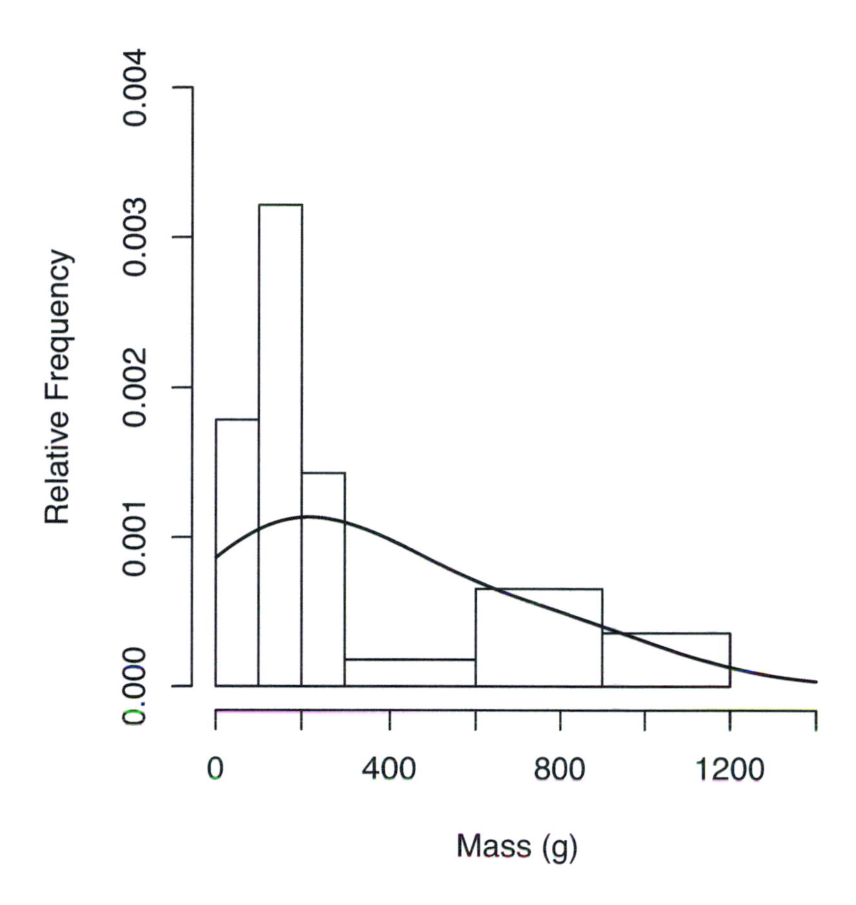

Figure 7.17 *Histogram of mass for a sample of 56 perch with variable class widths and Gram-Charlier type A density estimate*

function that nearly matches a reasonable approximation to the sampled distribution. Use Laguerre polynomials for a population with a distribution close to the exponential distribution and Legendre polynomials for a population with a distribution close to the uniform distribution.

The Box-Cox transformations are proposed in Box and Cox [12]. In addition to the Box-Cox transformations to normality, there are infinite series polynomial transformations that can be truncated and used as well—these are known as Cornish-Fisher expansions and are based on the Edgeworth expansions previously mentioned. Consult Kendall and Stuart [76] (pp. 172–185) for discussion of the Cornish-Fisher expansions and other transformations as well.

7.8 Exercises

1. Consider the data for the total compensation in 2008 received by chief executive officers employed by industrial companies listed in Table 4.3.
   (a) Produce a histogram for total compensation. Use Sturge's rule to determine the class width.
   (b) Estimate the mean and variance for total CEO compensation. Add a normal probability density function to your histogram of part (a). Do the data appear to be normal?
   (c) Produce a normal quantile plot or a normal probability plot. Add a normal distribution reference line to the plot. Do this by hand if your software package does not automatically provide one.
   (d) Which plot would you choose to include in a written media presentation? A histogram with a parametric density curve overlaid or a normal probability plot with a reference line? Justify your answer.
   (e) Which plot would you choose to include in an article for a business research journal? A histogram with a parametric density curve overlaid or a normal probability plot with a reference line? Justify your answer.
2. In 1882, Simon Newcomb measured the time required for light to travel from his laboratory on the Potomac River to a mirror at the base of the Washington Monument and back, a total distance of about 7400 meters. These measurements were used to estimate the speed of light. Table 4.4 contains the estimated speed of light for 66 trials. These data were reported in an article by Steven Stigler [115] in 1977 in *The Annals of Statistics.*
   (a) Produce a histogram for the data in Table 4.4 using the Freedman-Diaconis Rule.
   (b) Add a normal curve to the histogram you obtained for part (a).
   (c) Produce an outlier boxplot for the data in Table 4.4.
   (d) Comment on the impact of outliers when fitting a normal curve for a histogram.
3. On June 21st in 1798, Henry Cavendish [17] read before The Royal Society of London a paper concerning his experiments to determine the density of the earth. He modified an apparatus originally conceived by the Rev. John Michell. Inside a mahogany case was suspended a six-foot long slender rod from which two-inch diameter lead balls were suspended at either end. To estimate the density of the earth, Cavendish measured the displacement of the rod by the gravitational pull of a pair of eight-inch diameter lead balls positioned nearby and suspended by copper rods at the end of a wooden bar. Cavendish's experiments resulted in the 29 estimates for the density of the earth as given in Table 5.5.
   (a) Produce a histogram for Cavendish's estimates of the density of the earth relative to water. Use Doane's Rule to determine the class width.

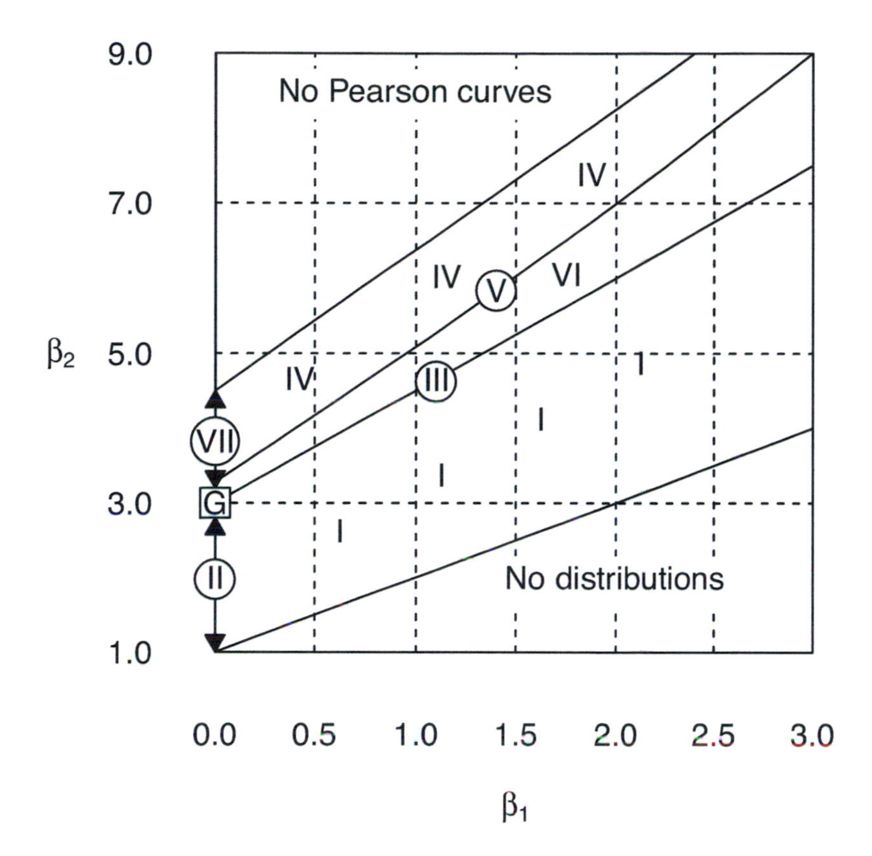

Figure 7.18 *Nomogram for Pearson's curves with horizontal axis expanded to 3 (note that any normal distribution corresponds to the point* $(0, 3)$, *and distributions of types II, III, V, and VII lie along lines or line segments as indicated)*

(b) Using the sample mean and sample variance, fit a normal curve to Cavendish's data and add this to your histogram of part (a). Compare the fit of the normal curve and Pearson's curve to the histogram.

4. Consider Cavendish's estimates for the density of the earth in Table 5.5.
   (a) Estimate the sample skewness β_1 and the sample kurtosis β_2 for Cavendish's density estimates. Plot the point (β_1, β_2) on the nomogram for Pearson's curves in Figure 7.18.
   (b) From the point alone plotted in part (b), determine the distribution. Fit this distribution to the data and then plot its probability density function on the histogram you produced in part (a). How good does the fit look? Comment.

(c) Compare the fit of the normal curve in the previous exercise and Pearson's curve to the histogram.

5. Consider Cavendish's estimates for the density of the earth in Table 5.5.
   (a) Produce a histogram for Cavendish's estimates of the density of the earth relative to water. Use the Freedman-Diaconis Rule to determine the class width.
   (b) Estimate the sample skewness β_1 and the sample kurtosis β_2 for Cavendish's density estimates. Plot the point (β_1, β_2) on the nomogram for Pearson's curves in Figure 7.9.
   (c) From the point alone plotted in part (b), determine the distribution. Fit this distribution to the data and then plot its probability density function on the histogram you produced in part (a). How good does the fit look? Comment.
   (d) Consider the variability of your estimate of the point (β_1, β_2) by using equations (7.21) and (7.22) to plot a rectangle around the point you plotted on the nomogram for Pearson's curves. Which other of Pearson's distributions might be the true underlying distribution for the data?
6. Consider Cavendish's estimates for the density of the earth in Table 5.5.
   (a) Produce a histogram for Cavendish's estimates of the density of the earth relative to water. Use Scott's Rule to determine the uniform class width.
   (b) Fit a Gram-Charlier expansion of type A to the data to fourth order. Report the estimates of the third- and fourth-order central moments.
   (c) Plot the Gram-Charlier expansion you obtained in part (b) on the histogram of part (a).
   (d) Add a parametric normal approximation to the histogram of part (a). Is there much difference between the normal curve and the Gram-Charlier expansion? Comment.
   (e) Add the third- and fourth-order components of the Gram-Charlier expansion to the histogram of part (a). Do either of these two terms add much to the Gram-Charlier expansion? Comment.
   (f) Based on the additions to the histogram in part (d), comment on which of the skewness or kurtosis components appears to have the greater impact on the fourth-order Gram-Charlier parametric density estimate.
7. Consider the data for the total compensation in 2008 received by CEOs employed by industrial corporations listed in Table 4.3.
   (a) Produce a normal quantile-quantile plot for total compensation. Do the data appear to be normally distributed? Discuss.
   (b) Consider a square root transformation of total compensation. Produce a normal quantile-quantile plot of the transformed data. Do the data appear to be normally distributed? Discuss.
   (c) Compare the two quantile-quantile plots for total compensation from your answers to parts (a) and (b). Which appears to be more normal:

Survival time (months)							
4.04	4.70	5.82	6.15	7.07	7.36	7.56	7.76
7.82	7.86	7.86	7.89	8.15	8.19	8.84	9.04
9.17	9.24	9.47	9.67	10.03	10.06	10.13	10.26
10.32	10.36	10.42	10.42	10.45	10.52	10.52	10.72
10.75	10.75	11.15	11.18	11.28	11.34	11.47	11.77
11.80	11.93	12.03	12.30	12.39	12.53	12.53	12.53
12.56	12.56	12.82	12.95	13.05	13.12	13.15	13.18
13.28	13.32	13.74	13.91	14.04	14.17	14.33	14.33
14.93	14.93	14.99	15.02	15.02	15.12	15.35	15.52
15.58	15.88	15.95	15.95	16.01	16.11	16.14	16.27
16.41	16.41	16.60	16.67	16.77	17.13	17.16	17.23
17.52	17.79	17.82	17.98	18.02	18.02	18.48	18.61
18.81	18.81	19.13	19.17	19.20	19.20	19.30	19.46
19.53	19.63	19.73	19.82	19.86	19.89	20.05	20.12
20.19	20.22	20.28	20.32	20.65	20.65	20.68	20.68
20.78	20.81	20.84	21.11	21.14	21.47	21.50	21.70
21.80	21.90	22.45	22.62	23.31	23.54	23.57	23.64
23.70	23.70	23.70	23.84	24.03	24.16	24.20	24.46
24.46	24.69	24.72	24.79	25.18	25.35	25.45	25.97
25.97	27.12	27.16	27.48	27.65	28.04	28.27	28.64
29.10	29.98	30.02	30.05	30.97	31.27	32.55	32.61
33.83	34.88	35.38	36.62	38.37	42.38	43.00	44.42
44.65	47.28	47.64	53.82	55.69	57.50	58.82	64.64

Table 7.1 *Survival time in months for lung cancer patients*

the original data or the data after a square root transformation? Justify your answer.

(d) Find an EDF goodness-of-fit test for normality in your statistical software package. The Shapiro-Wilk test is one such test. The Anderson-Darling test is another and is better. Test both total CEO compensation and its square root for normality. Which is closer to a normal distribution on this basis, the original data or the data after the square root transformation? Discuss.

8. Survival times in months for 184 patients who died from limited-stage, small-cell lung cancer are given in Table 7.1. These data have been provided by Patricia Tai, M.B., an oncologist with the Allan Blair Cancer Center in Regina, which is located in the Canadian province of Saskatchewan. The data in Table 7.1 involve cases diagnosed in Saskatchewan between 1981 and 1998 with follow-up until 2002. The data were the subject of an article by Dr. Tai and colleagues [119] that was published in the *International Journal of Radiation Oncology, Biology, and Physics* in 2003.

(a) Produce an outlier boxplot for survival times in Table 7.1. Comment on the shape and whether there are any outliers.
(b) Produce a stemplot for survival times in Table 7.1. Comment on the shape and whether there are any outliers.
(c) Are your characterizations of the distribution of survival time based upon the stemplot in part (b) different from those for the outlier boxplot in part (a)? Discuss.
(d) Based upon your answers to parts (a) and (b), is it reasonable to fit a normal probability density function to the data? Justify your answer.

9. Survival times in months for 184 patients who died from limited-stage, small-cell lung cancer are given in Table 7.1. The data in Table 7.1 involve cases diagnosed in Saskatchewan between 1981 and 1998 with follow-up until 2002.
   (a) Produce a histogram for survival times in Table 7.1. Use Doane's Rule to determine class width.
   (b) The distribution in your answer to part (a) is clearly not normal. One approach to finding a parsimonious parametric model for a nonnormal probability density is to find a suitable Box-Cox transformation. Do this.
   (c) Add the density distribution corresponding to your estimate of the transformation parameter λ in part (b) to the histogram given in your answer to part (a).
   (d) Is $\lambda = 0$ close to your answer in part (b)? Discuss.

10. Survival times in months for 184 patients who died from limited-stage, small-cell lung cancer are given in Table 7.1. The data in Table 7.1 involve cases diagnosed in Saskatchewan between 1981 and 1998 with follow-up until 2002.
   (a) Produce a kernel density estimate for survival times in Table 7.1.
   (b) The nonparametric density estimate in your answer to part (a) is clearly not normal. The challenge then is to find a probability distribution that does describe survival times for patients with limited-stage, small-cell lung cancer. Estimate the sample skewness β_1 and the sample kurtosis β_2 for the survival times in Table 7.1. Plot the point (β_1, β_2) on the nomogram for Pearson's curves with the expanded horizontal axis in Figure 7.18. Which type of Pearson's distributions does this point suggest?
   (c) Consider the variability of your estimate of the point (β_1, β_2) by using equations (7.21) and (7.22) to plot a rectangle about the point you plotted on the nomogram for Pearson's curves. Which other of Pearson's distributions might be the true underlying distribution for the data?
   (d) Is the gamma distribution included in your answer to part (c)? To check how well the gamma distribution fits the data, produce a gamma quantile-quantile plot. Comment.

11. The gamma distribution and the lognormal distribution are frequently used to model lifetime data. The gamma distribution is the Type III distribution in the system of Pearson's curves. The lognormal distribution is obtained when the logarithm of a random variable is approximately normal. Often the choice of model for lifetime data comes down to a choice between these two models.

    (a) Produce a step plot of the EDF for the survival time data in Table 7.1 for limited-stage, small-cell lung cancer.

    (b) Estimate the parameters of a gamma distribution for the survival time data in Table 7.1 and add the corresponding cumulative distribution function to the EDF plot of part (a).

    (c) Estimate the parameters of a normal distribution for the logarithm of survival time given in Table 7.1 and add the corresponding cumulative distribution function to the EDF plot of part (a).

    (d) Based upon the cumulative distribution functions plotted on the EDF plot, which of the gamma distribution and the lognormal distribution function best fits the data? Discuss.

12. North America's first universal medical insurance plan was enacted in the Canadian province of Saskatchewan on 1 July 1962. The province of Saskatchewan also has one of the oldest cancer registries in the world, dating to 1932. The survival times presented in Table 7.1 result not from a random sample of patients in Saskatchewan but a census. No census is perfect. But the Saskatchewan Cancer Registry comes close. The data reported in Table 7.1 are nearly complete, with just 2% lost to follow-up [119]. Consequently, the jagged step plot of your answer to part (a) in the previous exercise can be considered that of the actual distribution function and not an empirical approximation. The smooth curves in parts (b) and (c) of the previous exercise for the gamma and lognormal distributions are, however and ironically, approximations. Propose a justification for seeking a simple parametric approximation to a known distribution and find this approximation.

Part IV

Two Variables

Chapter 8

Depicting the Distribution of Two Discrete Variables

8.1 Introduction

The majority of situations involving statistical analysis involves not one variable but two or more. An important challenge in the process of data analysis for two or more variables is discovering whether it is possible to discriminate among different populations. The next step is then to find explanations for observed patterns. At each step of the way, graphical displays are essential.

The simplest multivariable setting is that of two discrete random variables. There is no lack of options for depicting the distribution of two discrete random variables. The discussion of graphical methods for this situation begins with a graphical display that is not as popular as a few other displays to be discussed later in this chapter. But this display has much in its favor and perhaps someday it will be more widely adopted.

8.2 Learning Outcomes

When you complete this chapter, you will be able to do the following.

- Be able to execute a grouped dot chart for two discrete variables with either the `graphics` package or the `ggplot2` package.
- Be able to draft a grouped dot-whisker chart for two discrete variables with either the `graphics` package or the `ggplot2` package.
- Be able to prepare a two-way dot chart for two discrete variables with either the `lattice` package or the `ggplot2` package.
- Be knowledgeable about the multi-valued dot chart.
- Be able to execute a side-by-side bar chart for two discrete variables with either the `graphics` package or the `ggplot2` package.

	Hair Color			
Eye Color	**Black**	**Brunette**	**Red**	**Blond**
Brown	68	119	26	7
Hazel	15	54	14	10
Green	5	29	14	16
Blue	20	84	17	94

Table 8.1 *Hair color and eye color in a convenience sample*

- Be able to draft a side-by-side bar-whisker chart with either the `graphics` package or the `ggplot2` package.
- Be able to prepare any form of a side-by-side stacked bar chart for two discrete variables with either the `graphics` package or the `ggplot2` package.
- Even though it is not recommended, be knowledgeable about the side-by-side pie chart.
- Be knowledgeable about the mosaic plot and be able to draft it using either the function `mosaicplot` in the `graphics` package or the function `geom_mosaic` in the `ggmosaic` extension package for `ggplot2`.

8.3 The Grouped Dot Chart

The first example to begin the discussion of how best to depict the distribution of two discrete variables concerns the joint distribution of eye color and hair color in humans. The inheritance of hair and eye color is not governed by simple Mendelian genetics. Multiple interacting genes are involved. But is it true that light eye-color tends to occur more often with light hair-color and similarly for dark eyes and dark hair?

Data were reported by Snee [112] in 1974 and relate to a convenience sample of 592 students collected as part of a class project by students in an elementary statistics course taught by Ronald Snee at the University of Delaware. The data are reproduced in Table 8.1. One depiction of these data is given in Figure 8.1.

The ordering for both hair color and eye color in Table 8.1 and Figure 8.1 is from dark to light. This choice was made based upon an intent to consider the question as to whether hair color and eye color tend to be both dark or both light. Other ordering schemes, based on the alphabet or frequency count, were considered but not adopted. Snee [112] apparently chose to order hair color based on darkness and eye color based on frequency. These choices appear to be haphazard.

The *grouped dot chart* was first presented in print in a peer-reviewed publication by Cleveland and McGill [25]. The year of publication was 1984. Figure 8.1

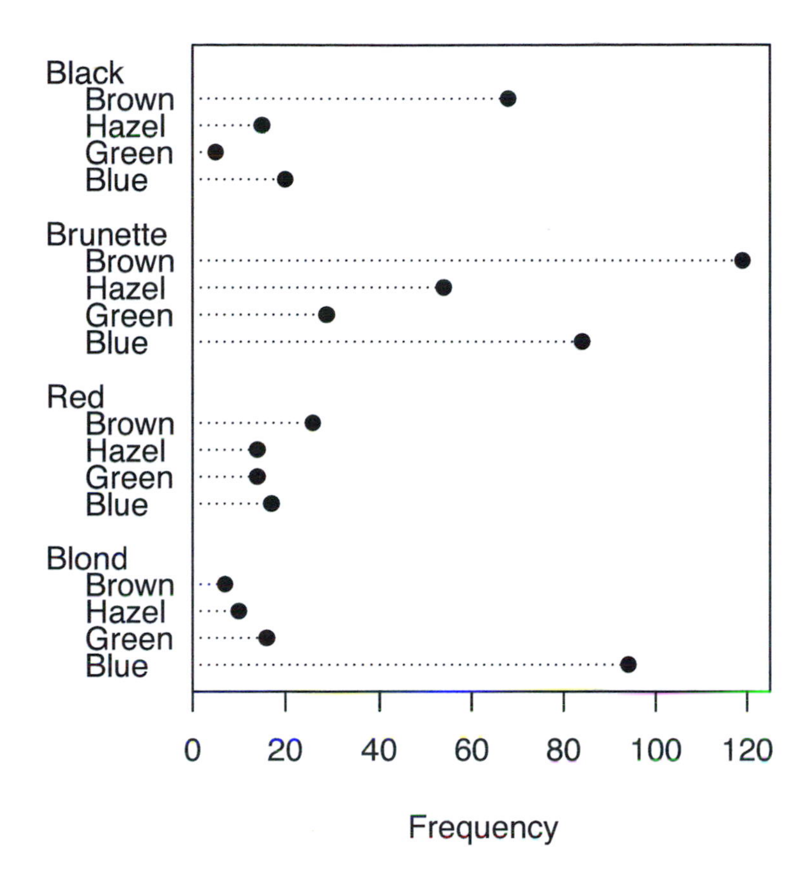

Figure 8.1 *Grouped dot chart of eye-color frequency grouped within hair color*

follows their plotting convention closely with the exception of not reporting the total frequency for each hair color. The justification for this omission is two-fold. The emphasis is on the bivariate distribution of the data. Adding information on the marginal distribution for hair color would detract from this. The other reason is that the variation amongst the hair colors would become compressed and harder to distinguish should a broader scale be used. This would be necessary in order to depict the counts for each hair color. Cleveland's [21] plotting convention of 1985 dispenses with plotting the frequency for the grouping variable.

Figure 8.1 is a thing of simplicity and elegance. With respect to the ACCENT rule, the grouped dot chart meets all requirements for a well-executed graphic. In Figure 8.1, the bivariate distribution of the data is clearly portrayed. With this figure, one can get down to work and explore the relationships between hair color and eye color.

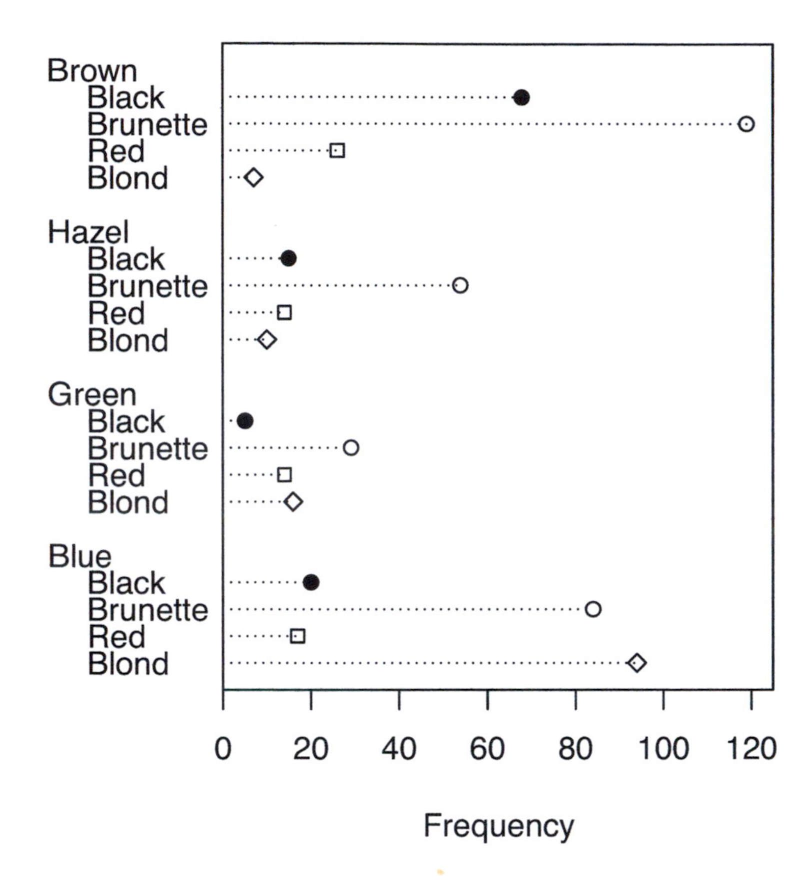

Figure 8.2 *Grouped dot chart of hair-color frequency grouped within eye color*

The distribution of eye color is similar among black-haired individuals, brunettes, and redheads. Blonds emerge with a distinct distribution. Brown eyes are associated with the darker hair colors of black and shades of brown but also with red. Although it is not possible with this data to determine whether gentlemen prefer blonds, Figure 8.1 shows that blonds are overwhelmingly blue eyed in the data reported by Snee [112].

Population geneticists have suggested a reason for this. As already stated, eye color is not a simple Mendelian recessive trait, but it acts like one. So it has been conjectured that blue-eyed individuals prefer blue-eyed mates. This way, if a blue-eyed female commits an indiscretion with a man whose eyes are not blue, the evidence will be apparent in the birth of a child. In the golden age of the Vikings, this likely would have been lethal to both the woman and her newborn. A brown-eyed woman fathering a child by a blue-eyed blond Viking would have no such worries. The striking excess of blue eyes among blonds in

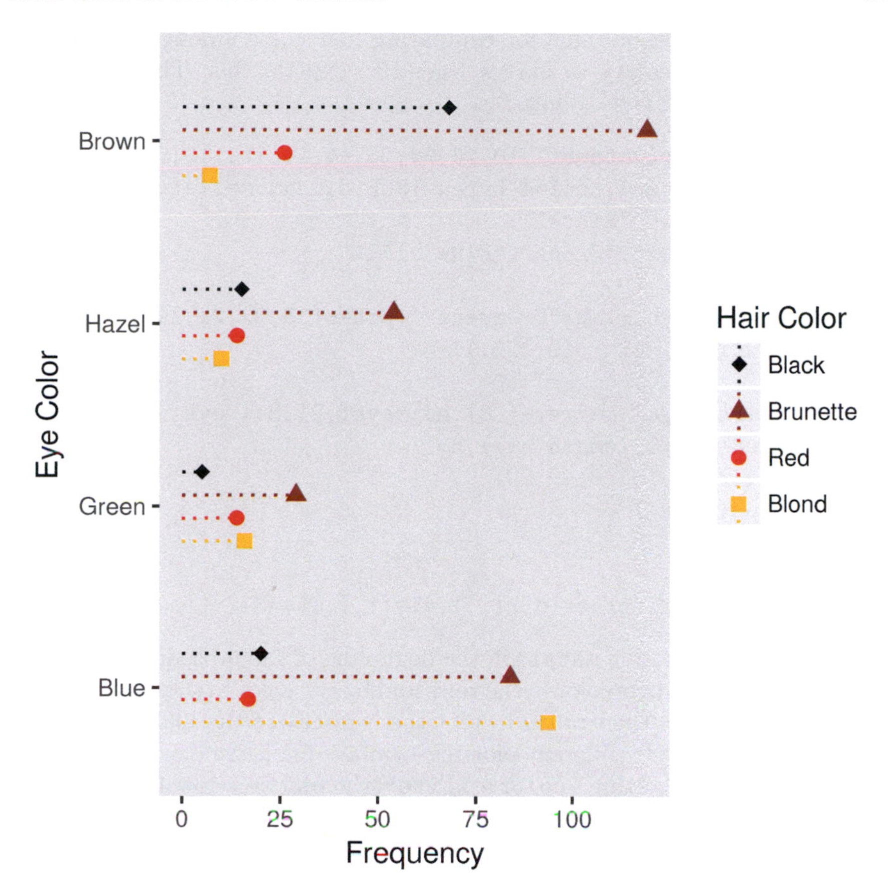

Figure 8.3 *Grouped dot chart of hair-color frequency grouped within eye color using* `ggplot2`

Figure 8.1 is consistent with this theory. So too is the fact that blue eyes are second most common, after brown eyes, among individuals with black, brown, or red hair.

Apparent in the bivariate distribution depicted in Figure 8.1 is the impression of the distribution of the combinations of the various hair and eye colors. Darker hair and eyes tend to predominate. With hair color as the grouping variable in Figure 8.1, a stronger impression of the relative numbers for hair color emerges. Figure 8.2 redrafts the grouped dot chart with eye color as the grouping variable.

Incorporated in Figure 8.2 is something that Cleveland [21] added in 1985 to the plotting convention for the grouped dot chart: distinct plotting symbols instead of a dot for terminating each dotted line segment. Although each line is labeled, using one distinct plotting symbol for each hair color in Figure 8.2

adds an additional visual clue for comparing hair color among the four eye colors. It is not necessary to have a legend to explain this. The R code for producing Figure 8.2 is as follows.

```
haireye<-matrix(data=c(7,10,16,94,26,14,14,17,119,54,29,84,
68,15,5,20),nrow=4,ncol=4,byrow=TRUE,dimnames=list(c("Blond",
"Red","Brunette","Black"),
c("Brown","Hazel","Green","Blue")))

dotchart(haireye,xlab="Frequency",pch=c(23,22,21,19),
lcolor="white",xlim=c(0,125))

he<-c(haireye[,4],haireye[,3],haireye[,2],haireye[,1])
dlx<-cbind(rep(0,length(he)),he)
ys<-c(1:4,7:10,13:16,19:22)
dly<-cbind(ys,ys)
ldl<-length(he)

for (i in 1:ldl) lines(dlx[i,],dly[i,],lty=3)
```

The call to the R function `matrix` in the beginning of the preceding R code arranges the counts by hair color in the rows and by eye color in the columns. The function `dotchart` is then called to plot the contents of the matrix `haireye`. The argument `pch` sets different plotting symbols for hair color to terminate the dotted lines. By setting `lcolor="white"`, the function `dotchart` does not plot the dotted lines. This is done with the remaining R code, the final line of which uses the function `lines` within a `for` loop to do the actual plotting of the dotted lines.

A `ggplot2` version of Figure 8.2 given in Figure 8.3 is produced by the following R code.

```
haireye<-matrix(data=c(7,10,16,94,26,14,14,17,119,54,29,84,
68,15,5,20),nrow=4,ncol=4,byrow=TRUE,dimnames=list(c("Blond",
"Red","Brunette","Black"),
c("Brown","Hazel","Green","Blue")))

haircol<-unlist(dimnames(haireye)[1])
eyecol<-unlist(dimnames(haireye)[2])

hair<-rep(" ",16)
eye<-rep(" ",16)
freq<-rep(0,16)

n=0

for (i in 1:4){ for (j in 1:4){
```

```
n<-n+1
hair[n]<-haircol[i]
eye[n]<-eyecol[j]
freq[n]<-haireye[i,j]
}}

hair<-factor(hair,levels=haircol,ordered=TRUE)
eye<-factor(eye,levels=rev(eyecol),ordered=TRUE)

haireyef<-data.frame(hair=hair,eye=eye,freq=freq)

figure<-ggplot(haireyef,aes(x=eye,ymin=0,ymax=freq,
y=freq)) +
geom_pointrange(aes(shape=factor(hair,
levels=haircol,ordered=TRUE),
color=factor(hair,levels=haircol,ordered=TRUE)),
position=position_dodge(width=0.5),linetype=3)+
scale_shape_manual(values=15:18) +
scale_color_manual(values=rev(c("black","brown",
"red","gold"))) +
coord_flip() +
labs(x="Eye Color",y="Frequency",
shape="Hair Color",color="Hair Color")+
guides(shape=guide_legend(reverse=TRUE),
color=guide_legend(reverse=TRUE)) +
theme(panel.grid=element_blank())

print(figure)
```

Additional code is required so that the data is saved in the format needed by `ggplot2`. This also requires the data to be saved in a `data.frame` object. Rather than manipulate `ggplot2` themes to render Figure 8.3 in grayscale more suitable for printing on paper, it was decided to render the figure in color so it would be more appealing for someone reading an electronic version of this second edition. The RGB color space used on computer monitors is additive and so essentially all three colors are needed to get a white background. This can be overwhelming and tiring on the eyes. The default theme in `ggplot2` instead produces a gray background that is less overpowering when viewed on a computer monitor.

Note that colors chosen for Figure 8.3 correspond to hair color. The R color `"yellow"` was found to be too stark for blond hair color so `"gold"` was substituted instead. Regardless of whether Figure 8.3 is rendered on paper or a computer monitor, the R color `"gold"` does not faithfully reproduce the appearance of real gold. Note that a legend was chosen for Figure 8.3 as this required less effort than placing labels after using the `ggplot2` function

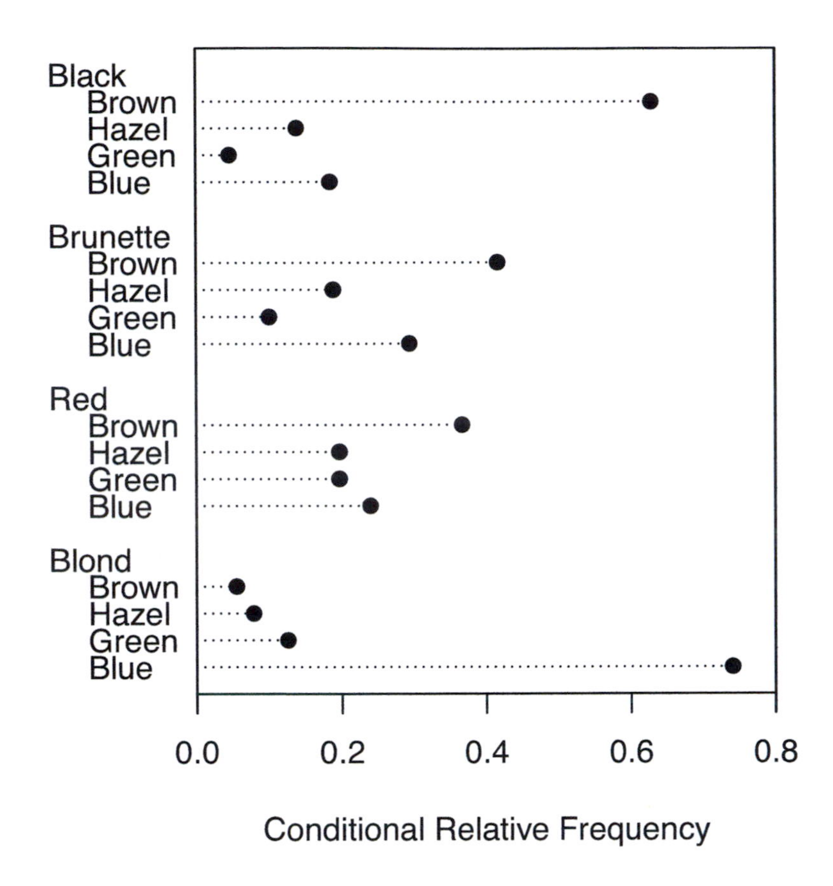

Figure 8.4 *Grouped dot chart of eye-color relative frequency conditioned on hair color (relative frequencies sum to one for each hair color)*

`position_dodge` to stagger (or dodge) the different hair colors for each of four eye colors.

From Figures 8.2 and 8.3, one gets the impression that blue eyes are not so rare. Blue eyes rival brown eyes for sheer numbers. While brown-eyed blonds are rare, the rarest combination is green eyes with black hair in this convenience sample.

To compare the distribution of eye color while treating hair colors as subpopulations, there is another version of the grouped dot chart better suited. Figure 8.4 portrays the relative frequency of eye color conditional on hair color. The relative frequencies of eye color for each hair color sum to one.

Figures 8.1 and 8.2 depict a bivariate distribution. Figure 8.4 depicts four univariate distributions, one for each hair color. With the grouped dot chart

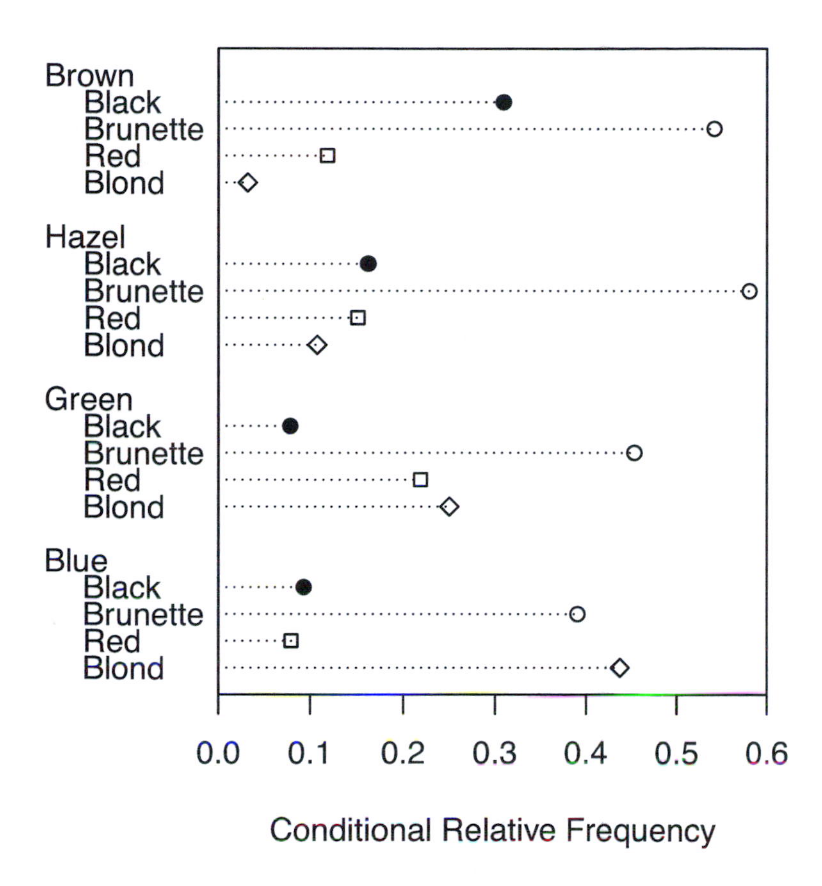

Figure 8.5 *Grouped dot chart of hair-color relative frequency conditioned on eye color (relative frequencies sum to one for each eye color)*

of Figure 8.1 it is still possible to compare the distribution of eye color conditional on hair color but more effort is required because it is necessary to compensate for the difference in frequency counts among hair colors when making comparisons. This compensation has already been made for the viewer in Figure 8.4.

It is just as easy to produce a grouped dot chart of hair-color relative frequency conditioned on eye color. This is done in Figure 8.5 where the use of distinct terminating plot symbols introduced in Figure 8.2 for hair color has been retained. In Figure 8.5, the similarity of the conditional distribution of hair color given eye color is clearly visible. It is apparent that the conditional distributions of hair color do not fit this pattern for green or blue eyes. The conditional distributions of hair color given eye color are quite different for green and blue eyes.

Returning to the conditional relative frequency dot chart in Figure 8.5 for eye color given hair color, blonds show the most limited degree of diversity. All four eye colors are seen in blonds but blue is overwhelmingly predominant. This is consistent with tight linkage among hair and eye color genes or a high degree of assortative mating among blue-eyed individuals, or both. In the English language, the adjectives *blond* and *blue-eyed* are synonymous with being naive or not too bright. These stereotypes would appear not to be a coincidence from the viewpoint of population genetics.

The greatest diversity, that is, a tendency toward a uniform distribution, of eye color is seen in Figure 8.4 to be among the redheads. From the bivariate distribution in Figure 8.1, the rarest hair color of the four surveyed is red. This would appear to suggest that the genes responsible for red hair act recessively but are equitably distributed with respect to eye color. Figure 8.5 appears to convey the latter for all eye colors but green for which there is a greater proportion of individuals with red hair compared to the other three eye colors.

The grouped dot chart is a simple but effective graphical tool. It shall be the *gold standard* for comparison for the other graphical displays to be presented in this chapter. The grouped dot chart can be used to portray the bivariate distribution of discrete variables as originally introduced by Cleveland [25]. Or it can be used to depict conditional distributions side-by-side. It is important to inform an audience of which version they are viewing.

It is not necessary that an audience have in advance a technical understanding of a conditional distribution. With a carefully worded explanation, it is possible to give a nontechnical description explaining what is being illustrated.

There are not a lot of options with respect to displaying a grouped dot chart. Variation in each of color, size, and style for the line segments and terminating symbols are about all there is. Figures 8.1 through 8.5, have been drafted in black ink on a white background using the R function *dotchart*. Wise selections of color can produce more eye-catching figures than these.

8.4 The Grouped Dot-Whisker Chart

In the previous section, the hair and eye color data have been treated as if they were a product of a census with the graphical displays depicting no variation whatsoever. It appears that Snee [112] obtained the data through a convenience sample of students at the University of Delaware. In his article, he treated the data as if they were obtained from a random sample. This is commonly done in marketing research.

There was no good reason given by Snee [112] for not treating the data as a result of a representative random sample of students at the University of Delaware. Depending on the recruitment strategy for the University of

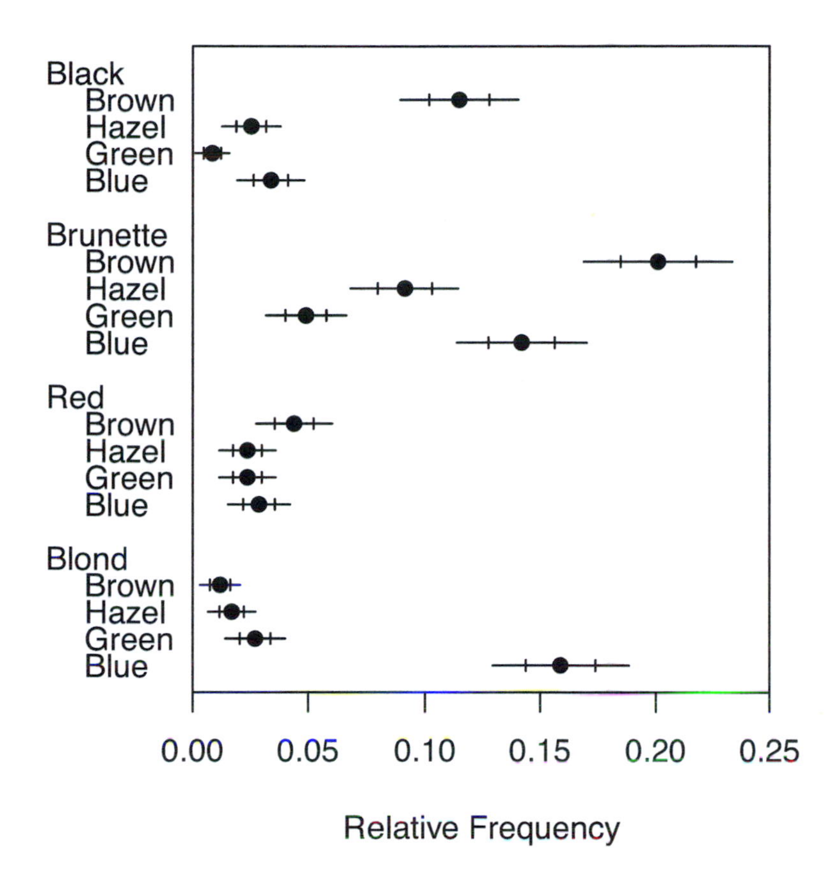

Figure 8.6 *Grouped two-tiered dot-whisker chart of eye-color relative frequency grouped within hair color (68% and 95% confidence intervals depicted)*

Delaware, it might even be reasonable to accept the data as being representative of the population in Delaware and its three neighboring states of Pennsylvania, Maryland, and New Jersey.

Treating the data as coming from a random sample and ignoring any finite population correction, Figure 8.6 gives a *grouped dot-whisker chart* of hair color and eye color. Note that this is a two-tiered dot-whisker chart with 68% and 95% confidence intervals depicted for each point estimate. This figure was drafted using the `dotchart` function in R with additional code written to depict the two confidence intervals. The code for doing this is as follows.

```
haireye<-matrix(data=c(20,84,17,94,5,29,14,16,15,
54,14,10,68,119,26,7),
nrow=4,ncol=4,byrow=TRUE,
dimnames=list(c("Blue","Green","Hazel","Brown"),
```

```
c("Black","Brunette","Red","Blond")))

haireye<-haireye/sum(haireye)

dotchart(haireye,xlab="Relative Frequency",
pch=19,lcolor="white",xlim=c(0,0.25))

he<-c(94,16,10,7,17,14,14,26,84,29,54,119,20,5,15,68)
nhe<-length(he)
hesum<-sum(he)
he<-he/sum(he)
sd<-sqrt((1.-he)*he/hesum)
hlo<-he-sd
hhi<-he+sd
sd2<-sd*qnorm(0.975)
hlo2<-he-sd2
hhi2<-he+sd2

ii<- -2
for (i in 1:nhe) {
ii<- ii + ifelse ((i %% 4) == 1,3,1)
lines(c(hlo2[i],hhi2[i]),c(ii,ii))
lines(c(hlo[i],hlo[i]),c(ii-0.2,ii+0.2))
lines(c(hhi[i],hhi[i]),c(ii-0.2,ii+0.2))
}
```

The grouped dot chart counterpart to Figure 8.6 is Figure 8.1. A comparison of these two figures reveals an important advantage of the dot-whisker plot: the depiction of variation. With this comes the ability to interact more fully with the data. For blonds, there doesn't seem to be much difference in relative frequency among green, hazel, and brown eye colors. For red hair, the relative frequency could be the same for all eye colors. This was noted in the previous section with the grouped dot chart. But discussions with point estimates in terms of data from a random sample are not as convincing.

The ggplot2 color counterpart to Figure 8.6 is Figure 8.7, which is produced by the following R code.

```
haireye<-matrix(data=c(7,10,16,94,26,14,14,17,119,54,
29,84,68,15,5,20),nrow=4,ncol=4,byrow=TRUE,
dimnames=list(c("Blond","Red","Brunette","Black"),
c("Brown","Hazel","Green","Blue")))

haircol<-unlist(dimnames(haireye)[1])
eyecol<-unlist(dimnames(haireye)[2])
```

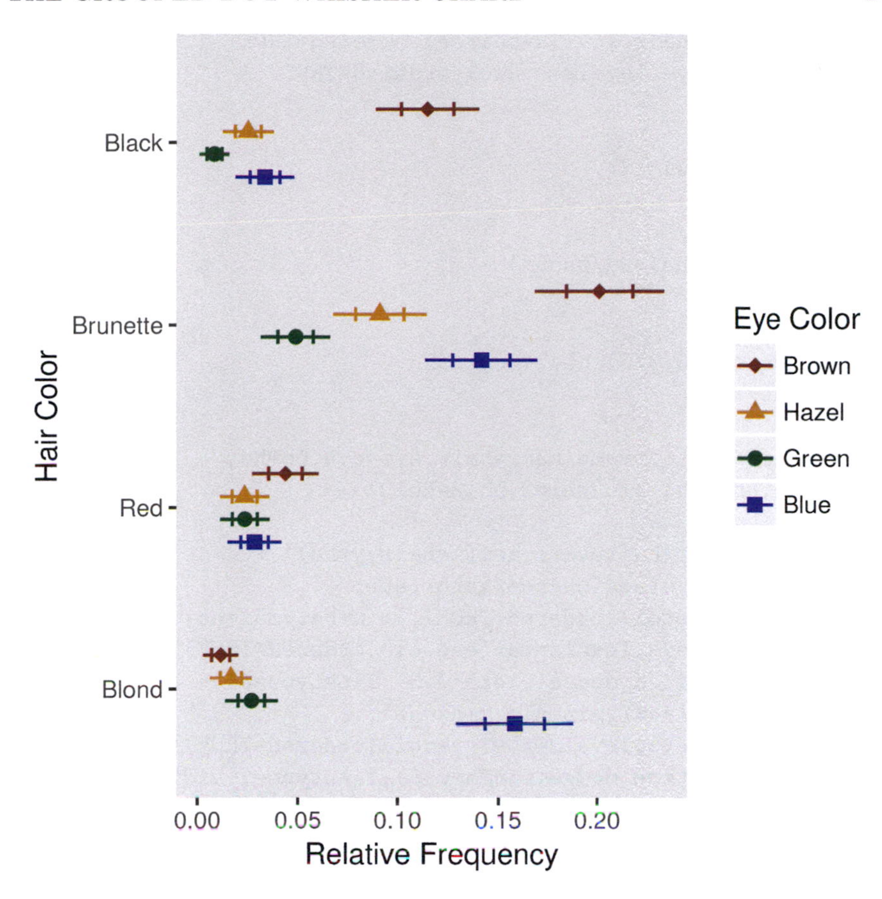

Figure 8.7 *Grouped two-tiered dot-whisker chart of eye-color relative frequency grouped within hair color (68% and 95% confidence intervals depicted) using* `ggplot2`

```
hair<-rep(" ",16)
eye<-rep(" ",16)
freq<-rep(0,16)

n=0

for (i in 1:4){ for (j in 1:4){
n<-n+1
hair[n]<-haircol[i]
eye[n]<-eyecol[j]
freq[n]<-haireye[i,j]
}}
```

```
hair<-factor(hair,levels=haircol,ordered=TRUE)
eye<-factor(eye,levels=eyecol,ordered=TRUE)

he<-freq
nhe<-length(he)
hesum<-sum(he)
he<-he/sum(he)
sd<-sqrt((1.-he)*he/hesum)
hlo<-he-sd
hhi<-he+sd
sd2<-sdqnorm(0.975) hlo2<-he-sd2
hhi2<-he+sd2

haireyerf<-data.frame(hair=hair,eye=eye,he=he,
hlo=hlo,hhi=hhi,hlo2=hlo2,hhi2=hhi2)

figure<-ggplot(haireyerf,aes(x=hair,y=he)) +
geom_pointrange(aes(shape=factor(eye,
levels=rev(eyecol),ordered=TRUE),ymin=he,ymax=he,
color=factor(eye,levels=rev(eyecol),ordered=TRUE)),
position=position_dodge(width=0.5),linetype=0) +
geom_errorbar(aes(ymin=hlo,ymax=hhi,
color=factor(eye,levels=rev(eyecol),ordered=TRUE)),
position=position_dodge(width=0.5),linetype=1,width=0.3)+
geom_errorbar(aes(ymin=hlo2,ymax=hhi2,
color=factor(eye,levels=rev(eyecol),ordered=TRUE)),
position=position_dodge(width=0.5),linetype=1,width=0)+
scale_shape_manual(values=15:18) +
scale_color_manual(values=c("blue","green4",
"darkgoldenrod3","brown4")) +
labs(x="Hair Color",y="Relative Frequency",
shape="Eye Color",color="Eye Color") +
guides(shape=guide_legend(reverse=TRUE),
color=guide_legend(reverse=TRUE)) +
theme(panel.grid=element_blank()) +
coord_flip()

print(figure)
```

Parallelling the two calls to the `graphics` function `lines` to produce Figure 8.6 are the two calls to the `ggplot2` function `geom_errorbar` for Figure 8.7. The call `scale_shape_manual(values=15:18)` selects specific filled end-of-line symbols and the call to the function `scale_color_manual` switches off the default colors and instead uses colors that match the actual eye colors.

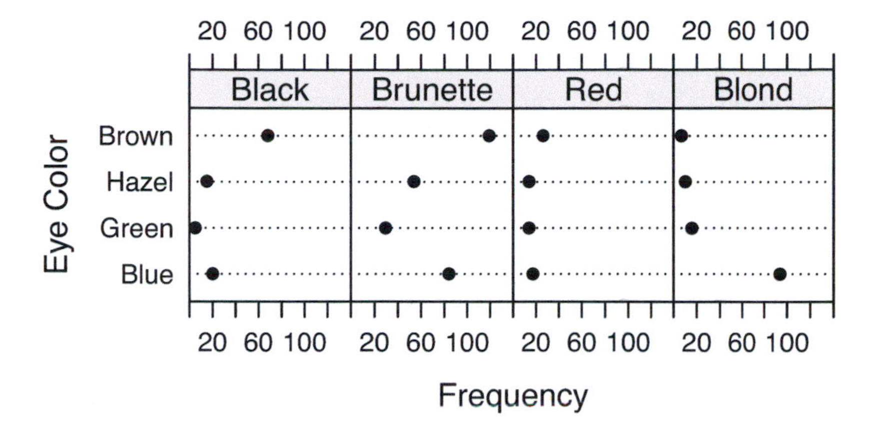

Figure 8.8 *Two-way dot chart in horizontal layout of eye-color frequency grouped within hair color*

Different end-of-line symbols and the same eye color order for each hair color, as in the legend, has been done to assist those with color deficient vision.

For brunettes, the relative frequencies for all four eye colors are widely spaced but there is a tiny bit of overlap of whiskers for adjacent estimates in Figures 8.6 and 8.7. For black hair, whiskers overlap for blue, green, and hazel eye colors but brown eye colors appear to have a distinct higher relative frequency.

It is also possible to use the grouped dot-whisker plot to depict side-by-side conditional distributions. This is left to an exercise for this chapter.

There is one disadvantage associated with a grouped dot-whisker plot. The use of the concept of a confidence interval will likely limit its use to audiences who have been exposed to this concept.

8.5 The Two-Way Dot Chart

The *two-way dot chart* was proposed by Cleveland [21] in 1985. An example of the two-way dot chart for the hair and eye color data of Table 8.1 is given in Figure 8.8. Note that the layout is horizontal with dot charts for the four hair colors side by side. Cleveland [21] could have just as easily named this the side-by-side dot chart.

The R code using the `lattice` package function `dotplot` for producing Figure 8.8 is as follows.

```
require(lattice)
```

```
trellis.device(color=FALSE)

graphics.off()
windows(width=4.5,height=2.5,pointsize=12)
par(fin=c(4.45,2.45),pin=c(4.45,2.45),
mai=c(0.85,0.85,0.25,0.25))

trellis.par.set("color",FALSE)

haireye<-matrix(data=c(20,84,17,94,5,29,14,16,15,54,
14,10,68,119,26,7),nrow=4,ncol=4,byrow=TRUE,
dimnames=list(c("Blue","Green","Hazel","Brown"),
c("Black","Brunette","Red","Blond")))

figure<-dotplot(haireye,xlab="Frequency",
ylab="Eye Color",as.table=TRUE,xlim=c(0,140),
groups=FALSE,stack=FALSE,layout=c(4,1),
col.line="black",lty="dotted",lwd=1,
scales=list(alternating=3))

print(figure)
```

The code to produce any figure rendered by R in the second edition, is embedded in a function. The call `require(lattice)` serves to execute `library(lattice)` if such a call has not been already made. The `lattice function` call `trellis.device(color=FALSE)` is followed by the second obsessive-compulsive call `trellis.par.set("color",FALSE)` to make sure that Figure 8.8 is rendered only in grayscale.

The `lattice` package shares a similarity with the `ggplot2` package in that calls to plotting functions within a function (inclusive of `for` and `while` loops) will not result in a plot as with the `graphics` package. Instead, with both the `lattice` and `ggplot2` packages, it is necessary to save the graph as an object to a variable and then use the generic `print` or `plot` functions to render the graph on the current plotting device. While this represents an extra step for both the `lattice` and `ggplot2` packages, the point is made that R is indeed an object-oriented environment for statistical and graphical analysis.

The `ggplot2` version of Figure 8.8 is Figure 8.9, which is produced by the following R code.

```
require(ggplot2)

graphics.off()
windows(width=4.5,height=2.0)
```

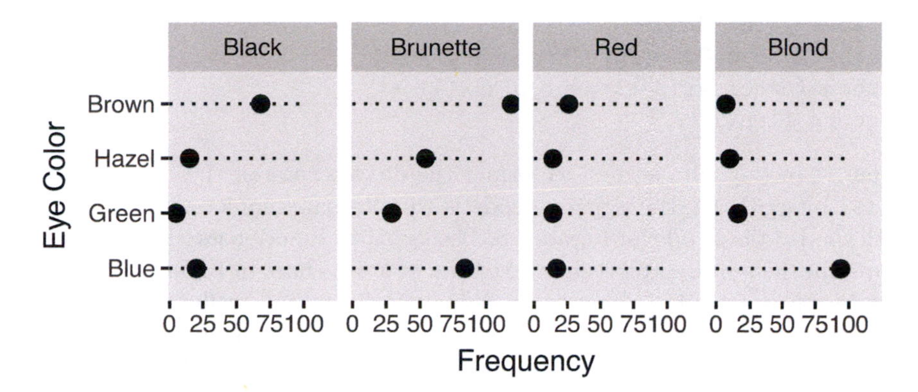

Figure 8.9 *Two-way dot chart in horizontal layout of eye-color frequency grouped within hair color using* ggplot2

```
haireye<-matrix(data=c(7,10,16,94,26,14,
14,17,119,54,29,84,68,15,5,20),nrow=4,ncol=4,
byrow=TRUE,dimnames=list(c("Blond",
"Red","Brunette","Black"),
c("Brown","Hazel","Green","Blue")))

haircol<-unlist(dimnames(haireye)[1])
eyecol<-unlist(dimnames(haireye)[2])

hair<-rep(" ",16)
eye<-rep(" ",16)
freq<-rep(0,16)

n=0

for (i in 1:4){ for (j in 1:4){
n<-n+1
hair[n]<-haircol[i]
eye[n]<-eyecol[j]
freq[n]<-haireye[i,j]
}}

hair<-factor(hair,levels=rev(haircol),ordered=TRUE)
eye<-factor(eye,levels=rev(eyecol),ordered=TRUE)

haireyef<-data.frame(hair=hair,eye=eye,freq=freq)
figure<-ggplot(haireyef,aes(x=eye,y=freq)) +
geom_pointrange(ymin=0,ymax=100,linetype=3) +
```

```
facet_grid(~hair)+
labs(x="Eye Color",y="Frequency") +
theme(panel.grid=element_blank()) +
coord_flip()
```

Other than the code needed to format the `data.frame` in the manner anticipated by `ggplot2`, the `ggplot2` code is equally parsimonious in comparison with the `lattice` code for Figure 8.8. The `ggplot2` function `geom_pointrange` to obtain the required plotting convention of a dot chart in Figure 8.9 and the call `facet_grid(~hair)` serves to request the required trellis layout for the variable `hair`. Finally, there is a call to the `ggplot2` function `coord_flip` to obtain a horizontal orientation for the dot chart.

There are a few problems with respect to the two-way dot chart that are apparent in the examples of Figures 8.8 and 8.9. Comparison of the counts for eye color amongst the hair colors is made more difficult in the two-way dot chart because there is not a common scale for frequency. In comparison, the grouped dot chart, proposed by Cleveland [21], is superior because there is one common scale for frequency. This requires less work on the part of the viewer when comparing counts.

In an attempt to ease comparisons, artistic license has been taken and labeled scales have been added along the top of the graph. This is not part of Cleveland's [21] original plotting standard for the two-way dot chart. While this duplication of labeled scales might constitute an improvement in apprehension on the part of the viewer, it does reduce the data-ink ratio, which detracts from efficiency.

A major problem with the horizontal layout in Cleveland's [21] original plotting standard is that the frequency axis is compressed. This negatively impacts clarity and forces the viewer to work harder. Figure 8.8 was executed using the function `dotplot` in the `lattice` package for the R statistical software system. It is a simple matter to change the horizontal orientation when using the function `dotplot`. This has been done in the following R code to produce Figure 8.10.

```
haireye<-matrix(data=c(20,84,17,94,5,29,14,16,
15,54,14,10,68,119,26,7),nrow=4,ncol=4,byrow=TRUE,
dimnames=list(c("Blue","Green","Hazel","Brown"),
c("Black","Brunette","Red","Blond")))

figure<-dotplot(haireye,xlab="Frequency",
ylab="Eye Color",as.table=TRUE,groups=FALSE,
stack=FALSE,layout=c(1,4),scales=list(alternating=3))

print(figure)
```

The argument `layout=c(1,4)` in the preceding call of the `lattice` function

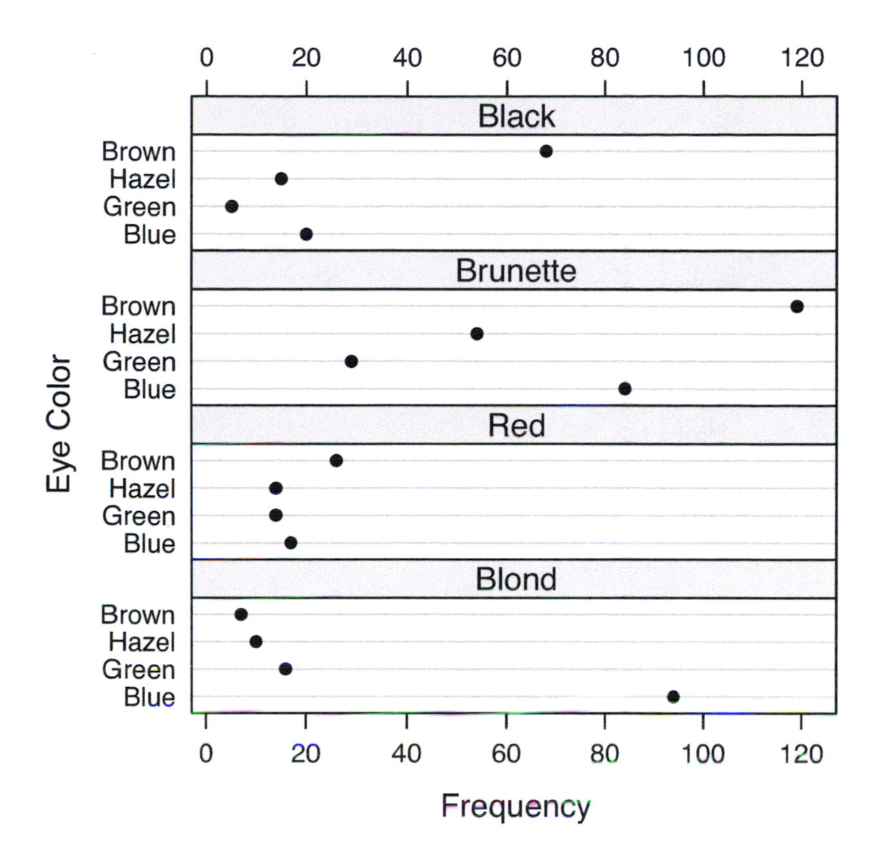

Figure 8.10 *Grouped dot chart of eye-color frequency grouped within hair color (alternative using* `dotplot` *in the R* `lattice` *graphics package)*

`dotchart` produced the vertical layout in Figure 8.10. The horizontal layout in Figure 8.8 was produced by setting `layout=c(4,1)`.

Note that Figure 8.10 is not, by Cleveland's [21] definition, a two-way dot chart. Comparison of Figure 8.10 with Figure 8.1 reveals that Figure 8.10 is an alternate version of the grouped dot chart. Note in Figure 8.10 that hair color is presented in a centered label within a shaded band as opposed to being offset to the left of the eye color labels as in Figure 8.1. While the hair color labeling in Figure 8.10 interrupts the flow of the dotted line segments it does make the hair-color distinctions more apparent.

Figures 8.8 and 8.10 give an exposure to Trellis graphics in the simplest of multivariable settings. Trellis graphics is implemented by the package `lattice` in the R statistical software package. *Lattice* is a play on the word *Trellis*. In 1996 the first article on Trellis graphics appeared. This was authored by Rick

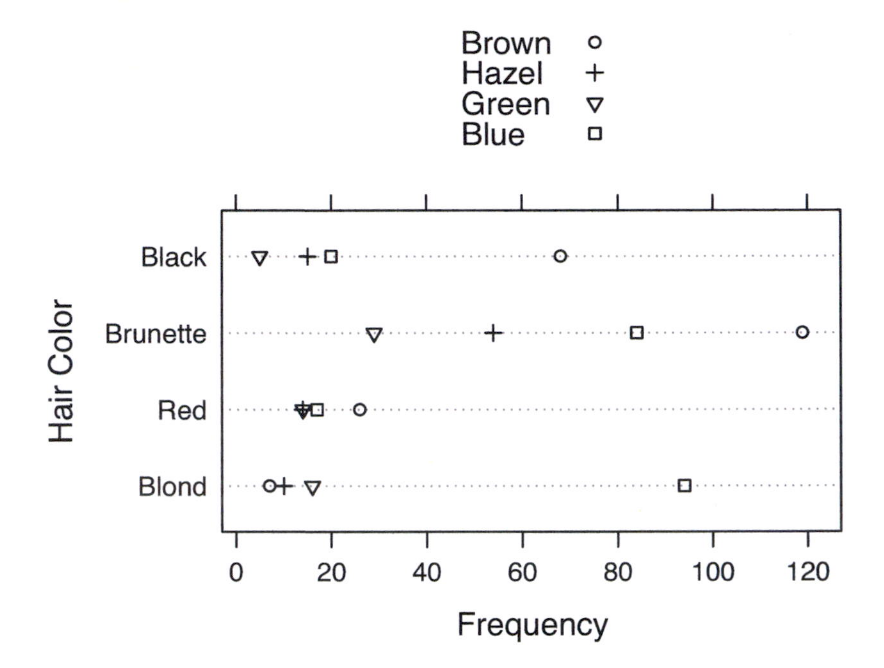

Figure 8.11 *Multi-valued dot chart in horizontal layout of eye-color frequency grouped within hair color*

Becker, Bill Cleveland, and Ming-Jen Shyu [10]. A manual authored by Becker and Cleveland [9] was published in the same year.

The `lattice` package requires the `grid` graphics package for R. Paul Murrell [84] developed the `grid` package. For a discussion of this package, including an introduction to the `lattice` package, see Murrell [84].

Although the two-way dot chart falls short of the gold standard, the grouped dot chart, the Trellis graphics package `lattice` is an important tool for data visualization. It will be encountered again in later chapters when three or more variables are involved.

8.6 The Multi-Valued Dot Chart

A third version of the dot chart for two discrete variables introduced by Cleveland [21] is the *multi-valued dot chart*. An example is given in Figure 8.11. This also was produced using the `dotchart` function in the `lattice` package for R.

There is a common scale for frequency in Figure 8.11 but interpretation of Figure 8.11 is slowed by the need to refer to the legend for eye color above the chart. Cleveland [21] stated on page 153 that it is reasonable to execute a multi-valued dot chart "provided one of the two groupings has a small number of categories...." This is the case for the hair and eye color data with four categories for each. But keeping track of the four different symbols for hair color takes a bit of work.

There is also the need for the symbols in a multi-valued dot chart to be separated enough to be distinguishable. Note the overprinting of symbols for frequency of green and hazel eyes for the redheads. Adding whiskers to illustrate variation in the point estimates of relative frequency in a multi-valued dot chart would only add more confusion if there was not sufficient separation.

The multi-valued dot chart is not as good as the corresponding grouped dot chart in Figure 8.1. Of the three versions of the dot-chart for depicting the distribution of two discrete variables presented by Cleveland [21], the grouped bar chart appears to be the winner.

8.7 The Side-by-Side Bar Chart

The *side-by-side bar chart* has been with us for some time. Figure 8.12 depicts the hair and eye color data of Snee [112] in a side-by-side bar chart. In many respects this graphical display is quite similar to the grouped dot chart in Figure 8.1.

There are a few features in Figure 8.12 that could be improved. The label for hair color and the hair colors themselves are vertical and not horizontal. The legend for eye color is set in a box inside the figure and this interferes with apprehension. Possibly the legend could be placed above or below the side-by-side bar charts. Figure 8.13 presents solutions to both these problems.

Note in Figure 8.13 that the labels for hair color are horizontal. Even the added labels for each bar for eye color are horizontal. Grayscale shading of the bars has been retained from Figure 8.12, but no separate legend is required in Figure 8.13.

For comparative purposes, Figure 8.14 presents a side-by-side bar chart for the relative frequency of eye color conditional on hair color. The grouped dot chart counterpart to this figure is Figure 8.4.

In comparison with the grouped dot chart, the side-by-side bar chart uses more ink for the same data so it is not as efficient. Yet the popularity of the bar chart persists as being higher than that of the dot chart.

8.8 The Side-by-Side Bar-Whisker Chart

Figure 8.15 illustrates a *side-by-side bar-whisker chart* for the hair and eye color data. In this figure the length of the bars denotes the relative frequency.

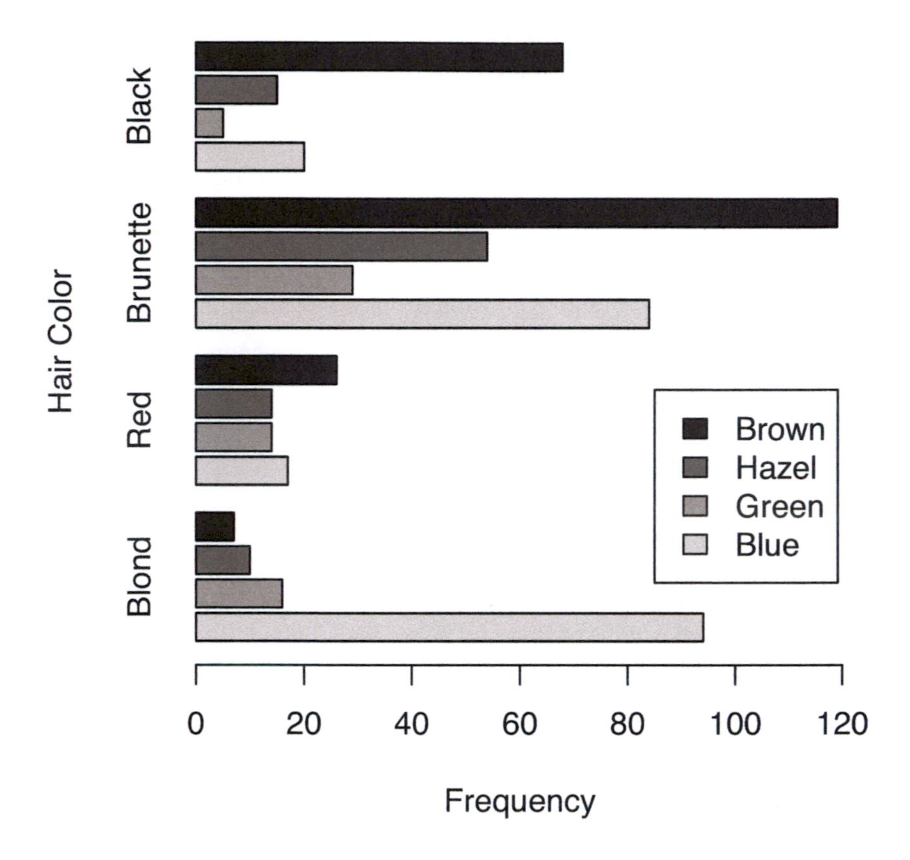

Figure 8.12 *Side-by-side bar chart of eye-color frequency grouped within hair color*

The distance between the ends of the whiskers is two standard deviations of the estimate of relative frequency denoted by the right end of each bar. The bars in Figure 8.15 have been plotted using the R function `barplot` and the whiskers have been manually added to the bar chart with calls to the R function `lines`.

Figure 8.16 is a color version of Figure 8.15. The `ggplot2` package was used to produce Figure 8.16. It uses colors corresponding to actual eye colors for the bars and whiskers. For individuals with deficient color vision, it need be noted that order of eye color within each hair color is the same as given in the legend of Figure 8.16.

For the sake of clarity in the **ACCENT** rule, there is no shading or coloring for the bars in Figure 8.15. The side-by-side bar-whisker chart and the grouped dot-whisker chart of Figure 8.6 are nearly equally effective in conveying both the point estimate of relative frequency and its error. The edge in performance

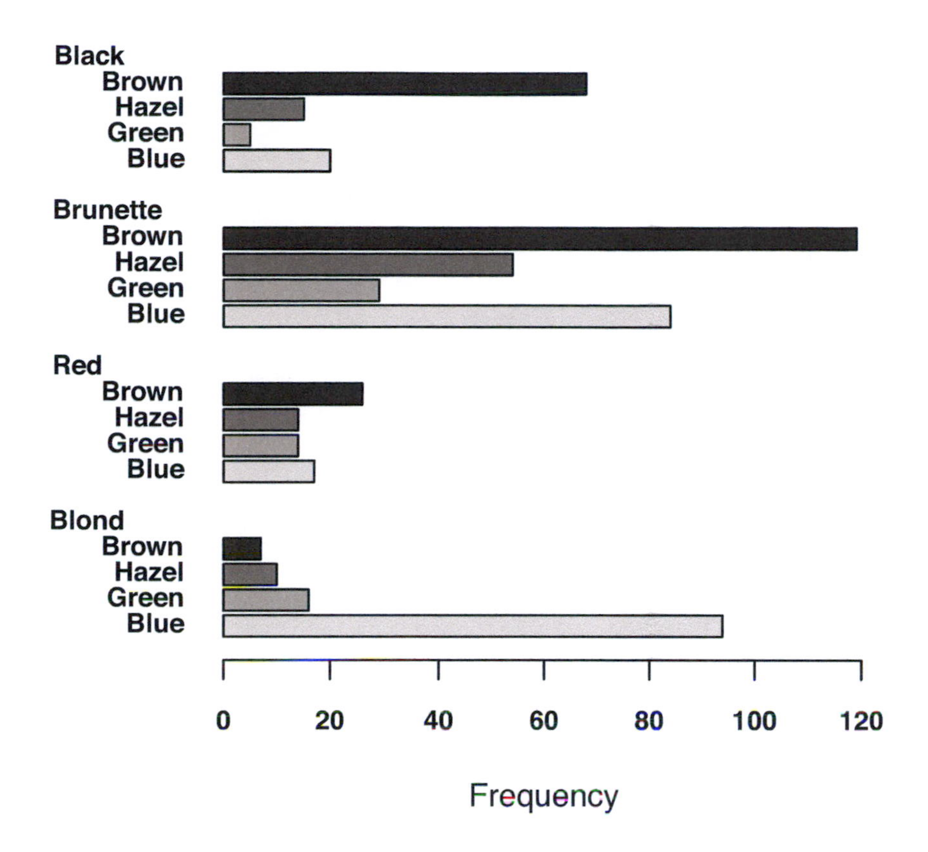

Figure 8.13 *Side-by-side bar chart of eye-color frequency grouped within hair color—with improved layout*

based on the better data-ink ratio is with the grouped dot-whisker chart. It could be claimed that this is splitting hairs. But more advocacy is needed before the dot chart replaces the bar chart.

8.9 The Side-by-Side Stacked Bar Chart

The final variation on the bar chart to be explored is the *side-by-side stacked bar chart*. The example given in Figure 8.17 was produced using the function `barplot` in the `graphics` package. The problems discussed with regard to a stacked bar chart for a single variable in Chapter 3 also hold for the side-by-side stacked bar chart, except the problem is compounded by there being more bars.

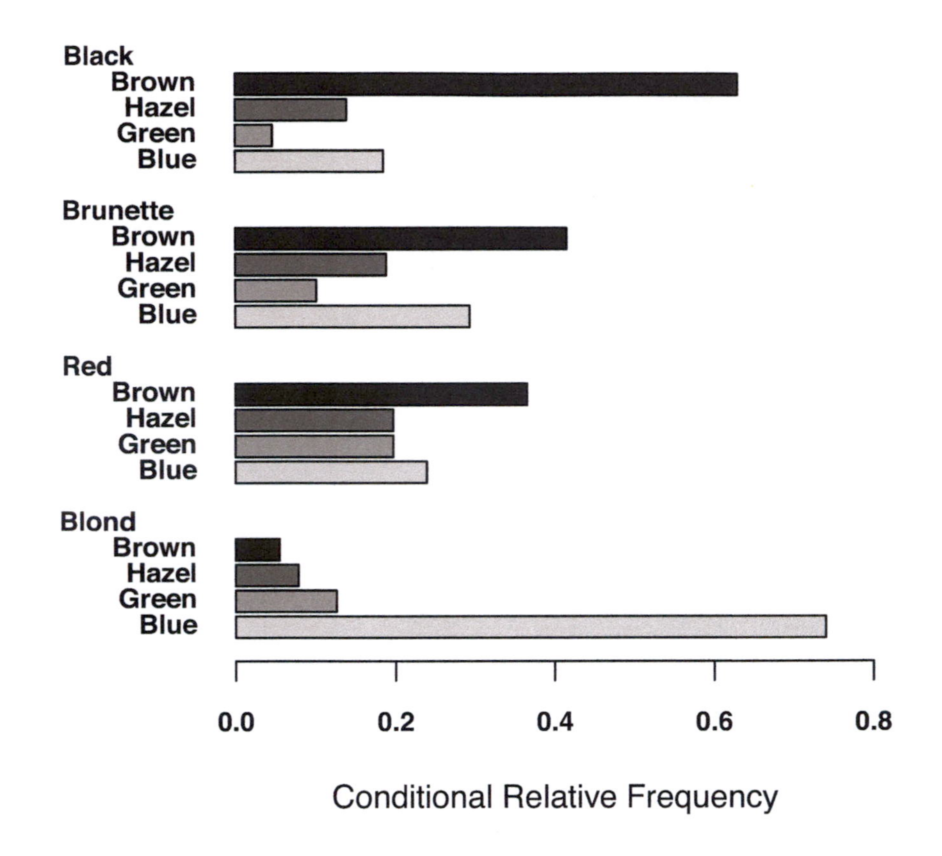

Figure 8.14 *Side-by-side bar chart of conditional frequency for eye-color grouped within hair color*

A `ggplot2` color version of Figure 8.17 is given in Figure 8.18. Again, the colors have been chosen to match the actual eye colors.

A further example is given in Figure 8.19 of the side-by-side bar chart using the `graphics` function `barplot` but in this instance with the roles of hair color and eye color exchanged in the layout of Figure 8.19 compared to Figure 8.17. Although there is one horizontal axis in each of Figures 8.17 and 8.19, there is not a common scale for comparing the lengths of the bars. There is also the chance that a viewer might attempt to make comparisons based upon area, which takes more effort than comparing lengths.

Estimates of conditional probability for eye color as a function of hair color are depicted with a side-by-side stacked bar chart in Figure 8.20. A color version of Figure 8.20 drafted by `ggplot2` is given in Figure 8.21.

Estimates of conditional probability for hair color as a function of eye color

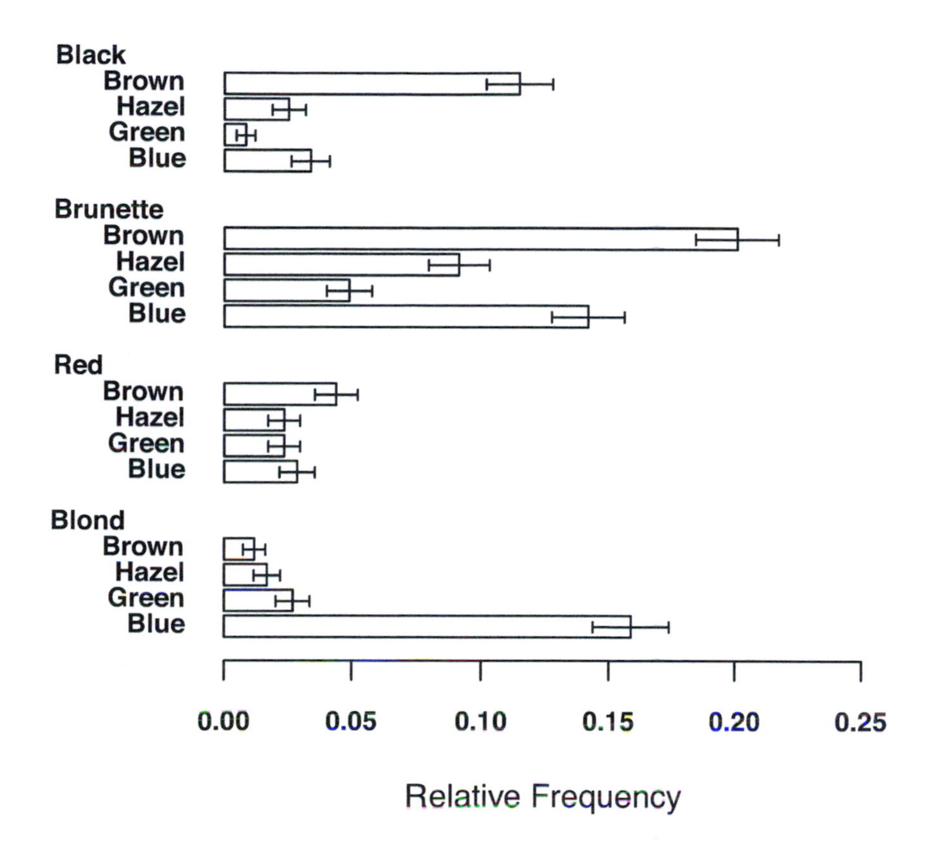

Figure 8.15 *Side-by-side bar-whisker chart of eye-color frequency grouped within hair color*

are depicted with a side-by-side stacked bar chart in Figure 8.22. A visual clue that these two figures are depicting a conditional distribution is given by the fact that all stacked bars are of equal length and terminate at a unit length.

In Figure 8.20, we see that most blonds are blue-eyed and most black-haired individuals have brown eyes. In Figure 8.22, we see the proportion of blonds monotonically increasing with the lightening of eye color while the proportion of black hair color monotonically decreases as eye color lightens.

The problem for the viewer of whether to use area or bar length for comparison with the side-by-side stacked bar chart is an issue. Cleveland and McGill [25] empirically demonstrated the superiority of the bar chart over the stacked bar chart. It is reasonable to infer that this result holds in the more complex case of more bars.

It was possible to add whiskers to the side-by-side bar chart and the grouped

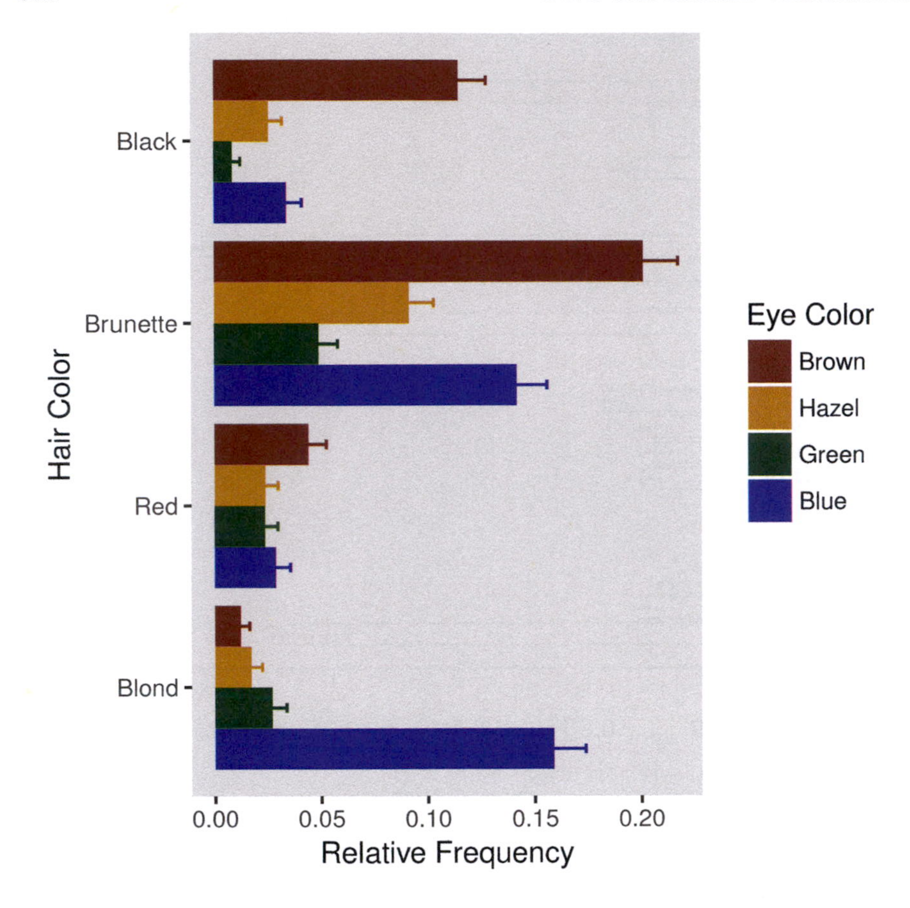

Figure 8.16 *Side-by-side bar-whisker chart of eye-color frequency grouped within hair color using* `ggplot2`

dot chart for the purpose of illustrating variability associated with each point estimate of relative frequency. It is hard to conceive how variation in point estimates could be illustrated in a side-by-side stacked bar chart. The side-by-side stacked bar chart is not recommended and the grouped dot chart remains preferred.

8.10 The Side-by-Side Pie Chart

Examples of *side-by-side pie charts* are given in Figures 8.23 and 8.24 using the hair and eye color data. Because of circular symmetry, a two-by-two layout has been chosen for the four pie charts in each of the two figures. A horizontal or vertical layout in the same space would require smaller pies and cramped labeling for the pie sectors.

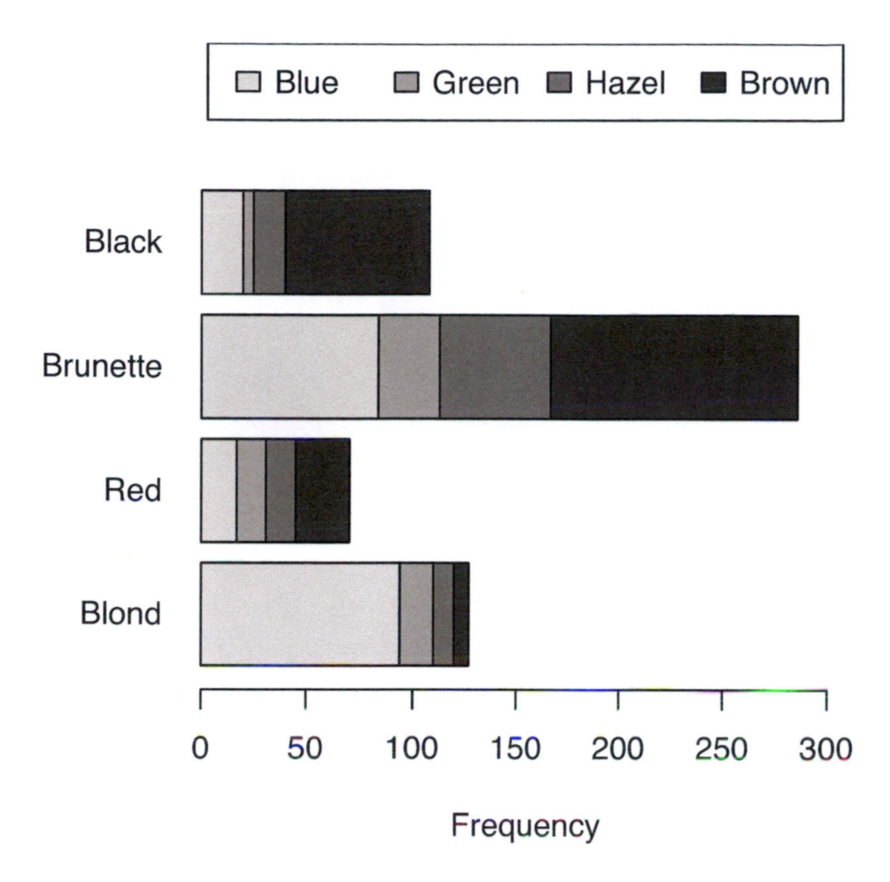

Figure 8.17 *Side-by-side stacked bar chart of eye-color frequency grouped within hair color*

Figure 8.23 illustrates the conditional distribution of eye color given hair color. Side-by-side pie charts cannot be used to depict the bivariate distribution of discrete variables if the pies are the same size.

To illustrate the bivariate distribution of hair and eye color, Figure 8.24 adjusts the size of the pie to depict the relative count for each eye color. This has been done so that the comparison is based upon area as the measure of size and not radius. A mistaken choice of making radius proportional to the marginal counts for eye color could lead to exaggeration of the differences among the marginal counts.

There are only four categories for each hair and eye color so there are only four wedges in each pie. There aren't so many wedges that color illustration is needed. Grayscale shading works fine. The lightest color starts at twelve o'clock with progressively darker colors being added on a clockwise basis for

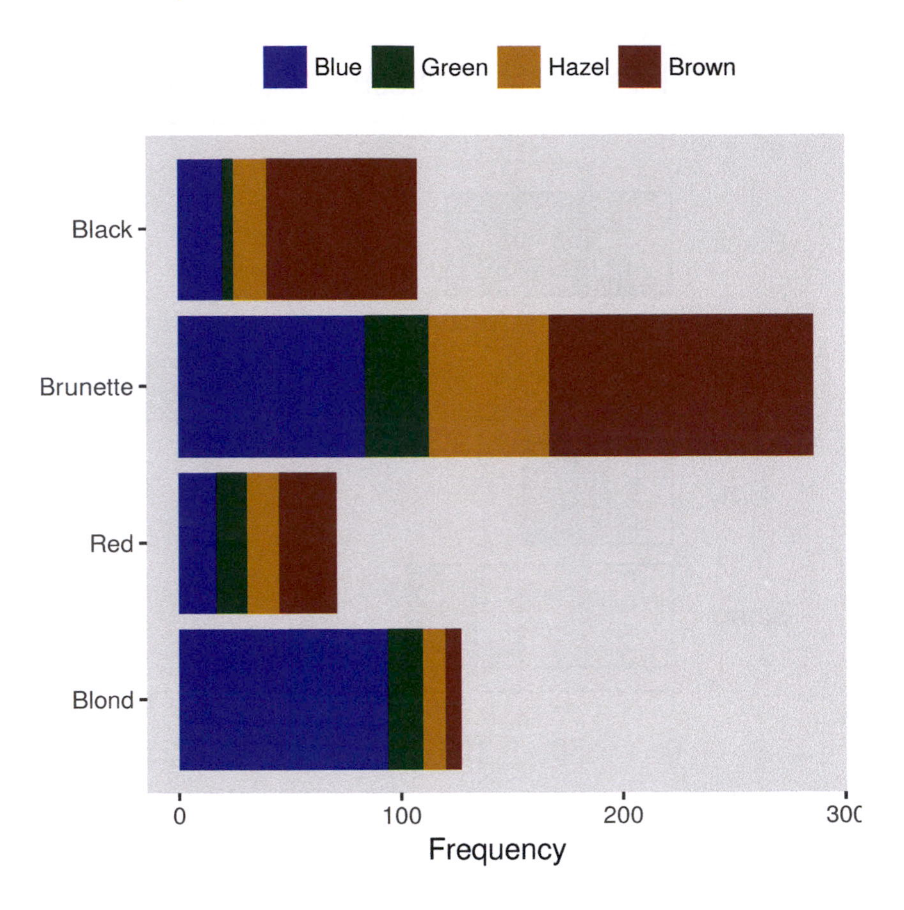

Figure 8.18 *Side-by-side stacked bar chart of eye-color frequency grouped within hair color using* `ggplot2`

each pie. But there is no common scale for comparison. Are comparisons to be made among sectors based upon area, arc, or internal angle? Not all viewers will make the same choice. This implies that some viewers will need to work harder than others.

The issue of comparison in Figure 8.24 is further complicated because the pies are not the same size. It is a complex task to compare wedges in pies of different radii.

Tufte [123] (pp. 178) wrote "A table is nearly always better than a dumb pie chart; the only worse design than a pie chart is several of them, for then the viewer is asked to compare quantities located in spatial disarray both within and between pies...." On Tufte's advice, it would have been better to stick with Table 8.1 than draft either Figure 8.23 or Figure 8.24.

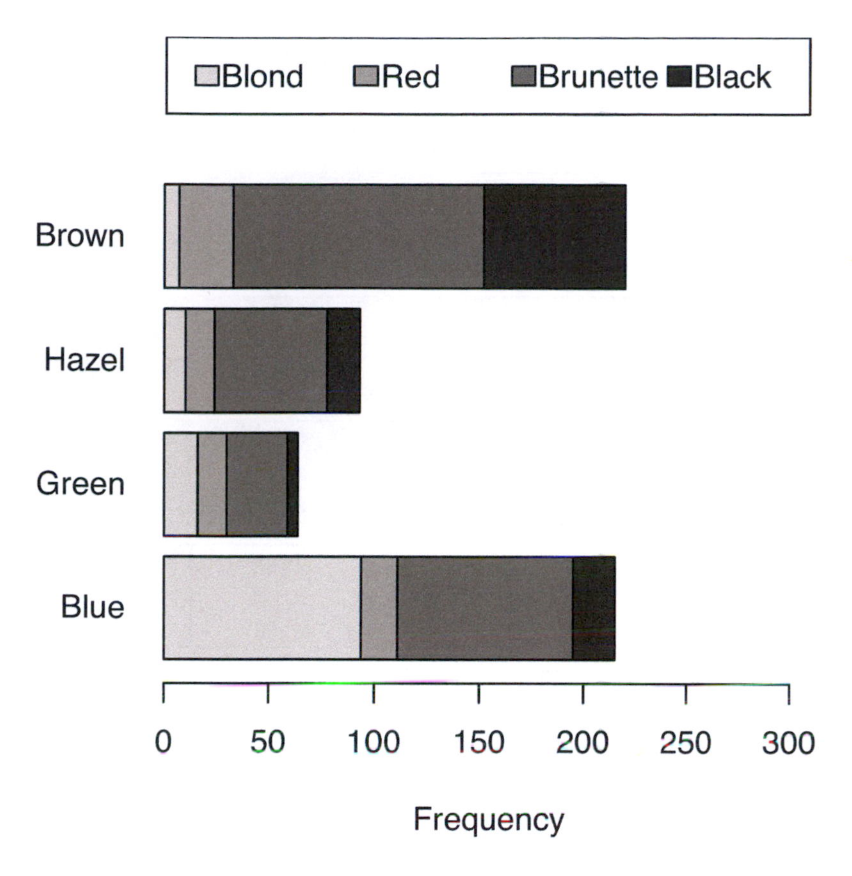

Figure 8.19 *Side-by-side stacked bar chart of hair-color frequency grouped within eye color*

A reasonable question is whether there is anything worse than side-by-side pie charts. According to Cleveland and McGill [25], there is. Based upon their perception experiments with bar charts, stacked bar charts, and pie charts, it can be inferred that the side-by-side stacked bar chart is worse. Side-by-side pie charts and side-by-side stacked bar charts share in common with each other the limitation that variation in relative frequencies cannot be depicted.

When it comes to choosing between the side-by-side pie chart and the gold standard, the choice is clear: go with the grouped dot chart.

8.11 The Mosaic Chart

A *mosaic chart* is a graphical display of the distribution of two discrete variables in which each count is represented by a rectangle of area proportional

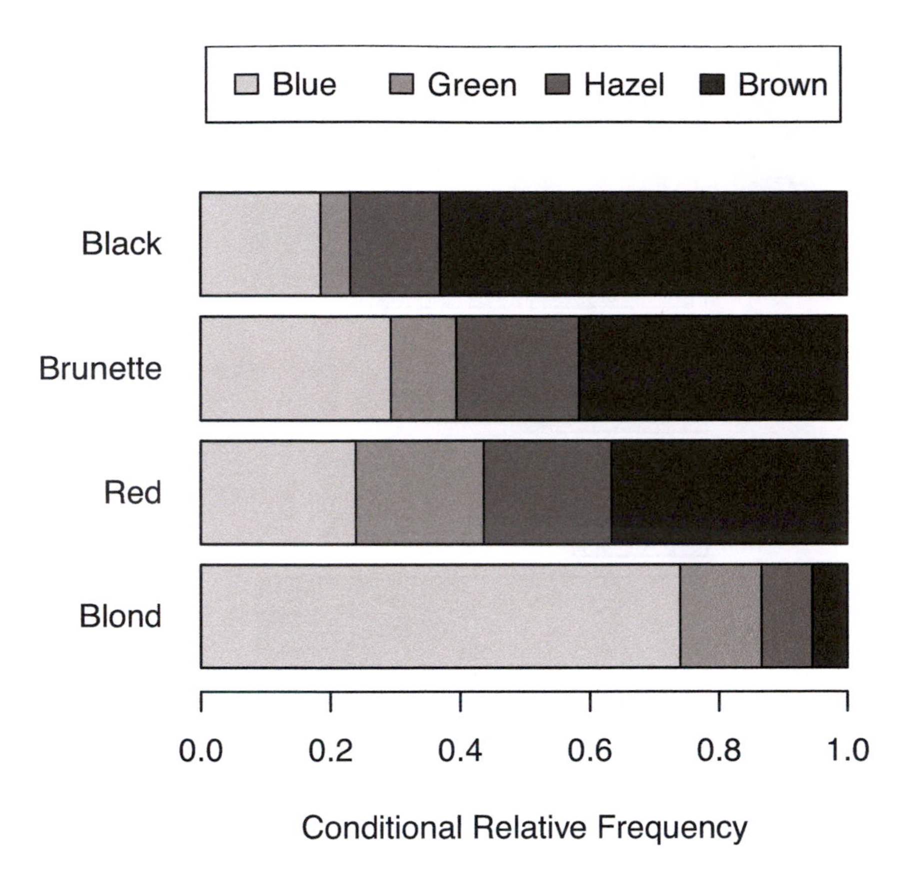

Figure 8.20 *Side-by-side stacked bar chart of eye-color relative frequency grouped within hair color*

to the count. An example of a mosaic chart for the hair and eye color data of Snee [112] is given in Figure 8.25.

Gray shading has been used in Figure 8.25 to make the mosaic plot more eye catching. The same gray tone has been used for each tile in the mosaic. Note the vertical axis corresponds to hair color in this figure. The horizontal axis denotes eye color. A good feature of the mosaic chart is that viewers are forced to focus their attention on the bivariate nature of the data. This nature can be overlooked at first glance in the grouped dot chart of Figure 8.1.

A problem with the mosaic chart of Figure 8.25 is that comparisons must be made on the basis of area. In the hierarchy of comparisons, this is more difficult and time consuming than comparing lengths of line segments in a grouped bar chart.

Figure 8.25 was executed using the `mosaicplot` function in the R statistical

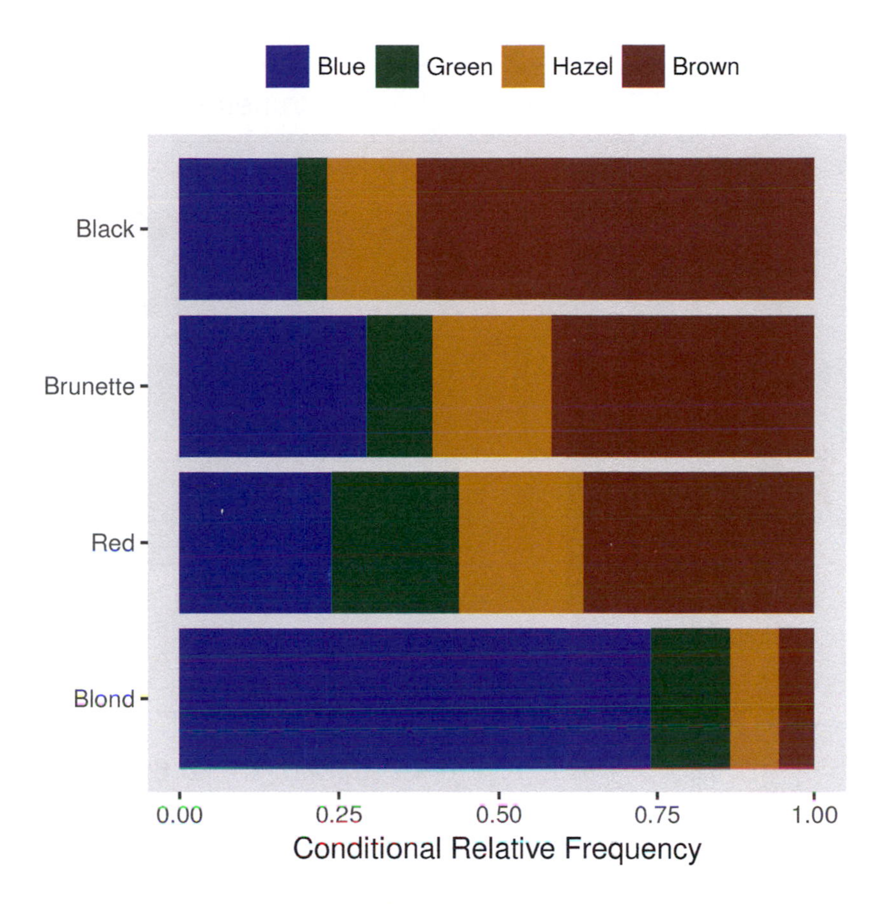

Figure 8.21 *Side-by-side stacked bar chart of eye-color relative frequency grouped within hair color using* ggplot2

analysis package. Figure 8.26 is a re-draft of Figure 8.25 using an option that allows the selection of gray shades as a function of the vertical axis, which is hair color. The R code for producing Figure 8.26 is as follows.

```
haireye<-matrix(data=c(20,5,15,68,84,29,54,
119,17,14,14,26,94,16,10,7),
nrow=4,ncol=4,byrow=TRUE,
dimnames=list(c("Black","Brunette","Red","Blond"),
c("Blue","Green","Hazel","Brown")))

mosaicplot(t(haireye),main=" ",las=1,
cex=0.75,color=TRUE)
```

Note that the transpose of the matrix `haireye` is the data argument passed

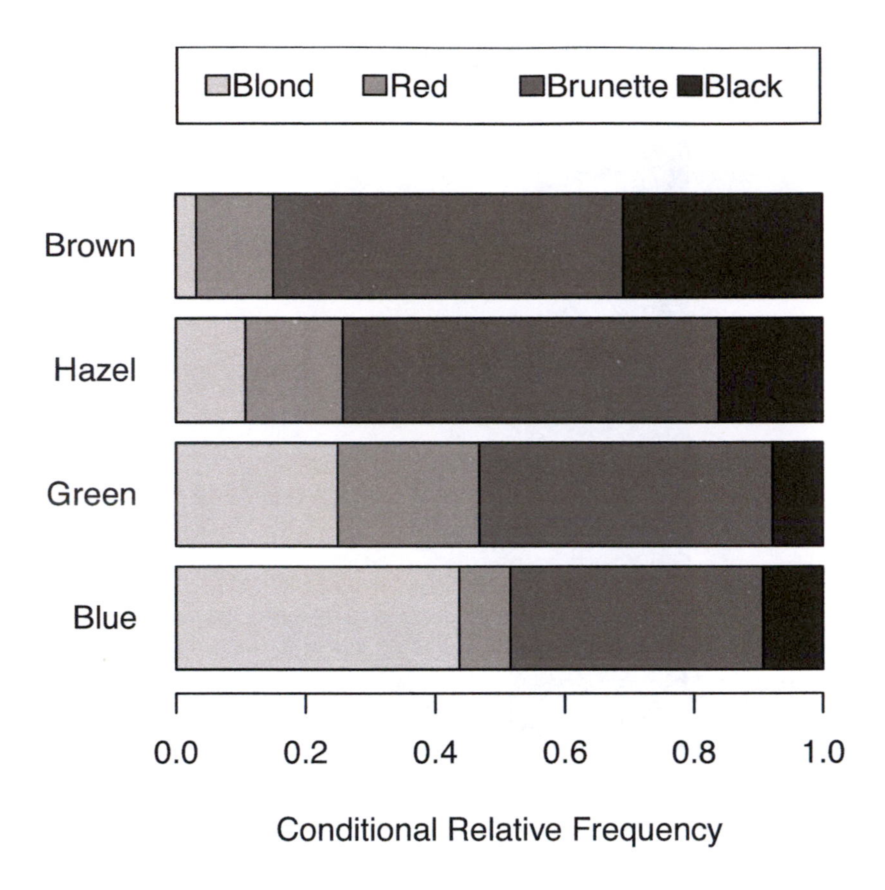

Figure 8.22 *Side-by-side stacked bar chart of hair-color relative frequency grouped within eye color*

to the R function `mosaicplot`. Setting `las=1` forces the axes labels to be horizontal for both rows and columns. The size of the color labels is reduced to 75 percent of usual size by setting `cex=0.75` in the preceding call. This has been done so that the eye color labels do not overrun tile width. The gray shading according to row category, hair color, is obtained by setting `color=TRUE`.

Figure 8.27 is a color version of Figure 8.26. The `ggplot2` package does not have a function for producing a mosaic plot. The R package `ggmosaic`, however, is an extension package for `ggplot2` written to produce mosaic plots. The `ggplot2` and `ggmosaic` code for producing Figure 8.27 is as follows.

```
require(ggplot2)
require(ggmosaic)
```

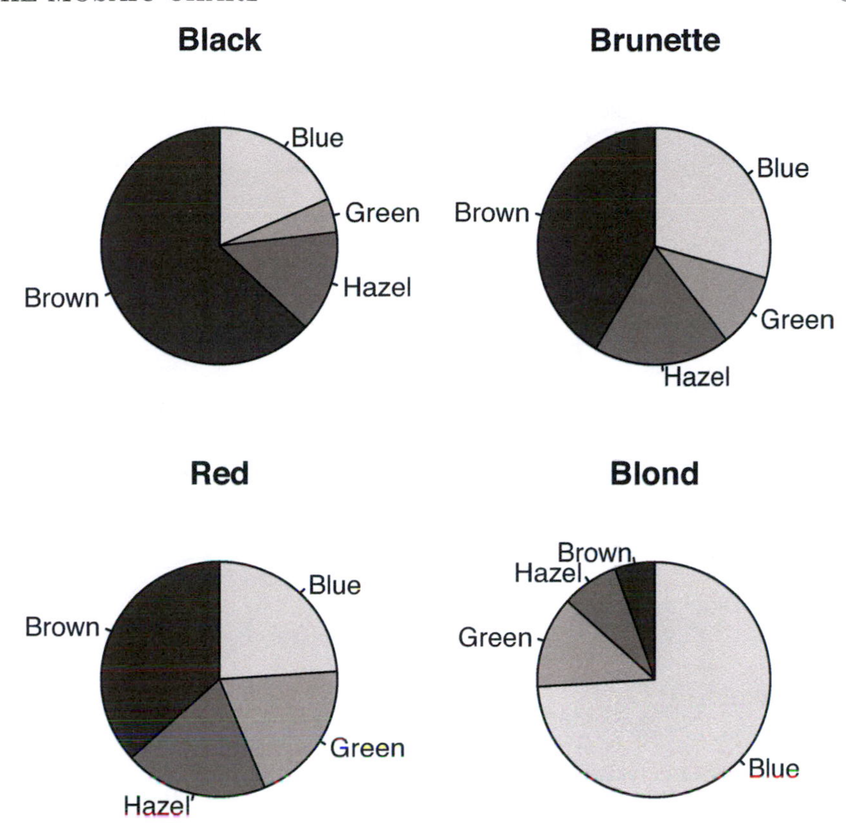

Figure 8.23 *Side-by-side pie chart of conditional eye-color relative frequency grouped within hair color*

```
graphics.off()
windows(width=4.5,height=4.5)

haireye<-matrix(data=c(68,15,5,20,119,54,29,
84,26,14,14,17,7,10,16,94),
nrow=4,ncol=4,byrow=TRUE,
dimnames=list(c("Black","Brunette","Red","Blond"),
c("Brown","Hazel","Green","Blue")))

haircol<-unlist(dimnames(haireye)[1])
eyecol<-unlist(dimnames(haireye)[2])

hair<-rep(" ",16)
eye<-rep(" ",16)
```

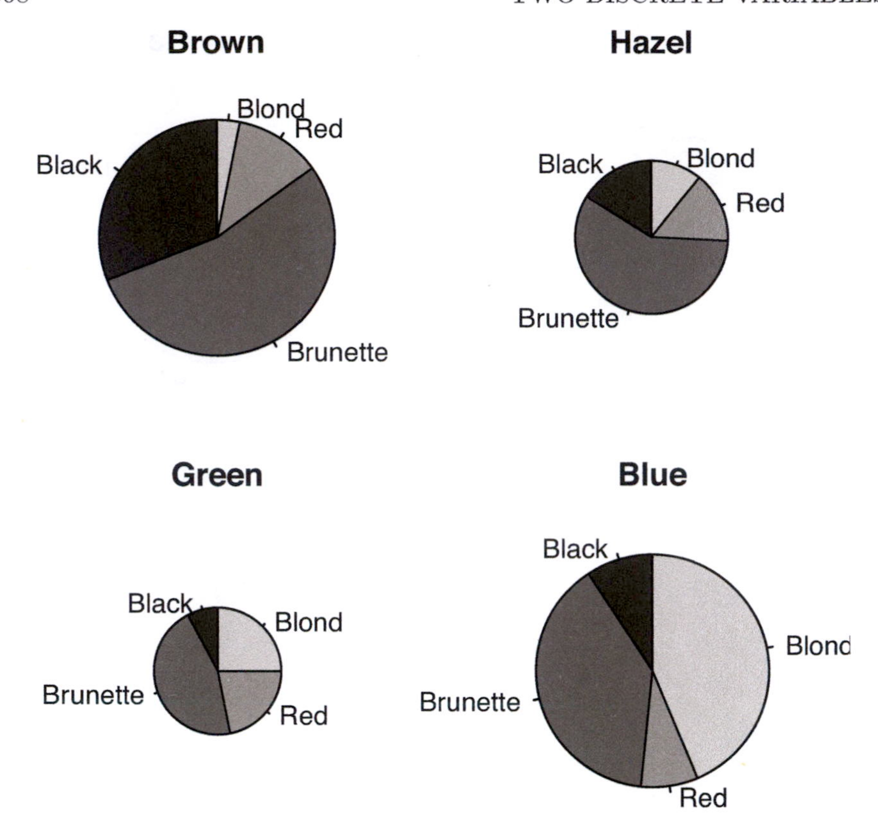

Figure 8.24 *Side-by-side pie chart of relative frequency for hair and eye color (pie area is proportional to eye color count)*

```
freq<-rep(0,16)

n=0

for (i in 1:4){ for (j in 1:4){
n<-n+1
hair[n]<-haircol[i]
eye[n]<-eyecol[j]
freq[n]<-haireye[i,j]
}}

haireyemf<-data.frame(hair=hair,eye=eye,freq=freq)
figure<-ggplot(haireyemf,aes(weight=freq,
```

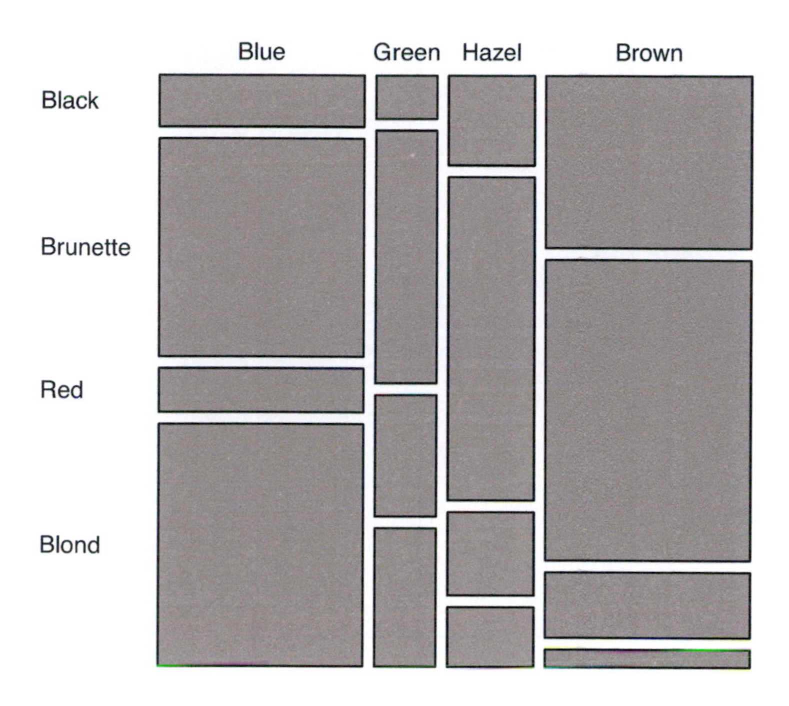

Figure 8.25 *Mosaic chart of eye and hair color*

```
x=product(factor(eye,levels=rev(eyecol),ordered=TRUE)),
fill=factor(hair,levels=rev(haircol),ordered=TRUE))) +
geom_mosaic(show.legend=FALSE,alpha=1) +
theme(panel.background=element_blank(),
panel.grid=element_blank(),
axis.ticks=element_blank()) +
scale_y_continuous(breaks=c(0.25,0.475,0.7,0.95),
label=(rev(c("Black","Brunette","Red","Blond")))) +
labs(x="",y="") +
scale_fill_manual(values=rev(c("black","brown",
"red","gold")))

print(figure)
```

The `ggmosaic` function for producing a mosaic plot with `ggplot2` is `geom_mosaic` and it is manipulated as an object as with any other native `ggplot2` plotting function. If colors for fill are manually set by a subsequent call to `scale_fill_manual`, then when saving the `ggmosaic` figure to an .eps file the mosaic itself will not be saved due to an issue with transparency not

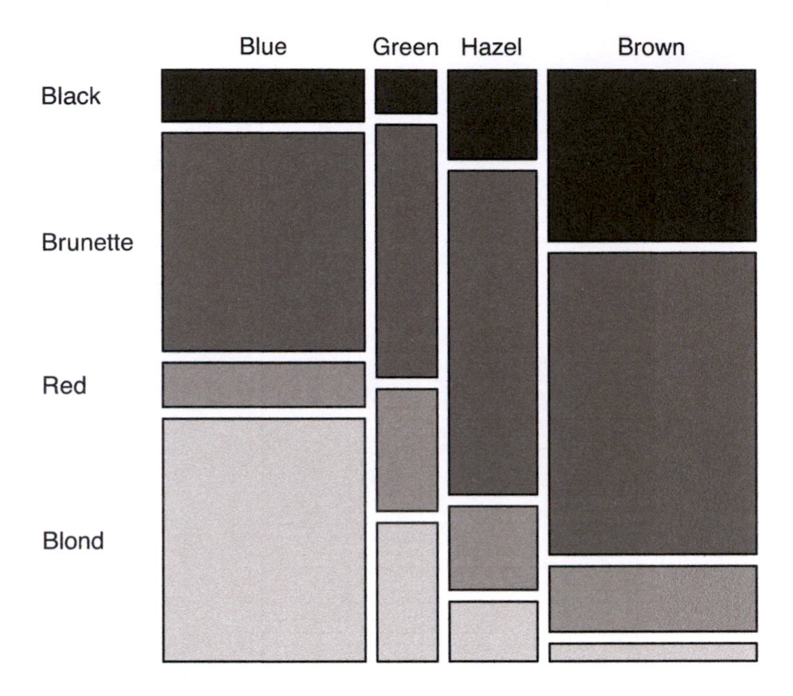

Figure 8.26 *Mosaic chart of eye and hair color with grayscales for hair color*

being supported by R for .eps files. Setting `alpha=1` when calling `geom_mosaic` forces there to be no use of color transparency in the mosaic figure produced. Note also that the legend for the mosaic plot in Figure 8.27 was switched off by passing `show.legend=FALSE` when calling `geom_mosaic`.

It is apparent from Figures 8.26 and 8.27 that the most common pairings of hair and eye color are brown-eyed brunettes followed by blue-eyed blonds. The rarest pairings are black hair with green eyes and blonds with brown eyes.

Shading in a mosaic plot can also be helpful in making conditional comparisons. It is apparent from Figure 8.26 that the proportion of blonds decreases with darkness of eye color. An increase in the proportion of black-haired individuals is associated with darkness of eye color. These same observations can be made with Figure 8.28, which exchanges the axes and associates gray shading with eye color.

The first appearance of a mosaic chart in a refereed journal can be attributed to Hartigan and Kleiner [61] in 1984. Its use was first proposed in 1981 by Hartigan and Kleiner [60]. They refer to the plot simply as a "mosaic." The examples of Kleiner [61] actually involve three and four discrete variables.

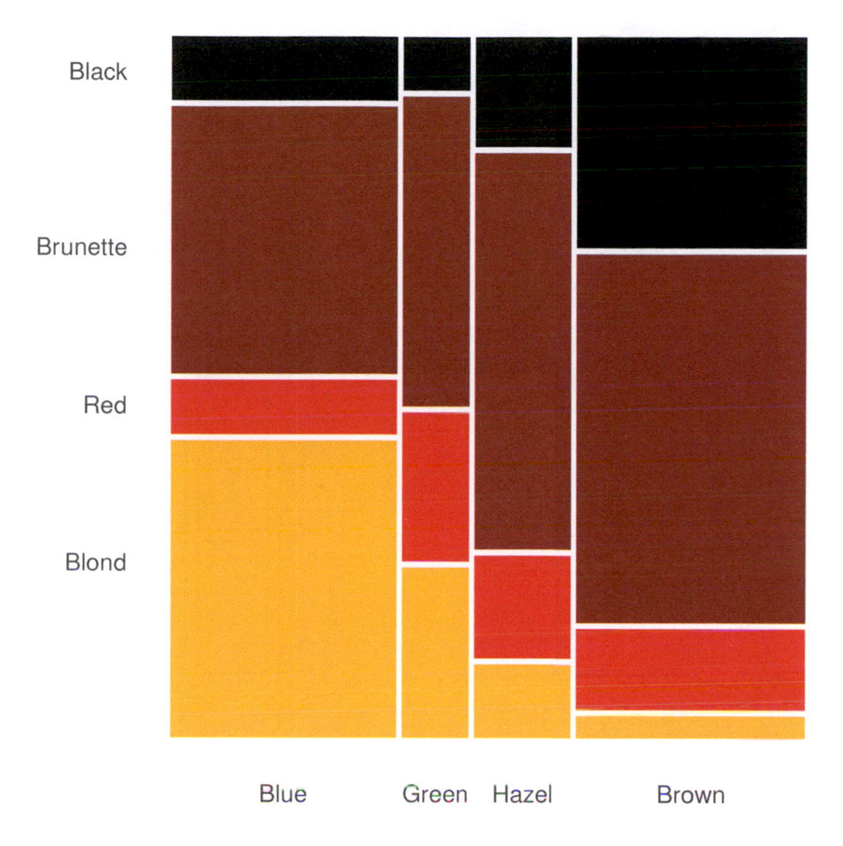

Figure 8.27 *Mosaic chart of eye and hair color with colors for hair color using* `ggplot2` *and* `ggmosaic`

With respect to a recommendation for using the mosaic chart, there are no comparative experiments with the grouped dot chart that can be cited. On the basis of the hierarchy of comparative tasks, because the mosaic chart relies on area comparisons and the gold standard relies upon length comparisons on a common horizontal scale, it must be concluded that the grouped dot chart is to be preferred. From the perspective of an eye-catching change from the commonplace, the mosaic chart is a temptation.

8.12 Conclusion

This chapter began with a discussion of the grouped dot chart. It was asked that this be accepted as the gold standard. With perhaps the exception of the side-by-side bar chart, the grouped dot chart was followed by a parade

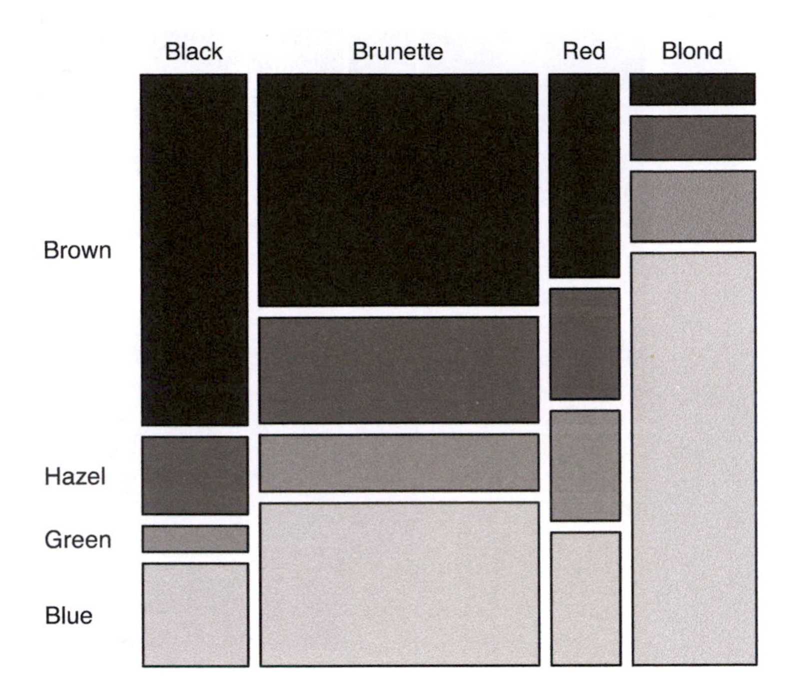

Figure 8.28 *Mosaic chart of eye and hair color with grayscales for eye color*

of graphics that did not measure up. Many were hideous. Other books on graphics, such as Cleveland's [21], do not subject their readers to this abuse.

The presentation of poor charts in this chapter was done because these charts are still used in practice. It was not necessary to give examples of side-by-side stacked bar charts with moiré effects, side-by-side pie charts in three-dimensional perspective, and so on. These topics have been dealt with in previous chapters. To discourage the reader, the R codes for these plots have not been included in this chapter. Examples of R codes have been provided in this chapter for the grouped dot chart and the grouped two-tiered dot-whisker chart, which are advocated, and the mosaic chart, which is a temptation. The R scripts for all the figures of this chapter are available for download from the website for this book.

This book has been written to get the message on effective graphics out there. A few recent issues of any scientific or medical journal will provide additional examples of poor graphics illustrated in this chapter. Grouped dot charts and grouped dot-whisker charts will be rare sightings.

In this chapter, a single data set has been taken as a case study and drafted a variety of different ways. The chapter began with the gold standard, and other

	Type of Individual		
Class	**Men**	**Women**	**Children**
First	173	144	5
Second	160	93	24
Third	454	179	76
Crew	875	23	0

Table 8.2 *Population on board the ocean liner Titanic at time of sinking*

graphical displays were executed and compared to this standard and to other displays. This book can be used as a reference if a boss expects a side-by-side stacked bar chart and gets a grouped dot chart instead. This book can be opened to the discussion as to why a grouped dot chart ought to be used.

The final recommendations with respect to displaying distributions of two discrete variables are the grouped dot chart and the grouped dot-whisker chart. Without being overly pedantic, side-by-side bar charts and side-by-side bar-whisker charts also get the job done but at the cost of a bit more ink.

8.13 Exercises

1. On the night of 14 April 1912, the ocean liner Titanic struck an iceberg and sank within three hours early on 15 April 1912. The ship had a lifeboat capacity of 1,178 despite being able to carry 3,547 people. Questions have been raised as to whether passengers and crew were evacuated to the life boats on the basis of "women and children first." It has been suggested that social class was the determining factor for survival. Before this can be examined, it is necessary to analyze the number of people with respect to age, gender, and class. Table 8.2 reports the number of men, women, and children among the three ticket classes and the crew, as compiled by the British Wreck Commissioner's Inquiry [13]. According to Table 8.2, a total of 2206 souls were embarked. Others report the number on board when the iceberg was struck as 2201, 2223, and 2240. The number of persons aboard a vessel should be determined by a census and not a random sample. Ignore any possible undercount or errors in recording and treat Table 8.2 as being fully accurate and without error.

   (a) Produce a grouped dot chart for Table 8.2. Use class as the grouping variable.

   (b) Produce a grouped dot chart for Table 8.2. Use type of passenger as the grouping variable.

   (c) Using the grouped dot charts prepared for parts (a) and (b), what can be summarized about the distribution of passengers and crew on board the Titanic at the time of sinking?

2. You are working with a colleague on an article of interest to both of you: the sinking of the Titanic. You both intend to submit the manuscript to a popular science magazine. You have decided to include a depiction of Table 8.2 in the manuscript. You suggest a grouped dot chart depicting counts with grouping according to type of individual. But your colleague disagrees.
   (a) Write a brief paragraph that could be enclosed in an email message to your colleague justifying the grouped dot chart and listing your reasons for this choice.
   (b) Execute the graphical display you proposed in part (a).
   (c) Create a label for your graphical display of part (b). Write a paragraph of a few sentences to accompany your graphic.
3. You are interested in comparing the distribution of men, women, and children among those embarked aboard the Titanic across the ticket classes and among the crew.
   (a) Execute a conditional relative frequency dot chart given class or crew status. Arrange the chart so that the crew appears after third class. Verify that the conditional relative proportions within a class sum to one, or one hundred if using percent rather than proportion.
   (b) Write a brief paragraph describing what you see in your chart for part (a).
   (c) Draft a second conditional grouped dot chart but condition on type of passenger instead of class.
   (d) What, if anything, is gained by considering the chart of part (c) in addition to the chart of part (b)? Discuss.
4. The original 1984 plotting convention for a grouped dot chart of Cleveland and McGill [25] included plotting the total frequency for each group. This was not done in Figure 8.1.
   (a) Using the data of Table 8.1, create a version of the grouped dot chart in Figure 8.1 according to the 1984 plotting standard.
   (b) The 1985 plotting convention of Cleveland [21] does not include plotting a line segment and dot for a group total. Speculate as to why this might have occurred and the reasons for it.
5. In 1985, Cleveland [21] added distinct different symbols for terminating line segments in the grouped dot chart.
   (a) Does the addition of different symbols constitute an important change from the 1984 plotting convention of Cleveland and McGill [25] for the grouped dot chart? Discuss.
   (b) The side-by-side bar chart of Figure 8.12 is a direct counterpart to the grouped dot chart of Figure 8.1. Produce a side-by-side bar chart for the data of Table 8.1 using only white fill within a black border for each.

Antibiotic Resistance	Tetracycline Treatment	Azithromycin Treatment	Untreated Control
Tetracycline	39	28	15
Azithromycin	0	5	0
Neither	87	130	76

Table 8.3 *Distribution of antibiotic resistance in nasopharyngeal S. pneumoniae*

(c) Compare your answer in part (b) to Figure 8.12. Comment on whether the difference between these figures is substantial enough to justify always using matching shading, or color, for the second variable.

(d) Based upon your answer to part (c) with respect to side-by-side bar charts, comment on the importance of discriminating plotting symbols for the second variable in a grouped dot chart.

6. In 2005, the *British Journal of Opthalmology* published an article reporting greater bacterial resistance when topical antibiotics are administered for control of the eye infection known as trachoma. The sites for the trial were four villages in Nepal. The researchers involved were from the University of California at San Francisco, the Geta Eye Hospital in Nepal, the Department of Public Health, and the Centers for Disease Control in Atlanta. The study involved children aged 1 to 10 years. Children in one village received oral azithromycin. Children in another village received topical tetracycline. Children were recruited as untreated controls in the other two villages. There were 194 children treated with oral azithromycin, 143 treated with topical tetracycline, and 107 controls. Table 8.3 is taken from Gaynor *et al.* [55]. This reports antibiotic resistance in those children aged 1–7 years found to be carriers of *Streptococcus pneumoniae* 6 months after the three-year treatment regime was completed as well as in children from the two control villages. Children were determined to be resistant to *S. pneumoniae* based on nasal swabs.

(a) Gaynor *et al.* [55] reported for the 3×3 contingency table in Table 8.3 a P-value of 0.004 for Fisher's exact test of no difference with respect to antibiotic resistance among the three types of villages. Confirm this by your own calculation.

(b) Create a grouped dot chart for the data in Table 8.3. Note that the data are not the result of random sample whereby children were classified according to two variables. The voluntary participants, however, can be treated as a random sample within the control villages, the tetracycline-treated village, and the azithromycin-treated village. So the grouped dot chart ought to be executed with a conditional relative frequency axis with separate groups based upon treatment.

(c) The grouped dot chart for your answer to part (b) may be accepted by a medical journal for publication. A grouped dot-whisker chart would be better. Produce this chart.

(d) Explain the results in part (a) using the grouped dot-whisker chart of part (c).

7. Consider the data in Table 8.3 regarding antibiotic resistance after treatment for trachoma in children in Nepal. A conventional graphical display in a medical journal for this table would be a side-by-side bar-whisker plot. This would be in the form of conditional relative frequency for type of infection given treatment. You believe the corresponding grouped dot-chart would be a better option.
   (a) Produce a side-by-side bar-whisker plot for the data in Table 8.3.
   (b) If you haven't already done so, produce a relative frequency dot-whisker chart conditional on treatment type.
   (c) Prepare a letter to the associate editor of an ophthalmology journal for which you plan to attach the graphs produced in parts (a) and (b) stating your preference to use a dot-whisker plot and explaining the reasons for your preference.

8. Pick a scientific, engineering, or technical journal of your own choice and examine twelve issues.
   (a) Enumerate the number and type of graphical displays for depicting the distribution of two discrete variables.
   (b) Depict your results from part (a) in a dot chart for a single discrete variable.
   (c) Summarize your findings from part (a) and its depiction in part (b) in a brief paragraph.

9. Pick a daily newspaper or online news outlet of your choice and examine fifty articles or stories.
   (a) Enumerate the number and type of graphical displays for depicting the distribution of two discrete variables.
   (b) Summarize your findings from part (a) in a brief paragraph. Use an appropriate graphical display if necessary.

10. It is not easy to miss the fact that two variables are being portrayed in a mosaic chart. This graphical display is rarely seen outside of statistics journals and textbooks. A mosaic chart is one of those graphical displays that does not lend itself to a quick look. It takes time for a viewer to digest.
    (a) In your own discipline, would you consider using a mosaic chart for a presentation or a written document? Discuss.
    (b) Would you still draft and use a mosaic chart for your use if you decided not to share it with another viewer or a larger audience? Discuss.

11. Portray the Titantic data given in Table 8.2 in two different mosaic plots. Which of the two is the more effective? Discuss.

12. Portray the data in Table 8.3 regarding antibiotic resistance after treatment for trachoma in children in Nepal in two different mosaic plots. Which of the two is the more effective? Discuss.

Chapter 9

Depicting the Distribution of One Continuous Variable and One Discrete Variable

9.1 Introduction

Most applications of statistical methods involve the comparison of two or more populations. For example, an agricultural experiment may be performed to determine whether the application of a new fertilizer results in higher yields compared to the control treatment. A new drug developed to reduce serum cholesterol levels may be studied in a medical experiment against other drugs already in use. The distribution of personal income in different nations may be studied by survey sampling. In these three examples, there is a quantitative variable of interest indexed by a single qualitative variable.

Any of the graphical displays in this chapter are typically the first step in data analysis, or at least they ought to be. The purpose of these graphical displays is to examine for differences in location and spread among the distributions.

To continue the use of a convention previously adopted for the distribution of a single continuous variable, a graphical display introduced in this chapter will be referred to as a *plot*. The term *chart* is reserved for one or more discrete variables.

9.2 Learning Outcomes

When you complete this chapter, you will be able to do the following.

- Be able to execute a side-by-side dotplot with either the `graphics` package or the `ggplot2` package.
- Be able to execute a side-by-side boxplot, with or without notches, with either the `graphics` package or the `ggplot2` package.

Name				
Scientific	**Finnish**	**Swedish**	**English**	**Sample Size**
Abramis bjoerkna	Parkki	Bjoerknan	Silver Bream	11
Abramis brama	Lahna	Braxen	Bream	34
Esox lucius	Hauki	Jaedda	Pike	17
Leusiscus idus	Sikka	Iiden	Ide	6
Osmerus eperlanus	Norssi	Norssen	Smelt	14
Perca fluviatilis	Ahven	Abborree	Perch	56
Rutilus rutilis	Saerki	Moerten	Roach	20

Table 9.1 *Species and sample sizes of fish caught in Längelmävesi*

- Be able to overlay side-by-side dotplots and side-by-side boxplots with either the `graphics` package or the `ggplot2` package.
- Be knowledgeable about back-to-back and side-by-side stemplots.
- Be able to execute a side-by-side dot-whisker plot with either the `graphics` package or the `ggplot2` package.
- Optionally, be able to prepare a trellis kernel density estimate with either the `lattice` package or the `ggplot2` package.
- Optionally, be knowledgeable about how to prepare a trellis plot of dotplots, boxplots, histograms, or quantile-quantile plots of one continuous variable for multiple values of a discrete variable with either the `lattice` package or the `ggplot2` package.

9.3 The Side-by-Side Dotplot

In Chapter 3, a source of data was from a research trawl on the lake known as Längelmävesi near Tampere, Finland. The first dotplot examples were for the mass of 56 perch. There are other fish in Längelmävesi that are also important to a commercial freshwater fishery. Six species in addition to perch are listed in Table 9.1 with their scientific, Finnish, Swedish, and English names. Table 9.1 reports the number of fish caught for each species.

Five of the fish species listed in Table 9.1 are not native to North America. In Canada and the United States, the English names for some of the species differ from those used in England because there are North American counterparts. *Abramis brama* is known as the carp bream. *Perca fluviatilis* is known as the European perch. *Osmerus eperlanus* is known as the European smelt.

Two of the species listed in Table 9.1 are found in North America. Ide (or orfe) was imported by the United States Fish Commission in the nineteenth century as a commercial and ornamental species but escaped pond containment. *Esox*

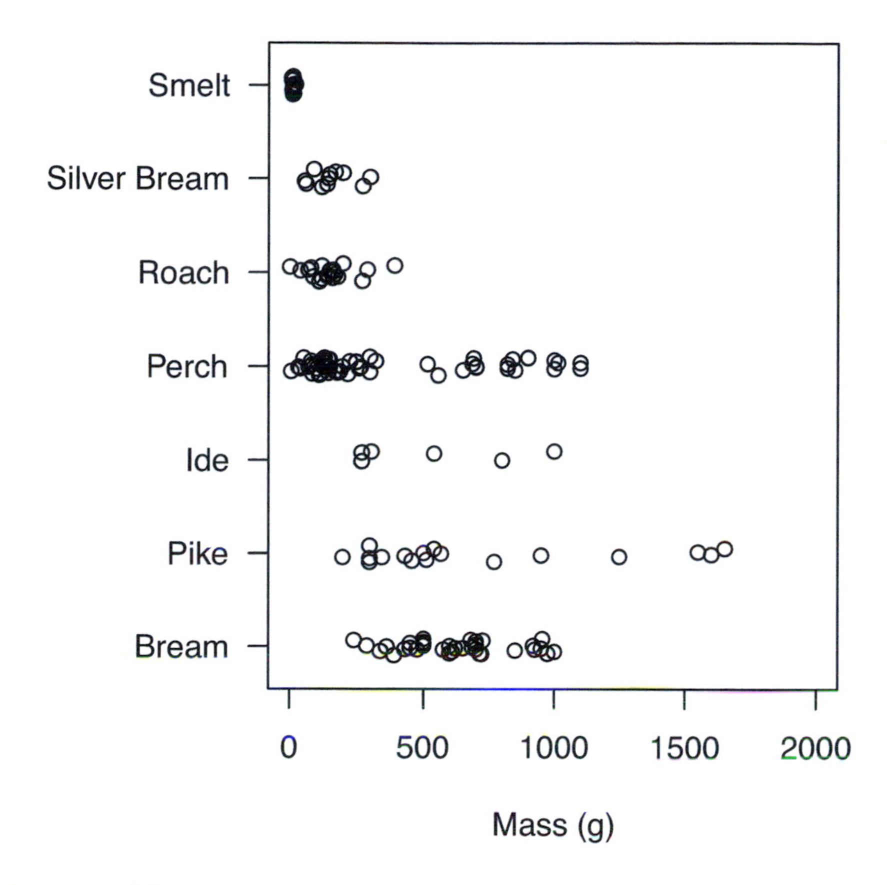

Figure 9.1 *Side-by-side dotplot of mass (g) of seven species of fish caught in Längelmävesi*

lucius is found throughout the northern hemisphere. In North America it is known as the Northern Pike, with the exception of parts of western Canada and the southern United States where the common name is jackfish or, simply, jack.

The data are complete for mass with the exception of one roach. The *side-by-side dotplot* of Figure 9.1 plots mass, in grams, for each of 158 fish in the seven species. All are game fish for the sports fisher, even the tiny smelt.

Figure 9.1 was plotted using the R function `stripchart` in the following R code.

```
fc<-Fishcatch

fc$Species<-ordered(fc$Species,
```

```
levels=c(1,6,2,7,3,4,5),
labels=c("Bream","Pike","Ide","Perch",
"Roach","Silver Bream","Smelt"))

set.seed(475)

stripchart(fc$Weight  fc$Species,pch=21,
method="jitter",cex=1.0,xlim=c(0,2000),
xlab="Mass (g)",group.names=rep(" ",7),
vertical=FALSE)

axis(2,at=1:7,labels=levels(fc$Species),las=1)
```

The ordering of smallest fish at the top of Figure 9.1 to largest fish at the bottom of the side-by-side dotplot was achieved by the first five lines of the preceding R code. This is not the numerical ordering in the original R list `Fishcatch`, which is left unchanged by assigning its contents to the new R list `fc`. So the reordering has been done with a call to the R function `ordered` that sets the desired ordering in the list `fc` and matches labels to the numerical codes of the fish species. The species' labels are applied to Figure 9.1 not by the call to `stripchart` but by a separate call to the lower-level function `axes`. Printing the labels in the call `stripchart` were suppressed by setting `group.names=rep(" ",7)`.

The points have been randomly jittered because of duplication and close proximity for many of the masses for each species. The seed for the default pseudo-random-number generator has been set to 475 by the code `set.seed(475)`. The number 475 was selected after a bit of experimentation to determine a seed that produced the best separation between overlapping data points.

At a cost to truthfulness, jittering has not been completely successful in achieving clarity in Figure 9.1. But it is still possible to make some comparative comments about the seven species.

The smallest species is the smelt. They are a food source for the other species. The pike is clearly the apex freshwater predator in Längelmävesi. Pike are also known by the translation of their Latin name: water wolf. Seeing one while scuba diving in a clear freshwater lake, don't be surprised if it comes nose to nose. They are fearless freshwater barracuda. Insects, amphibians, reptiles, fish, birds, and even small mammals are all food for the pike. They will eat anything with a pulse, including other pike of the same size.

Pike can grow to 147 centimeters in length and 31 kilograms in mass. To be considered a master angler in Finland, one must catch a ten-kilogram pike. By law in the Province of Manitoba, in Canada, a pike when caught must be a minimum of 104 centimeters in length.

Because of their relative abundance, the most important commercial fish reported in Figure 9.1 are perch and bream. The silver bream, also known as

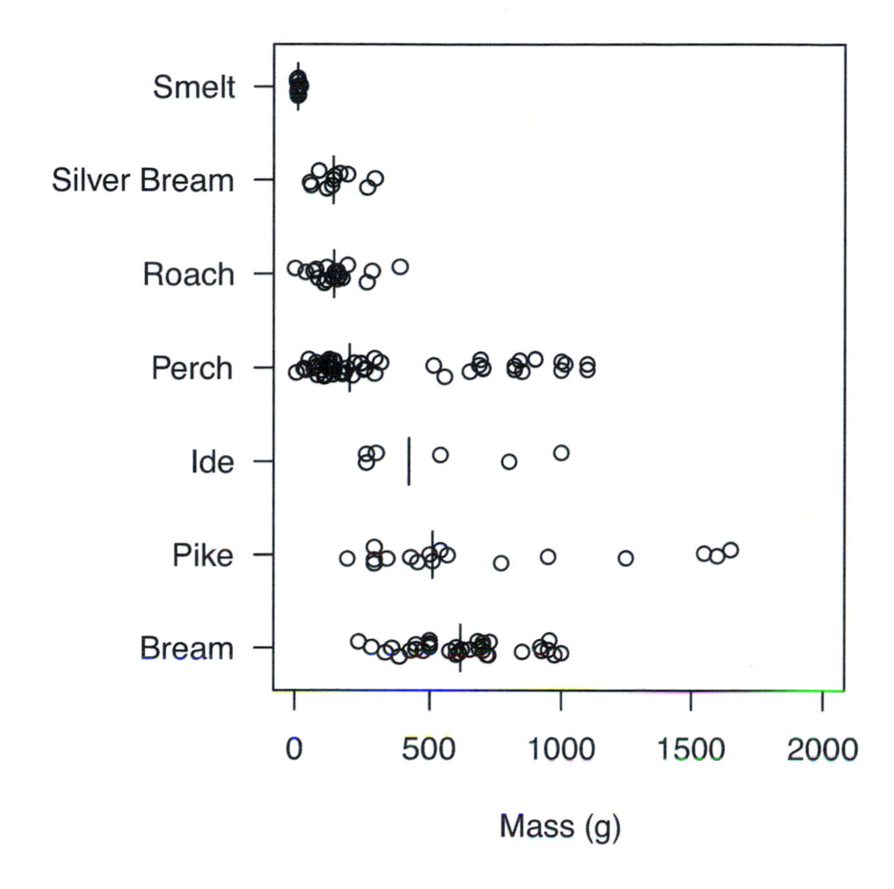

Figure 9.2 *Side-by-side dotplot of mass (g) of seven species of fish caught in Längelmävesi (the line segment for each species locates its sample median)*

white bream, are not as highly desired as the bream. Note that the mass of silver bream is narrower in spread and smaller in location than the bream, also known as the common bream.

The order of species in Table 9.1 is readily apparent as alphabetical by scientific name. The order in Figure 9.1 is not obviously apparent. The order is based on the sample median for each species. In Figure 9.2 the sample median for each species has been added to the dotplot. In the next section, this embellishment is taken to the next logical step.

9.4 The Side-by-Side Boxplot

Figure 9.3 gives an example of the *side-by-side boxplot*. These are outlier boxplots for mass for seven species of fish. A common plotting symbol has

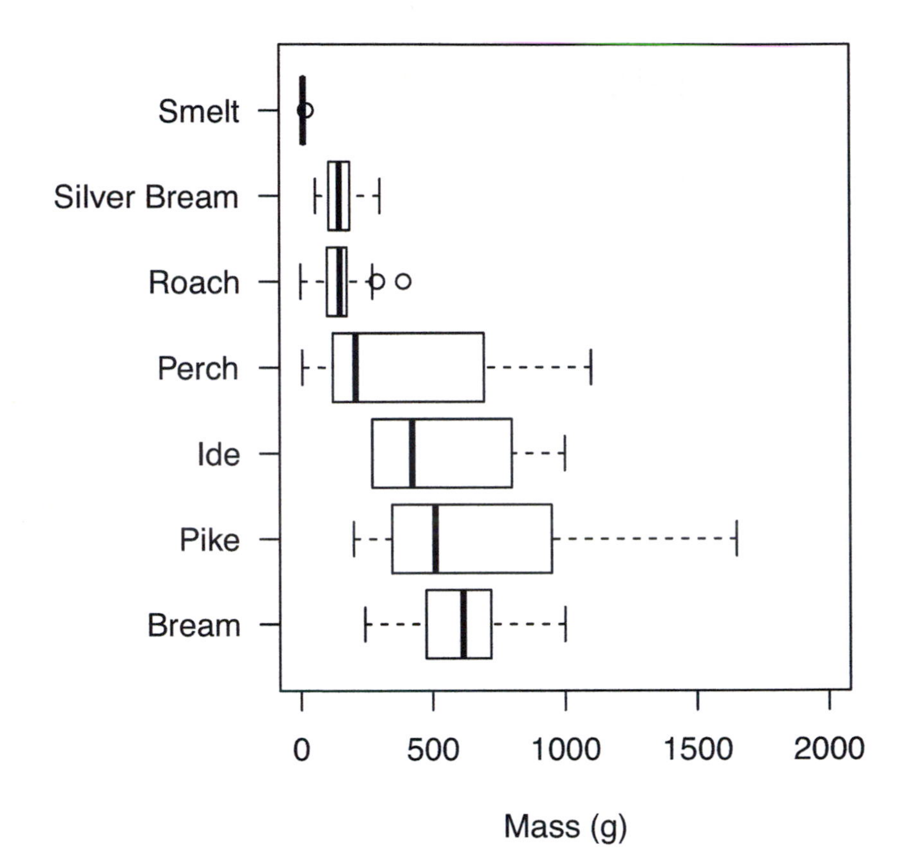

Figure 9.3 *Side-by-side outlier boxplot of mass (g) of seven species of fish caught in Längelmävesi*

been used for both outliers and extreme outliers. This is an unalterable feature of the R function `boxplot` that has been used to produce Figure 9.3. The R code is given below.

```
boxplot(Weight ~ Species,data=fc,ylim=c(0,2000),
xlab="Mass (g)",horizontal=TRUE,las=1)
```

Because of the scale required to accommodate the whiskers for pike, little is discernible for the sample of 14 smelt other than location and the presence of an outlier. They all start from a small egg, but it is the pike that has the heaviest weight according to the upper quartile in addition to being the species of the heaviest fish caught overall at 1650 grams.

One might infer from Figure 9.3 that the distribution of mass is rightward skewed for perch, ide, and pike. But we know from previous analyses, and evi-

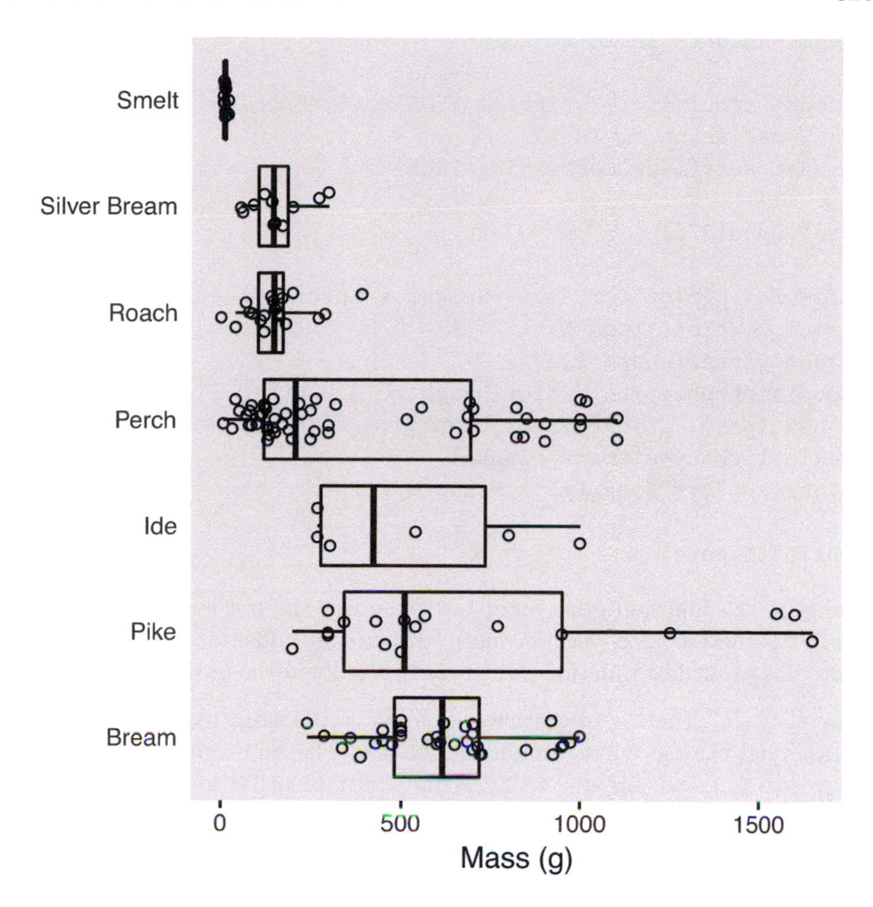

Figure 9.4 *Side-by-side outlier boxplot with overlaid dotplot of mass (g) of seven species of fish caught in Längelmävesi using* `ggplot2`

dent in the side-by-side dotplot Figure 9.3, that a bimodal normal distribution is a good fit for perch. There is evidence of rightward skewness in Figure 9.3 for silver bream as well, but the outlier boxplot in Figure 9.3 appears to be symmetric for bream.

Figure 9.4 uses `ggplot2` to overlay a dotplot on the side-by-side dotplot for each of seven fish species. The R code for doing this is as follows.

```
fc<-Fishcatch

fc$Species<-ordered(fc$Species,
levels=c(1,6,2,7,3,4,5),
labels=c("Bream","Pike","Ide","Perch",
"Roach","Silver Bream","Smelt"))
```

```
fcf<-as.data.frame(fc)

meds<-tapply(fc[,"Weight"],INDEX=fc[,"Species"],
FUN=median,na.rm=TRUE)
meds<-sort(meds,decreasing=TRUE)

set.seed(475)

figure<-ggplot(fcf,aes(y=Weight,x=Species)) +
geom_boxplot(alpha=0) +
geom_jitter(shape=1,
position=position_jitter(height=0.2,width=0.2)) +
theme(panel.grid=element_blank(),
axis.ticks.y=element_blank()) +
labs(x="",y="Mass (g)") + coord_flip()

print(figure)
```

The ggplot2 function `geom_boxplot` generates the outlier boxplot while the ggplot2 function `geom_jitter` more efficaciously adds the dotplot rather than `geom_dotplot`. The function `coord_flip` results in the horizontal orientation.

Note that the ordering of the lower quartile is the same as for the median. In commercial fishing, what remains in the net is based on mesh dimensions. In the research trawl, bream is the second most plentiful and has the heaviest lower quartile.

The economic rate of return in commercial fishing is closely tied to the mass of fish sold. Making species comparisons based on a measure of location is reasonable. Given long tails and bimodality, comparison for location based upon median is better than that based upon mean. Although Figure 9.3 illustrates variability for each species by illustrating the distribution of each species with side-by-side outlier boxplots, it does not depict the variability of the estimate of the median. This is done in the plot of the next section.

An early example of the side-by-side outlier boxplot can be found in 1977 on page 100 of Tukey [127].

9.5 The Notched Boxplot

The *notched boxplot* adds to the plotting convention of the outlier boxplot a visual means to conduct hypotheses tests regarding no difference in medians. Figure 9.5 gives side-by-side notched boxplots for mass for the seven species of fish caught in the research trawl on Längelmävesi in Finland around 1917. The notches in Figure 9.5 were obtained by adding the argument `notch=TRUE` when calling the R function `boxplot`.

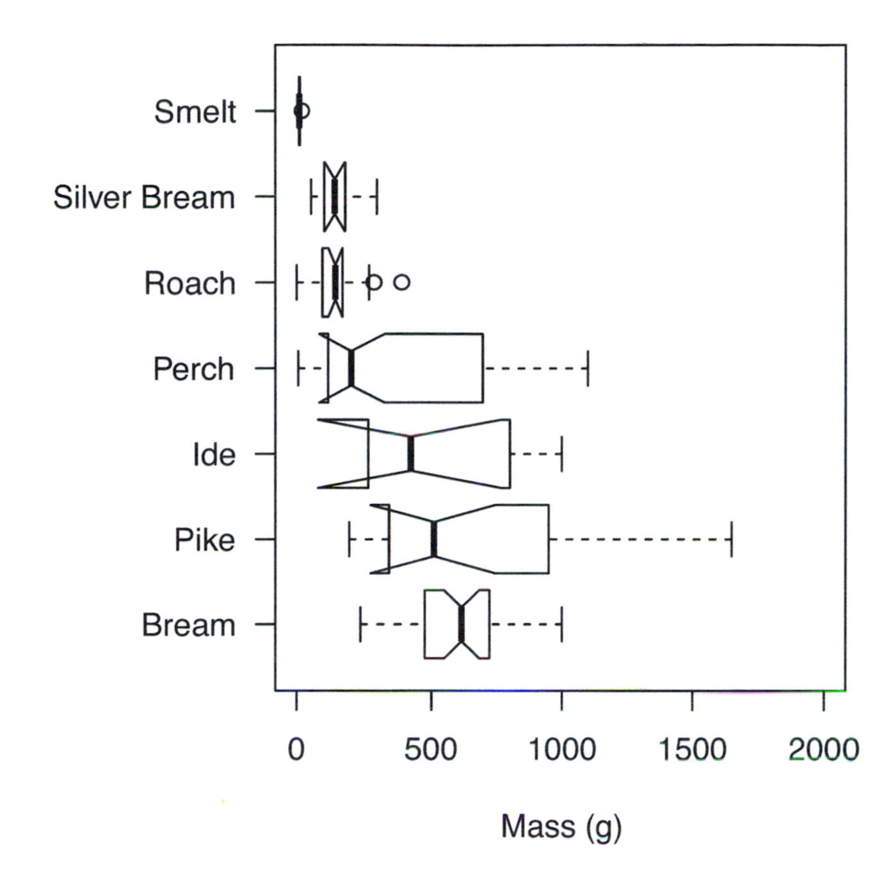

Figure 9.5 *Side-by-side notched boxplot of mass (g) of seven species of fish caught in Längelmävesi (this is an outlier boxplot with a common plotting symbol ○ for outliers and extreme outliers)*

The notches surrounding the medians provide a measure of the rough significance of the difference between a pair of values. If the notches about two medians do not overlap, then the medians are roughly different.

It is not possible to discern the notched box for smelt in Figure 9.5 because of the scale of the horizontal axis. But there is evidently no overlap in notches between smelt and any other species. It is possible for the notch to exceed the limits of the box at either the lower or upper quartile. This is seen in Figure 9.5 for the species of perch, ide, and pike. The notches for perch overlap with all species except smelt at the lower end of the spectrum and bream at the upper end.

The notched boxplot was introduced by McGill, Tukey, and Larsen [83]. It is not necessary to use the adjective side-by-side because according to the plotting convention of McGill *et al.* [83] there must be at least two boxplots to compare. Note that notched outlier boxplots are depicted in Figure 9.5. The examples given in McGill *et al.* [83] apply notches to quantile boxplots but the authors note that the type of boxplot can be switched to an outlier boxplot if desired.

The notched outlier boxplots in Figure 9.5 were produced by the R function `boxplot`. The notches are centered about the sample median M and extend to

$$M \pm 1.58\frac{IQR}{\sqrt{n}}. \tag{9.1}$$

The coefficient 1.57 instead of 1.58 appears on p. 62 of Chambers, Cleveland, Kleiner, and Tukey [18] in 1983. Formula (9.1) doesn't actually appear in McGill *et al.* [83] who report instead

$$M \pm 1.7\frac{1.25(IQR)}{1.35\sqrt{n}}. \tag{9.2}$$

Note that $1.7 \times 1.25/1.35 \approx 1.5741$.

The justification given in McGill *et al.* [83] for formula (9.2) is that the asymptotic normal distribution of the sample median M has an approximate standard deviation

$$s_M = \frac{1.25(IQR)}{1.35\sqrt{n}}. \tag{9.3}$$

This asymptotic property is insensitive to the underlying distributions of the samples.

The notch around the median is then calculated by

$$M \pm Cs_M \tag{9.4}$$

where the value C is determined from the standard normal distribution.

McGill *et al.* [83] note that an appropriate value for comparing two medians would be $C = 1.96/\sqrt{(2)} \approx 1.386$. This assumes that the standard deviations of the medians are close in value. McGill *et al.* [83], however, do not recommend the value $C = 1.386$. They choose $C = 1.7$, which they "empirically selected as preferable." Presumably a Monte Carlo simulation was done but this is not documented in McGill *et al.* [83].

Looking at Figure 9.5, it is apparent from the length of the notches that there are wide variations in the estimates of the standard deviation of the median over the seven species. Given the similarity of the lengths of the notches for roach and silver bream, it is fair to infer that these species are similar with respect to median mass. The standard deviations for the median for perch and bream are close, so it is fair to conclude the medians for these two species are different at an approximate 95% level of significance. Any other pairwise comparison of medians is subject to challenge.

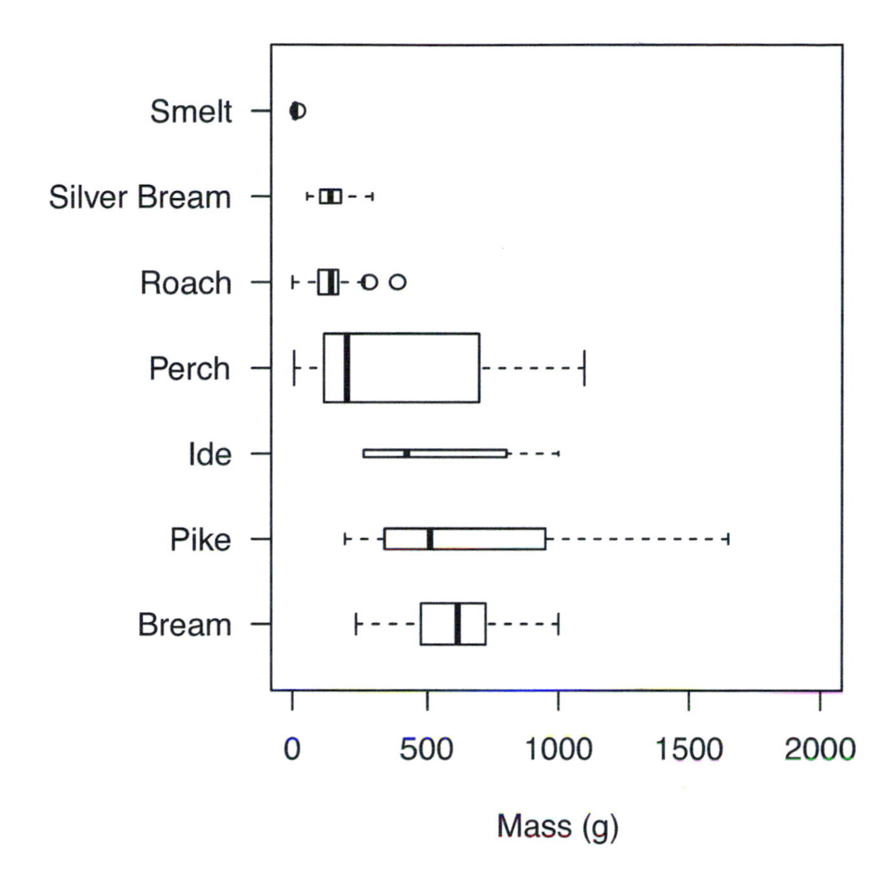

Figure 9.6 *Variable-width outlier boxplot of mass (g) of seven species of fish caught in Längelmävesi with box width proportional to sample size*

With software it is possible to draft a solitary notched boxplot. In this case to illustrate variation in the estimate of the median for a single population, the multiplier C ought to be 1.96 and not $1.96/\sqrt{2}$. In some statistical software packages, it is not possible to change the value of C. This value is fixed at $1.96/\sqrt{2}$ in the R function `boxplot`.

9.6 The Variable-Width Boxplot

The *variable-width boxplot* was introduced in 1978 by McGill *et al.* [83]. An example for the mass of the seven species of fish caught in the research trawl around 1917 is given in Figure 9.6. Note that these are side-by-side outlier boxplots. The width of the boxes are proportional to the number of observations for each species, as reported in Table 9.1. This is one approach to varying box widths discussed in McGill *et al.* [83]. The call to the R function `boxplot` for producing Figure 9.6 is as follows.

```
fc<-Fishcatch

fc$Species<-ordered(fc$Species,
levels=c(1,6,2,7,3,4,5),
labels=c("Bream","Pike","Ide","Perch",
"Roach","Silver Bream","Smelt"))

fc$Count<-fc$Weight
fc$Count[fc$Count >= 0]<-1
fc$Count[is.na(fc$Count)== TRUE]<-0
sums<-tapply(fc[,"Count"],
INDEX=fc[,"Species"],FUN=sum,na.rm=TRUE)
sums<-sums/max(sums)

boxplot(Weight ~ Species,data=fc,
ylim=c(0,2000),xlab="Mass (g)",
horizontal=TRUE,las=1,width=sums)
```

The vector `sums` is proportional to the sample size and this is passed to the argument `width` in the call to the function `boxplot`.

Another approach to varying box widths is based upon the asymptotic formula (9.3) for the standard deviation S_M of the sample median. Note the denominator involves the square root of the sample size n. The denominator of the variance of the standard deviation of the sample mean also is a function of the square root of the sample size n. Figure 9.7 depicts variable-width notched boxplots for mass with box widths proportional to the square root of the sample size for each species. This is accomplished by adding the argument `varwidth=TRUE` when calling the R function `boxplot`. The whiskers in Figure 9.7 are those for outlier boxplots.

As noted by McGill *et al.* [83], if the intent is to minimize visual impact of the differences among samples, then square roots will do the job. There were 6 ide caught and 17 pike. The nearly three-fold difference is apparent in Figure 9.6 with the linear scale for sample size. This difference is muted with square roots in Figure 9.7. The difference in box width is similarly muted between pike and bream. There were 34 bream caught in the research trawl. So if the absolute differences in sample size are in the teens, then square root scale results in roughly comparable box widths.

Figure 9.8 is a `ggplot2` representation of Figure 9.7. The code for producing Figure 9.8 is as follows.

```
fc<-Fishcatch
```

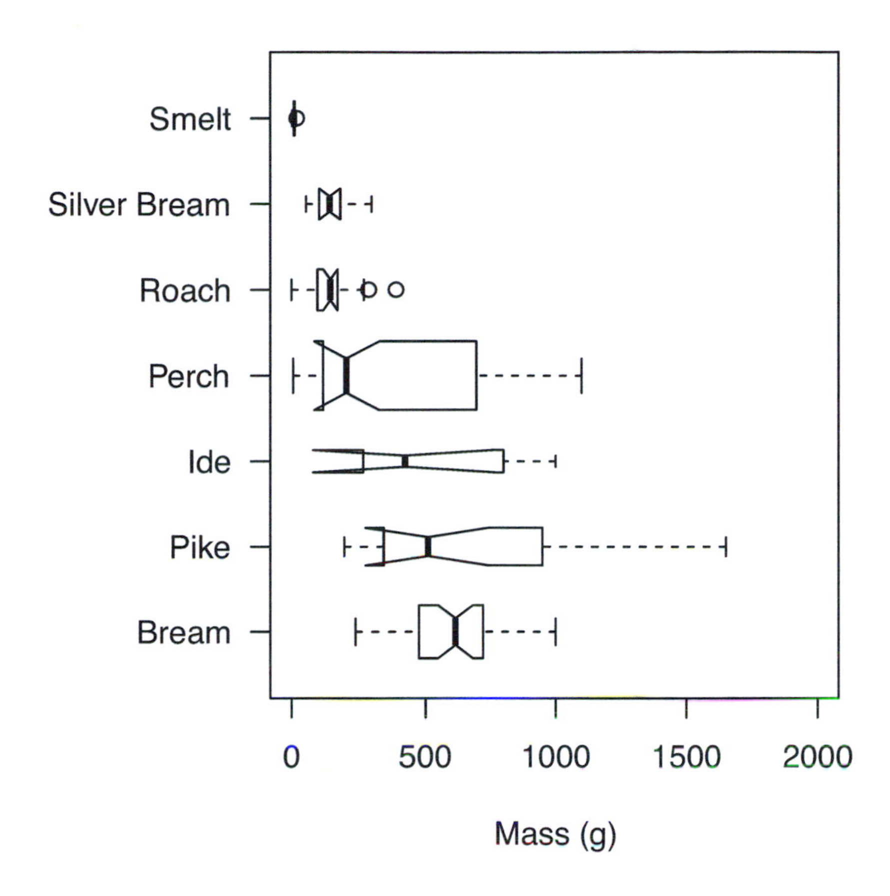

Figure 9.7 *Variable-width notched boxplot of mass (g) of seven species of fish caught in Längelmävesi with box width proportional to square root of sample size*

```
fc$Species<-ordered(fc$Species,
levels=c(1,6,2,7,3,4,5),
labels=c("Bream","Pike","Ide","Perch",
"Roach","Silver Bream","Smelt"))

fcf<-as.data.frame(fc)

meds<-tapply(fc[,"Weight"],INDEX=fc[,"Species"],
FUN=median,na.rm=TRUE)
meds<-sort(meds,decreasing=TRUE)

figure<-ggplot(fcf,aes(y=Weight,x=Species)) +
geom_boxplot(alpha=0,notch=TRUE,varwidth=TRUE) +
```

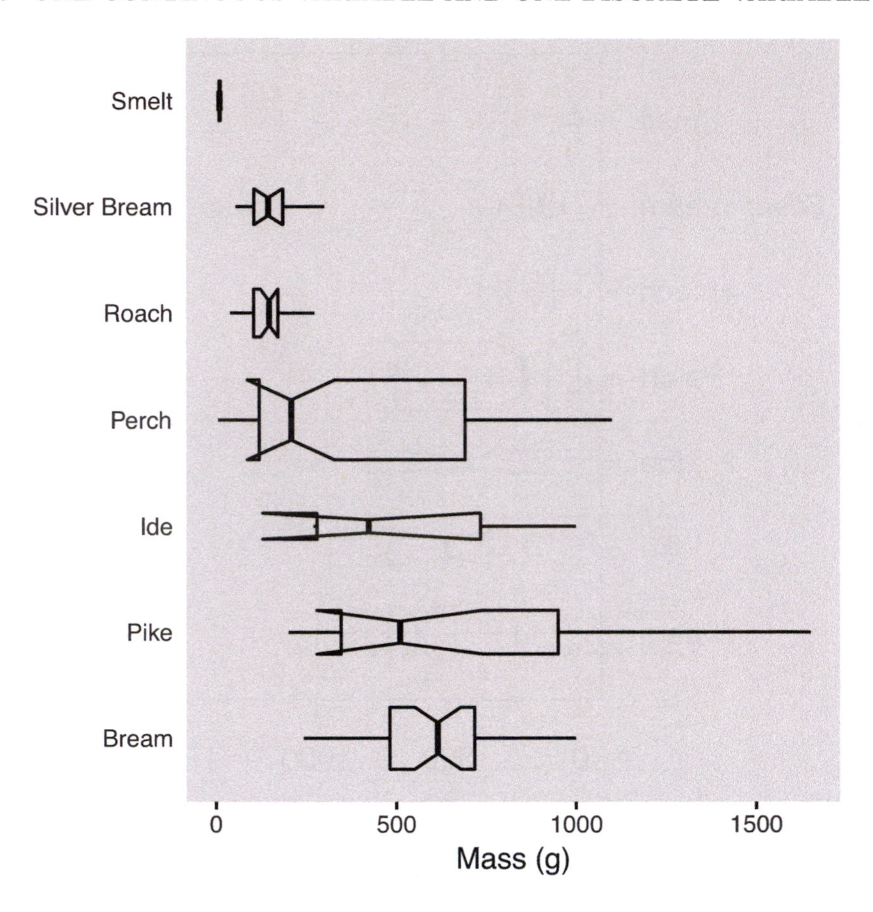

Figure 9.8 *Variable-width notched boxplot of mass (g) of seven species of fish caught in Längelmävesi with box width proportional to square root of sample size using* `ggplot2`

```
theme(panel.grid=element_blank(),
axis.ticks.y=element_blank()) +
labs(x="",y="Mass (g)") + coord_flip()

print(figure)
```

The parameter passed to the `ggplot2` function `geom_boxplot` to obtain box widths proportional to the square root of the sample size is `varwidth=TRUE`, which is identically the same as that passed to the `graphics` function `boxplot`. So while the packages `ggplot2` and `graphics` are different graphics languages with different grammars, these two packages do have some nouns in common.

Taking another look at Figures 9.7 and 9.8, it is evident that there is a lot of information being portrayed. There is information on location, spread, and

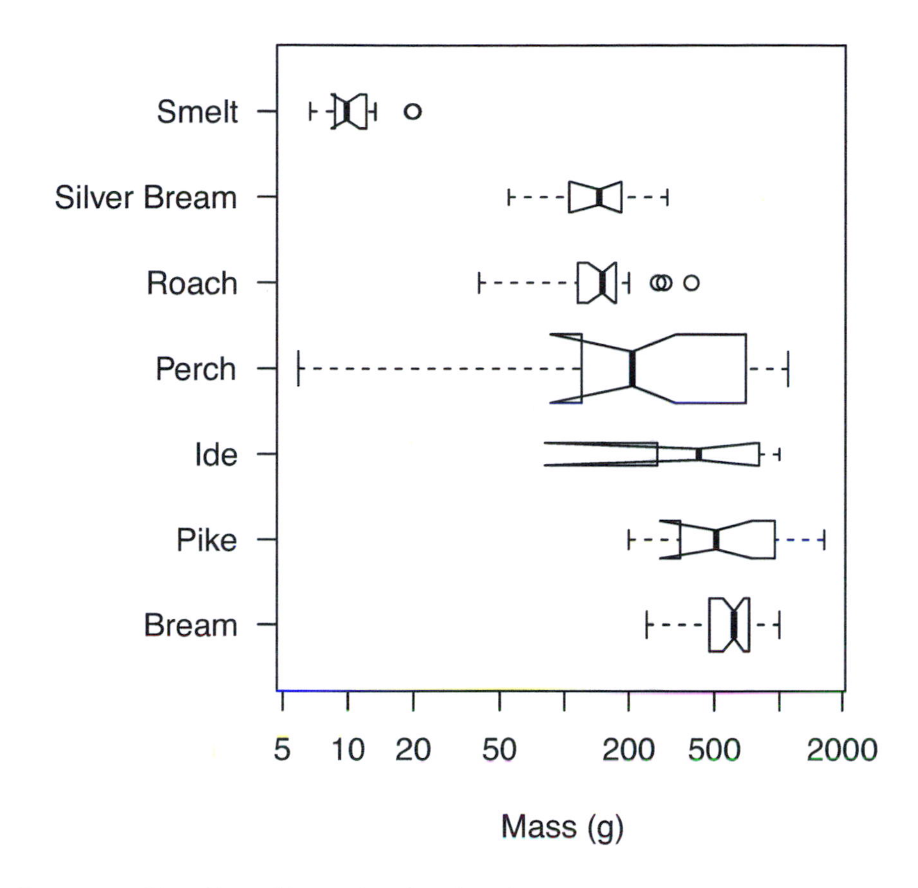

Figure 9.9 *Variable-width notched boxplot of seven species of fish caught in Längelmävesi with logarithmic scale for mass*

skewness. Pairwise differences of location can be made with the notches. Sample size is illustrated. For some viewers this definitely can be information overload. But a graphical display like Figure 9.7 can be a beauty in the eye of the beholder if used as a tool for data analysis and not just an illustration of summary statistics. Figure 9.7 and its `ggplot2` counterpart Figure 9.8 are figures to be savored slowly.

There is a glaring problem with Figure 9.8 and the four previous boxplots. For smelt, there is only a smudge. This is solved in Figure 9.9 with a logarithmic scale for mass generated by adding the syntax `log="x"` to the call to the R function `boxplot` as given below.

```
boxplot(Weight ~ Species,data=fc,xlab="Mass (g)",
horizontal=TRUE,las=1,notch=TRUE,varwidth=TRUE,
log="x"))
```

```
1 | 2: represents 120
leaf unit: 10

        Bream      Silver Bream

               |  0  | 569
               |  1  | 24457
            94 |  2  | 07
           964 |  3  | 0
          7553 |  4  |
         70000 |  5  |
       8852100 |  6  |
        221000 |  7  |
             5 |  8  |
         75522 |  9  |
             0 | 10  |
```

Figure 9.10 *Back-to-back stemplot for mass (g) of bream and silver bream*

In the course of producing Figure 9.9 it was discovered that fish #47, a roach, has a weight of zero grams. This could be the result of a coding error or a rounding down with a fluke catch of a minnow less than 0.5 g as roach appeared to have been weighed to within one gram. Because a value of zero cannot be plotted on a logarithmic scale, fish #47 was omitted to produce Figure 9.9.

The lognormal distribution is one possible right-skewed distribution. From Figure 9.7, there is certainly the impression that the distribution of mass for pike is right skewed.

With the logarithmic scale in Figure 9.9, the mass of pike appears to be symmetric. That is, the distribution of pike mass has about the same amount of skewness as would be expected for a random variable with a lognormal distribution.

9.7 The Back-to-Back Stemplot

The *back-to-back stemplot* is an ingenious approach to depicting two stemplots when comparing a continuous variable for two distributions. It does not generalize if the discrete variable indexing the continuous variables has three or more distinct values.

The bream and silver bream are easily confused with each other. This is particularly true in the younger stages. Figure 9.10 gives the back-to-back stemplot for weight for bream and silver bream. Note that the values of the leaves increase to the right for silver bream but increase to the left for bream on the left side of the back-to-back stemplot.

Stemplots are not high-resolution graphical displays. While common in introductory statistics textbooks, software for the back-to-back stemplot is hard to find. There is no function to do this in the base packages (`graphics` and `stats`) in R. Figure 9.10 was drafted manually.

9.8 The Side-by-Side Stemplot

An example of a *side-by-side stemplot* for weight for all seven species reported from the research trawl on Längelmävesi is given in Figure 9.11. Note the vertical common axis for mass for all seven species of fish.

In the upper right corner of Figure 9.11 are the observations for the smallest species: smelt; roach; and silver bream. These three species appear to be unimodal and skewed to the right. The mass of perch is bimodal. There are only six observations for ide. The two biggest species, bream and pike, are on the left of the side-by-side stemplot. Four of the pike are the largest predators caught in this research trawl.

Figure 9.11 was pieced together manually from a call to the R function `stem` for each of the seven species. The plotting convention originated in Tukey [127] on page 100. This was back in 1977. A stemplot is basically a nonparametric density estimate. In Section 9.9, a more modern alternative is given.

9.9 The Side-by-Side Dot-Whisker Plot

The *side-by-side dot-whisker plot* is visually austere. An example is given in Figure 9.12. In this version of the side-by-side dot-whisker plot an estimate for location of the distribution of mass is given for each species by the sample mean. Variation for each population is depicted by a whisker to the left and right of the dot. The length of each whisker is one sample standard deviation.

Figure 9.12 is depicting the location and spread of the distribution of mass for each species. Because the species are either rightward skewed or bimodal with the minor mode on the right, the dot-whisker plots in Figure 9.12 are a poor choice. Any of the plots considered previously in this chapter are better for this sort of data. But if the data are symmetric, or better yet, normal, then side-by-side dot-whisker plots depicting the mean and standard deviation for each sample are acceptable. Figures like Figure 9.12 are sometimes drafted deliberately for what they can conceal rather than reveal.

Note that the order of the species in Figure 9.12 is different from that in Figures 9.1 through Figure 9.9, inclusive. The reason for this is that the order

```
1 | 2: represents 120
leaf unit: 10

     Bream      Pike    Ide  Perch                  Silver  Roach         Smelt
                                                    Bream

 0 |          |       |    | 034577888            | 569   | 04678       | 00000000111111
 1 |          |       |    | 0111222333344557889  | 24457 | 12244566668 |
 2 | 49       | 0     | 77 | 125566               | 07    | 079         |
 3 | 469      | 0004  | 0  | 002                  | 0     | 9           |
 4 | 3557     | 35    |    |                      |       |             |
 5 | 00007    | 0146  | 4  | 15                   |       |             |
 6 | 0012588  |       |    | 589                  |       |             |
 7 | 000122   | 7     |    | 00                   |       |             |
 8 | 5        |       | 0  | 2245                 |       |             |
 9 | 22557    | 5     |    | 00                   |       |             |
10 | 0        |       | 0  | 0001                 |       |             |
11 |          |       |    | 00                   |       |             |
12 |          | 5     |    |                      |       |             |
13 |          |       |    |                      |       |             |
14 |          |       |    |                      |       |             |
15 |          | 5     |    |                      |       |             |
16 |          | 05    |    |                      |       |             |
```

Figure 9.11 *Side-by-side stemplot for mass (g) of seven species of fish*

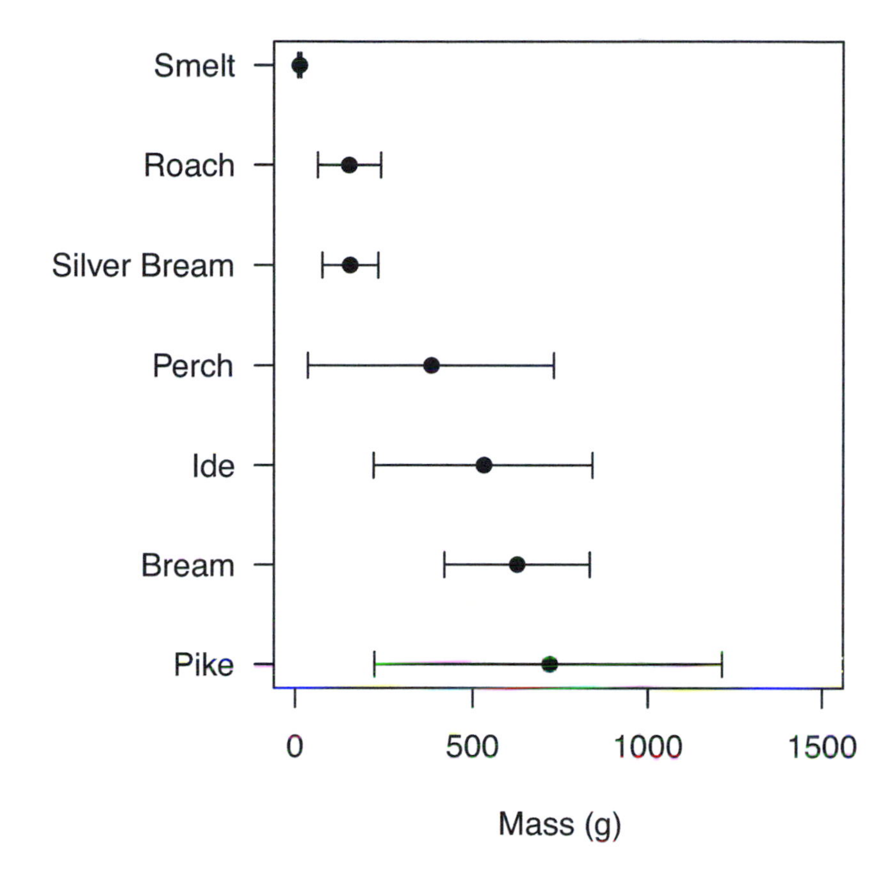

Figure 9.12 *Side-by-side dot-whisker plots for mass (g) of seven species of fish (the dot gives the location of the mean and each whisker is one standard deviation from the mean for each species)*

of fish species for Figure 9.12 is based on the mean of mass rather than the median. Compared to the median, the mean is not robust to outliers. Moreover, the mean is the best choice of estimate for location if the distribution is skewed or bimodal, as mass apparently is for some of the species. Nevertheless, it cannot be expected that the order of means and medians will be identical. The order selected for the side-by-side dotplots and boxplots in this chapter is based on the medians while the order for the side-by-side dot-whisker plots is based on the mean with the plots shifting to the right when following down from the top of each figure.

Instead of a standard deviation s, the standard error $s/\sqrt{n}$ is usually chosen for whisker length. In this situation, the variation being depicted is not that of the distribution but rather that of the sample mean. This is the usual choice.

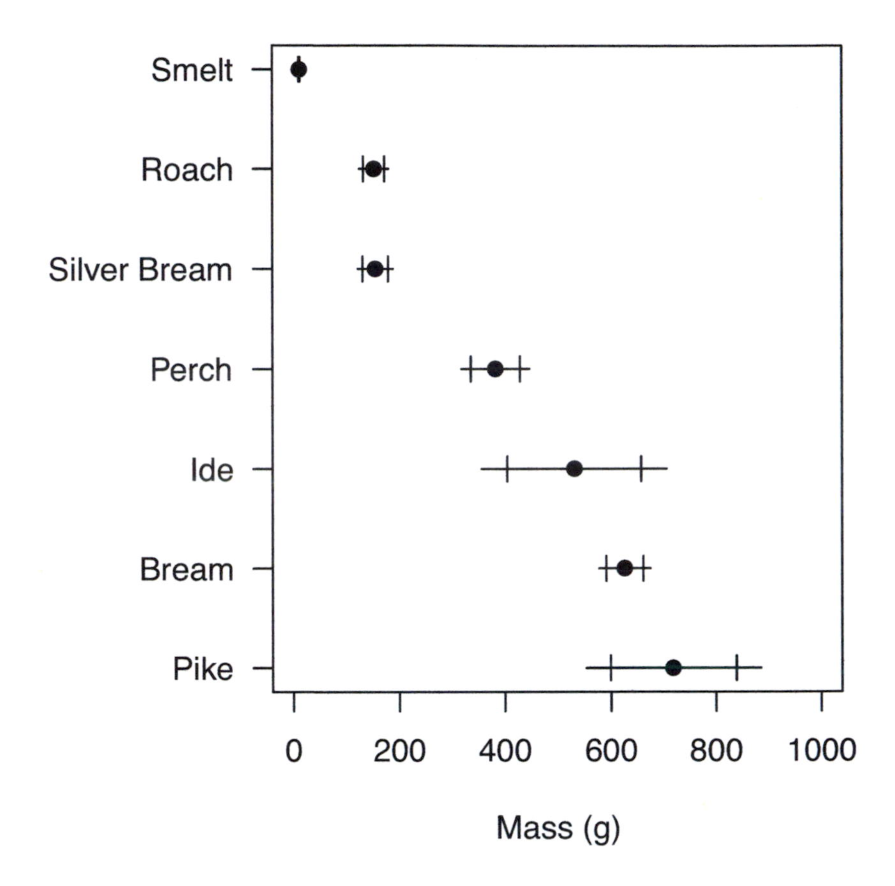

Figure 9.13 *Side-by-side dot-whisker plots for mass (g) of seven species of fish (the dot gives the location of the sample mean and the bars terminating the whisker indicate the length of one standard error, and overlapping whiskers for a pair of species indicate no difference in sample means at an approximate 95% confidence level)*

An example is given in Figure 9.13 for the mass of seven species of fish caught in Längelmävesi. In addition to depicting the standard error, the full length of the whiskers can be used to assess whether pairs of means are significantly different.

Overlapping whiskers in Figure 9.13 for a pair of species indicate no difference in sample means at an approximate 95% confidence level. While this has been done to mimic the notched boxplot, this is typically not done in a side-by-side dot-whisker plot. If anything else is done, then the full length of a whisker on each side corresponds to $1.96s/\sqrt{n}$ for a 95% confidence interval for the mean. It is important to check the text accompanying the side-by-side dot-whisker plot to find out what exactly is being depicted. When drafting a side-by-side dot-whisker plot, also place this information in the caption, or a legend.

In the following R code for drafting Figure 9.13, the R function `tapply` is used to compute the means and standard deviations of mass for each species of fish. The third call to `tapply` computes the number of nonmissing masses for each species.

```
fc<-Fishcatch

fc$Species<-ordered(fc$Species,
levels=c(6,1,2,7,4,3,5),
labels=c("Pike","Bream","Ide","Perch",
"Silver Bream","Roach","Smelt"))

means<-tapply(fc[,"Weight"],INDEX=fc[,"Species"],
FUN=mean,na.rm=TRUE)
sdf<-tapply(fc[,"Weight"],INDEX=fc[,"Species"],
FUN=sd,na.rm=TRUE)
fc$Iswna<-!is.na(fc$Weight)
counts<-tapply(fc[,"Iswna"],INDEX=fc[,"Species"],
FUN=sum)
se<-sdf/sqrt(counts)
sds<-sdf/sqrt(2*counts)

qf<-qnorm(0.975)

stripchart(fc$Weight   fc$Species,col="white",
xlim=c(0,1000),xlab="Mass (g)",
group.names=rep(" ",7),vertical=FALSE)

for (i in 1:7) points(means[i],i,pch=19)
for (i in 1:7) lines(c(means[i]-qf*sds[i],means[i]+qf*sds[i]),
c(i,i))
for (i in 1:7) lines(rep(means[i]-se[i],2),c(i-0.125,i+0.125))
for (i in 1:7) lines(rep(means[i]+se[i],2),c(i-0.125,i+0.125))

axis(2,at=1:7,labels=levels(fc$Species),las=1)
```

By setting `qf<-qnorm(0.975)`, the critical value for the 95% confidence intervals is found.

The function `stripchart` is called only to provide a skeleton for plotting the dots and their whiskers. Setting `col="white"` plots the dots in white so that they are invisible. The following four `for` loops plot the points and the whiskers. Setting `group.names=rep(" ",7)` forces a blank to be printed for

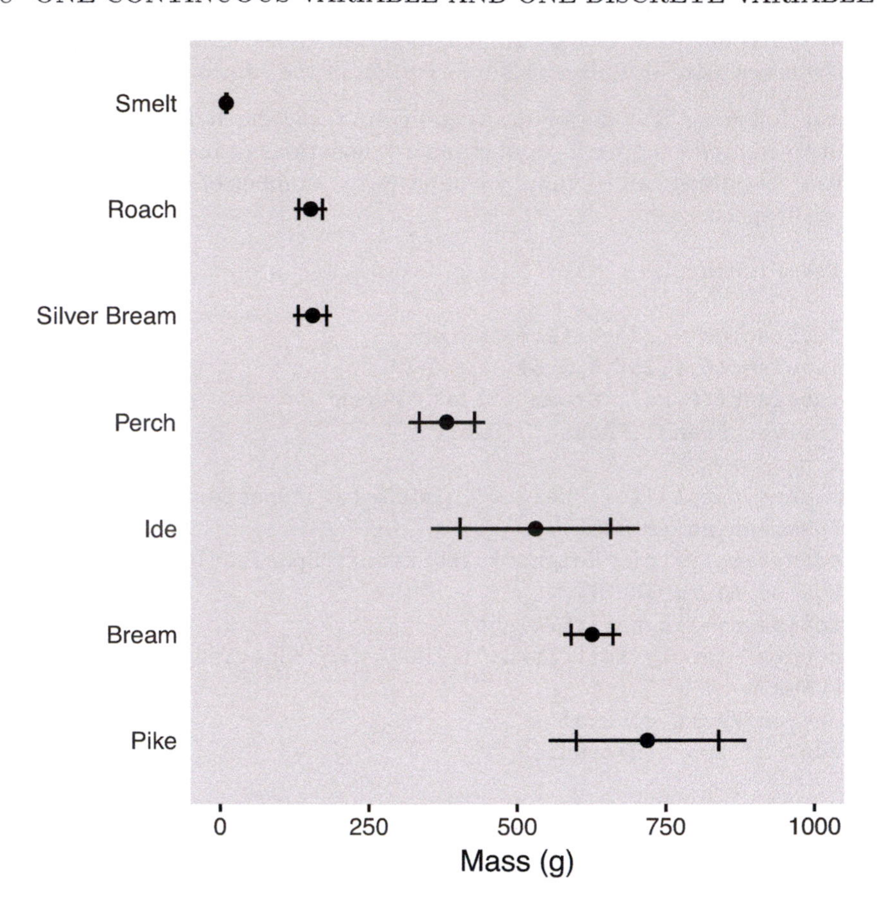

Figure 9.14 *Side-by-side dot-whisker plots for mass (g) of seven species of fish (the dot gives the location of the sample mean and the bars terminating the whisker indicate the length of one standard error, and overlapping whiskers for a pair of species indicate no difference in sample means at an approximate 95% confidence level) with* `ggplot2`

each species label. The final call to the function `axis` prints the vertical axis with the names of the fish species.

Figure 9.14 is a version of side-by-side dot-whisker plots rendered by the following `ggplot2` code.

```
fc<-Fishcatch

fc$Species<-ordered(fc$Species,
levels=c(6,1,2,7,4,3,5),
labels=c("Pike","Bream","Ide","Perch",
"Silver Bream","Roach","Smelt"))
```

```
means<-tapply(fc[,"Weight"],INDEX=fc[,"Species"],
FUN=mean,na.rm=TRUE)
sdf<-tapply(fc[,"Weight"],INDEX=fc[,"Species"],
FUN=sd,na.rm=TRUE)
fc$Iswna<-!is.na(fc$Weight)
counts<-tapply(fc[,"Iswna"],INDEX=fc[,"Species"],
FUN=sum)
se<-sdf/sqrt(counts)
qf<-qnorm(0.975)
sds<-qf*sdf/sqrt(2*counts)

fcsf<-data.frame(Species=levels(fc$Species),
means=means,se=se,sds=sds)

figure<-ggplot(fcsf,aes(y=means,
x=factor(fcsf$Species,levels=fcsf$Species))) +
geom_point(fill="black",size=2) +
geom_errorbar(aes(ymin=means-se,ymax=means+se),
width=0.2) +
geom_errorbar(aes(ymin=means-sds,ymax=means+sds),
width=0) +
ylim(0,1000) +
theme(panel.grid=element_blank(),
axis.ticks.y=element_blank()) +
labs(x="",y="Mass (g)") + coord_flip()

print(figure)
```

The function geom_point is called to plot the mean for each species of fish. Two calls to geom_errorbar produce the error bars. Finally coord_flip is called to obtain a horizontal orientation.

In Figures 9.13 and 9.14, we see that the estimate of mean mass for ide overlaps with those for perch, bream, and pike. Roach and silver bream form a similar group. The smelt are by themselves. Because of scale, there is no meaningful information available in these two figures regarding confidence intervals for mean mass of smelt.

The central limit theorem states that the distribution of the sample mean approaches normality as the sample size increases. This is true for the vast majority of distributions, regardless of whether the distribution is long tailed or bimodal. A sample size of 30 is generally accepted as sufficiently large, but the approximation can be quite good even for long-tailed distributions with sample sizes as small as ten. If in doubt, use the notched boxplot instead.

Because the normal distribution is symmetric and the focus in Figures 9.13 and

9.14 is on the sample mean, these figures do not mislead with respect to lack of symmetry in the data as Figure 9.12 does. Figures 9.13 and 9.14 ought to be presented with an additional graphic display, say, the trellis kernel density estimates of Figure 9.15, which do present information regarding skewness and modality.

9.10 The Trellis Kernel Density Estimate*

The *side-by-side kernel density estimates* in Figure 9.15 were produced in the following R script using the `densityplot` function in R's `lattice` package implementation of Trellis graphics.

```
figure<-densityplot(~ Weight | Species,data=fc,
xlab="Mass (g)",ylab="Relative Density",layout=c(1,7),
scales=list(y=list(at=(0:3)*0.001,limits=c(0.0,0.003))),
type="density",kernel="gaussian",bw=139.86,na.rm=TRUE,
from=0,to=2000)

print(figure)
```

This is a more modern approach to depicting the distribution of a continuous variable for several values of a discrete variable. The origins of Trellis graphics can be traced to Cleveland's [23] book on visualizing data. This book was published in 1993, sixteen years after Tukey's [127] book that defined exploratory data analysis.

The order of the species for the figures in this section will revert to that originally used in Figure 9.1, which was used for all the dotplots and boxplots.

In Chapter 6, nonparametric density estimates for perch mass were studied. After considering different choices of kernel function and bandwidth, the final kernel density estimate was created with a Gaussian kernel and the bandwidth of 139.86 grams determined by Silverman's rule of thumb. Taking perch as the reference species, this same kernel and bandwidth was used for the other six species in producing Figure 9.16. It is possible to have different kernels and bandwidths for each species when using the R function `densityplot`.

So that there is a common axis for the mass for the seven species, the nonparametric kernel density estimates have been stacked one on top of the other in Figure 9.15. The vertical scales for the seven nonparametric kernel density estimates are identical. Note tick marks on both vertical sides of the box in Figure 9.15 with the labels for the tick marks alternating between the left and right sides.

According to the kernel density estimates in Figure 9.15, the smallest and

*This section can be omitted without loss of continuity.

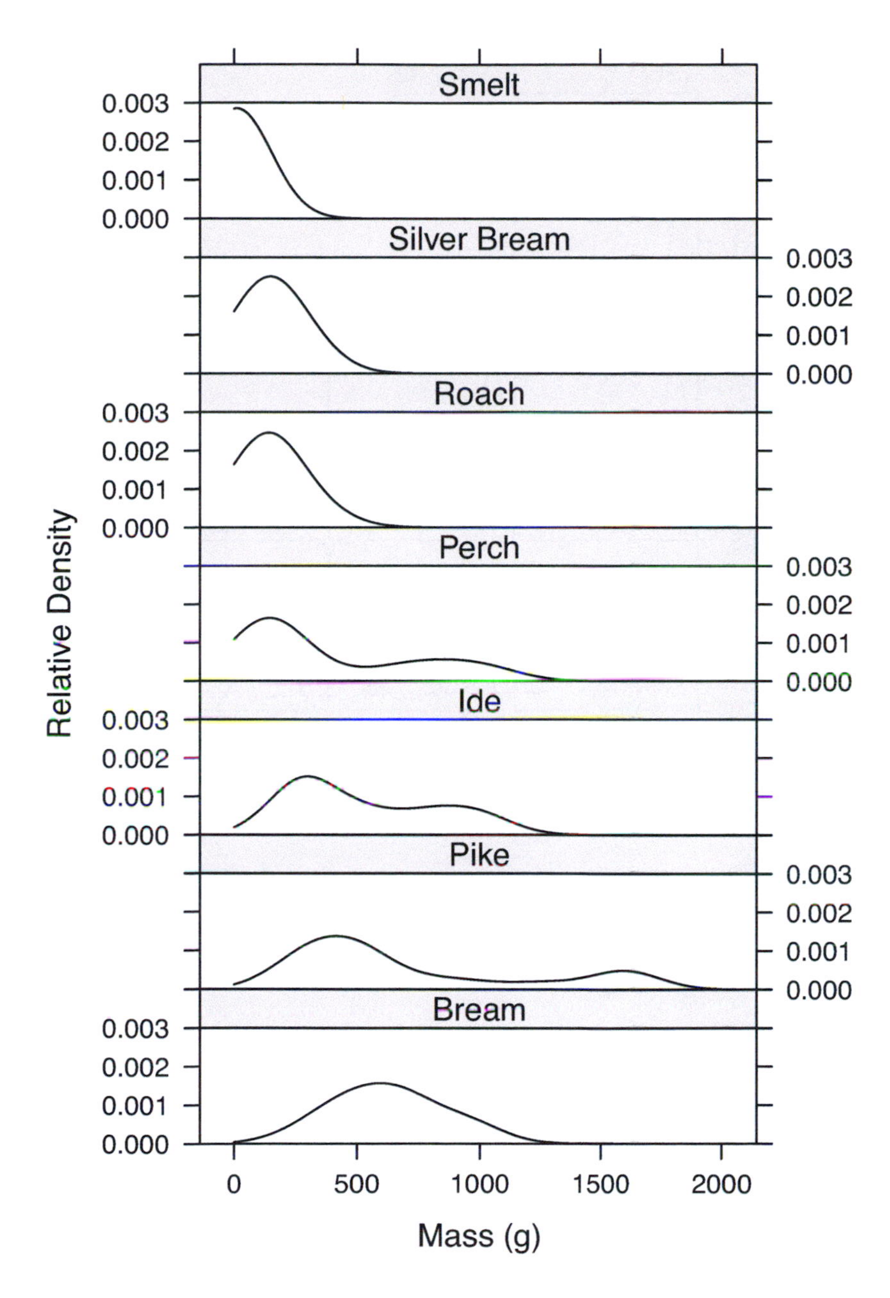

Figure 9.15 *Side-by-side nonparametric kernel density estimates for mass (g) of seven species of fish (Gaussian kernel density used with a bandwidth of 139.86 grams for each species)*

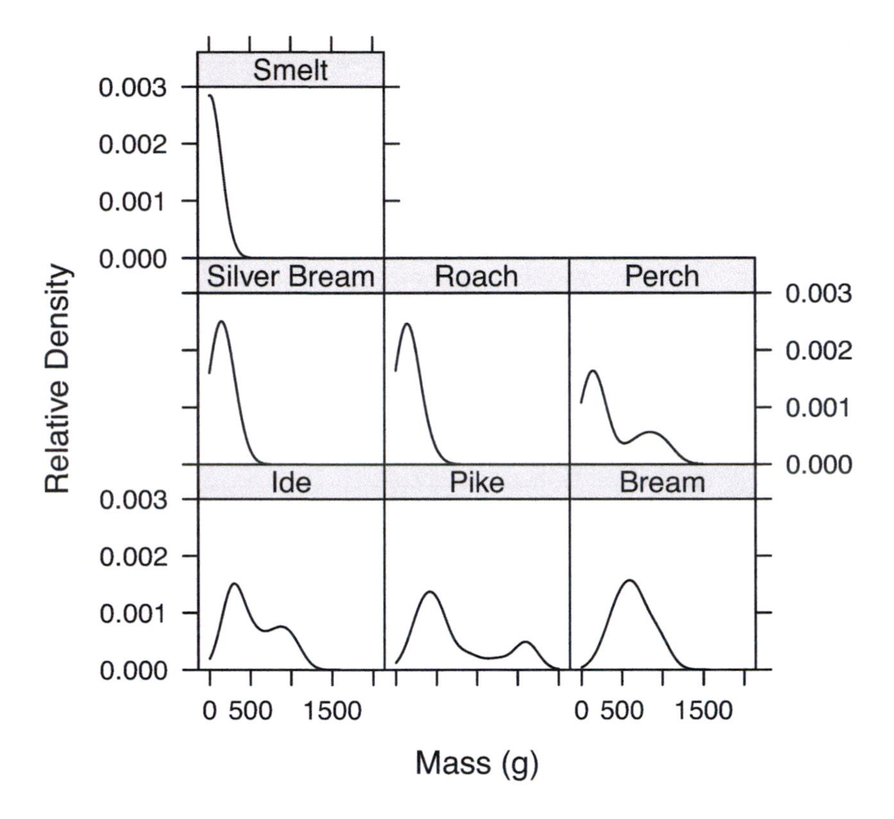

Figure 9.16 *Panel-layout of nonparametric kernel density estimates for mass (g) of seven species of fish (Gaussian kernel density used with a bandwidth of 139.86 grams for each species)*

most preyed upon species appear to be smelt, roach, and silver bream. The distribution of mass for perch, ide, and pike appear to be bimodal. Bream appears to be unimodal but has the greatest value for the major mode of all seven species. By virtue of the secondary mode for pike, this would appear to be the apex predator in Finland's Längelmävesi.

A criticism of Figure 9.15 is that the vertical scale is compressed so discerning the secondary modes is made more difficult. Figure 9.16 is an attempt to solve this problem. The kernel density estimates in Figure 9.16 are laid out in a tile pattern. This is to be read row by row, starting at the upper left-hand corner to achieve the same ordering as in Figure 9.16. The vertically stacked layout of kernel density estimates in Figure 9.15 was produced with the argument `layout=c(1,7)` in the call of the `lattice` function `densityplot` while the layout of Figure 9.16 was produced with the argument `layout=c(3,3)`. Also added to the call to `densityplot` was the argument `par.strip.text=list(cex=0.9)` to reduce the font size so that the species name “Silver Bream” was not too close to the edges of the strip box.

The multiple modes for perch, ide, and pike are easily discerned in the kernel density estimates of Figure 9.16. The high narrow peaks for the kernel density estimates of mass for smelt, roach, and silver bream convey the impression that these are small fry compared to the lower broader kernel density estimates for the four other species. The biggest gap between modes occurs for pike.

The technical sophistication in the trellis kernel density estimates in Figure 9.16 is far and away greater than that required for the side-by-side stemplots in Figure 9.11. Ironically, considerably more effort and time are required to study Figure 9.11 compared to Figure 9.16. This makes the trellis kernel density estimates in Figure 9.16 the better choice for an oral presentation to an audience.

Total biomass for a given species is the main consideration for conservation with respect to commercial and sports fisheries. Total biomass is estimated from research trawls. It is a function of the distribution of mass and the total number of fish caught. Fish tagging and a second research trawl are further required to determine the fraction of the population sampled in the first trawl. There is additional information in the side-by-side stemplots of Figure 9.11 that is not available in the trellis kernel density estimates of Figure 9.16, that is, the relative abundance of each species. In the side-by-side stemplot each leaf represents one fish caught.

Figure 9.17 is a `ggplot2` version of Figure 9.16 produced by the following code.

```
figure<-ggplot(fcwf,aes(x=Weight)) +
geom_density(bw=139.86) +
facet_wrap(~ Species,as.table=FALSE) +
scale_x_continuous(limits=c(0,2000),
breaks=(0:4)*500,
labels=c("0","500","","1500","")) +
scale_y_continuous(limits=c(0,0.003)) +
theme(panel.grid=element_blank()) +
labs(x="Mass (g)",y="Relative Density")

print(figure)
```

The `ggplot2` function `facet_wrap` extends the density plot to the seven species of freshwater fish. Additional `scale` commands are there to produce axis tick marks and labels similar to the `lattice` output in Figure 9.16.

Kernel density estimation is not the only choice in R’s `lattice` and `ggplot2`

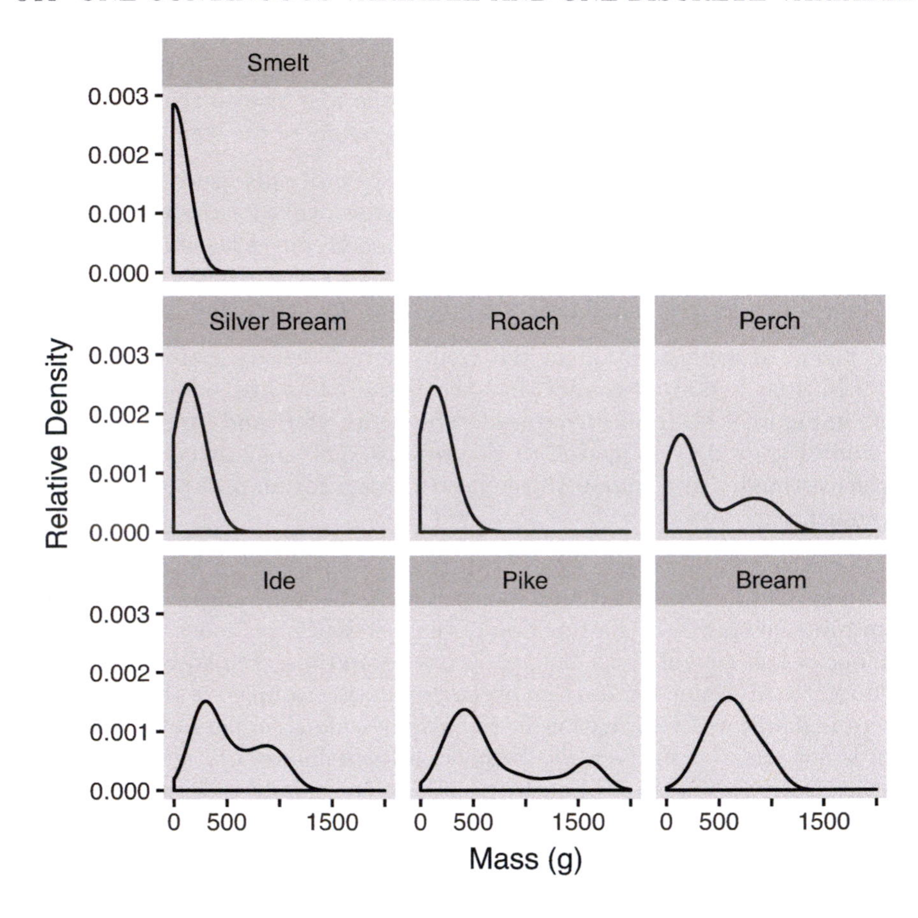

Figure 9.17 *Panel-layout of nonparametric kernel density estimates for mass (g) of seven species of fish (Gaussian kernel density used with a bandwidth of 139.86 grams for each species) using* `ggplot2`

packages for depicting the distribution of a continuous variable and a discrete variable. Other choices include boxplots (without notches), histograms, quantile-quantile plots, and dotplots (without whiskers). The trellis kernel density estimates in this section are the tip of the iceberg. In Chapter 13, Trellis graphics will be shown to be a powerful tool for visualizing multidimensional data.

9.11 Conclusion

The purpose of plotting the graphical displays side-by-side is to examine for differences in location and spread among the distributions of a single continuous variable that are indexed by the values of a discrete variable. The approach has been to select one exemplary data set and then use the various different graphical displays to depict the data. The rationale behind this approach is to provide a case study of the different graphical displays with data that becomes increasing familiar to the reader.

The first graphical display in this chapter was jittered dotplots of mass depicted side-by-side for the seven species of fish caught in a research trawl on Finland's lake Längelmävesi. This allows the viewer to see the data. The remaining graphical displays invite the viewer to become engaged in the process of statistical analysis with the displays themselves as tools of analysis. For example, in the side-by-side boxplot of Figure 9.3, a statistical test for outliers is performed and these are found for smelt and roach.

The material in this chapter has been presented in a somewhat condensed format. For all the various types of side-by-side boxplots in this chapter, there is not one example with the plotting convention for a quantile boxplot. All the illustrations in this chapter are black ink on a white background with the exception of the trellis graphics. The trellis kernel density estimates in Figures 9.15 and 9.16 only use grayscale shading for the pre-panel areas listing fish species. The default is for color in trellis graphics and this was turned off for Figures 9.15 and 9.16. Experiment with different colors for fish species and see whether this can produce more eye-catching graphics.

In this chapter, Trellis graphics in R's `lattice` and `ggplot2` packages was only used to produce kernel density estimates. If you use the R statistical system, then try dotplots, boxplots, histograms, and quantile-quantile plots for depicting the distribution of a continuous variable and a discrete variable.

Take a look at the popular online media and scientific journals and see what is currently being used. With respect to medical journals, one is likely to encounter frequent use of side-by-side quantile boxplots. Less frequently, one encounters side-by-side dotplots and versions of side-by-side dot-whisker plots. Trellis kernel density estimates would be quite rare.

9.12 Exercises

1. In Chapter 4, we considered the distribution of the width of perch in addition to the distribution of the mass of perch. Width is expressed as a percentage and given by the ratio of the maximal width of a fish to its length from nose to the end of the tail, as opposed to the notch in the tail. The full data set for the research trawl on Längelmävesi around 1917 is available from this book's website. It can be downloaded either as a text file or as the R workspace file `Fishcatch.RData`. The data set and accompanying descriptive text can also be obtained from various sites on the world wide web by using a web browser.
   (a) Produce a side-by-side dotplot for the width ratio for each of the seven species.

(b) Produce a side-by-side boxplot for the width ratio for each of the seven species. Use the plotting convention for the outlier boxplot.

(c) Write a short paragraph commenting on the width of fish species based upon your answers to parts (a) and (b).

2. Consider the distribution of mass for the seven species caught in the research trawl on Längelmävesi. These data are available in the full data set available from this book's website and other locations on the world wide web.

   (a) All the examples of boxplots in this chapter use the outlier boxplot convention. Create a counterpart to Figure 9.3 using quantile boxplots.

   (b) Compare your answer to part (a) with Figure 9.3. What would be your preference for the style of boxplot: outlier or quantile? Justify your answer.

   (c) Actually, the quantile boxplot is the more likely of the two plotting conventions to be seen in a scientific journal. Comment on why this is the case.

3. Refer to Exercise 1 for the discussion of fish width.

   (a) Construct a notched boxplot for the width percentage using all seven species of fish.

   (b) Construct a variable-width notched boxplot for the width percentage using all seven species of fish.

   (c) Compare your answers to parts (a) and (b). Which of the two would you routinely use to analyze the distribution of a continuous variable and a discrete variable? Justify your answer.

   (d) Which of the two plots produced for parts (a) and (b) would you feel comfortable showing to an audience in your own area of specialization? Justify your answer.

4. Find a user-contributed package in R that can produce a back-to-back stemplot.

5. Consider the side-by-side stemplot.

   (a) Verify whether there is a user-contributed package in R that can produce a side-by-side stemplot.

   (b) Can you get by without a side-by-side stemplot? Discuss.

6. Consider the maximal width of fish expressed as a percentage of length from nose to the end of the tail as discussed in Exercise 1 for the research trawl on Längelmävesi.

   (a) Produce a side-by-side dot-whisker plot depicting the mean and standard deviation for each species' sample.

   (b) Using your answer from part (a), compare and contrast the species.

   (c) Given the nature of the distribution of maximal percentage width, is the request in part (a) reasonable? Discuss.

7. Consider fish length from the tip of the nose to the notch of the tail as discussed in Exercise 6 for the research trawl on Längelmävesi.
   (a) Produce a side-by-side dot-whisker plot depicting the mean with one and two standard errors for each species' sample.
   (b) Using your answer from part (a), compare and contrast the species.
   (c) Develop catch-and-release regulations for sports fishers in Finland based on your answer to part (a).
8. In some jurisdictions, such as the Province of British Columbia in Canada, catch-and-release regulations are stipulated according to fish length measured from the tip of the nose to the notch in the tail. Minimum and maximum lengths are listed in provincial regulations for many species of freshwater fish. Some of these regulations are specific to individual lakes. The data collected for each fish caught in the research trawl on Längelmävesi include this length. It is the variable `Length2` in the data set. The unit of measure is the centimeter. Refer to Exercise 1 for directions on downloading the data set that includes this variable.
   (a) Produce a trellis kernel density estimate for length determined from the tip of the nose to the notch of the tail for the seven species of freshwater fish.
   (b) Based only on your answer to part (a), develop catch-and-release fishing regulations for the lake Längelmävesi in Finland.
9. Continue with the data of Exercise 8.
   (a) Produce a trellis violin plot for length determined from the tip of the nose to the notch of the tail for the seven species of freshwater fish.
   (b) Compare the trellis violin plot of part (a) with the trellis kernel density estimate of part (a) of Exercise 8. Which of the two do you prefer?
10. Consider the distribution of mass for the seven species caught in the research trawl on Längelmävesi. These data are available in the full data set available from this book's website and other locations on the world wide web.
    (a) Produce a trellis normal quantile-quantile plot for the mass of all seven species of fish.
    (b) Do any of the species appear to have a distribution for mass that is vaguely normal? Discuss.
11. Is there any advantage of a trellis boxplot over a side-by-side boxplot for depicting a continuous variable as a function of different levels of a discrete variable? Discuss.
12. Of all the different plots illustrated in this chapter, which one best addresses the issue of depicting the joint distribution of a continuous variable and a discrete variable? Discuss.

Chapter 10

Depicting the Distribution of Two Continuous Variables

10.1 Introduction

Quite possibly the earliest depiction of the distribution of two continuous variables is due to Francis Galton [51]. This was published in 1886 in the *Journal of the Anthropological Institute of Great Britain and Ireland*. Galton [51] examined the inheritance of stature among humans.

Given that in Galton's day, a computer was a human being, Galton's approach to presenting this data was to produce tabular summaries. At some point in time, the original data was lost. Or rather, filed and forgotten. James Hanley [59] took an interest in finding the original data. With the help of Rebecca Fuhrer, her postgraduate student Beverley Shipley at University College London (UCL), and Kate Lewis and Gill Furlong of UCL Library Services, Hanley was able to track down a booklet with the heights of family members. Photographs of the pages of this booklet can be downloaded from the internet. These were used by the author of this book to create a data set of heights for nuclear family members. This data set is slightly different and more complete than the one used by Hanley [59] in his own analysis of the data in 2004. But the differences are minor.

In this chapter, several different plots are presented for depicting the distribution of two continuous variables. But only one of these appeared in 1886 in Galton's [51] article.

10.2 Learning Outcomes

When you complete this chapter, you will be able to do the following.

- Be able to execute a scatterplot with either the `graphics` package or the `ggplot2` package.

- Understand how to read and be able to execute a sunflower plot for two continuous variables with the `graphics` package.
- Understand how to read and be able to execute a bagplot for two continuous variables using the R package `aplpack`.
- Be knowledgeable about the two-dimensional histogram.
- Be able to produce a two-dimensional histogram as either a scatterplot or a sunflower plot using the `graphics` package.
- Be able to use a levelplot to illustrate a two-dimensional histogram with either the `graphics` package or the `ggplot2` package.
- Be knowledgeable about the cloud plot and be able to use the `lattice` package to depict a two-dimensional histogram as a cloud plot.
- Optionally, be able to depict a two-dimensional kernel density estimate for two continuous random variables in a contour plot using either the `graphics` package or the `ggplot2` package.
- Optionally, be able to depict a two-dimensional kernel density estimate for two continuous random variables in a wireframe plot using the `graphics` package.

10.3 The Scatterplot

Early in 1884, Galton sent letters to the editor of *The Times* of London and the president of the Royal Statistical Society (which appeared in the *Journal of the Royal Statistical Society*) soliciting families to participate in a study collecting data on family members in exchange for a chance to win cash prizes [46, 49]. Confidentiality was promised by Galton. On May 15, 1884, the *Times* published a letter from Galton [47] confirming "150 good records of different families" chiefly from "the upper and middle classes of society." In fact, 205 families were ultimately entered into Galton's Records of Family Faculties.

Galton [51] was concerned with the inheritance of height between two generations. Because the full data set used by Galton [51] is available, analyses can be done that have only appeared as a numerical summary without the benefit of graphical displays. The first example of a *scatterplot*, given in Figure 10.1, was produced using the `plot` function in the R statistical package as follows.

```
plot(Galton$Mother+60,Galton$Father+60,
xlab="Wife's Height (inch)",
ylab="Husband's Height (inch)",xlim=c(55,75),
ylim=c(60,80))
```

Because Galton's notebook lists height offset from 60 inches, it is necessary to add 60 to the heights of both husbands and wives. (Note that 1 inch equals 2.54 centimeters.) The limits of the horizontal and vertical axes are set by `xlim=c(55,75)` and `ylim=c(60,80)`, respectively.

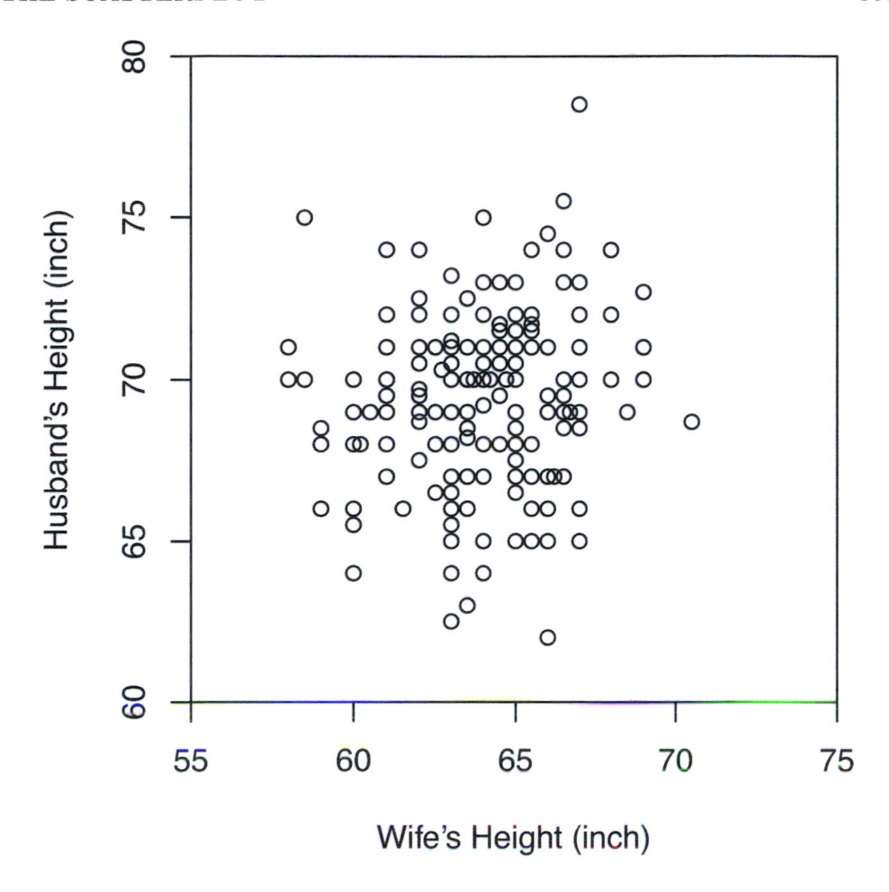

Figure 10.1 *Scatterplot of husband's height versus wife's height*

The code for producing the ggplot2 version of Figure 10.1 given in Figure 10.2 is as follows.

```
figure<-ggplot(Galton,aes(x=Mother+60,
y=Father+60)) +
geom_point(shape=1) + theme_linedraw() +
theme(panel.grid=element_blank(),
axis.line=element_line(),
axis.title=element_text(size=12),
axis.text=element_text(size=12)) +
scale_x_continuous(breaks=55+5*(0:4),
limits=c(55,75)) +
scale_y_continuous(breaks=60+5*(0:4),
limits=c(60,80)) +
labs(x="Wife's Height (inch)",
```

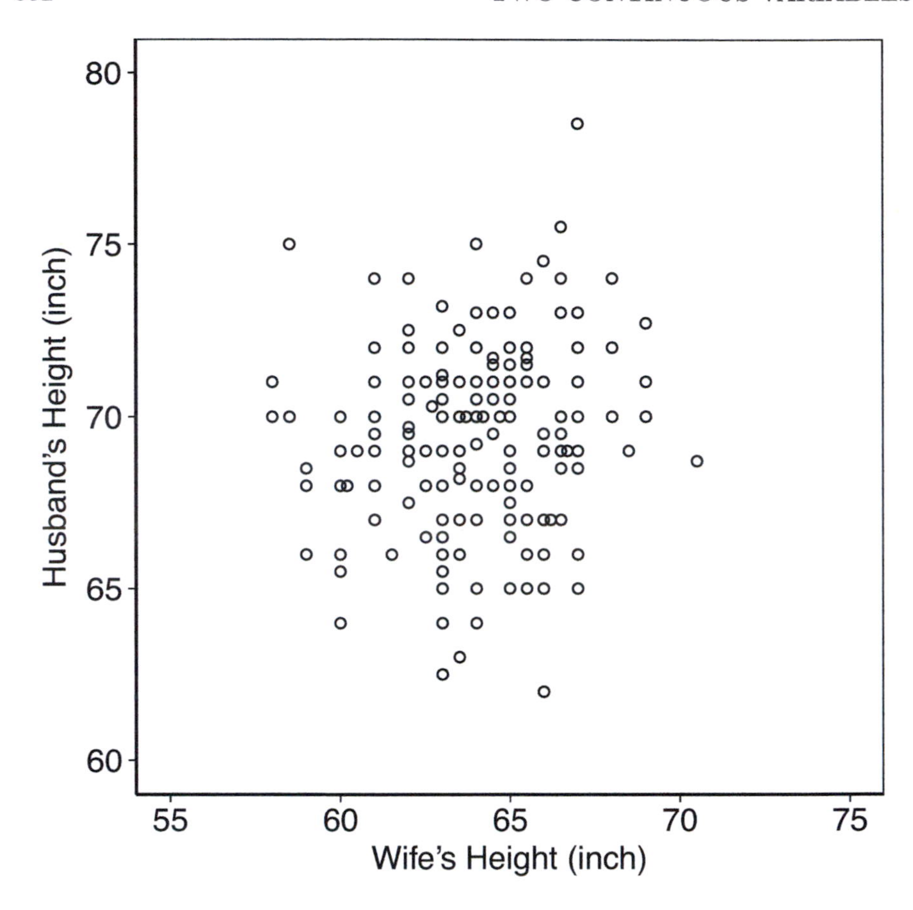

Figure 10.2 *Scatterplot of husband's height versus wife's height using* `ggplot2`

```
y="Husband's Height (inch)")

print(figure)
```

The `linedraw` theme `ggthemes` has been used to dispense with the light gray background and figure-framing box in the `ggplot2` default theme. The background grid has been switched off with the parameter `panel.grid=element_blank()` passed to the function `themes` yet the amount of code for `ggplot2` for Figure 10.2 still exceeds that required for the `graphics` function `plot` to achieve a similar set of axis ticks, axis labels, and axis titles. This is similar to human languages. It is well known to bilingual Canadians that French compared to English requires more syllables when speaking and more letters when writing beyond what is required for the additional syllables. It can be argued that neither length of spoken speech nor length of written text need be an impediment to acquiring and using another language, be that a human language or a synthetic graphics language.

The sampling unit represented by each open circle in Figure 10.1 is the nuclear family that consists of a mother, a father, and their children. Figure 10.1 only depicts the height of a husband and wife. The horizontal axis corresponds to the height of the wife and the vertical axis corresponds to the height of the husband. The units of measure are those used by Galton: Imperial inches.

Figure 10.1 depicts one generation, that of the parents. This figure can be used to assess whether assortative mating with respect to height is occurring, that is, whether tall men tend to marry tall women and short men tend to marry short women. No such pattern appears in Figure 10.1.

The capacity of a scatterplot to convey information among researchers accounts for its popularity. This graphical icon is learned through repeated exposures in textbooks early in the educational process. This experience is reinforced later by exposure in technical journals. But the scatterplot tends to be underused in briefings and in popular publications.

A plotting convention for the scatterplot ought to stipulate that:

(i) no points are omitted from the graph;

(ii) the extreme points are plotted just inside the axis lines (or frame if drawn);

(iii) no points are plotted directly on the axes; and

(iv) the plotting symbol for each point is of adequate size and color for viewing.

It is good practice to select plotting symbols so that nearby points are not obscured. Open circles were selected for Figure 10.1 for this reason. However, this does not provide a solution to the situation of multiple observations at a single point in the Cartesian coordinate plane. One solution would be to jitter the location of the plotting symbol by jittering one or both variables. But this does not provide a solution that is completely truthful. The graphical plot of the next section provides a better solution.

10.4 The Sunflower Plot

Cleveland and McGill [26] created the *sunflower plot* to solve the inadequacy of the scatterplot when overplotting occurs. An example of a sunflower plot for Galton's data for the heights of husbands and wives in nuclear families is given in Figure 10.3.

The plotting convention of 1984 by Cleveland and McGill [26] for the sunflowers is as follows. A single dot is a count of 1, a dot with two line segments is a count of 2, a dot with three line segments is a count of 3, and so forth.

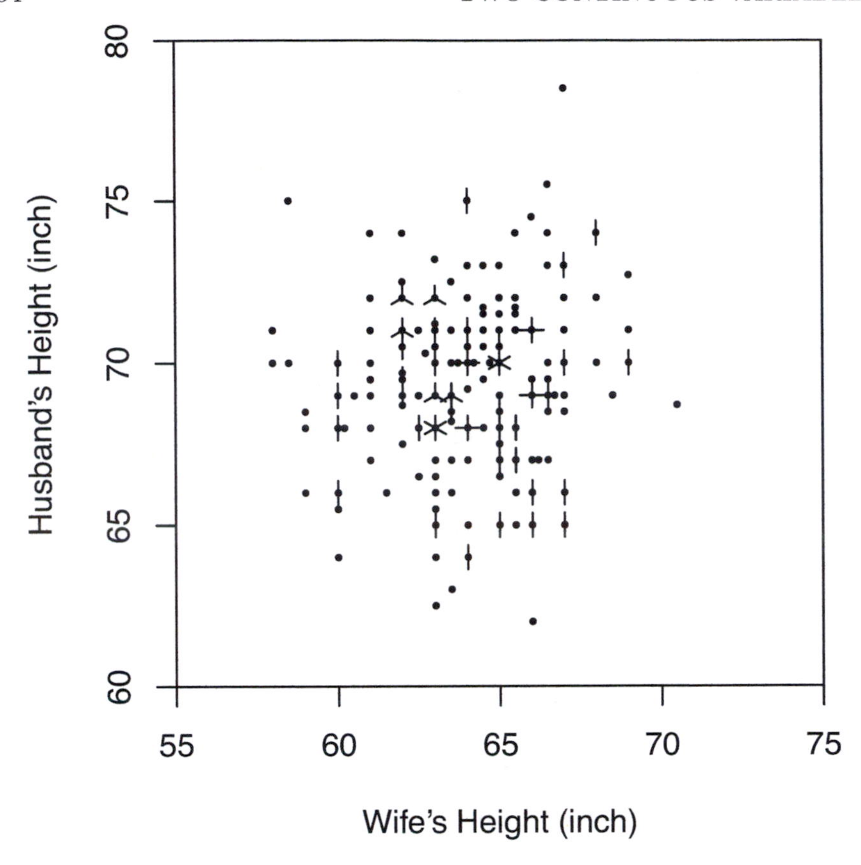

Figure 10.3 *Sunflower plot of husband's height versus wife's height*

With a consistent orientation of the petals for each count of 2 or more, there is a chance of artifactual visual impressions. This can be minimized by using random jitter to assign the direction of the first petal. This has been done in Figure 10.4.

Figures 10.3 and 10.4 have been drafted using the function `sunflowerplot` in R. In 1994 Schilling [103] noted that the plotting convention of Cleveland and McGill [26] did not use ink in proportion to the number of observations represented by a sunflower. Schilling [103] suggested instead that a petal be used for each observation.

Schilling [103] also suggested using a logarithmic sequence for the angles of the petals. The successive petals of a sunflower ought to be drawn so that the kth petal makes an angle of $2\pi\,[\log_2(2k-1)(\text{mod } 1)]$ measured counter-clockwise from the vertical. These variations have not been implemented in the `sunflowerplot` function.

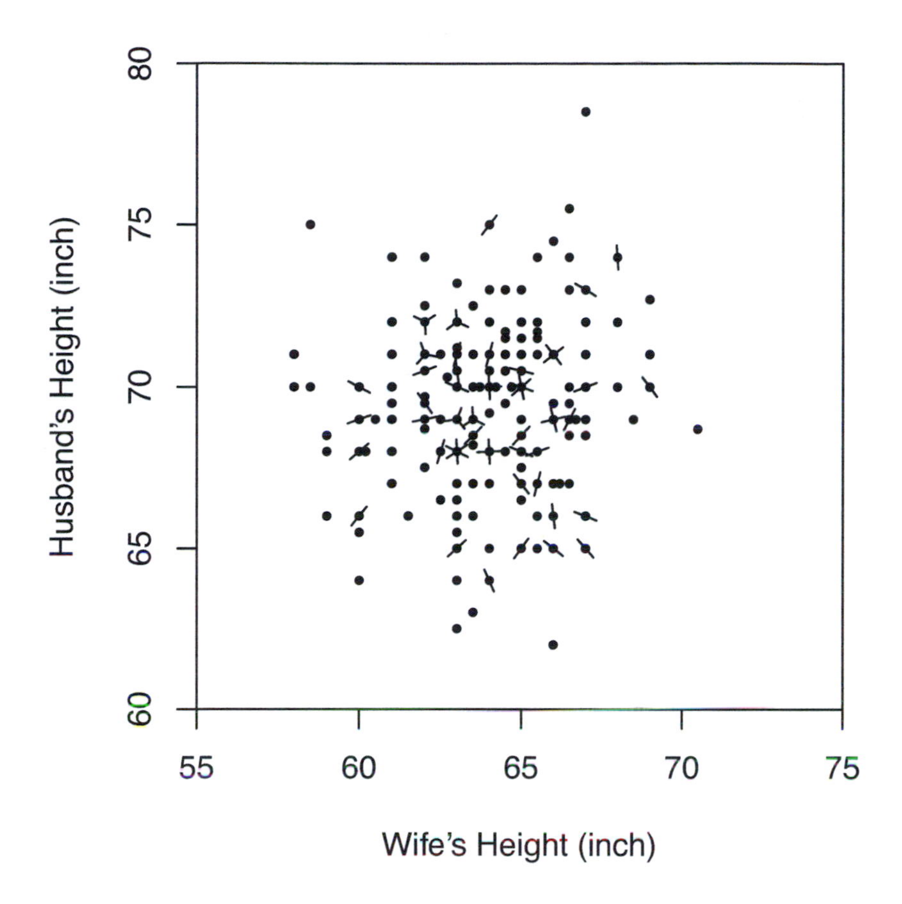

Figure 10.4 *Sunflower plot of husband's height versus wife's height with random orientation of petals*

Figure 10.5 is a variation on Figure 10.4 with a couple of changes. One change in Figure 10.5 is with respect to the plotting symbol. A small solid black circle is used to denote a single observation and the location of the center when two or more petals are drawn. This was done to achieve better clarity over the presentation in Figure 10.4. The other change is the addition of an ellipse to Figure 10.5 compared to that of Figure 10.4. The R script for producing Figure 10.5 is as follows.

```
set.seed(345)

sunflowerplot(Galton$Mother+60,Galton$Father+60,
xlab="Wife's Height (inch)",
ylab="Husband's Height (inch)",
xlim=c(55,75),ylim=c(60,80),pch=20,
```

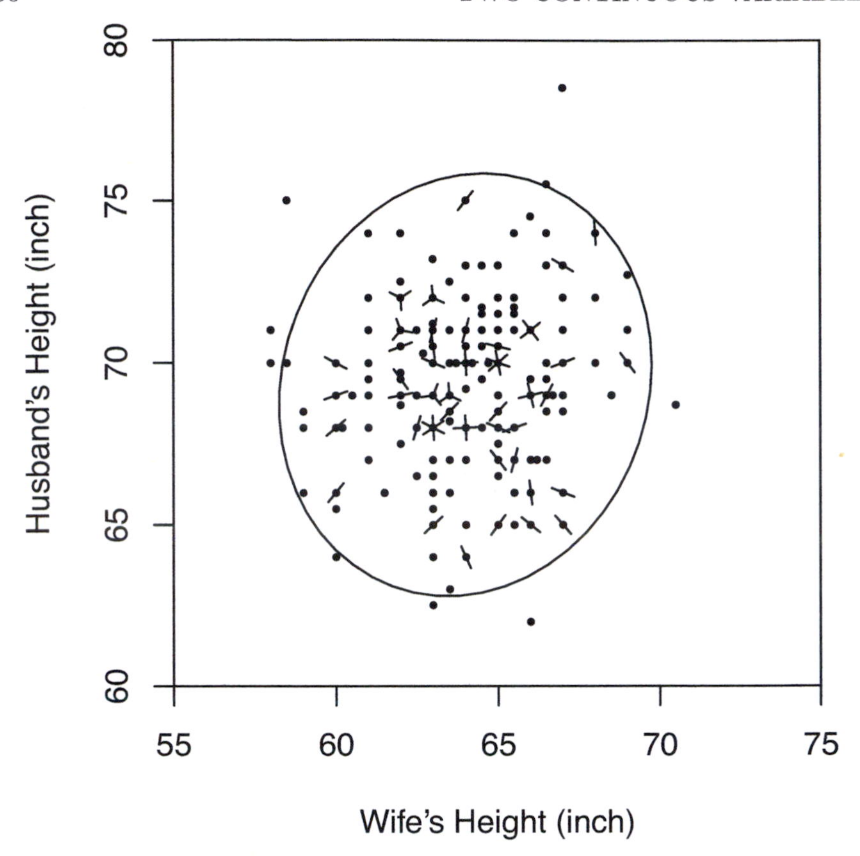

Figure 10.5 *Sunflower plot of husband's height versus wife's height with random orientation of petals and a 0.95 probability ellipse for a bivariate normal distribution*

```
seg.col="black",cex=0.6,cex.fact=1,
seg.lwd=1.,size=1/16,rotate=TRUE)

dataEllipse(Galton$Mother+60,Galton$Father+60,
add=TRUE,center.pch=NULL,col="black",lwd=1.0,
levels=0.95,plot.points=FALSE)
```

The function `dataEllipse` is found in the R package `car`, which was written to accompany the book *An R and S-PLUS Companion to Applied Regression* written by John Fox [43].

In the call to the function `sunflower`, `pch=20` sets the plotting symbol for a single observation to a closed black circle, `seg.col="black"` makes the petals black, `cex=0.6` shrinks the plotted center symbols to 60 percent of size, `cex.fact=1`, the default, sets the shrinking factor for the center points when there are petals, and `size=1/16` sets the length of the petals to one-sixteenth of an inch.

Setting `rotate=TRUE` in the call to the function `sunflower` randomly rotates the original direction for petals for each point. The first function call `set.seed(345)` in the preceding R script is for the purpose of producing the same plot for each call. The value 345 was selected, after a bit of experimentation, for the purpose of clarity in Figure 10.5.

The theory of using an ellipse to represent a region of constant density for a bivariate normal distribution can be traced to Francis Galton and Hamilton Dickson [54] in 1886. Their article was entitled "Family Likeness in Stature" and it was published in the *Proceedings of the Royal Society* (of London). The first depiction of this was given later that same year by Galton [51] in the article "REGRESSION *towards* MEDIOCRITY *in* HEREDITARY STATURE." published in the *Journal of the Anthropological Institute of Great Britain and Ireland.*

The ellipse depicted in Figure 10.5 encloses 95% of the probability according to a bivariate normal distribution. The method used for determining the ellipse in Figure 10.5 is not that developed by Dickson for Galton. Sunflower plots did not originate until nearly one hundred years after Galton's two publications in 1886.

10.5 The Bagplot

The *bagplot* was proposed in 1999 by Peter Rousseeuw, Ida Ruts, and John Tukey [101]. They characterized it as a bivariate generalization of the univariate boxplot. An early version of the bagplot without a discussion of how to produce it was given by Tukey [126] in 1972.

Like the boxplot, the bagplot is a robust graphical tool. An example of a bagplot is given in Figure 10.6 for the heights of husbands and wives for the 205 nuclear families collected by Galton in 1884. This figure was produced by the function `bagplot` in the R package `aplpack`. The R code for generating this figure is as follows.

```
bagplot(Galton$Mother+60,Galton$Father+60,
xlab="Wife's Height (inch)",
ylab="Husband's Height (inch)",
xlim=c(55,75),ylim=c(60,80),
show.whiskers=FALSE)
```

The main components of a bagplot are a *bag* that contains 50% of the data points, a *fence* that separates inliers from outliers, and a *loop* indicating the points outside the bag but inside the fence. In Figure 10.6, the bag is the inner polygon drawn with black line segments, with a dark blue interior. The fence

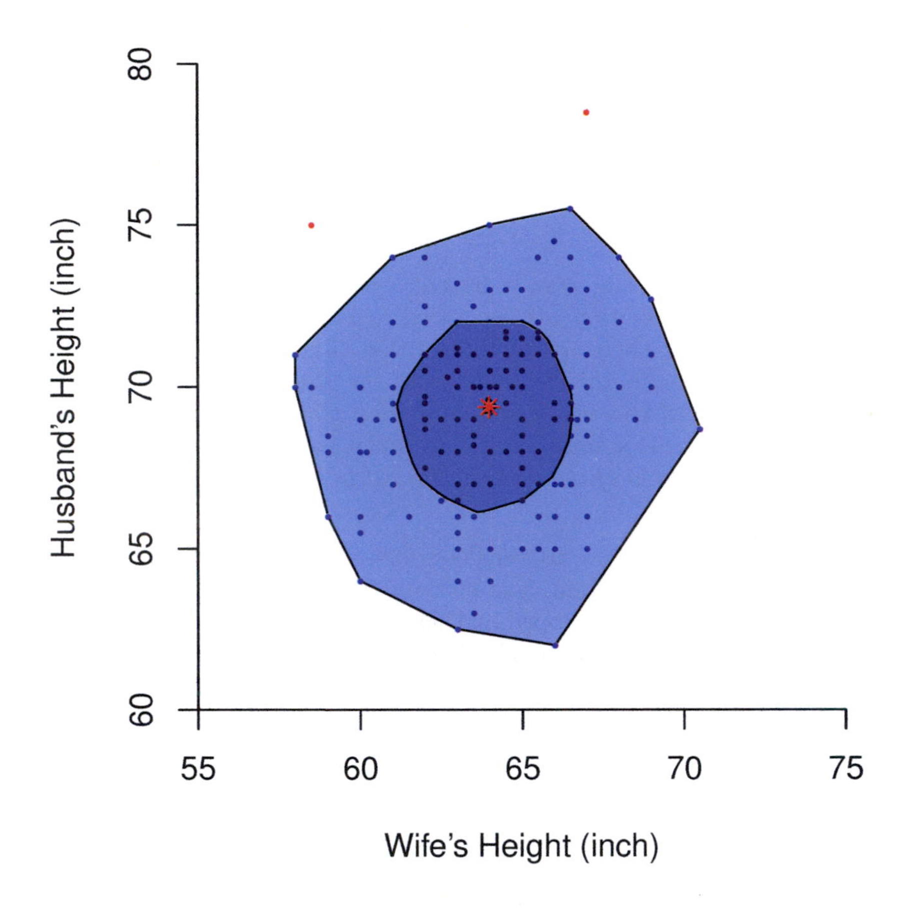

Figure 10.6 *Bagplot of husbands' and wives' heights produced by the function* `bagplot` *in the R package* `aplpack` *for 205 nuclear families collected by Francis Galton in 1884*

is the outer polygon. The loop is the light blue area. Two outliers appear as red dots outside the fence.

The *halfspace location depth* $\text{ldepth}(\theta, Z)$ of some point $\theta \in R^2$ relative to a bivariate data cloud $Z = \{\mathbf{z}_1, \mathbf{z}_1, \ldots \mathbf{z}_m\}$ is the smallest number of $\{\mathbf{z}_i\}$ contained in any closed halfplane with a boundary line through θ.

The *depth region* D_k is the set of all θ with $\text{ldepth}(\theta, Z) \geq k$. The depth regions are convex polygons such that $D_{k+1} \subset D_k$. The *depth median* $\mathbf{T}^*$ is defined as the θ with the highest $\text{ldepth}(\theta, Z)$ if uniquely defined; otherwise, it is the center of gravity of the deepest region. The depth median is noted by a red asterisk in Figure 10.6.

The bag B is constructed as follows. Let $\#D_k$ denote the number of data points contained in D_k. Determine the value k for which $\#D_k \leq \lfloor n/2 \rfloor <$

$\#D_{k-1}$ then interpolate linearly between D_k and D_{k-1} relative to the depth median $\mathbf{T}^*$ to obtain the set B. The bag B is a convex polygon.

The fence in Figure 10.6 was obtained by inflating B relative to the depth median $\mathbf{T}^*$ by a factor 3. This is the default value in the R function `bagplot`. It is possible to set the inflation factor to other values. The choice of 3 is based on simulations done by Rousseeuw and Ruts [100].

It is possible to incorporate a 95% confidence region for the depth median in the bagplot. This is done in Figure 10.6, using formula (2) in Rousseeuw, Van Aelst, and Hubert [102], but it is hard to see. The red confidence region is mostly obscured by the large red asterisk for the depth median.

In analogy to the notch of the one-dimensional boxplot, the confidence region for the depth median in a bagplot is called a *blotch*. Figure 10.6, in fact, depicts a *blotched bagplot*.

If the pairs of points fall on a straight line, the function `bagplot` will produce an outlier boxplot at a 45-degree angle. A test of this is given in Figure 10.7 by calling the function `bagplot` with wives' heights for both variables.

Like the univariate boxplot, the bagplot displays several characteristics of the data. The opportunity will now be taken to study the bivariate distribution of height for husbands and wives. The depth median gives a measure of location in two dimensions. The size and orientation of the bag measures spread. The shapes of both the bag and the loop provide information on skewness. The outliers and points near the outer boundary of the loop inform on the tails.

The depth median in Figure 10.6 suggests that the location for the heights of husbands is nearly 6 inches greater than that of their wives. The size of the blotch indicates that the two-dimensional measure of location is quite precise and that it is reasonable to conclude that the location for the heights of husbands is significantly greater than the location for heights of wives.

The shape of the bag in Figure 10.6 is nearly a circular disk. The light blue loop looks like the outline of a donut. Thus the spread of stature is similar between husbands and wives and the correlation between the heights of husbands and wives appears negligible.

There is no indication of skewness in any direction in the Cartesian coordinate plane of Figure 10.6. There are only two outliers.

Based on the bagplot, there is no reason not to model Galton's height data with a bivariate normal distribution. There are no concerns with respect to using the usual measures of location and spread in such a situation. The sample mean height of husbands is 69.316 inches and that for wives is 64.002 inches. The sample standard deviation is 2.647 inches for men and 2.333 inches for women. These two estimates of spread are quite close.

The sample correlation coefficient is 0.09690. A 95% confidence interval for

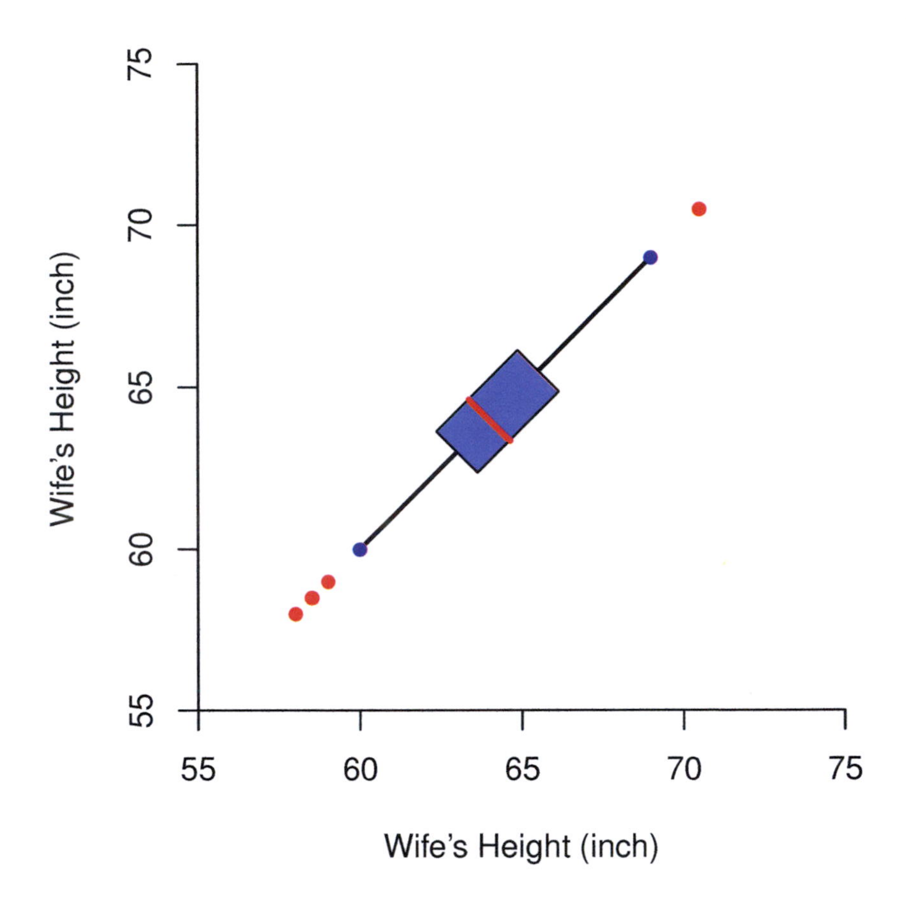

Figure 10.7 *One-dimensional bagplot of wives' heights produced by the function* `bagplot` *in the R package* `aplpack`

the correlation between heights of husbands and wives is $[-0.0407, 0.23087]$. The hypothesis test that the correlation coefficient is zero versus it is not has a p-value of 0.1669. This is evidence against assortative mating. The publications by Galton in 1886 were in advance of his creation of the concept of the correlation coefficient.

Galton did consider the possibility of assortative mating in British society. As President of the Anthropology section of the British Association for the Advancement of Science in 1885, he was obliged to give a presidential address. This was published by Galton [50] a year in advance of his two aforementioned publications in 1886. Galton [50] referred to the publication of a longer report in the *Journal of the Anthropological Institute of Great Britain and Ireland* to appear in November 1885. This is a source of confusion in the literature as this article by Galton [51] did not appear until 1886. The topic of assortative mating was printed in the report of his Presidential Address in 1885 and the subsequent paper in 1886.

Galton's [50] assessment was that "marriage selection takes little or no account of shortness or tallness." As to the statistical method used, Galton [50] wrote the following.

> This is by no means my only inquiry into this subject, but, as regards the present data, my test lay in dividing the 205 male parents and the 205 female parents each into three groups—tall, medium, and short (medium being taken as 67 inches and upwards to 70 inches), and in counting the number of marriages in each possible combination between them. The result was that men and women of contrasted heights, short and tall or tall and short, married just about as frequently as men and women of similar heights, both tall or both short; there were 32 cases of one to 27 of the other. In applying the law of probabilities to investigations into heredity of stature, we may regard the married folk as couples picked out of the general population at [sic] haphazard.

His opinion was that the 205 married couples could be considered as selected at random from the general population. Galton [51] repeats his conclusion that mating does not appear to be assortative with respect to stature again in 1886.

What Galton was engaged in was quantitative genetics. He did not know this at the time. This appellation came later. Stature is a quantitative trait. The conventional wisdom in quantitative genetics is that mating is assortative with respect to height. This even appears in textbooks. Falconer and Mackay [35] on page 174 write the following.

> Mating in human populations is assortative with respect to many metric characters, such as stature and IQ score, though not necessarily by deliberate choice of mates.

This quote is taken from the 1996 edition of the *Introduction to Quantitative Genetics*. The textbook was first published in 1960. Yet the first quantitative geneticist, Galton, thought differently and had the data to back up his statistical inference. Falconer and Mackay [35] provided no citation to support their statement that mating in human populations is assortative with respect to height.

Although Francis Galton came up with the concept of the correlation coefficient in 1888, it was Karl Pearson [91] who gets the credit in 1896 for the estimator known as Pearson's product-moment correlation coefficient. Pearson [91] writes that he hopes eventually to collect 1000–2000 families.

> Meanwhile, Mr. Galton, with his accustomed generosity, has placed at my disposal the family data on which his work on 'Natural Inheritance' was based.

For reasons unexplained, Pearson [91] selects 200 out of the 205 families for his computations. He estimates the mean and standard deviation for the height of husbands to be 69.215 inches and 2.628 inches. For wives, the corresponding estimates are 63.869 inches and 2.325 inches. All four estimates are quite close to those for the full set of 205 families.

The first product-moment estimate of the simple correlation coefficient between two variables ever published is that for husband's height and wife's height. Pearson [91] reports a value for r of 0.0931 with a standard error of 0.0473. The estimate for all 205 families is 0.0969 with a standard error of 0.0162. Pearson's estimate of the correlation coefficient is quite close but his estimate of its standard error is off by a factor of 2.92.

On the topic of assortative mating, Pearson [91] writes the following.

> Although the probable error (Table II.) is about half the coefficient of correlation, it is unlikely that the latter can really be zero, and although we must not lay very great stress on the actual value r, still we are justified in considering there is a definite amount of assortative mating with regard to height going on in the middle classes.

This is the likely origin of the misconception that mating in human populations is assortative with respect to height. As previously noted, a p-value of 0.1669 for the hypothesis test that the correlation coefficient is zero, versus the alternative that it is not, does not support a conclusion that mating is assortative for height.

To be fair to Pearson, he was missing two pieces of the puzzle. The sampling distribution of his product-moment estimator of the correlation coefficient was not known until 1921 when it was derived by Ronald Aylmer Fisher [38]. Pearson did not have the benefit of Fisher's [41] comments on the p-value and its interpretation because these were not published until 1926. Pearson and Galton can be faulted for not demonstrating that they had looked at the data on the heights of husbands and wives. If seeing is believing, there is nothing in the scatterplot of Figure 10.1 through to the sunflower plot of Figure 10.5 and the bagplot of Figure 10.6 to suggest that mating in Galton's sample of 1884 is assortative with respect to stature.

10.6 The Two-Dimensional Histogram

10.6.1 Definition

Given that in Galton's time all drafting was done by human hand, producing a scatterplot with 205 points would have been considered overly burdensome. Producing a bagplot with a bag and loop overlaid on a scatterplot would have been computationally unfeasible. The approach instead would have been to produce a summary tabulation first. This is done in Table 10.1 for the heights of husbands and wives.

Table 10.1 appears very much like the tables in Galton and Dickson [54] and Galton [51]. But a version of Table 10.1 does not appear in either of these publications of 1886. Galton's pre-occupation was with the inheritance of stature. This will be considered in the next chapter. The case study for this chapter concerns assortative mating.

Data tables, such as Table 10.1, and narrative descriptions of bivariate relationships typically take the place of plots in presentations for administrators and the public. Galton [51] in 1886 managed to go beyond this approach. Figure 10.6 presents a graphical display very much similar to that of Plate X in Galton [51]. But Galton was more interested in heritability of height and his Plate X concerned parents and offspring. He never did produce a table like Table 10.1 or a figure like Figure 10.6 for husbands and wives. If not for the success in 2004 of Jim Hanley in tracking down Galton's original data, it would not have been possible to produce Table 10.1 or any of the figures of this chapter.

Figure 10.8 takes the frequencies in Table 10.1 and plots them at the cell midpoints in a scatterplot. A 0.95 probability ellipse for a bivariate normal distribution has been added. So it is possible to imitate a graphical display introduced by Galton in 1886 in a fraction of the time that he would have required. But it is possible to do better. Figure 10.9 takes the frequencies in Table 10.1 and depicts each as a sunflower.

The ellipses for the bivariate normal distribution in Figure 10.5, 10.8, and 10.9 were produced with the original data for the 205 couples. The ellipses in these three figures are identical. The center is given by the mean heights for husbands and wives. The shape is determined from unbiased estimates of the variances for these two variables and their covariance. The ellipses were plotted by the function `data.ellipse` in the R package `car`.

Overplotting of symbols is still a problem for sunflower plots such as Figures 10.3, 10.4 and 10.5. It just so happened for this data from Galton's Records of Family Faculties that the problem wasn't particularly troublesome. Cleveland and McGill [25] proposed dividing a square plotting region into square subregions, or cells, of equal area. The number of observations in each cell is then plotted with a sunflower, as in Figure 10.9.

Sunflower plots go beyond coping with overlapping data points in a scatterplot. Figure 10.9 is in essence a nonparametric estimate of the bivariate density function for the heights of husbands and wives. A *two-dimensional histogram* is depicted in Figure 10.9. To this has been added a parametric density estimate enclosing 95% of the values according to a bivariate normal distribution. In the next section, an alternative nonparametric density estimate and its graphical display will be presented.

Height of Husband (inch)	Height of Wife (inch) 57.5	58.5	59.5	60.5	61.5	62.5	63.5	64.5	65.5	66.5	67.5	68.5	69.5	70.5
78.5	0	0	0	0	0	0	0	0	0	1	0	0	0	0
77.5	0	0	0	0	0	0	0	0	0	0	0	0	0	0
76.5	0	0	0	0	0	0	0	0	0	0	0	0	0	0
75.5	0	0	0	0	0	0	0	0	0	1	0	0	0	0
74.5	0	1	0	0	0	0	2	0	1	0	0	0	0	0
73.5	0	0	0	1	1	1	0	0	1	1	2	0	0	0
72.5	0	0	0	0	1	0	2	2	0	3	0	1	0	0
71.5	0	0	0	1	3	4	1	4	3	1	1	0	0	0
70.5	1	0	0	1	5	6	4	5	5	1	0	1	0	0
69.5	1	1	2	2	3	2	7	9	1	4	1	2	0	0
68.5	0	1	2	2	3	4	6	3	4	6	0	1	0	1
67.5	0	1	2	2	1	8	4	4	2	0	0	0	0	0
66.5	0	0	0	1	0	3	2	3	3	2	0	0	0	0
65.5	0	1	3	0	1	2	1	0	3	2	0	0	0	0
64.5	0	0	0	0	0	2	1	2	3	2	0	0	0	0
63.5	0	0	1	0	0	1	2	0	0	0	0	0	0	0
62.5	0	0	0	0	0	1	1	0	0	0	0	0	0	0
61.5	0	0	0	0	0	0	0	0	1	0	0	0	0	0

Table 10.1 *Number of married couples for various statures in 205 nuclear families from Galton's Records of Family Faculties (the heights listed above are mid-points of intervals each one inch in width, and inclusive of the right endpoint)*

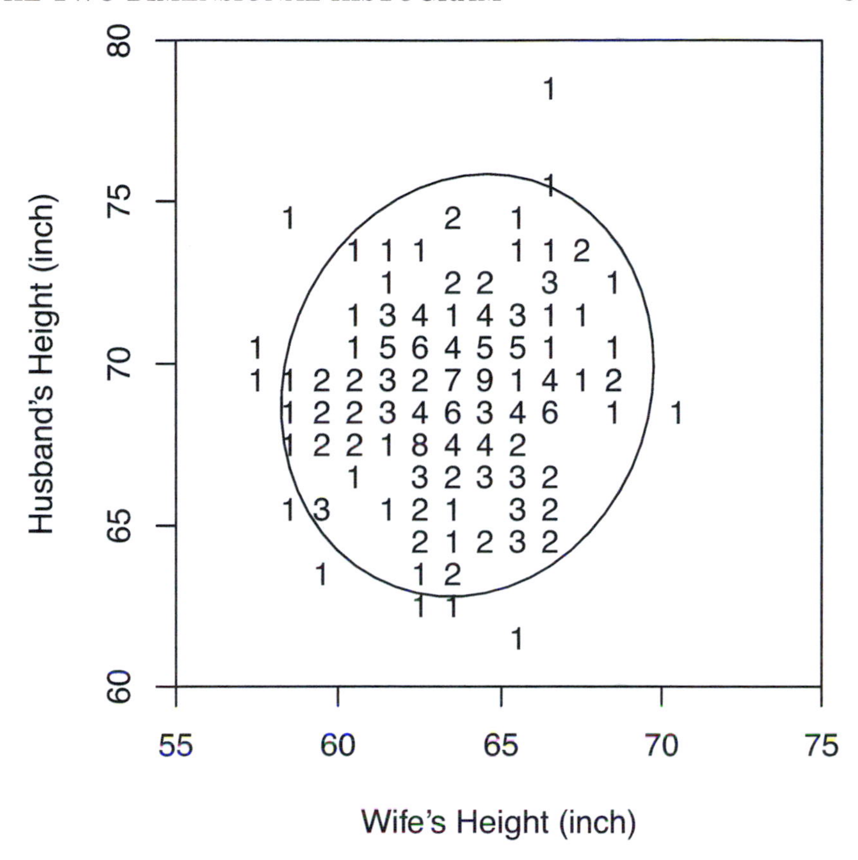

Figure 10.8 *Scatterplot of frequency for combinations of husband's height and wife's height with a 0.95 probability ellipse for a bivariate normal distribution*

10.6.2 The Levelplot

In using the sunflower plot in Figure 10.9 to visualize the bivariate density for the joint distribution for the heights of man and wife, a two-dimensional graphical display is being used to depict a three-dimensional entity.

An alternative to the use of petals to depict frequency counts is the use of color or depth of gray shade. This is done in the grayscale *levelplot* for sample density of the heights of husbands and wives in Figure 10.10.

The function `image` in the standard R `graphics` package was used to draft the levelplot in Figure 10.10. An alternative in R would have been to use the function `levelplot` in the `lattice` package (which implements Trellis graphics in R). The function `image` was chosen because the axis scale can be set to the actual scale.

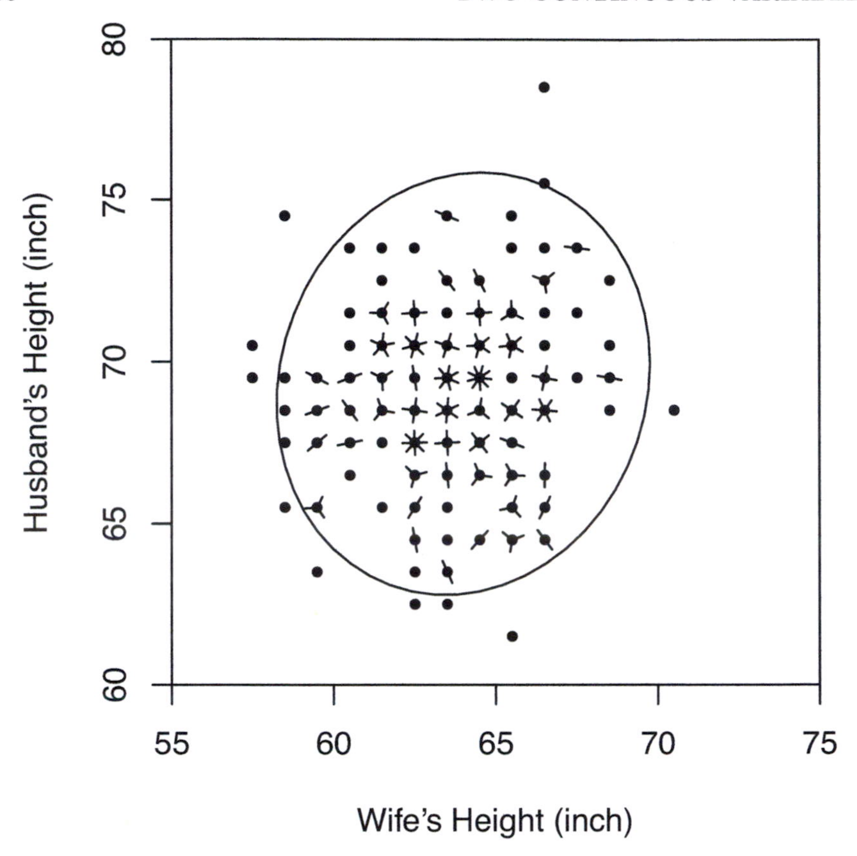

Figure 10.9 *Sunflower plot of frequency for combinations of husband's height and wife's height with a 0.95 probability ellipse for a bivariate normal distribution*

The `lattice` function `levelplot` naturally uses the interval $[0, 1]$ for both axes. The ticks and numbers for each in Figure 10.10 were added by two separate calls to the function `axis`.

The legend for the frequency count and the corresponding shade of gray was produced by placing to the left of the levelplot a separate plot generated by the `barplot` function in R.

The outline of the levelplot in Figure 10.10 is rectangular. But a careful look will reveal that this is to accommodate the wider range of husbands' heights as compared to wives. The cells for heights are one inch on edge and appear as squares in Figure 10.10.

For comparison with the sunflower plot of Figure 10.9, an ellipse for a bivariate normal density has been added to the levelplot in Figure 10.11.

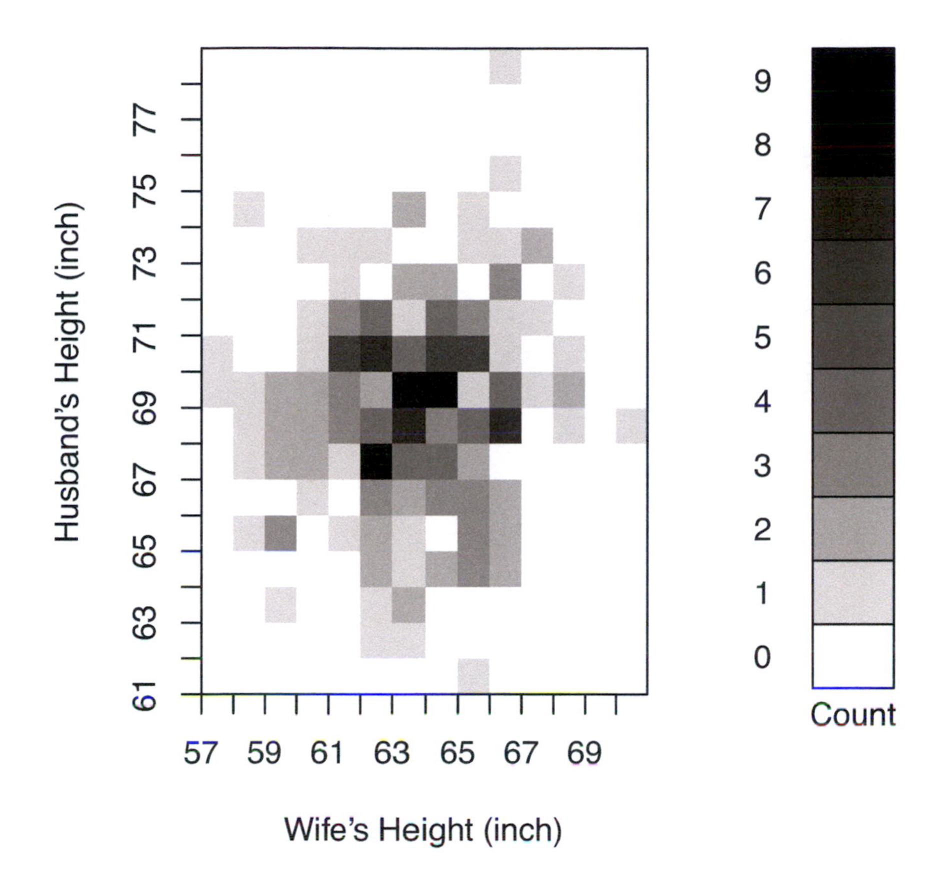

Figure 10.10 *Levelplot of frequency for combinations of husband's height and wife's height*

Instead of tones of gray in Figures 10.10 and 10.11, colors could have been used. Several different color palettes are available in R for this purpose. One palette uses the colors of the rainbow. Another uses heat colors from red hot (coolest) through yellows to white hot (hottest). There are two different palettes for coloring topographic maps.

Colors have been avoided in Figures 10.10 and 10.11 for a couple of reasons. Color blindness in viewers of levelplots is a concern. This is a greater concern for male viewers than female viewers.

The other reason is that while the darkness of gray for each level in Figures 10.10 and 10.11 increases linearly with the count, this cannot be said to be the case for colors.

Shades of gray are ordered along a single axis: darkness. Colors can be ordered

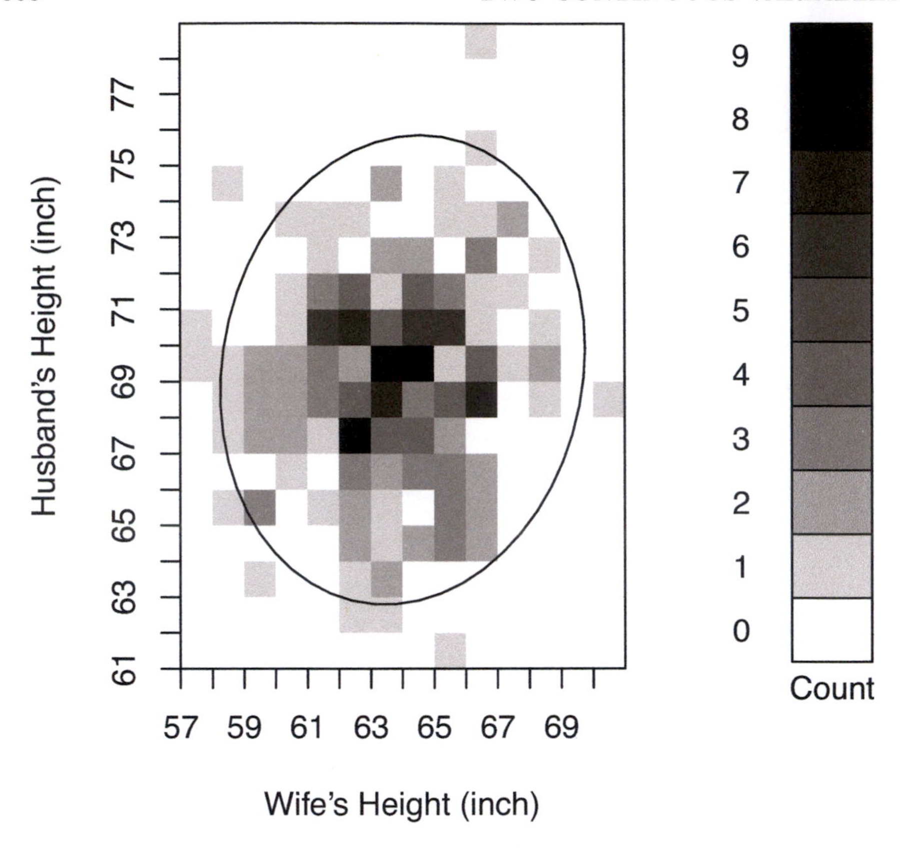

Figure 10.11 *Levelplot of frequency for combinations of husband's height and wife's height with a 0.95 probability ellipse for a bivariate normal distribution*

along three different axes. Moreover, there are at least three sets of axes in common use.

One set of color axes consists of hue, saturation, and value. Another of hue, chroma, and luminance. Yet another of intensity for each of the three primary colors of red, green, and blue. (For a full discussion see Appendix A.)

The choice of color palette can be deliberately made by the drafter of a levelplot to produce a desired apprehension by the viewer. This has been done in Figure 10.12, which has been drafted using the functions `geom_bin2d` and `stat_ellipse` in the `ggplot2` package to obtain the levelplot and the confidence ellipse, respectively, and the following call to the function `daltonize` in the `LSD` package to obtain a version of the 10 colors in the R topographic palette that is adjusted for color-vision deficient individuals who have deuteranopia.

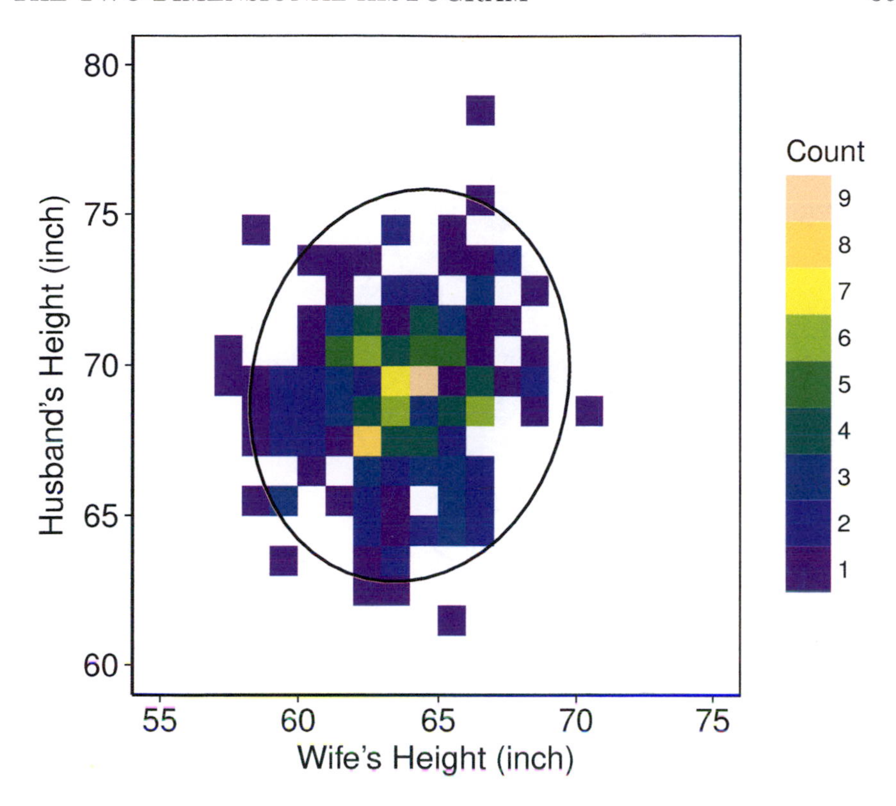

Figure 10.12 *Levelplot of frequency for combinations of husband's height and wife's height with a 0.95 probability ellipse for a bivariate normal distribution using* `ggplot2`

```
topod10<-daltonize(topo.colors(10),cvd="d",show = FALSE)[[2]]
```

The function `daltonize` also offers palettes enhanced for protanopia or tritanopia but of the three forms of color deficiency, deuteranopia is the most common. (See Appendix A for a fuller discussion and prevalence estimates.)

Not everyone agrees that a rectangular grid for a levelplot is optimal. Some would advise using a hexagonal grid to reduce sampling bias due to edge effects. It is thought by some that the bias is greater for higher perimeter:area ratios. There is also the issue that the distance between centroids is the same for all six nearest neighbors of a hexagon but this is not the case for a square for which the distance is a factor $\sqrt{2}$ greater for diagonal neighbors. There is also the issue that the hexagon is the most complex regular polygon that can cover a plane and not leave any gaps. Figure 10.13 is a levelplot with a hexagonal grid executed in `ggplot2` by merely replacing the call `geom_bin2d(binwidth=1)` used for Figure 10.12 with `stat_bin_hex(binwidth=1)`.

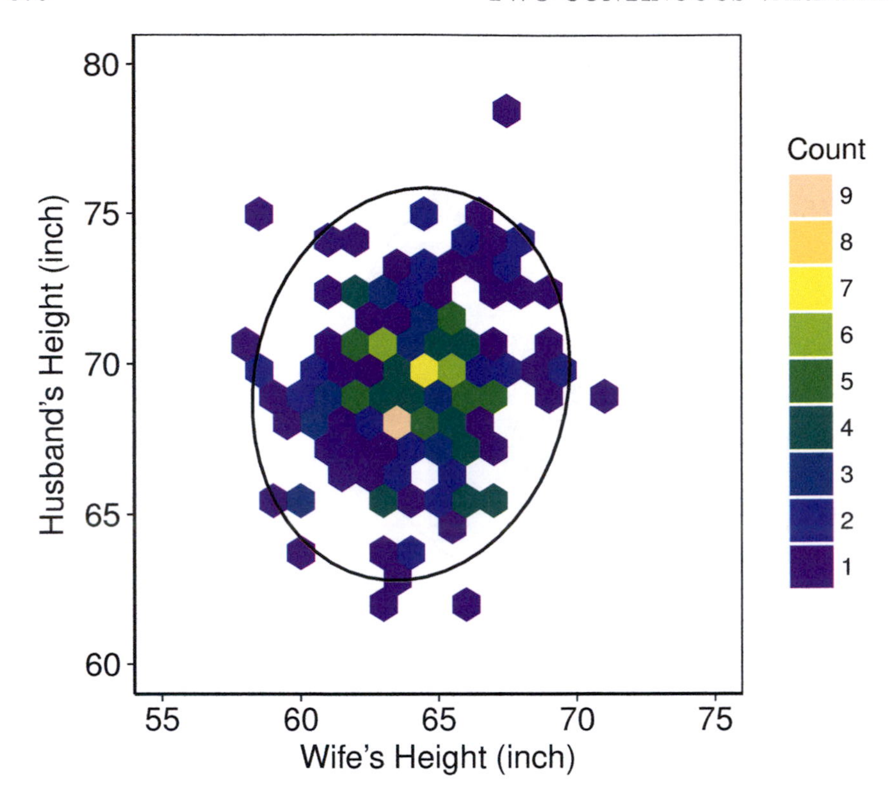

Figure 10.13 *Levelplot of frequency using a hexagonal grid for combinations of husband's height and wife's height with a 0.95 probability ellipse for a bivariate normal distribution using* ggplot2

The ggplot2 function stat_bin_hex relies on the R package **hexbin** to do the heavy work needed for the hexagonal grid. The package **hexbin** will be loaded automatically by ggplot2 without intervention provided that the **hexbin** package has already been installed. The package **hexbin** has its own plotting function gplot.hexbin that was used to produce Figure 10.14. Unfortunately, the car package function dataEllipse cannot overlay a confidence ellipse on a plot produced by the **hexbin** plotting function gplot.hexbin so this is omitted in Figure 10.14.

10.6.3 The Cloud Plot

The levelplots of Figures 10.10 and 10.11 use two dimensions for a histogram that could be displayed in three dimensions with the count of frequencies as the third dimension. The *cloud plot* in Figure 10.15 is intended to be a three-dimensional portrayal of the two-dimensional histogram for heights of

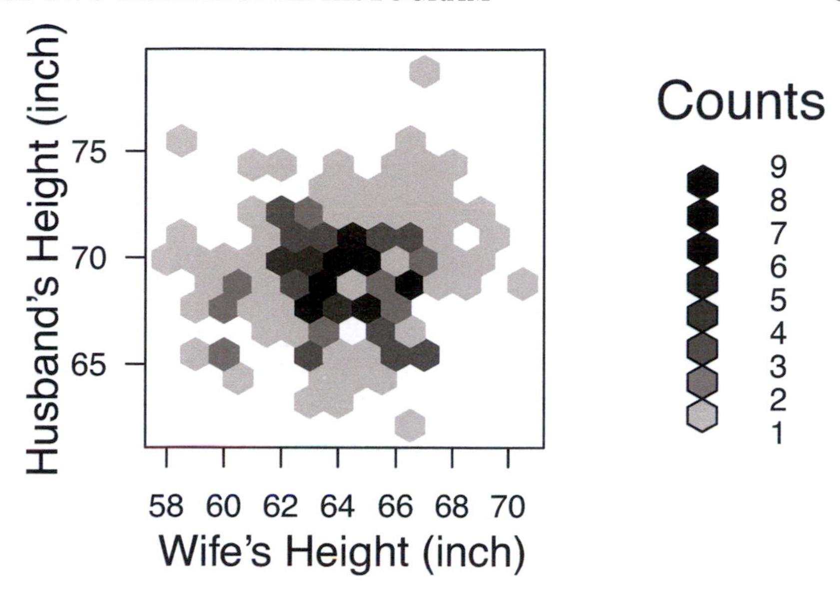

Figure 10.14 *Levelplot of frequency using a hexagonal grid for combinations of husband's height and wife's height without a probability ellipse for a bivariate normal distribution using* `hexbin`

husbands and wives. The cloud still is constrained to two dimensions but uses a drafted perspective to gain the third dimension.

The R code for producing Figure 10.15 is given below and uses the function `cloud` in the `lattice` package.

```
figure<-cloud(prop.table(MF),type=c("h"),
screen=list(x=-10,y=7,z=10),
xlab="Wife",ylab="Husband",
zlab=" Count",par.box=FALSE,
col="black",lwd=1.5,
scpos=list(x=1,y=8,z=12))

print(figure)
```

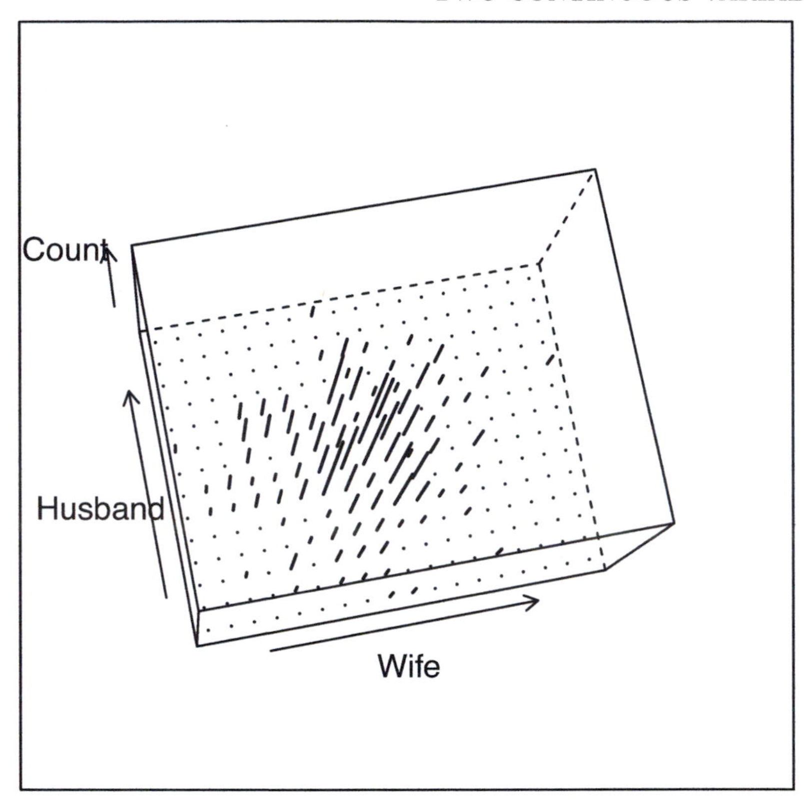

Figure 10.15 *Cloud plot of two-dimensional histogram of husbands' and wives' heights*

The function `prop.table` is in the `base` package of R and is being used in this instance to convert the counts in the matrix `MF` into proportions.

As a static graphical display, the cloud plot of Figure 10.15 is not as effective as the other plots of this section. But a cloud plot dynamically displayed on a computer screen with the viewer able to change the orientation as well as the axis and rate of rotation is much more useful. A flat image with motion gives a better portrayal of three dimensions.

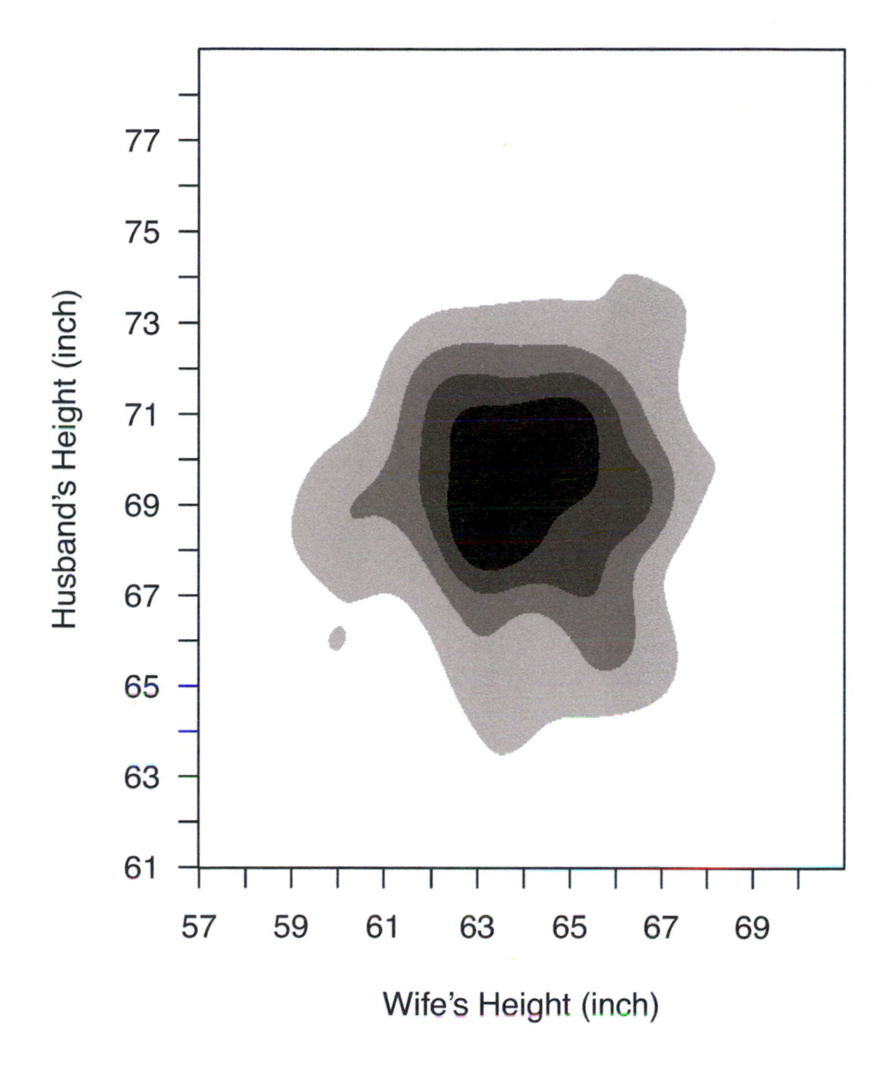

Figure 10.16 *Levelplot of a two-dimensional kernel density estimate for the joint distribution of husband's height and wife's height (the darker the shade of gray, the greater the estimate of relative frequency)*

10.7 Two-Dimensional Kernel Density Estimation*

10.7.1 Definition

A levelplot depicting a *two-dimensional kernel density estimate* for the distribution of husbands' height and wives' height based on Galton's data from

*This section can be omitted without loss of continuity.

1884 is given in Figure 10.16. The function `kde2d` in the R package `MASS` was used to produce the kernel density estimate displayed in this figure. (The `MASS` package was developed to accompany the textbook *Modern Applied Statistics with S* by William Venables and Brian Ripley [128].) The two-dimensional kernel in the R function `kde2d` is

$$K_G(x, y) = \frac{1}{2\pi} e^{-\frac{x^2+y^2}{2}} \tag{10.1}$$

for variables x and y. This kernel is Gaussian and axis aligned. The kernel density estimate produced by the function `kde2d` is determined over a square grid.

The bandwidth used for the kernel density estimate depicted in Figure 10.16 was estimated for each variable by

$$\hat{\lambda}_{2D} = 4 \times 1.06 A n^{-1/5} \tag{10.2}$$

where

$$A = \min(s, IQR/1.34) \tag{10.3}$$

with the standard deviation s and the interquartile range IQR estimated from the bivariate sample. Note that the bandwidth $\hat{\lambda}_{2D}$ is a composite of the simple normal reference bandwidth $\hat{\lambda}_{SNR}$ and Silverman's Rule of Thumb estimator $\hat{\lambda}_{SROT}$.

The levelplots in Figures 10.10 and 10.11 have one shade of gray for each distinct whole number. The levelplot in Figure 10.16 depicts a continuous variable representing a continuous probability density for two random variables. A single shade of gray in Figure 10.16 represents a range of real numbers. The levelplot is not the best option for presenting a continuous surface above a two-dimensional plane. The plot of the next section does a better job.

A more visually dynamic graphical display is given in Figure 10.17, which overlays contours on a colored version of the grayscale level plot of Figure 10.16. The R code for generating Figure 10.17 is as follows.

```
bwmf<-c(bandwidth.nrd(Galton$Mother+60),
bandwidth.nrd(Galton$Father+60))

kdemf<-kde2d(Galton$Mother+60,Galton$Father+60,
h=bwmf,n=500,lims=c(57,71,61,79))

image(kdemf,col=terrain.colors(6),
breaks=(0:6)*0.005,axes=FALSE,
xlab="Wife's Height (inch)",
ylab="Husband's Height (inch)")

contour(kdemf,xlab="Wife's Height (inch)",
```

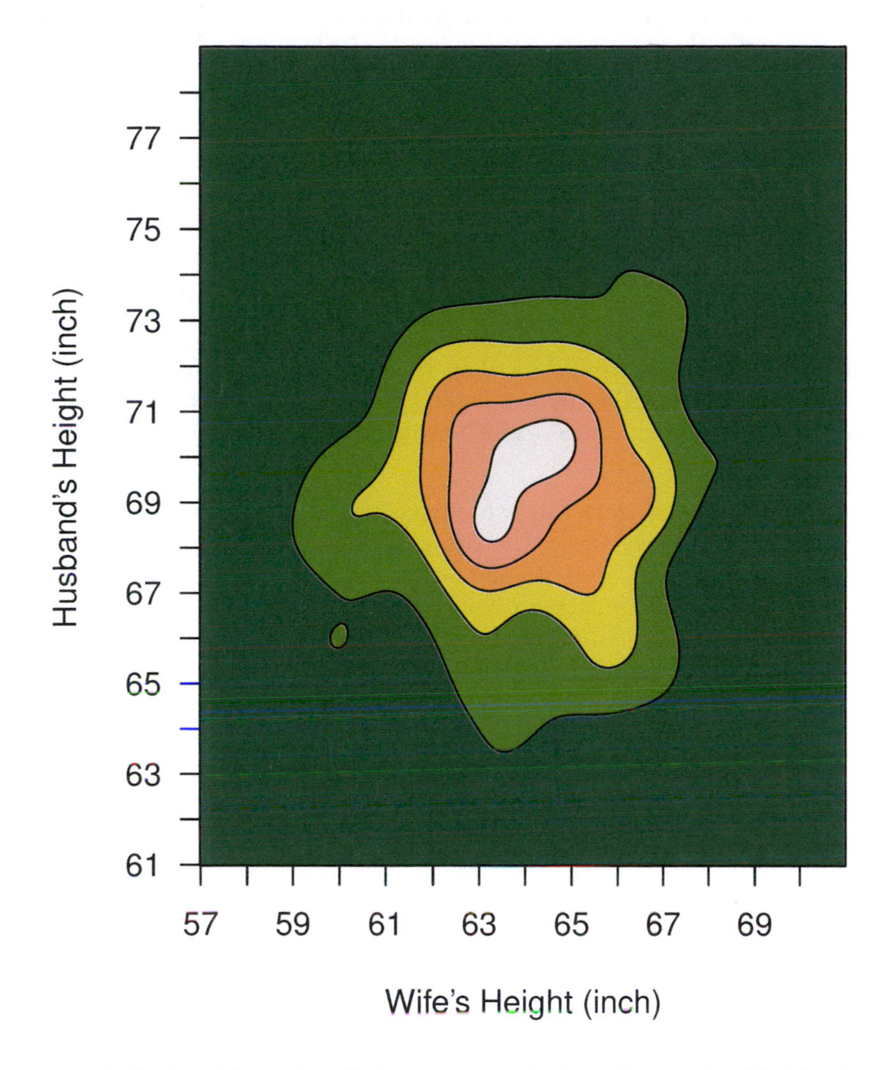

Figure 10.17 *Levelplot with added contours of husbands' and wives' heights for 205 nuclear families collected by Francis Galton in 1884*

```
ylab="Husband's Height (inch)",axes=FALSE,
nlevels=9,labcex=0.6,add=TRUE,drawlabels=FALSE)

axis(1,at=57.+0:13)
axis(2,at=(61.+0:17),
labels=c("61"," ","63"," ","65"," ","67"," ",
"69"," ","71"," ","73"," ","75"," ","77"," "))
box("plot")
```

The call to the MASS function bandwidth.nrd sets the bandwidth for heights of mothers and fathers to that of the normal reference distribution:

$$\hat{\lambda} = 4 \cdot 1.06 \cdot \min(s, IQR/1.34)n^{-1/5}. \tag{10.4}$$

Then the MASS function kde2d is executed to compute the kernel density estimate, which is stored in the R list kdemf. The graphics package function image then creates the contour plot from kdemf. The two subsequent calls to the lower-level function axis then add the tick marks and their labels for the horizontal and vertical. Finally, a call to the function box completes the black box along all four sides of the contour plot.

The impression from Figure 10.17 is that the joint density can be characterized as unimodal. Although the highest-value contour, corresponding to a 0.025 probability, shows a roughly 45-degree orientation, this pattern is not repeated for other contours.

The overall impression from the two-dimensional kernel density estimate in Figure 10.17 and the levelplot of the less sophisticated nonparametric density estimate given by the two-dimensional histogram in Figure 10.10 is that there is not a tight connection in stature for married couples. Mating in the upper and upper-middle classes of Great Britain during the Victorian era does not appear to be assortative with respect to height.

Figure 10.18 is a ggplot2 version of Figure 10.17 produced by the following code.

```
figure<-ggplot(Galton,aes(x=Mother+60,y=Father+60)) +
stat_density_2d(aes(fill = ..level..),
geom = "polygon") +
theme_linedraw() +
theme(panel.grid=element_blank(),
axis.line=element_line(),
axis.title=element_text(size=12),
axis.text=element_text(size=12)) +
guides(fill=guide_legend(title="Density",
reverse=TRUE))+
scale_x_continuous(breaks=seq(57,71,2),
limits=c(57,71)) +
scale_y_continuous(breaks=seq(61,79,2),
limits=c(61,79)) +
labs(x="Wife's Height (inch)",
y="Husband's Height (inch)")

print(figure)
```

The ggplot2 function stat_density_2d also relies on the MASS function kde2d to do the heavy work. The parameters aes(fill = ..level..) and

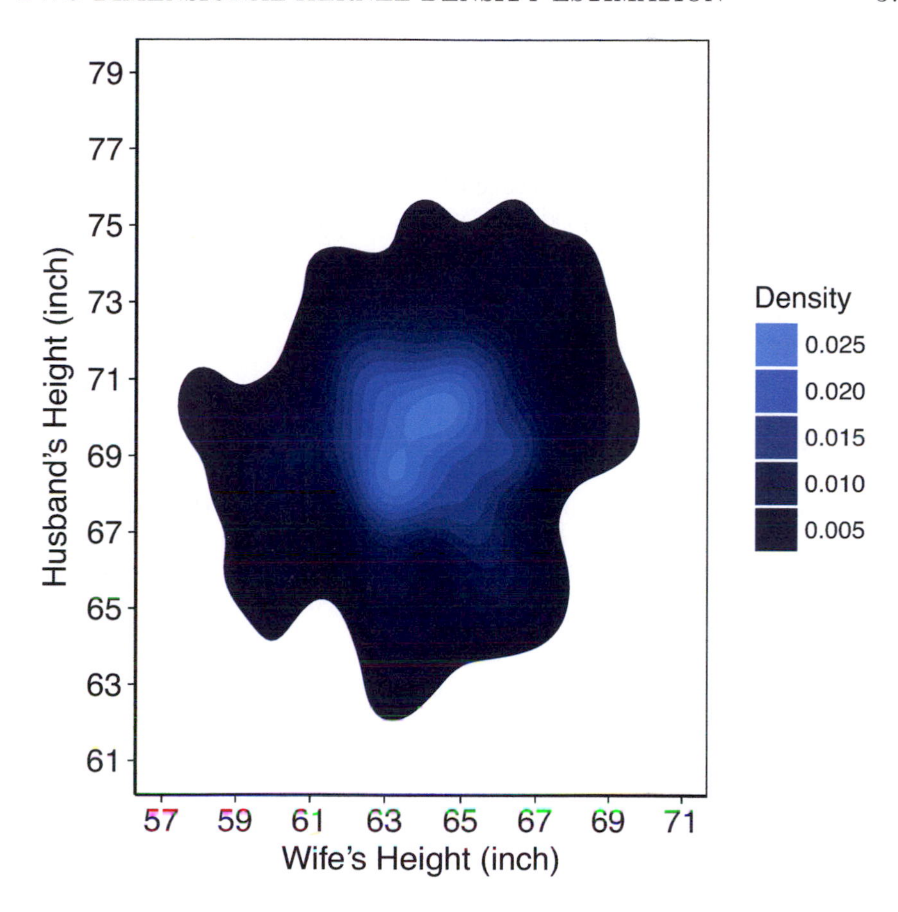

Figure 10.18 *Levelplot with added contours of husbands' and wives' heights for 205 nuclear families collected by Francis Galton in 1884 using* `ggplot2`

`geom = "polygon"` passed to the function `stat_density_2d` are responsible for the colored contours. The shades of blue, and in fact the number of shades of blue, seen in Figure 10.18 are by default only. The result is quite stunning.

Francis Galton never published a graphical display similar to Figure 10.18 for human stature. If it were not for the lack of computation power in the 1880s, Galton might have produced Figure 10.18. Galton actually did produce something quite similar but in another discipline of application: meteorology. In 1863, Macmillan of London published Galton's [48], *Meteorographica*. Galton illustrated this book with more than 600 diagrams to depict the weather during the month of December 1861 over Great Britain and a large part of Europe.

His main pre-occupation in this book was to introduce the synoptic surface weather chart with its pictographs for depicting wind velocity and cloud cover at weather observatories. But in the appendix to this work, one finds for the morning, afternoon, and evening of each day in December 1861, separate charts for barometric pressure, temperature, and wind streams over Great Britain and Western Europe.

These maps were hand drawn to save costs both in production and printing; the colored regions are stippled (lots of small ink dots), vertically hatched, and cross-hatched. Galton [48] used the levelplot of the previous section with added contours for temperature and pressure. The results were published in 1863 for a month of weather in 1861. There is no evidence to support that he used the next plot to be considered.

10.7.2 The Contour Plot

A common two-dimensional plot for depicting a surface in three dimensions is the *contour plot*. The two-dimensional kernel density estimate for the joint distribution of heights of husbands and wives is depicted in the contour plot of Figure 10.19. The R function `contour` was used in the following code to produce this graphical display.

```
bwmf<-c(bandwidth.nrd(Galton$Mother+60),
bandwidth.nrd(Galton$Father+60))

kdemf<-kde2d(Galton$Mother+60,Galton$Father+60,
h=bwmf,n=500,lims=c(57,71,61,79))
contour(kdemf,xlab="Wife's Height (inch)",
ylab="Husband's Height (inch)",axes=FALSE,
nlevels=9,labcex=0.8)

axis(1,at=57.+0:13)
axis(2,at=(61.+0:17),
labels=c("61"," ","63"," ","65"," ","67"," ",
"69"," ","71"," ","73"," ","75"," ","77"," "))
box("plot")
```

The call to the function `box` completes the frame around the contour plot.

A ggplot2 version of Figure 10.19 is given in Figure 10.20 and is produced by the following code.

```
figure<-ggplot(Galton,aes(x=Mother+60,
y=Father+60)) +
stat_density_2d(bins=6) +
theme_linedraw() +
theme(panel.grid=element_blank(),
axis.line=element_line(),
```

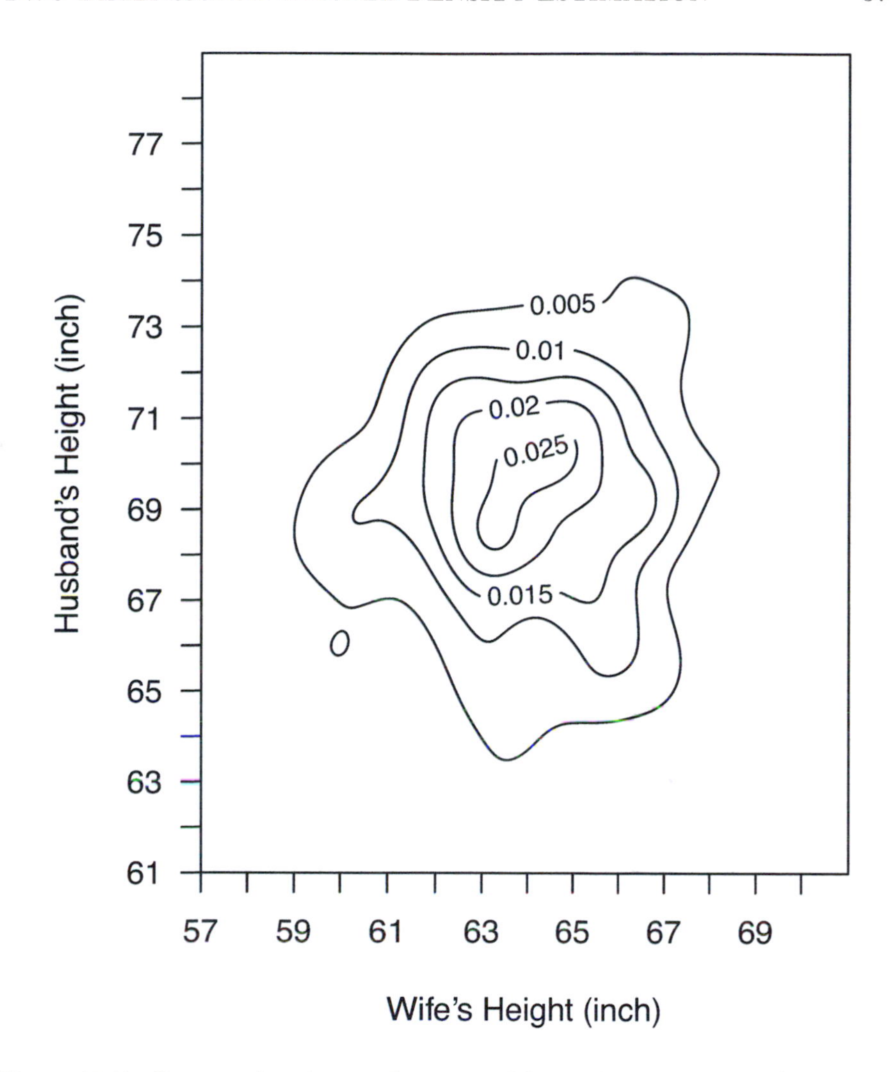

Figure 10.19 *Contour plot of a two-dimensional kernel density estimate for the joint distribution of husband's height and wife's height*

```
axis.title=element_text(size=12),
axis.text=element_text(size=12)) +
scale_x_continuous(breaks=seq(57,71,2),
limits=c(57,71)) +
scale_y_continuous(breaks=seq(61,79,2),
limits=c(61,79)) +
labs(x="Wife's Height (inch)",
y="Husband's Height (inch)")
```

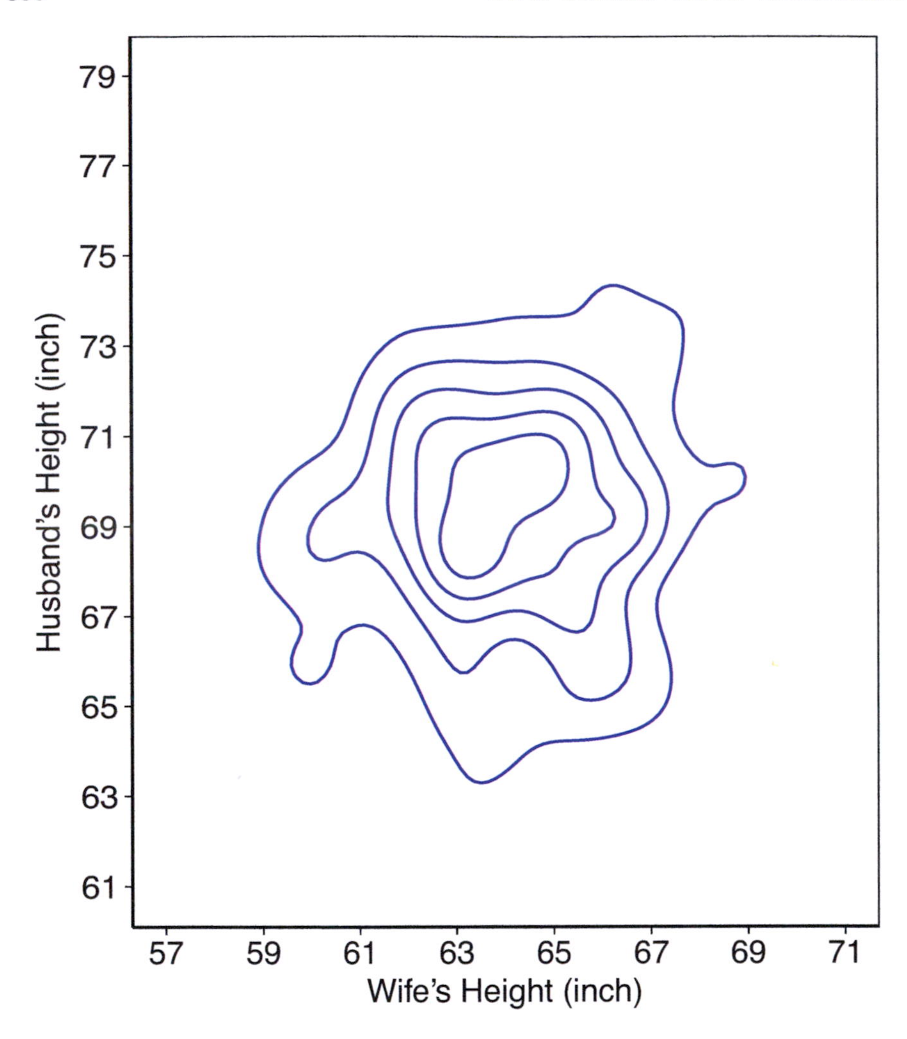

Figure 10.20 *Contour plot of a two-dimensional kernel density estimate for the joint distribution of husband's height and wife's height using* `ggplot2`

```
print(figure)
```

Passing the parameter `bins=6` results in 5 contours being drawn as in Figure 10.19. The bandwidth, kernel, and contour color (blue) are a result of relying on defaults so the appearance of Figure 10.20 is slightly different from Figure 10.19. Also the `graphics` function `contour` provides labels for the contours, whereas these would need to be manually added using the `annotate` function for the `ggplot2` version in Figure 10.20.

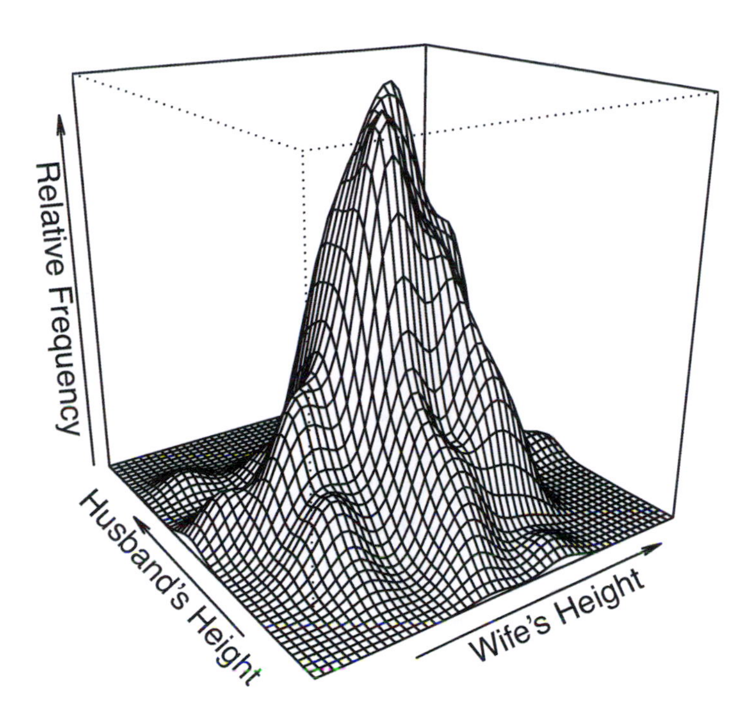

Figure 10.21 *Wireframe plot of a two-dimensional kernel density estimate for the joint distribution of husband's height and wife's height for 205 nuclear families collected by Francis Galton in 1884*

10.7.3 The Wireframe Plot

The *wireframe plot* in Figure 10.21 depicts the two-dimensional kernel density estimate for the joint distribution of husbands' height and wives' height previously displayed with level and contour plots. The name *wireframe* is obvious from the figure. The wire mesh is used to impart a perspective of three dimensions from a two-dimensional surface.

Figure 10.21 was created using the R function `persp` found in the `graphics` package, which is the package for graphics that is commonly loaded when starting R. The corresponding function in the `lattice` package for R is `wireframe`. This has been used to incorporate topographical colors on the

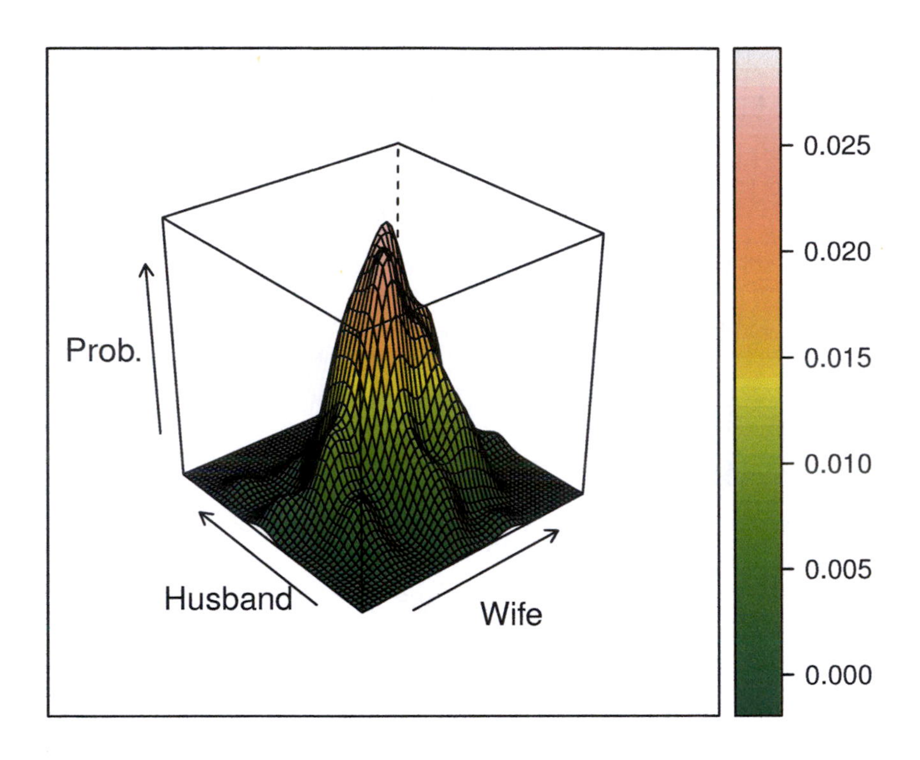

Figure 10.22 *Wireframe plot, with topographic colors, of a two-dimensional kernel density estimate for the joint distribution of husband's height and wife's height for 205 nuclear families collected by Francis Galton in 1884*

wire mesh of the kernel density estimate in Figure 10.22. The R code for generating the color figure is as follows.

```
> mfd<-expand.grid(x=kdemf$x,y=kdemf$y)
> mfd$z<-c(kdemf$z)
> #
> wireframe(z ~ x * y,data=mfd,drape=TRUE,
+ col.regions=terrain.colors(100),
+ xlab="Wife",ylab="Husband",zlab="Prob.")
```

Unfortunately, the list structure `kdemf` for the two-dimensional kernel density estimate produced by the `MASS` function `kde2d` cannot be used as is by the `lattice` function `wireframe`. The `base` R function `expand.grid` places the horizontal and vertical coordinates in a form that can then be used when calling `wireframe`. The argument `drape=TRUE` in the call to `wireframe` causes the colors given by `col.regions=terrain.colors(100)` to be literally draped over the wireframe.

For both wireframe figures, there is a loss of information in a static image because some of the features can be obscured behind "high terrain." An interactive wireframe plot displayed on a computer with a mouse used to determine the flight path as over mountainous terrain is a lot more revealing.

The use of a static wireframe plot in conjunction with a contour plot is worthy of consideration if there is a feature in the two-dimensional kernel density estimate that is important enough to be highlighted.

Figure 10.23 is another wireframe plot of a two-dimensional kernel density estimate, but it depicts the joint distribution of length, as measured from the tip of the nose to the notch of the tail (cm), and mass (g) for each of 56 perch caught during a research trawl on Längelmävesi as reported in 1917. There are a couple of differences between Figure 10.22 and 10.23. Figure 10.23 has been draped with heatmap colors instead of the topographical colors of Figure 10.22. There is also a further difference in Figure 10.23: the heatmap colors have been adjusted for improved perception by individuals with anomalous deuteranopic (red-green) color-vision using the function `daltonize` in the `LSD` package (see Appendix A).

A version of Figure 10.22 appeared in heatmap colors as one of a stereographic pair on the cover of the first edition of this book. Found on the cataloguing page in both editions are instructions on how to view a stereographic pair as a single three-dimensional image. To do this, cross one's eyes until four images appear, then relax gradually to allow the images to converge to a set of three. Focus on the center image. It might take a bit of practice and patience, but the center image will appear to be in three dimensions. Figure 10.23, without the color key, is the right image of the stereographic pair on the cover of the second edition. Heatmap colors have been chosen for the cover figures to match the red cover scheme for the series *Texts in Statistical Science*.

From the two-dimensional kernel density estimate in Figure 10.22, there appears to be a simple linear relationship between the heights of husbands and wives. The modeling of height in Galton's data by simple linear models is further examined in Chapter 11. On the other hand, a close examination of the two-dimensional kernel density estimate in Figure 10.23 reveals a curvilinear relationship between length and mass of perch caught on Längelmävesi in addition to two modes. The perch data will be revisited in Chapter 12 during discussion of polynomial regression. As for a biological explanation for the two modes: the major mode in this bimodal distribution corresponds to juvenile fish (prey) and the minor mode corresponds to mature fish (predator).

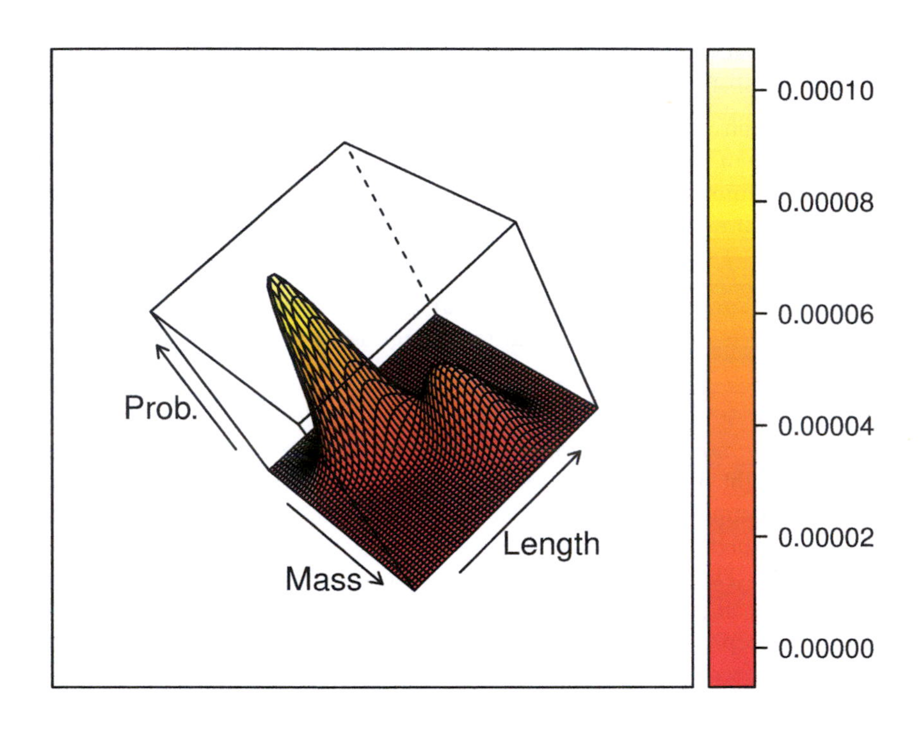

Figure 10.23 *Wireframe plot, with heatmap colors adjusted for improved perception by individuals with anomalous deutanopic (red-green) color-vision, of a two-dimensional kernel density estimate for the joint distribution of length as measured from the tip of the nose to the notch of the tail (cm) and mass (g) for each of 56 perch caught during a research trawl on Längelmävesi as reported in 1917*

10.8 Conclusion

Although scatterplots are a staple of science, the sunflower plot is encountered much less frequently. Neither appear that often in the popular media. The bagplot and levelplot are rare birds indeed.

The cloud plot for depicting a two-dimensional histogram is really only used in a dynamic interactive computer display. Two-dimensional kernel density estimates do make appearances in journals in disciplines with a high degree of mathematical and computational sophistication. The chance that these plots would be sighted in a medical journal, for example, would be slight.

For the R statistical software system, there are no less than three packages that support some form of two-dimensional kernel density estimation. These tend to be connected with textbooks. The `KernSmooth` package is associated with Wand and Jones [131] published in 1995. The function `bkde2D` in this package was evaluated for this book. But the decision was made to use the function `kde2d` in the `MASS` package. The letters `MASS` stand for *Modern Applied Statistics with S*, which was written by Venables and Ripley [128]. The fourth edition of this book was published in 2002.

The R package `np` was created by Hayfield and Racine [62]. This package supports kernel density estimation not just for two continuous variables but a multiple number of continuous or discrete variables. It is the newest of the three packages. It became available from the R package depository in June 2009.

An alternative to fitting bivariate distribution to a pair of continuous variables is fitting a straight line, or other simple curve, through data by the estimation of parameters. This is the topic of the next chapter. The example for the next chapter is the first published case study for linear regression. This is Galton's case study of regressing children's heights on their parents'. It has a more interesting bivariate distribution than that chosen as the case study for this chapter.

But the analysis of the joint distribution of heights of husbands and wives has its place. For one, it is worthwhile to have the example of a situation in which the correlation between two variables is low. Another reason for considering this distribution will be presented at the start of the next chapter.

There is some unfinished business with respect to Karl Pearson's own data on family faculties. In 1896, Pearson [91] thanked Francis Galton for access to the latter's Records of Family Faculties. The first published product-moment estimate for the correlation coefficient was 0.0931 for the correlation between husbands and wives. One might be curious as to why these individuals were not referred to as fathers and mothers in this chapter. The choice was determined by Pearson [92] in 1899 when he undertook a reanalysis of Galton's data.

In the reanalysis, pairs of mothers and fathers were repeated as many times as the number of children they had. While Pearson calculated the sample correlation coefficient of 0.0931 for 205 husbands and wives, the calculation of the sample correlation coefficient for 965 pairs of mothers and fathers was reported by Pearson [92] as being done by Mr. L. Bramley-Moore and assisted by Mr. K. Tressler. The value they found, as reported by Pearson [92], was 0.1783.

This is quite a jump in value, nearly a doubling of the original estimate just by changing the weight function from the unit family to the number of offspring. With the original data available now, thanks to Jim Hanley [59], it is possible to verify this calculation. In so doing, 963 pairs are found in Galton's data and not 965 as determined by Messrs. Bramley-Moore and Tressler. This is a slight difference. The sample correlation coefficient, however, is found to be 0.04818. This is about half the value of 0.09690 for the 205 pairs and considerably less than the value of 0.1783 that Pearson reported.

Pearson [92] concluded the following.

> We have practically *doubled* [the italics are Pearson's] the intensity of assortative mating by weighting the observations with fertility.

Actually, the sample correlation has been reduced by approximately one-half.

Pearson and co-author Alice Lee returned to this topic in 1903. Pearson and Lee [95] reported the results of a similar calculation from Pearson's own collection of family data: the Family Record Series. After 5 years, the Family Record Series consisted of approximately 1100 families with information on height, arm span, and length of left forearm on parent and as many as two sons and two daughters.

Based on 1000 to 1050 cases of husbands and wives, Pearson and Lee [95] reported a product-moment correlation of 0.2804. They declared this estimate to be in close agreement with the value of 0.1783 published by Pearson [92] in 1899. Pearson and Lee [95] wrote: "We could hardly want stronger evidence of the existence of assortative mating in man ... ."

Roberts, Billewicz, and McGregor [99] reported estimates from two communities of the polygynous Mandingo people in Gambia. Their sample correlation coefficients were 0.085 for a husband and each of his wives and 0.096 for a husband and the mean height of his wives, consistent with Galton's data but not Pearson's. A lingering question is whether there was a calculation error in the value of 0.2804 by Pearson and Lee with the Family Record Series collected by Pearson.

10.9 Exercises

1. Andrews and Herzberg [4] produced a book published in 1985 that consists only of multiple data sets and their stories. The data sets can be downloaded from the StatLib archive at http://lib.stat.cmu.edu/datasets/Andrews. For this exercise, consider the urine data in their Table 44.1. Consider only the controls without calcium oxalate crystals in their urine.
   (a) Produce a scatterplot for the calcium concentration and the urea concentration. Both concentrations are measured in millimoles per liter.
   (b) Discuss whether there is a pattern in your answer to part (a).
   (c) Is it necessary to produce a sunflower plot for these two variables? Discuss.
2. Refer to Exercise 1. Consider only the controls without calcium oxalate crystals in their urine.

(a) Produce a bagplot for the calcium concentration and the urea concentration. Both concentrations are measured in millimoles per liter.

(b) Are there any outliers in the bagplot?

(c) What proportion of the data points appear to be in the loop in your answer to part (a)?

(d) Comment on the overall pattern in the bagplot.

(e) Does the spread in urea concentration appear to conditionally depend on calcium concentration? Describe the pattern you see.

3. Refer to Exercise 1. Consider only the controls without calcium oxalate crystals in their urine.

(a) Produce a blotched bagplot for the calcium concentration and the urea concentration. Both concentrations are measured in millimoles per liter.

(b) What can be said about the spread of the estimate of depth median in your answer to part (a) ?

(c) Comment on any similarity or dissimilarity among the blotch, the bag, and the outer polygon for the loop in the blotched bagplot. What can be inferred from this?

4. The year 1888 saw the publication of an article entitled "Co-relations and their Measurement, chiefly from Anthropometric Data." by Francis Galton [52]. In this article, Galton defines the term *correlation* for the first time.

> Two variable organs are said to be co-related when the variation of the one is accompanied on the average by more or less variation of the other, and in the same direction. Thus the length of the arm is said to be co-related with that of the leg, because a person with a long arm has usually a long leg, and conversely.

The modern definition of correlation accepts a negative correlation and not just a positive. Galton collected data on 350 males of 21 years of age and older at his anthropometric laboratory at South Kensington. Presumably, the original data does still exist. But at time of printing all that is available is the summary table from Galton [52] that has been reproduced in Table 10.2. The quantities measured are the height (stature) and the distance from the tip of the middle finger to the elbow of the left arm (left cubit).

(a) Create a sunflower plot to depict the two-dimensional histogram in Table 10.2.

(b) Create a levelplot to depict the two-dimensional histogram in Table 10.2.

(c) Which of the two plots produced for parts (a) and (b) do you prefer? Discuss.

5. Refer to the previous exercise and the data in Table 10.2.

(a) Explore a dynamic cloud plot for the data in Table 10.2.

Stature in inches	Length of left cubit in inches, 348 adult males								Total
	Under 16.5	16.5 and under 17.0	17.0 and under 17.5	17.5 and under 18.0	18.0 and under 18.5	18.5 and under 19.0	19.0 and under 19.5	19.5 and above	
71 & above				1	3	4	15	7	30
70				1	5	13	11		30
69		1	1	2	25	15	6		50
68		1	3	7	14	7	4	2	48
67		1	7	15	28	8	2		61
66		1	7	18	15	6			48
65		4	10	12	8	2			36
64		5	11	2	3				21
Below 64	9	12	10	3	1				34
Total	9	25	49	61	102	55	38	9	348

Table 10.2 *Bivariate distribution of left cubit and stature in males (from Table II from Francis Galton, "Co-relations and their Measurements, chiefly from Anthropomorphic Data."* Proceedings of the Royal Society, *Vol. 45, pp. 135–145, 1888)*

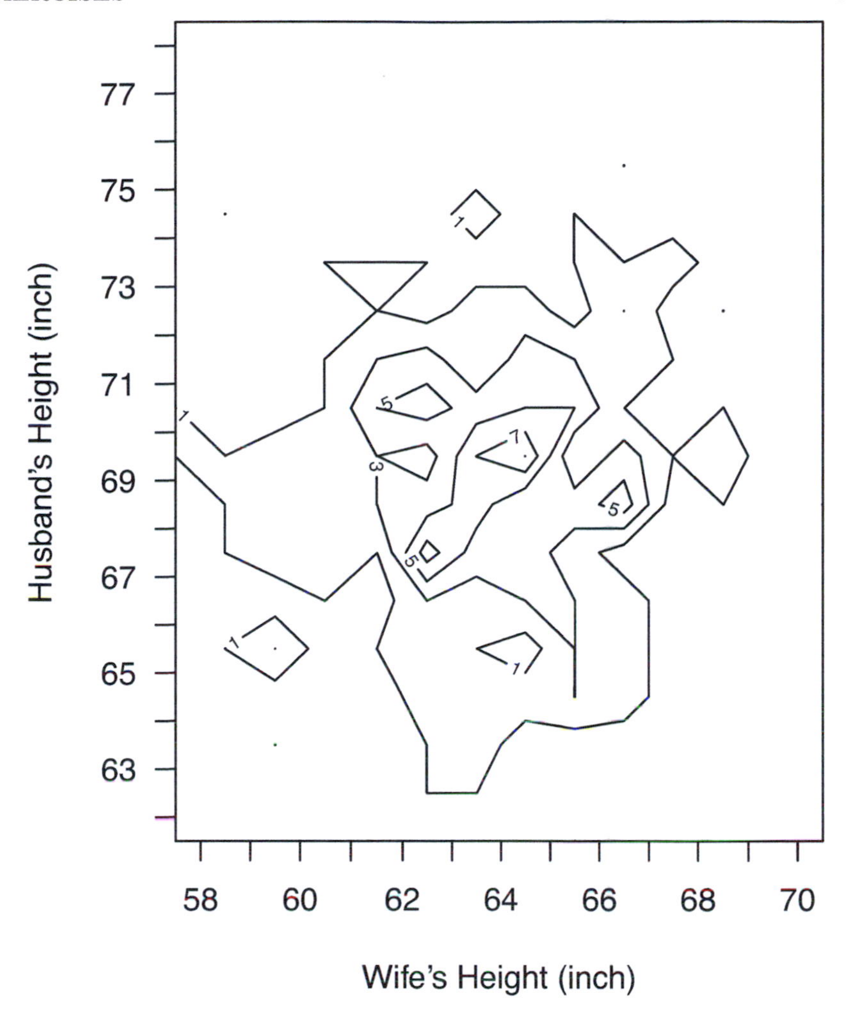

Figure 10.24 *Contour plot of two-dimensional histogram of husbands' heights and wives' heights for 205 families in Galton's data collected in 1884*

(b) Does the distribution of stature and cubit in Galton's data for 348 young men appear to be bivariate normal? Discuss.

6. In the section on two-dimensional histograms, an example of a contour plot was neither given nor discussed. This was not an oversight. Look at the contour plot in Figure 10.24 of the two-dimensional histogram for heights of husbands and wives from Galton's Record of Family Faculties. Critique Figure 10.24. Comment on the decision not to discuss the use of contour plots for depicting two-dimensional histograms in this chapter.

7. Among the levelplot, the contour plot, and the wireframe, which is best for depicting a two-dimensional kernel density estimate for a bivariate distribution of two continuous random variables? Justify your answer. Can it ever be worthwhile to use more than one of these plots? Discuss.

8. Refer to Exercise 1. Consider only the controls without calcium oxalate crystals in their urine.
   (a) Plot a two-dimensional kernel density estimate for calcium and urea concentrations.
   (b) Which plot did you select for part (a): levelplot; contour plot; or wireframe plot? Justify your choice.
   (c) Does the density estimate in your plot for part (a) have one or more modes? Discuss.

9. Consider the 56 perch caught around 1917 from Längelmävesi, a Finnish lake near Tampere, during a research trawl. Download the complete data from this book's website. Subset on the species variable to create a data set with the perch only. The two variables of interest are length from tip of the nose to the notch of the tail and width. Width is not given in the data set but must be calculated from width as a percentage of length as measured from the nose to the end of the tail.
   (a) Create a sunflower plot for the length and width of perch.
   (b) Create a bagplot plot for the length and width of perch.
   (c) Write a short report characterizing the joint distribution of length and width for perch using the plots produced for parts (a) and (b).

10. Refer to Exercise 9.
    (a) Create a levelplot for the joint distribution of width and length using a rectangular grid. Overlay a normal ellipse at the 95% level of confidence.
    (b) Create a levelplot for the joint distribution of width and length using a hexagonal grid. Overlay a normal ellipse at the 95% level of confidence.
    (c) Compare your answers to parts (a) and (b). Which is better: a rectangular grid or a hexagonal grid? Discuss.

11. Refer to Exercise 9.
    (a) Create a contour plot depicting a two-dimensional nonparametric kernel density estimate for the joint distribution of width and length.
    (b) Create a wireframe plot depicting a two-dimensional nonparametric kernel density estimate for the joint distribution of width and length.
    (c) Write a short report characterizing the joint distribution of width and length for perch using the plots produced for parts (a) and (b).

12. Some countries, and jurisdictions within countries, use weight to determine catch-and-release regulations for commercial and sports fisheries. Others use length. Pike (*Esox lucius*) is ubiquitous in freshwater lakes in the Northern Hemisphere. Using the methods of this chapter, study the bivariate distribution of weight and length defined from the tip of the nose to the

notch in the tail in the pike caught around 1917 from Längelmävesi, a lake near Tampere, Finland. Note that the research trawl only captured 17 pike. Should this be a concern?

Part V

Statistical Models for Two or More Variables

Chapter 11

Simple Linear Regression: Graphical Displays

11.1 Introduction

The first case study presented in this chapter will be the first case study ever published for the simple linear regression model. It will be Galton's [51] regression of the heights of adult children on their parents, which was published in 1886. Before working through this case study, there is some unfinished business from the previous chapter on the topic of heights of mothers and fathers.

With complete data from Galton's Records of Family Faculties for the heights of mothers and fathers, there are 205 pairs. Galton [51] gives no reason not to treat the families as a simple random sample. The caveat might be added by some that the sample is only representative of the upper middle class of Great Britain and Ireland during the Victorian era. With respect to statistical modeling, there is no reason to select response and explanatory variables from husband's height or wife's height. So this data was a good choice as a case study for the discussion in the previous chapter concerning the depiction of the distribution of two continuous variables.

In the case of regressing children's heights on their parents' heights, it is necessary to deal with the perception that men tend to be taller than women. Galton's [51] analysis proceeded by reducing the heights of the father and mother in each family to one summary statistic. In the previous chapter, measures of location for the heights of fathers and mothers were briefly discussed. The sample depth means and their blotch were discussed when the bagplot was presented. The sample mean for height was reported as 69.316 inches for husbands and 64.002 inches for wives. There was no statistical inference with respect to these measures of location. The discussion was limited to the context of exploratory data analysis.

A good starting point for statistical inference was the consideration in the previous chapter as to whether there was a correlation between the husband's and his wife's height. With a p-value of 0.1669 for the test of no correlation against the two-sided alternative, it can be concluded that there is no correlation. The genetic implication is that human mating is not assortative with respect to height. Pearson's product-moment correlation for the data was $r = 0.09690$. If the population correlation coefficient is, on the other hand, assumed to be nonzero, then the amount of variation in one parent's height due to a linear relationship with the other is $100 \times r^2 \approx 0.9\%$. This is quite low. It is virtually negligible. Population correlation coefficients, although quite small, can be quite real. If the data follow a normal distribution, then treating the two variables as stochastically independent when they are not can impact statistical inference concerning comparisons of means. A safer approach to using a Student's t-test assuming independence is to conduct a matched-pairs test.

For a matched-pairs t-test there must be a logical pairing of variables. This is the situation for Galton's data with married couples and their offspring. It is best for the matched-pairs t-test if the difference between the two variables is nearly normally distributed. It is not necessary that each variable is normally distributed or that the pair of variables is bivariate normal. Figure 11.1 presents a normal quantile-quantile plot together with an outlier boxplot. The difference in height is along the horizontal axis so that the outlier boxplot is depicted in a horizontal orientation. A quantile-quantile plot typically depicts the standard normal quantiles along the horizontal axis but Figure 11.1 does not.

A straight line corresponding to a normal distribution has been added as a visual reference. The line is drawn through the upper and lower quartiles by the R function `qqline`. Displaying a normal quantile-quantile plot together with an outlier boxplot is particularly efficient when examining data visually for the first time. If the data in the normal quantile-quantile plot do not appear to fall close to a straight line, then the outlier boxplot is available for ready reference.

For the difference formed by subtracting a wife's height from her husband's height, the data appear to be normally distributed. Two observations, one at each extreme end of the data, are ruled outliers according to the outlier boxplot in Figure 11.1. These correspond to one couple for which the wife is 4 inches taller than her husband and another couple for which the husband is 16.5 inches taller than his wife.

For the family with the mother 4 inches taller than her husband, her height was 5 feet 6 inches while his was 5 feet 2 inches. They had a son with height 5 feet 4 inches and two daughters with a height of 5 feet 2 inches for the elder and 5 feet 1 inch for the younger. This family (#5 in the Records of Family Faculties) appears otherwise unremarkable.

For the family in which the husband is 16.5 inches taller than the wife, his height is 6 feet 3 inches, hers is 4 feet 10.5 inches. They had 6 children. For the sons, from eldest to youngest, the heights are 6 feet, 5 feet 9 inches, and

Figure 11.1 *Normal quantile-quantile plot and outlier boxplot for the difference of husband's height less his wife's height for 205 families in Galton's Records of Family Faculties*

5 feet 8 inches. No wonder Galton observed the heights of sons *regressing* away from a tall father toward the mean height of men, or *mediocrity* as he phrased it. For daughters, the height of the eldest is 5 feet 6.5 inches, and the two younger are both 5 feet 2.5 inches. This family (#203 in the Records of Family Faculties) also appears otherwise unremarkable.

There are no reasons to justify excluding families numbered 5 and 203 from further statistical analysis, so they shall be retained despite height differences being flagged as outliers. The matched-pairs t-test using the difference in husband's and wife's heights allows testing the null hypothesis $H_O : \mu_F - \mu_M =$

0 versus the alternative hypothesis that $H_1 : \mu_F - \mu_M \neq 0$. The p-value for this test is less than 2.2×10^{-16} based on 204 degrees of freedom. The corresponding 95% confidence interval for the difference of the mean height of husbands minus the mean height of wives is $[4.85 \text{ inches}, 5.78 \text{ inches}]$. There is sufficient evidence to conclude that men are taller than women.

Of course, Student's t-test was not available to Galton in 1885. (This did not appear until 1908 when the sampling distribution for the t-statistic was determined by William Sealy Gosset to be a Pearson Type III distribution through simulation. Gosset published the result under the pseudonym Student [117]. The theoretical derivation had to wait until Ronald Aylmer Fisher [39] in 1925.) Galton [50] did conclude there was a difference in height between men and women. He also knew that he somehow had to adjust for this difference when attempting to study the heritability of stature.

The point estimate of the mean difference in heights of married couples in Galton's data is 5.31 inches. One possible approach to adjusting for the gender difference in height would be to add the mean difference to the height of each mother and daughter. This is not the approach that Galton [50, 51, 54] chose. Instead Galton applied a multiplicative factor of 1.08 to the heights of the females. Galton [50] wrote the following.

> In every case I transmuted the female statures to their corresponding male equivalents and used them in their transmuted form, so that no objection grounded on the sexual differences of stature need to be raise when I speak of averages. The factor I used was 1.08, which is equivalent to adding a little less than one-twelfth to each female height. It differs a very little from factors employed by other anthropologists, who, moreover differ a trifle between [sic] themselves; anyhow it suits my data better than 1.07 or 1.09. The final result is not of a kind to be affected by these minute details, for it happened that, owing to a mistaken direction, the computer to whom I first entrusted the figures used a somewhat different factor, yet the result came closely the same.

Not only did Galton estimate the factor to be 1.08 but he tested its robustness on account of mis-programming his computer.

But how did Galton arrive at the factor of 1.08? The hint in the quoted text above is the word "averages." Taking the mean height of fathers and dividing by the mean height of mothers, one obtains 1.0830. Upon rounding to two decimal digits, this is Galton's value of 1.08. Pearson's [91] estimate for the same data was 1.082. This value is listed in Table V on page 271 of Pearson [91]. In the footnote to Table V, and likely for Table IV immediately above, Pearson [91] reported that "Mr Galton excluded from his calculations the larger families, but it seems to me that large families form an essential feature of the community."

If Galton did exclude the larger families, it had no effect on the estimate of

1.08 for the ratio. By standard methods, the large-sample standard error of the ratio estimate is 0.0038. This calculation could have been done in Galton's and Pearson's time but was not. The large-sample 95% confidence interval for the ratio is [1.0756, 1.0905]. Pearson's [91] estimate for the same data of 1.082 falls within this interval. We also have Galton's [50] own comments regarding the robustness of this estimate.

Jim Hanley [59] in 2004, in the first re-analysis of Galton's original data since Pearson and Lee [95] in 1903, noted that he and students usually agree that the proportional scaling of Galton "is a more elegant and biologically appropriate adjustment than the additive one." However, it needs to be noted that scaling with height in animals is not quite so simple. An eight percent increase in height requires much more than an eight percent increase in muscle mass with commensurate increases in blood vessels, lungs, and heart to supply the added muscle mass. But as a first approximation, a constant scaling factor is not bad, and for historical consistency, Galton's scaling by a multiplicative factor of 1.08 will be used.

11.2 Learning Outcomes

When you complete this chapter, you will be able to do the following.

- Know the definition of the simple linear regression model in its most general form.
- Be able to execute a scatterplot and a sunflower plot for two continuous variables with the `graphics` package.
- Be able to fit a simple linear regression model using the linear model function `lm` in the `stats` package in R and extract the residuals.
- Be able to execute scatterplots of residuals versus response, explanatory, and other variables.
- Be able to produce a normal quantile-quantile plot of residuals.
- Know the definition of semistandardized residuals and be able to do a normal quantile-quantile plot of semistandardized residuals.
- Know the definition of standardized residuals and be able to do a normal quantile-quantile plot of standardized residuals.
- Be able to do a scatterplot of semistandardized or standardized residuals versus another variable.
- Know the definition of studentized residuals and be able to do a Student's t quantile-quantile plot of studentized residuals as well as a scatterplot of studentized residuals versus another variable.
- Know the definition of leverage, and be able to execute and interpret a scatterplot of leverage with another variable to determine whether an observation has high leverage.

- Know the definitions of the difference in fitted values and differences in fitted intercept and slope estimates (the betas), be able to execute and interpret a scatterplot of these with another variable to determine whether an observation has high leverage.
- Be able to use the `car` package plotting function `influencePlot` to find influential observations being fitted by a simple linear regression model.

With respect to executing scatterplots, boxplots, and quantile-quantile plots in the process of doing regression diagnostics, the choice of either the `graphics` package or `ggplot2` rests with the R user.

11.3 The Simple Linear Regression Model

11.3.1 Definition

Over a period of decades, the *simple linear regression model* has been defined to be given by

$$y_i = \beta_0 + \beta_1 x_i + \epsilon_i \tag{11.1}$$

where y_i is the response variable and x_i is the explanatory variable for $i = 1, 2, \ldots, n$. The random errors $\{\epsilon_i\}$ are assumed to be independently normally distributed with mean 0 and variance σ^2. The parameters β_0 and β_1 are the *regression coefficients.* All three parameters σ^2, β_0, β_1 are typically unknown and are estimated from a simple random sample of ordered pairs $\{(x_i, y_i)\}$.

11.3.2 The Scatterplot

The first step in analysis for a simple linear regression model is to plot the data. In the case of simple linear regression analysis this means using the familiar scatterplot of the ordered pairs $\{(x_i, y_i)\}$.

Table 11.1 reports cell counts in Table I on page 248 of Galton [51] as published in 1886. The term *mid-parent* was defined by Galton to be the average height of a father and mother after multiplying the mother's height by 1.08. Galton [50, 51] referred to the process of adjusting the heights of females in this way as "transmuting." This process has been applied to the height of each daughter, just as Galton had done.

Table 11.1 has been plotted on a Cartesian coordinate plane in Figure 11.2 with a 95% confidence ellipse added corresponding to a bivariate normal distribution. Figure 11.2 is very much like the first scatterplot for simple linear regression that appears in Plate X of Galton [51]. There are differences between Figure 11.2 and Galton's version.

The cells used in Figure 11.2 are not those of Galton's Table I but instead

Heights of Mid-Parents (inch)	Heights of Adult Offspring (inch)													
	Below	**62.2**	**63.2**	**64.2**	**65.2**	**66.2**	**67.2**	**68.2**	**69.2**	**70.2**	**71.2**	**72.2**	**73.2**	**Above**
Above												1	3	
72.5								1	2	1	2	7	2	4
71.5					1	3	4	3	5	10	4	9	2	2
70.5	1		1		1	1	3	12	18	14	7	4	3	3
69.5			1	16	4	17	27	20	33	25	20	11	4	5
68.5	1		7	11	16	25	31	34	48	21	18	4	3	
67.5		3	5	14	15	36	38	28	38	19	11	4		
66.5		3	3	5	2	17	17	14	13	4				
65.5	1		9	5	7	11	11	7	7	5	2	1		
64.5	1	1	4	4	1	5	5		2					
Below	1		2	4	1	2	2	1	1					

Table 11.1 *Number of adult children of various statures born of 205 mid-parents of various statures (all female heights have been multiplied by 1.08)*

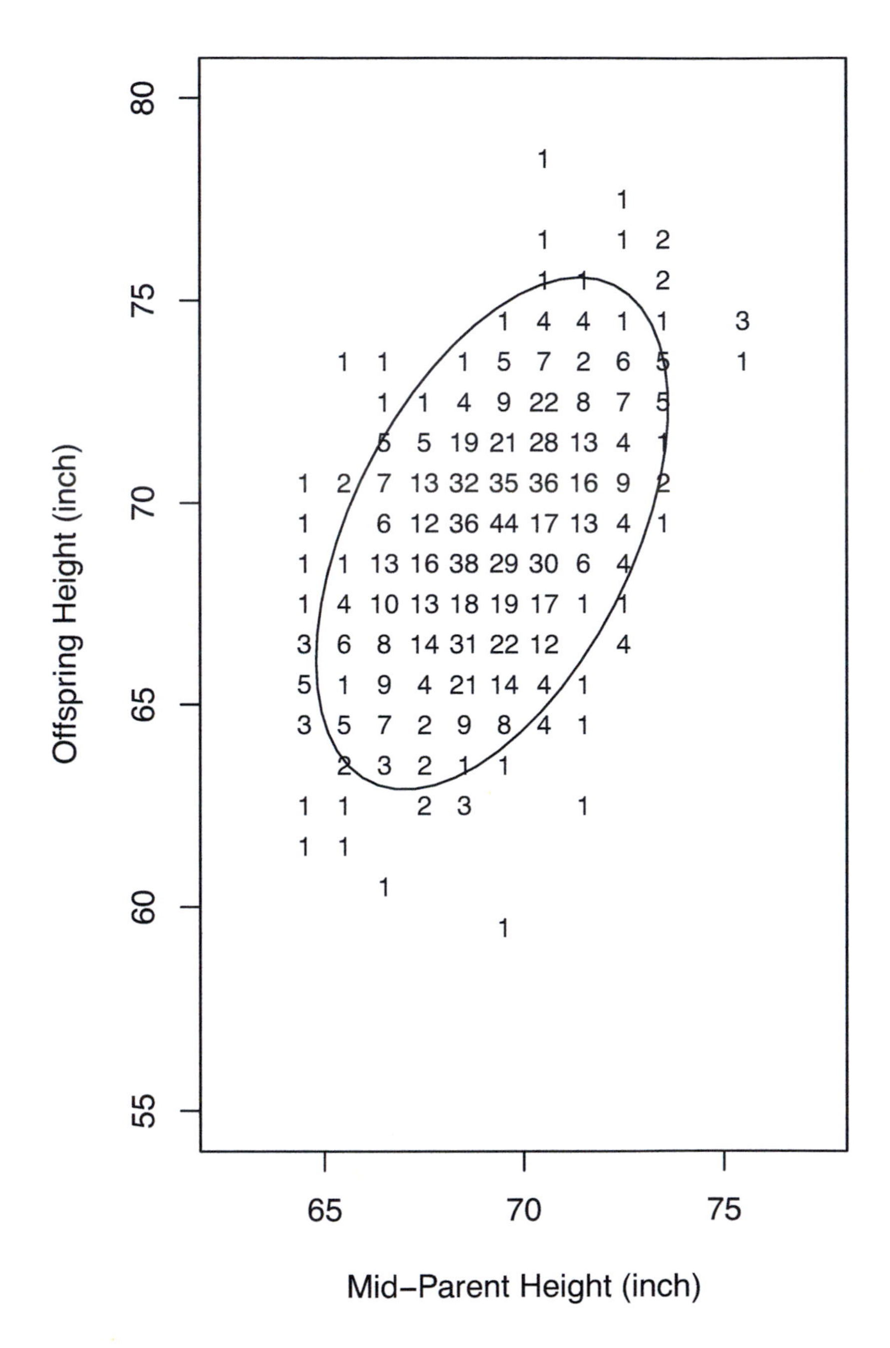

Figure 11.2 *Scatterplot of frequency for heights of mid-parents and offspring with a 0.95 probability ellipse for a bivariate normal distribution*

are a unit inch in length for both axes and inclusive of the upper end-point. Galton's original data were used to determine the frequency for each cell and these are plotted at the midpoint of the unit square.

Comparing Table 11.1 with Figure 11.2, discrepancies in frequencies are apparent. Galton [51] reports 928 pairs in his Table I, as reproduced in Table 11.1. But the actual count is 934. Note also in Table 11.1 that unit inch intervals for the offspring are centered at values like 62.2 inch, 63.2 inch, and so on.

The intervals for offspring in Figure 11.2 are centered at the half inch. Galton also chose the height of offspring for the horizontal axis. Figure 11.2 adopts the convention of plotting the response variable along the vertical axis. So the axes of Figure 11.2 are interchanged with respect to Galton's [51] Plate X.

Different approaches for drawing the bivariate-normal ellipses were used for Figure 11.2 and Galton's Plate X. The ellipse in Figure 11.2 is centered at the sample mean mid-parent and offspring heights with the orientation of the ellipse and its extent determined by the sample variances and covariance following Pearson's [91] approach of 1896. Pearson's denominators for the sample variances and covariance have the sample size n in the denominator. The ellipse in Figure 11.2 used $n - 1$ for the unbiased variants.

11.3.3 The Sunflower Plot

In Chapter 10, a better plot was found for depicting bivariate densities. This was the sunflower plot. Figure 11.3 is the *sunflower plot* counterpart to Figure 11.2. Note that when the petal count reaches numbers above 20, or so, they tend to become indistinguishable and the result is a black ink smudge for the sunflower. So there are limits to what can be depicted with the sunflower plot.

No doubt that Galton would have given his eyeteeth to have seen a scatterplot of all 934 pairs of mid-parent and offspring heights. This was prohibitively expensive in his day. He would have had to wait until 2004 when Hanley's [59] article was published in *The American Statistician.* A version of Hanley's [59] scatterplot is given in Figure 11.4, which is, in fact, a sunflower plot. The R code to produce this plot from the R list `Galton` transcribed from Galton's own notebook is as follows.

```
MidParent<-NULL
Offspring<-NULL

n<-length(Galton$Father)

for (i in 1:n) {
OS<-c(60+c(Galton$S1[i],Galton$S2[i],
Galton$S3[i],Galton$S4[i],Galton$S5[i],
```

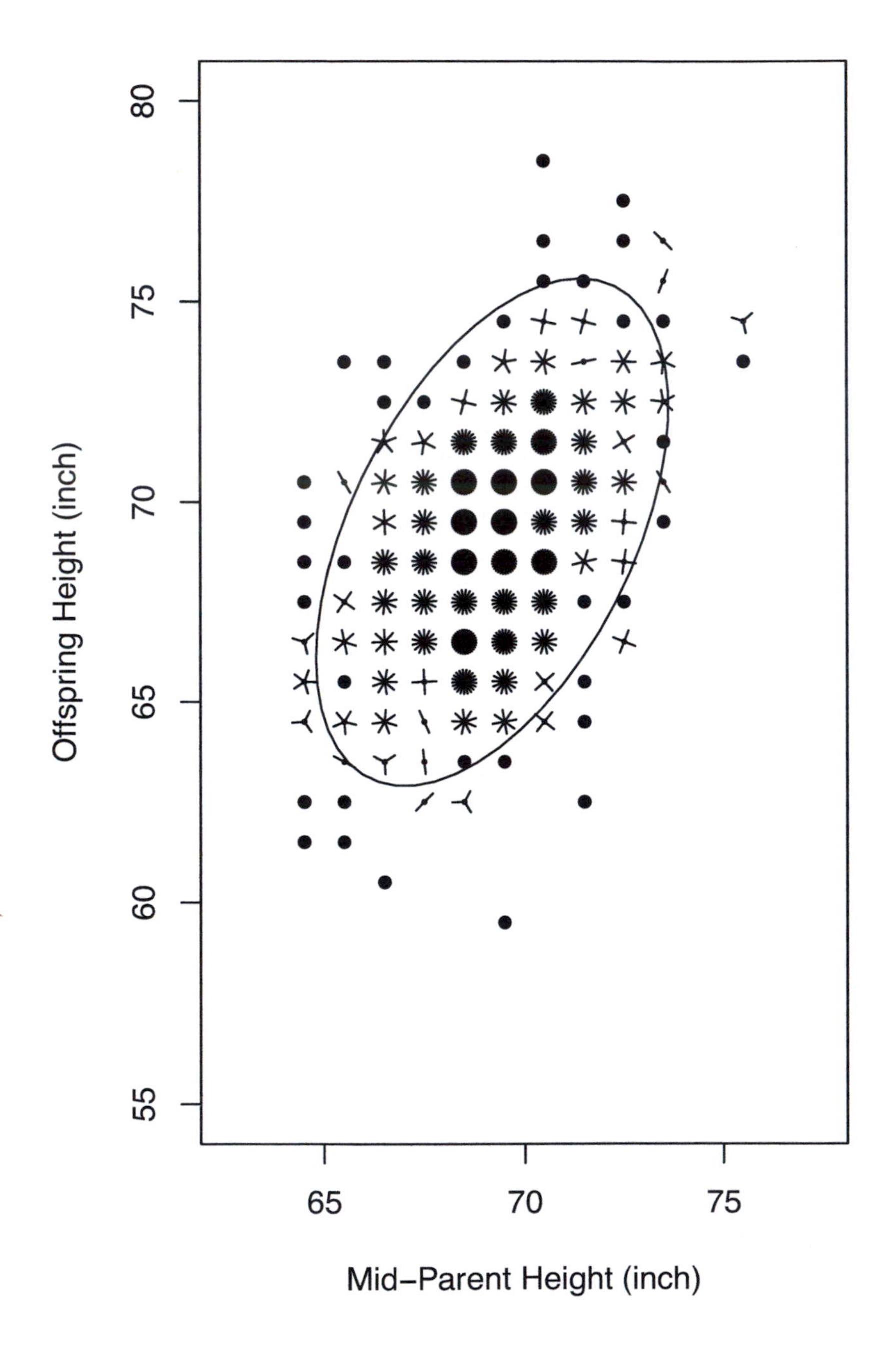

Figure 11.3 *Sunflower plot of frequency for heights of mid-parents and offspring with a 0.95 probability ellipse for a bivariate normal distribution*

```
Galton$S6[i],Galton$S7[i],Galton$S8[i],
Galton$S9[i],Galton$S10[i]),
1.08*(60+c(Galton$D1[i],Galton$D2[i],
Galton$D3[i],Galton$D4[i],Galton$D5[i],
Galton$D6[i],Galton$D7[i],Galton$D8[i],
Galton$D9[i],Galton$D10[i])))
for (j in 1:20) {
if (!is.na(OS[j])) {
MidParent<-c(MidParent,((60+Galton$Father[i])+
1.08*(60+Galton$Mother[i]))/2)
Offspring<-c(Offspring,OS[j])
}
}
}

graphics.off()
windows(width=4.5,height=6.8,pointsize=12)
par(fin=c(4.45,6.7),pin=c(4.45,6.75),
mai=c(0.85,0.85,0.25,0.25),xaxs="r",yaxs="r")

set.seed(345)

sunflowerplot(MidParent,Offspring,
xlab="Mid-Parent Height (inch)",
ylab="Offspring Height (inch)",xlim=c(62.5,77.5),
ylim=c(55,80),pch=21,seg.col="black",cex=0.5,cex.fact=1,
seg.lwd=1.,size=1/16,rotate=TRUE)

dataEllipse(MidParent,Offspring,add=TRUE,
center.cex=0.,col="black",lwd=1.0,
levels=0.95,plot.points=FALSE)
```

The R vector variables `MidParent` and `Offspring` are initialized as empty by assigning the value `NULL` to each. This is done so that the R function `c` can be used to add the first observation of each when the nested structure consisting of two `for` loops and one `if` statement are executed to produce offspring height matched with the mid-parent height from their family.

Note in the preceding code that 60 inches must be added to Galton's original heights, which were recorded as offset from 60 inches. Also, the female heights have been adjusted to bring them into equivalence with male heights by using Galton's multiplicative factor of 1.08. The R script ends with calls to the `graphics` function `sunflowerplot` and the `car` function `dataEllipse`.

Note that the petals in the sunflower plot are randomly rotated so there is a previous call to the function `set.seed` for the purpose of setting the pseudo-random-number seed to 345. This value was found by experimentation to improve clarity among nearby pairs of mid-parent and offspring heights with multiple observations.

Hanley [59] chose the horizontal axis for the mid-parents' heights as did Galton [51]. Also, Hanley's [59] scatterplot did not use sunflowers to denote repeated observations among the 934 pairs, nor did it include a 95% probability ellipse for a bivariate normal distribution. Small open circles in Figure 11.4 have been chosen to lessen, as much as possible, obscuration of nearby points.

There are actually very few petals in the sunflower plot of Figure 11.4. There is reasonably good separation among the 934 points plotted because the heights of family members were measured by Galton to the nearest one-half inch. Consequently, Hanley's [59] scatterplot with its black dots shows some smudging of the data points but not enough to be a concern.

There is another much more serious concern in Figure 11.4. The mid-parent height is an average of heights of each family's mother and father. It is hard to see, but data points for families with more than one child are arranged in a vertical line. This is because the ordered pairs for each family have the same horizontal coordinate. The problem is thus that the data portrayed in Figure 11.4 are not representative of a simple random sample yet this is an assumption of the simple linear regression model.

The matter is compounded by the genetic fact that each child within a family is correlated with their siblings. It is, therefore, not appropriate to apply the simple linear regression model of equation (11.1) to the data depicted in the sunflower plot of Figure 11.4.

There are two approaches to handling this problem. One approach would be to incorporate the pedigree structure for the families in a statistical model different from the simple linear regression model but taking into account that the families were obtained from a simple random sample.

The other approach is to take a page from Galton's book, which quite literally would be *Natural Inheritance*, published in 1889. (This book by Galton [53] summarizes his earlier peer-reviewed genetics articles, including those cited in this and the previous chapter.) Galton dealt with the problem of two parents by averaging their observations and then defined the result to be the mid-parent value. The *mid-offspring* value is similarly defined to be the average of the available heights for children in each family. The resulting 205 ordered pairs form a simple random sample for height from Galton's Records of Family Faculties from 1884 and are depicted in the scatterplot of Figure 11.5. Note that it is not necessary to use a sunflower plot when the plotting the mid-offspring heights versus the mid-parent heights.

It is not conventional to add a probability ellipse from a bivariate normal distribution to a scatterplot for which a simple linear model is contemplated. So this is not done in Figure 11.5. Note that the simple linear regression model in equation (11.1) has only a single random error term. But it is conventional

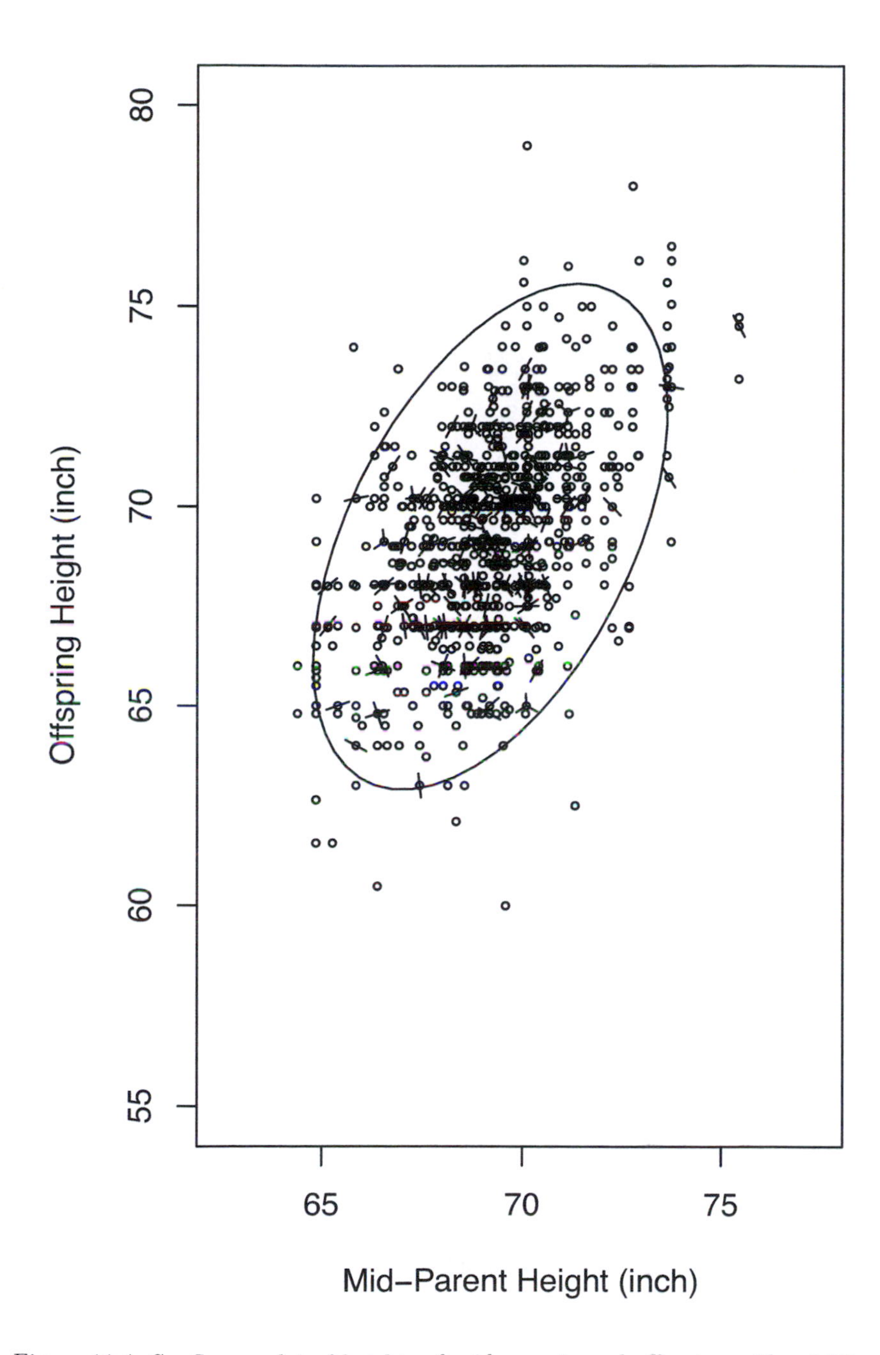

Figure 11.4 *Sunflower plot of heights of mid-parents and offspring with a 0.95 probability ellipse for a bivariate normal distribution*

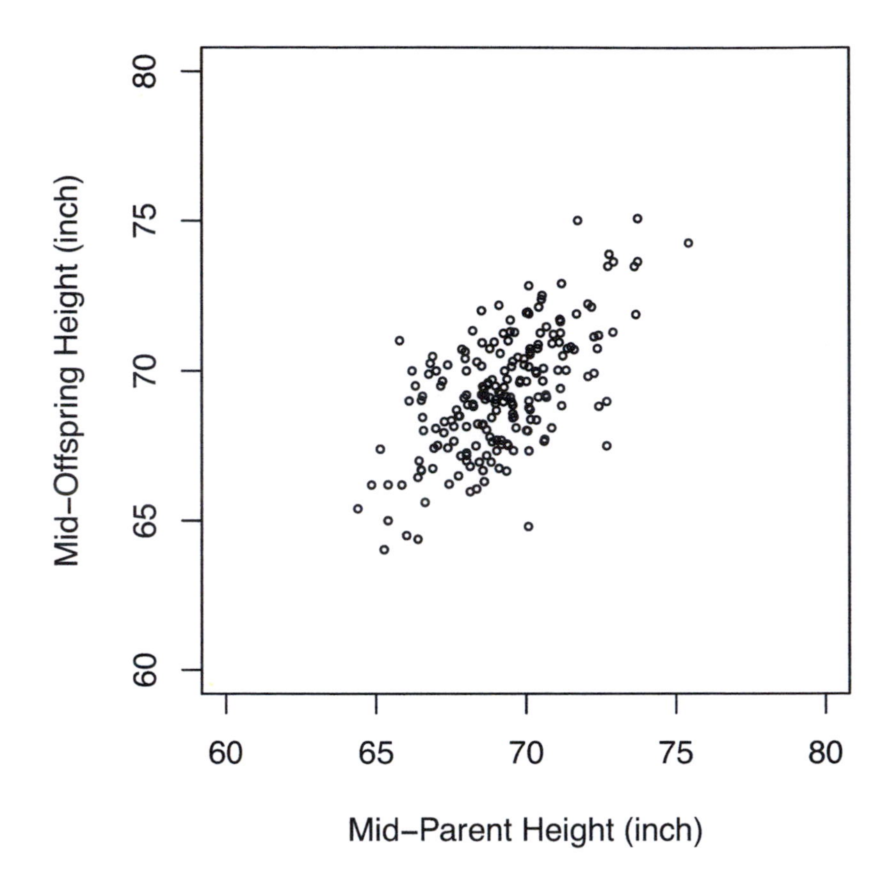

Figure 11.5 *Scatterplot of heights of mid-parents and mid-offspring*

to add the line representing the linear model. This is done in Figure 11.6, which is produced, in part, by the following R script.

```
n<-length(Galton$Father)

MidParent<-rep(0,n)
MidOffspring<-MidParent

for (i in 1:n) {
MidParent[i]<-mean(c(60+Galton$Father[i],
1.08*(60+Galton$Mother[i])),
na.rm=TRUE)
MidOffspring[i]<-mean(c(60+c(Galton$S1[i],Galton$S2[i],
Galton$S3[i],Galton$S4[i],
Galton$S5[i],Galton$S6[i],Galton$S7[i],
```

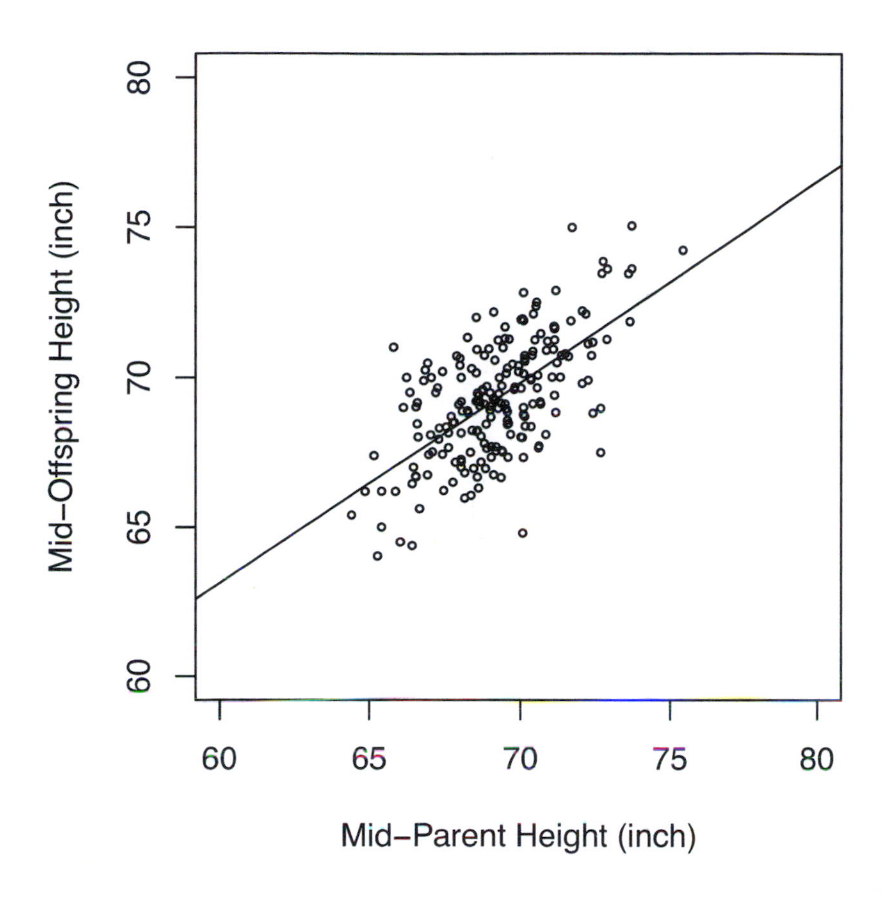

Figure 11.6 *Scatterplot of heights of mid-parents and mid-offspring with least-squares regression line*

```
Galton$S8[i],Galton$S9[i],Galton$S10[i]),
1.08*(60+c(Galton$D1[i],Galton$D2[i],
Galton$D3[i],Galton$D4[i],Galton$D5[i],
Galton$D6[i],Galton$D7[i],Galton$D8[i],
Galton$D9[i]))),na.rm=TRUE)
}

molm<-lm(MidOffspring ~ MidParent)

graphics.off()
windows(width=4.5,height=4.5,pointsize=12)
par(fin=c(4.45,4.45),pin=c(4.45,4.45),
mai=c(0.85,0.85,0.25,0.25),xaxs="r",yaxs="r")
```

```
plot(MidParent,MidOffspring,
xlab="Mid-Parent Height (inch)",
ylab="Mid-Offspring Height (inch)",
xlim=c(60,80),ylim=c(60,80),pch=21,cex=0.5)

abline(coef(molm))
```

The preceding R script starts with vector variables `MidParent` and `MidOffspring` being initialized to zero for all 205 of Galton's families. The subsequent `for` loop then calculates the mid-parent and mid-offspring height with the R function `mean`. The argument `na.rm=TRUE` is passed in each call of `mean` in order to eliminate any missing values in the calculation of the averages.

The function call to `lm` fits a simple linear model for mid-offspring height as a function of mid-parent height. The result is stored in the R list `molm`. Because of multiplicities for pairs of mid-parent and mid-offspring heights, the R function `sunflowerplot` is used to plot Galton's original data. The R function `abline` plots the least squares in Figure 11.6 after the regression coefficients stored in the list `molm` are extracted by the R function `coef`.

Note that the least-squares regression line in Figure 11.6 is banked at an angle of approximately 34° to the horizontal. Plate X in Galton and Dickson [51] in essence did the same. Galton and Dickson [51] determined their estimate of the regression coefficient by comparing the tangent of this angle to the tangent of 45°. That is, $\tan(34°) : \tan(45°) \approx 2 : 3$.

Banking to 45° was studied a century later in the measured-response experiments of Cleveland, McGill, and McGill [24] in 1988. They showed that estimation of the rate of change is most accurate when orientations are centered on 45°. Their work rested on the premise of slope estimation as a basis for pattern recognition. Later theoretical considerations by Cleveland [22] suggested this was not the case. But the empirical results of Cleveland, McGill, and McGill [24] remain valid and a basis for the recommendation of banking to 45° given in 1993 by Cleveland [23] in his book *Visualizing Data.*

Implementing the 45° *banking principle* in practice can be somewhat tedious because it involves manipulating the *aspect ratio* between the height of the vertical scale and the width of the horizontal scale. This has not been done in Figure 11.6, which sets the ratio between the vertical and horizontal scales to 1 inch : 1 inch as done by Galton [51].

The R function `lm` was used to fit the simple linear regression model to the mid-parent and mid-offspring height pairs according to Pearson's [91] principle of least squares. That is, the regression coefficients have been determined to give the line that minimizes the sum of the square distances between the points and the line of best fit. The explanatory variable is mid-parent height.

Regression Coefficient	Estimate	Standard Error	Two-Sided p-Value
β_0	22.9025	3.9682	2.91×10^{-8}
β_1	0.6703	0.0573	$< 2 \times 10^{-16}$

Table 11.2 *Summary of results for fit of simple linear regression model to mid-offspring height as a function of mid-parent height with the principle of least squares*

The values are denoted $\{x_i\}$ and are plotted along the horizontal axis. The response variable is mid-offspring height. The values are denoted $\{y_i\}$ and are plotted along the horizontal axis. The results of the least-squares fit are summarized in Table 11.2.

An estimate not reported in Table 11.2 is the sample correlation coefficient $r = 0.6345$ with 95% confidence interval $[0.5449, 0.7099]$. The sample correlation coefficient is not of much interest to a geneticist concerned with assessing the heritability of stature in a human population. Francis Galton did not know this in 1886. Karl Pearson did not know this in 1896. The world had to wait until 1918 when Ronald Fisher's [37] landmark paper was published.

Fisher's [37] model for a single quantitative trait affected by one or more loci is the result of an additive genetic effect, a dominant genetic effect, and everything else is due to an environmental effect. The variance of the quantitative trait, the so-called *phenotype*, is the sum of these variances. This is Fisher's [37] *analysis of variance.*

Since Fisher [37], there are two definitions of heritability. Heritability in the *broad sense* is the ratio of the variance of the total genetic effect divided by the phenotypic variance. Heritability in the *narrow sense* is the ratio of the variance of the additive genetic effect divided by the phenotypic variance. The symbol h^2 is used to denote the latter.

Fisher [37] showed that the narrow-sense heritability h^2 is given by one-half of the regression coefficient of a single offspring on a single parent. He also showed that the narrow-sense heritability h^2 is given by one-half of the regression coefficient of offspring on a single parent when a single offspring's value is replaced by the mean of the offspring's values. There are certain conditions that must be met in order to use this result. Firstly, the sample of families must be a simple random sample. This appears to be so for Galton's data.

Secondly, the population must be in Hardy-Weinberg equilibrium with respect to the quantitative trait of interest. This will not be satisfied if the population is undergoing assortative mating with respect to the quantitative trait of interest. It has been previously inferred from Galton's Records of Family Faculties that mating does not appear to be assortative with respect to height (p-value $= 0.1669$). Lastly, the phenotypic variance must be the same in men and women. This is usually accomplished by an F-test for comparing variances of husbands and wives after demonstrating that there is no evidence for assortative mating. This F-test has a p-value of 0.0719 for the 205 families in Galton's data.

What Fisher [37] did not demonstrate in 1918 was that when the mean of the parents is used instead of a single parent, the narrow-sense heritability is given by the regression coefficient itself. See pages 148–150 of Falconer and Mackay [35] for details. Note that the term *mid-offspring* is typically not used by quantitative geneticists, including Falconer and Mackay. The term *offspring* is used when the value of just one offspring is involved, or the mean of more. Geneticists rely on context to sort things out. Based upon the regression analysis for Galton's data as depicted in Figure 11.6, the estimate of narrow-sense heritability for stature in humans is $\hat{h}^2 = \hat{\beta}_1 = 0.6703$.

With a standard error of 0.0573 and Fisher's [40] theory for the t-distribution, published in 1925, a 95% confidence interval for heritability of stature is $[0.5573, 0.7833]$. This confidence interval includes the value of 0.654 for heritability of height reported in 1978 by Roberts, Billewicz, and McGregor [99] with a standard error of 0.052. Falconer and Mackay [35] report this point estimate from a polygynous population in the Gambia on page 162 of their book. Mackay did not have access in 1999 to the estimate of $\hat{h}^2 = 0.6703$ from Galton's original data.

Before accepting this result, there is more work to be done. In 1973, Francis Anscombe [5] objected to the practice, dating from Pearson [91] in 1896, of emphasizing numerical calculation almost to the exclusion of graphing data.

Anscombe [5] produced four fictitious data sets in an argument that graphical analysis ought not be considered "cheating." He was adamant that "A computer should make *both* calculations *and* graphs." The italics are Anscombe's. The four examples have come to be known as *Anscombe's Quartet*. They are depicted in Figure 11.7.

Note that each of the 4 data sets consists of exactly 11 ordered pairs of points. What is remarkable about these four vastly different data sets is that they all share a common least-squares regression line:

$$\hat{y} = 3.0 + 0.5x \tag{11.2}$$

with $r^2 = 0.67$ and a p-value of 0.022 for the ANOVA test of significance of the linear regression model.

The fact that the model only seems appropriate for the first data set could easily go undetected if one relied on the numerical results of the least-squares regression and failed to graph the scatterplot of the raw data for each data set.

The simple linear regression model (11.2) is appropriate for the data depicted in panel I of Figure 11.7. Panel II, however, is a perfect quadratic relationship and the linear model is not appropriate. In panel III, all but the tenth observation fall in perfect line. Looking at the data reveals an outlier for this data

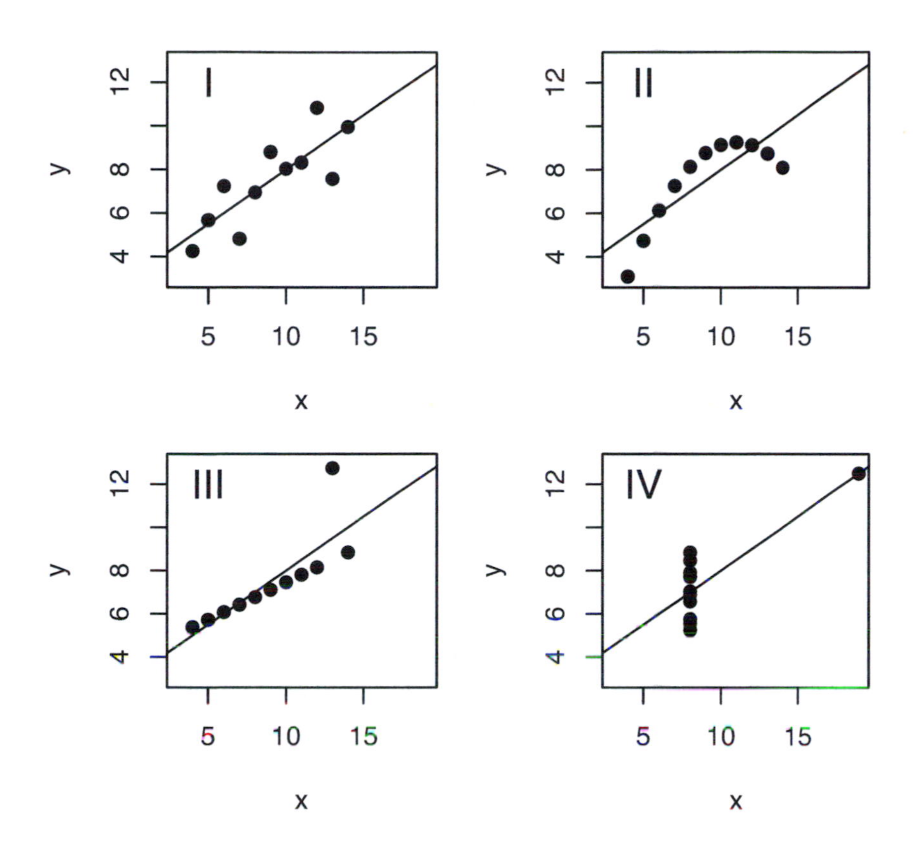

Figure 11.7 *Anscombe's Quartet*

set. In panel IV, all points but one fall in a vertical line. For the data of panel IV, the point $(19.0, 12.5)$ is influential as well as being an outlier.

Using a scatterplot to display data and the straight line of a simple linear regression model is an essential part of regression analysis. But it doesn't detect all problems. The graphical displays of the next section are important tools.

11.4 Residual Analysis

11.4.1 Definition

When the model is fitted and the least-squares estimates of regression coefficients $\hat{\beta}_0$ and $\hat{\beta}_1$ are obtained, the *least-squares regression line* is given by

$$\hat{y}_i = \hat{\beta}_0 + \hat{\beta}_1 x_i. \tag{11.3}$$

The estimate $\hat{y}_i$ is called the *fitted value* for the ith observation. An estimate of the error ϵ_i is given by the *residual*

$$\begin{aligned} e_i &= \text{Observed} - \text{Fitted} & (11.4)\\ &= y_i - \hat{y}_i. & (11.5) \end{aligned}$$

The estimate of the error variance σ^2 is given by the *mean square error*

$$MSE = \frac{\sum_{i=1}^{n} e_i^2}{n-2}. \tag{11.6}$$

Residuals for the simple linear regression model of mid-offspring height on mid-parent height is depicted in Figure 11.8. The residuals are depicted by the directed vertical line segments between the data points and the least-squares regression line. The magnitude of the residual is given by the length of the line segment. The sign of the residual is positive if the line segment is above the least-squares regression line and negative if below.

Residual analysis consists of executing various plots to verify the assumptions of the simple linear regression model. Sometimes the resulting action is to simply remove outliers from the data after investigation. But these plots can also be used prior to a second attempt at modeling the data by suggesting transformations of the explanatory variable or the response variable, or both.

If a logarithmic transformation, or another of the Box-Cox transformations, does not do the trick, then additional nonlinear terms or more explanatory variables might be the route to go.

11.4.2 Residual Scatterplots

There are a couple of problems with the depiction of residuals in Figure 11.8. Because of the proximity of some of the mid-parent heights, some line segments are smudged together and, thus, indistinct. Another potentially more troublesome issue is that comparison of the residuals must literally be made along a sliding scale caused by the least-squares regression line. The scatterplot of residuals versus mid-parent heights in Figure 11.9 solves this problem by introducing a common scale of comparison. Note that Figure 11.9 is, in

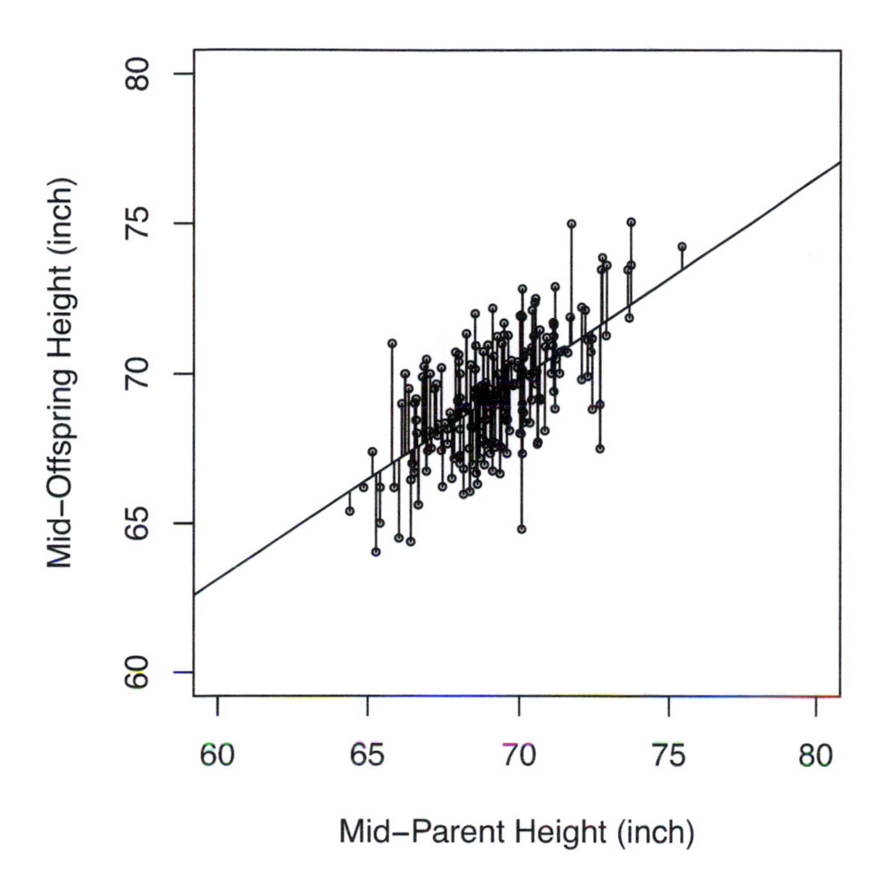

Figure 11.8 *Scatterplot of heights of mid-parents and mid-offspring with least-squares regression line (the residuals are depicted by the directed vertical line segments between the data points and the least-squares regression line)*

fact, a sunflower plot—but there is only one point with a sunflower of two petals.

The sum of the residuals must necessarily be zero, so a dashed reference line with slope and intercept both equal to zero have been drawn in Figure 11.9. Comparison of data to a line is a powerful visual metaphor. This was, of course, used in the scatterplot of mid-offspring height versus mid-parent height in Figure 11.6. It was also used in the first figure of this chapter in the normal quantile-quantile plot for the difference in height of husbands and wives.

A close alignment of the points to the line through the upper and lower quartiles in a normal quantile-quantile plot is taken to be confirmation of normality. In a residual scatterplot, one looks for patterns about the reference as an indication that model assumptions are not being met.

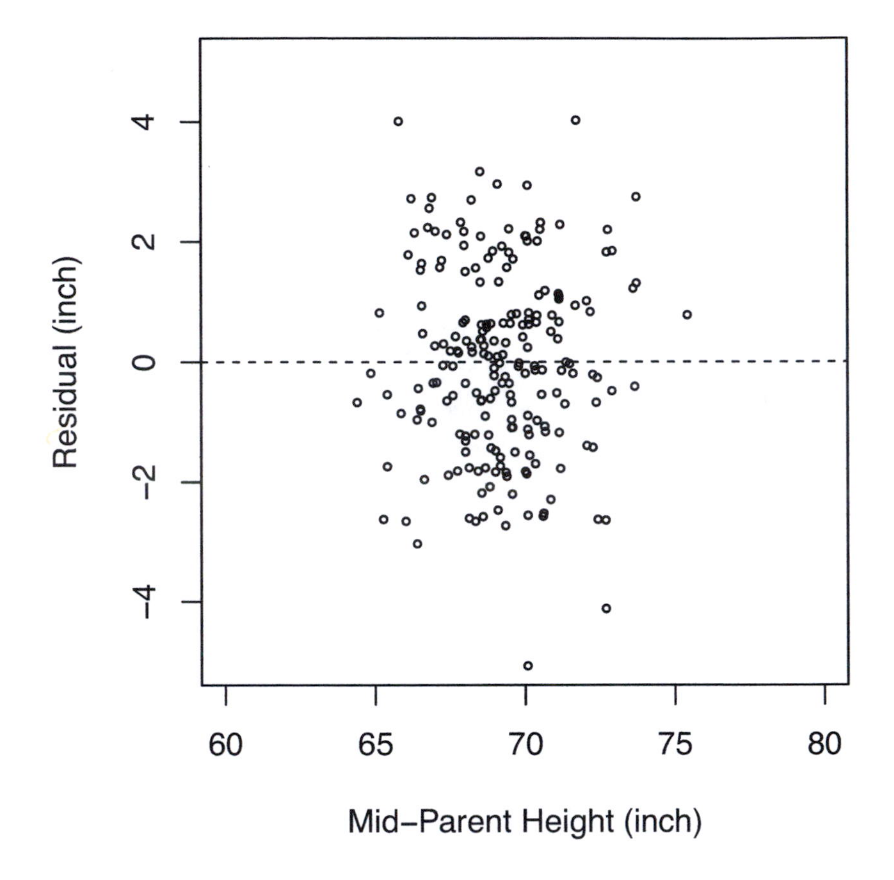

Figure 11.9 *Scatterplot of residuals versus heights of mid-parents*

In 1963, Francis J. Anscombe and John W. Tukey [6] recommended plotting residuals against fitted values as well as other potential explanatory variables such as time and geography. They recommended this as a standard procedure to find outliers or patterns in the data. In the context of the simple linear model, plotting residuals against the explanatory variable is equivalent to plotting residuals against fitted values because, by virtue of equation (11.3), the fitted values are a linear function of the explanatory values.

Taking a look at Figure 11.9, there appears to be no pattern. One typically looks for a curvilinear relationship in a residual scatterplot or an indication of the vertical spread of points varying with the explanatory variable. These patterns are not evident in Figure 11.9.

The situation is different when residuals are plotted against the height of offspring, as in Figure 11.10. The residuals appear to follow a linear pattern. This does not give great cause for concern. Such a pattern is not unusual if the

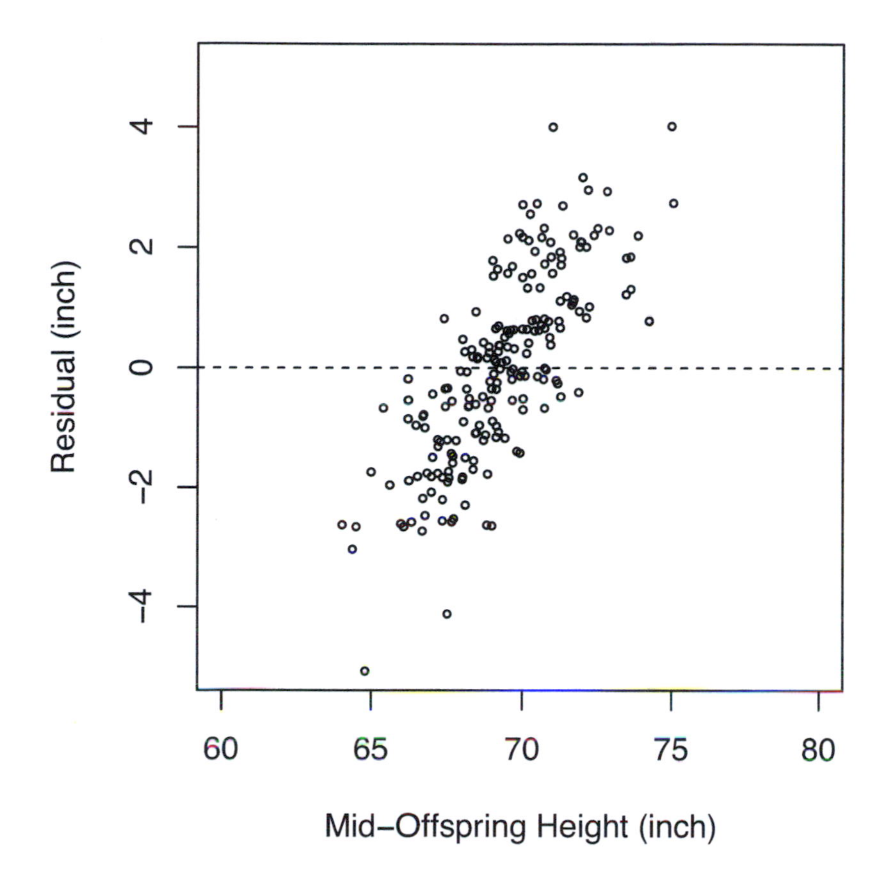

Figure 11.10 *Scatterplot of residuals versus heights of offspring*

absolute value of the sample correlation coefficient between the two variables of interest is not close to the value of one.

The principle of least squares finds the estimates of the regression parameters that minimize the sum of the square vertical distances between the data points and the line.

Minimizing the sum of the distances between the points and the line would probably do a better job of aligning the regression line with the major axis of an ellipse of constant density for a bivariate normal distribution.

The misfit between the major axis and the least-squares regression line is responsible for the pattern in Figure 11.10. Despite the group of points not having a horizontal orientation, if it were to be rotated horizontal, there would be no suspicious pattern to report.

No other explanatory variables have been provided with the data, so plots

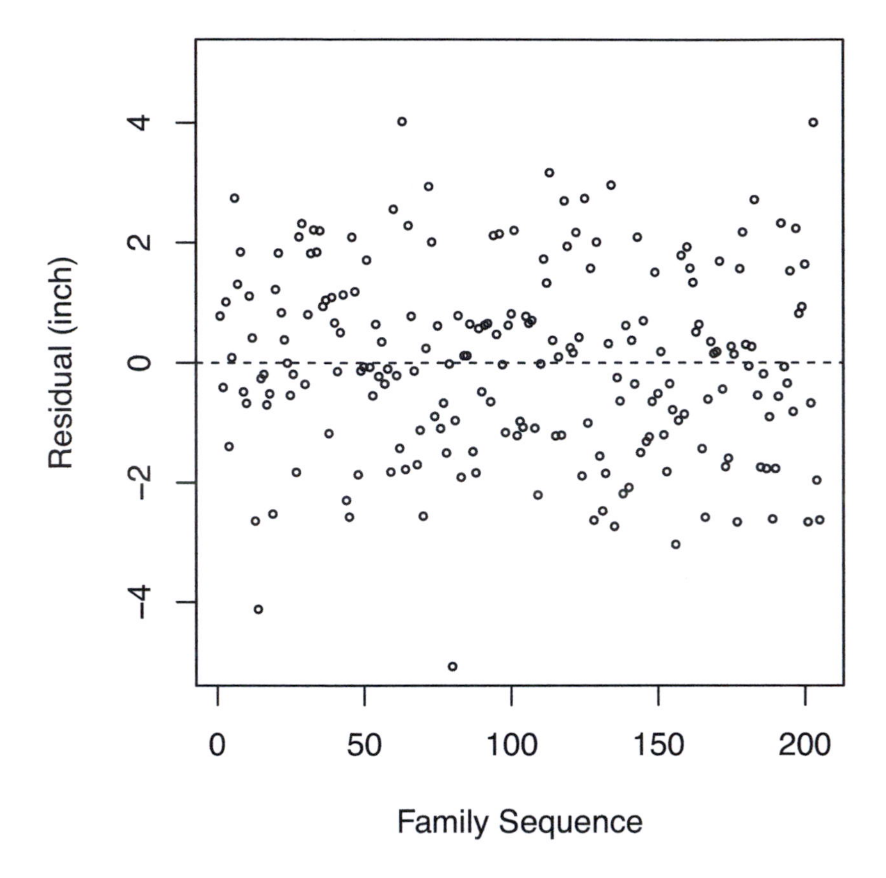

Figure 11.11 *Scatterplot of residuals versus sequence number in which families were recorded*

to check patterns in the residuals against these variables, as suggested by Anscombe and Tukey [6], are not possible.

On the chance that there could be something related to the sequence of families recorded in Galton's notebook, Figure 11.11 displays residuals against the sequence in which families were recorded. No pattern is apparent.

The point corresponding to residual values of approximately -5.079 could be an outlier from the overall pattern, but a graphical display of the type considered next is usually used to make this decision.

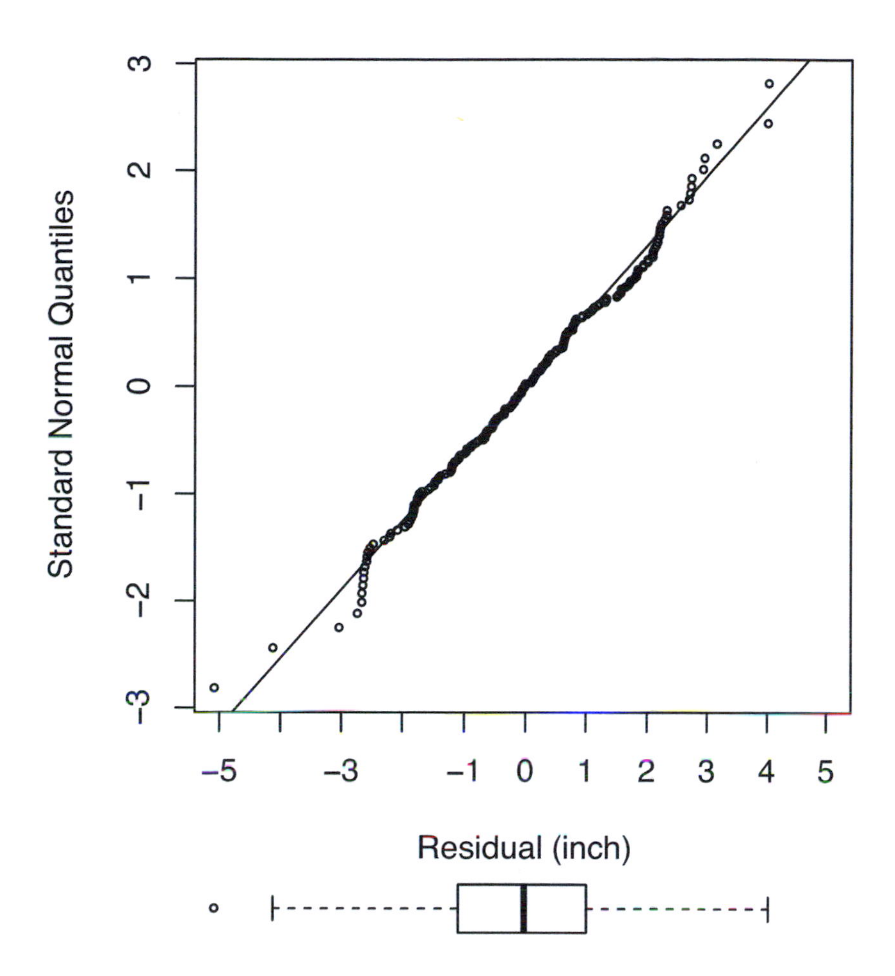

Figure 11.12 *Normal quantile-quantile plot and outlier boxplot of residuals from modeling mid-offspring height as a linear function of mid-parent height for Galton's Records of Family Faculties*

11.4.3 Depicting the Distribution of the Residuals

Normal quantile-quantile plots are typically used to verify the assumption of normality for the residuals obtained from fitting a simple linear regression model.

This is done for the residuals obtained from a simple linear model of mid-offspring height as a function of mid-parent height in Figure 11.12. The normal distribution appears to be a good fit.

There are some deviations from normality at either tail end in Figure 11.12, but these could just be a result of sampling variation.

Also presented in Figure 11.12 along the horizontal axis is an outlier boxplot. The distribution of the residuals appears to be symmetric from the box and whiskers, as one would expect for a normal distribution. The residual of −5.079, as previously noted, is judged to be an outlier.

The residual of −5.079 is associated with family #80. The father is 5 feet 10.5 inches in height, the mother is 5 feet 4.5 inches, and their only child, a daughter, is 5 feet even. This family is not remarkable based upon the heights of its members, so the family will not be omitted from the analysis.

The next topic of discussion concerns a variation on the residuals.

11.4.4 Depicting the Distribution of the Semistandardized Residuals

Recall that the sum of the residuals is zero when fitting a least-squares regression line to a simple random sample of ordered pairs. Hence, the mean of the residuals is zero.

The *semistandardized residual* for the ith observation is, therefore, given by

$$e_i^* = \frac{e_i}{\sqrt{MSE}}. \tag{11.7}$$

Note that some authors (see Neter, Kutner, Nachtsheim, and Wasserman [86]) refer to e_i^* as a *semistudentized residual* . A normal quantile-quantile plot and an outlier boxplot for the semistandardized residuals from the linear model of mid-offspring height on mid-parent height is given in Figure 11.13. The only difference between Figures 11.12 and 11.13 is a linear change of scale. Semistandardized residuals are dimensionless, whereas residuals are in units of the response variable. Given that there are semistandardized residuals, there must be standardized residuals. These are a topic in the next section.

11.5 Influence Analysis

11.5.1 Definition

An observation (x_i, y_i) is deemed *influential* in a linear model if its removal has a significant impact on the point estimates of the regression coefficients. A weakness in the application of the principle of least squares to obtain estimates of linear models is that an individual observation can have a large impact on the fitted model.

An influential data point need not be an outlier. So the residual analysis

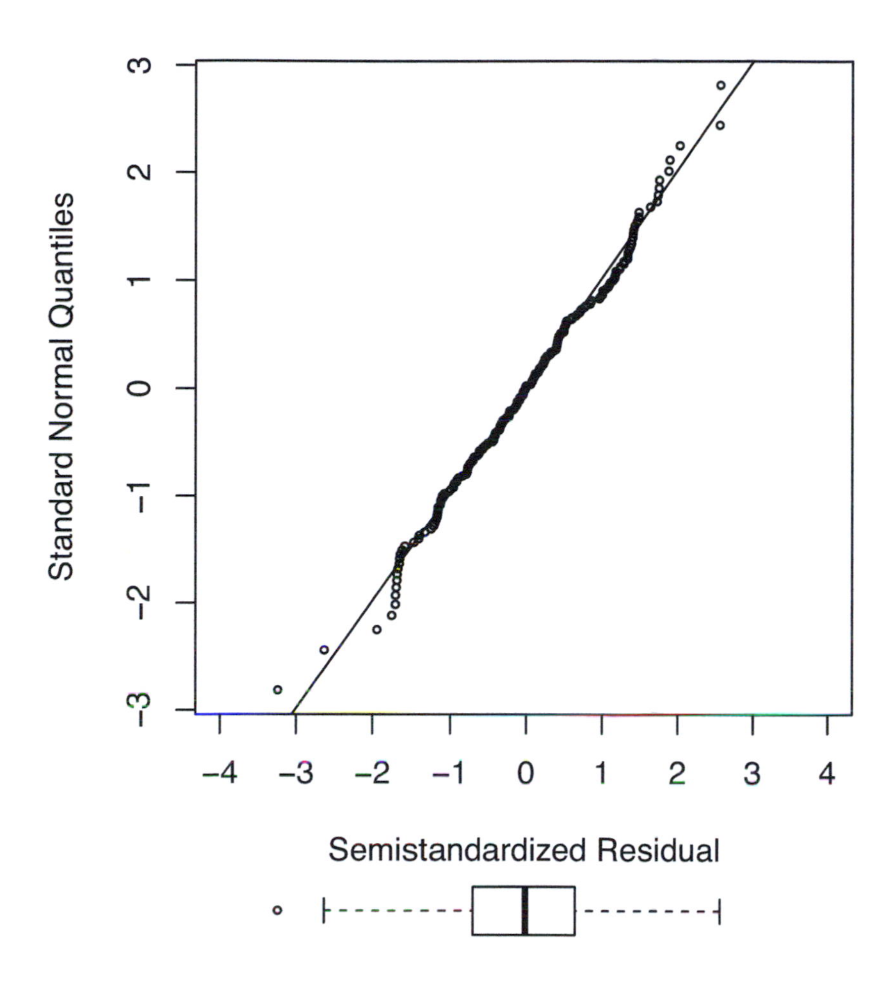

Figure 11.13 *Normal quantile-quantile plot and outlier boxplot of semistandardized residuals from modeling mid-offspring height as a linear function of mid-parent height for Galton's Records of Family Faculties*

of the previous section is insufficient for the task of identifying influential observations.

A number of different measures have been developed to identify influential observations. The presentation of this theory is best done using matrix notation.

This can be done by keeping the details to a minimum. Consider arranging the responses in a column vector:

$$\mathbf{y} = \begin{bmatrix} y_1 \\ y_2 \\ \vdots \\ y_n \end{bmatrix}. \tag{11.8}$$

Similarly, let the fitted values be arranged as

$$\hat{\mathbf{y}} = \begin{bmatrix} \hat{y}_1 \\ \hat{y}_2 \\ \vdots \\ \hat{y}_n \end{bmatrix}. \tag{11.9}$$

There exists a square $n \times n$ matrix $\mathbf{H}$ such that

$$\hat{\mathbf{y}} = \mathbf{H}\mathbf{y}. \tag{11.10}$$

The matrix $\mathbf{H}$ is called the *hat matrix*. In the following subsection, the hat matrix is shown to depend only on the explanatory values and not $\mathbf{y}$. It is the diagonal entries $\{h_{ii}\}$ of the hat matrix $\mathbf{H}$ that measure influence. The value h_{ii} is called the *leverage* of the observation (x_i, y_i).

The following discussion of matrix details can be skipped if so desired.

11.5.2 Matrix Notation for the Simple Linear Regression Model

Let the matrix of explanatory variables be given by

$$\mathbf{X} = \begin{bmatrix} 1 & x_1 \\ 1 & x_2 \\ \vdots & \vdots \\ 1 & \hat{x}_n \end{bmatrix}. \tag{11.11}$$

Arrange the regression coefficients in a column vector as follows:

$$\boldsymbol{\beta} = \begin{bmatrix} \beta_0 \\ \beta_1 \end{bmatrix}. \tag{11.12}$$

Let the column vector of random errors be given by

$$\boldsymbol{\epsilon} = \begin{bmatrix} \epsilon_1 \\ \epsilon_2 \\ \vdots \\ \epsilon_n \end{bmatrix}. \tag{11.13}$$

The simple linear regression model in matrix notion is

$$\mathbf{y} = \mathbf{X}\boldsymbol{\beta} + \boldsymbol{\epsilon}. \tag{11.14}$$

The *ordinary least-squares estimator* of $\boldsymbol{\beta}$ is

$$\hat{\boldsymbol{\beta}} = (\mathbf{X}^T\mathbf{X})^{-1}\mathbf{X}^T\mathbf{y}. \tag{11.15}$$

The column vector of *fitted values* is thus

$$\hat{\mathbf{y}} = \mathbf{X}(\mathbf{X}^T\mathbf{X})^{-1}\mathbf{X}^T\mathbf{y}. \tag{11.16}$$

The *hat matrix* $\mathbf{H} = [h_{ij}]$ is defined to be

$$\mathbf{H} = \mathbf{X}(\mathbf{X}^T\mathbf{X})^{-1}\mathbf{X}^T. \tag{11.17}$$

The column vector of residuals is given by

$$\mathbf{e} = (\mathbf{I} - \mathbf{H})\mathbf{y}. \tag{11.18}$$

The *leverage* is measured by the entries $\{h_{ii}\}$ along the main diagonal of the hat matrix $\mathbf{H}$.

11.5.3 Depicting Standardized Residuals

The *standardized residual* for the ith observation is

$$r_i = \frac{e_i}{\sqrt{MSE(1 - h_{ii})}}. \tag{11.19}$$

Figure 11.14 gives an illustration of the standardized residuals for mid-offspring height as a linear function of mid-parent height with a boxplot for the residuals given below the normal quantile-quantile plot.

That there is not much difference between Figure 11.14 and Figure 11.13 for the semistandardized residuals suggests that leverage is relatively uniform for Galton's data. The same outlier, as previously noted, is flagged and the residuals appear to follow a normal distribution.

Standardized or semistandardized residuals can also be plotted against the explanatory variable in a scatterplot. Because, in this example, the standardized values and semistandardized values are so close in value, only a scatterplot of standardized values versus mid-parent height is provided in Figure 11.15. Note that standardized and semistandardized residuals are free of dimension.

The standardized residuals depicted in Figures 11.14 and 11.15 were produced by the R function `rstandard` after using the function `lm` to fit the simple linear regression model. Note that Neter *et al.* [86] use the term *studentized residual* to refer to the *standardized residual.* The term *internally studentized residuals* is also used synonymously. Studentized residuals are the next topic.

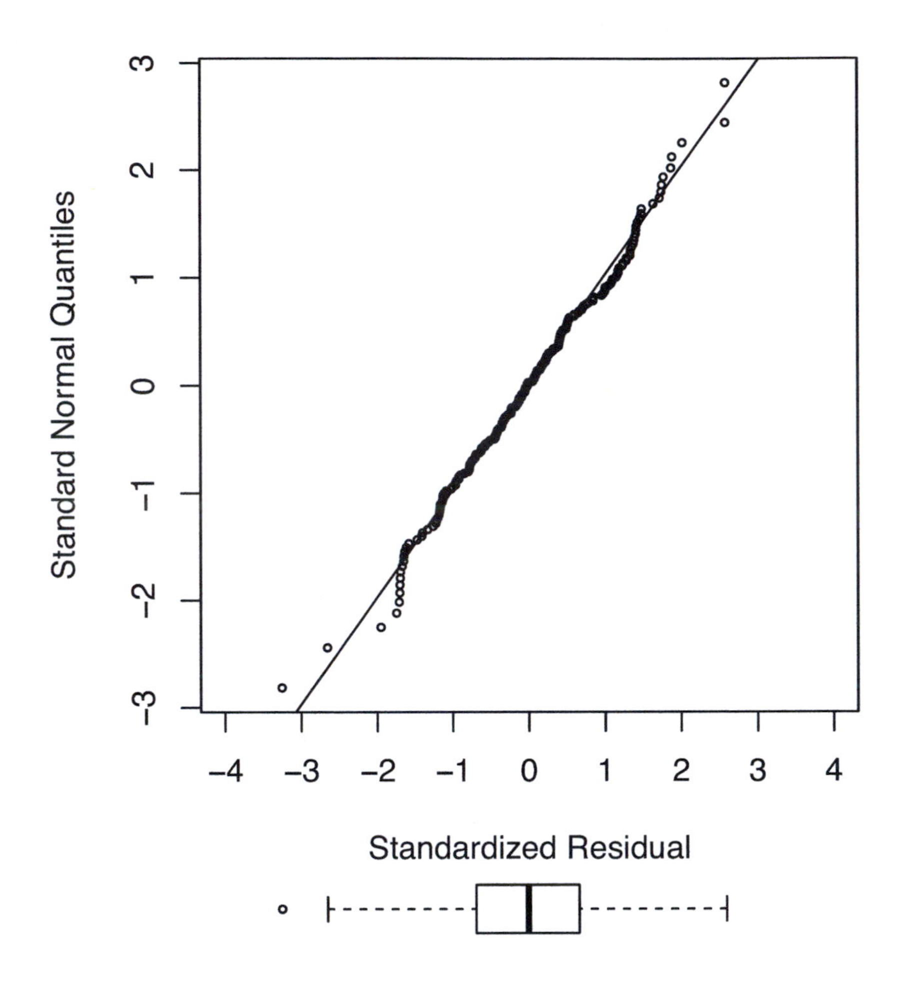

Figure 11.14 *Normal quantile-quantile plot and outlier boxplot of standardized residuals from modeling mid-offspring height as a linear function of mid-parent height for Galton's Records of Family Faculties*

11.5.4 Depicting the Distribution of Studentized Residuals

The next version of a residual requires working with subsets of the original data that are formed by deleting each observation in turn. In general, the procedure deletes the ith observation and fits the linear model to the remaining $n-1$ observations. The point estimate $\hat{y}_{i(i)}$ of the ith response is then obtained with the value of the explanatory variable x_i substituted. The *deleted residual* for the ith observation is

$$d_i = y_i - \hat{y}_{i(i)}. \tag{11.20}$$

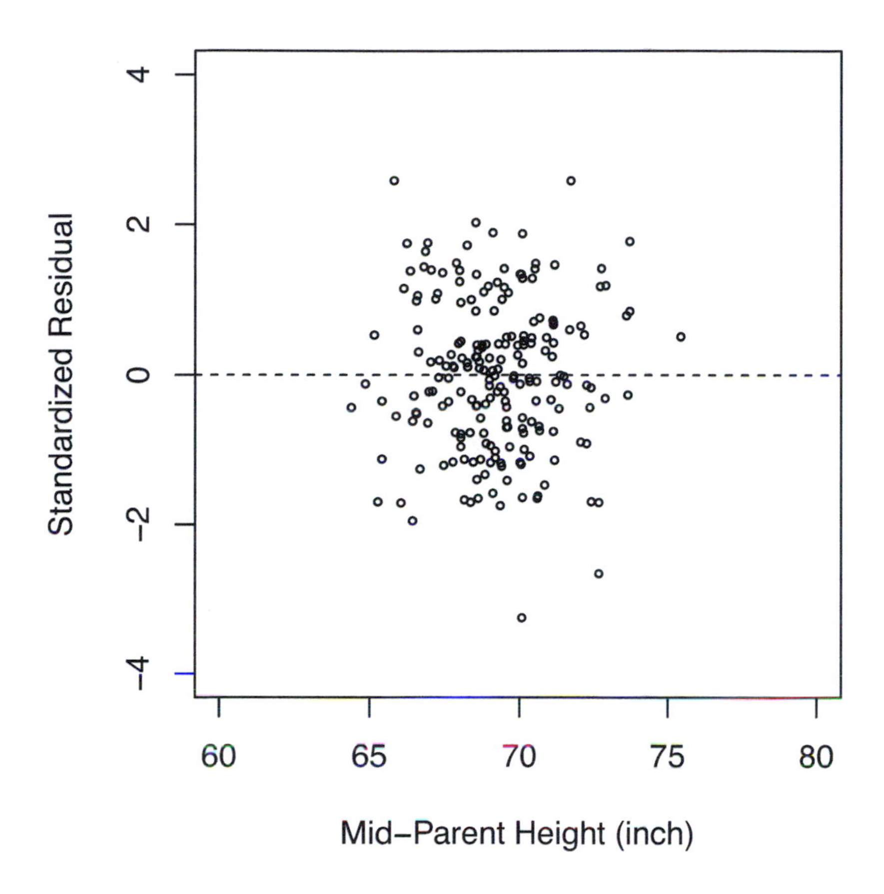

Figure 11.15 *Scatterplot of standardized residuals versus mid-parent height*

An algebraically equivalent expression for d_i that does not require n recomputations is

$$d_i = \frac{e_i}{1 - h_{ii}}. \tag{11.21}$$

If $MSE_{(i)}$ denotes the mean square error for the model with the ith observation deleted, it can be shown that the *studentized residual*

$$t_i = \frac{e_i}{\sqrt{MSE_{(i)}(1 - hii)}} \tag{11.22}$$

follows Student's t-distribution with $n - 3$ degrees of freedom for the simple linear regression model. Neter *et al.* [86] refer to t_i as the *studentized deleted residual.*

Figure 11.16 gives an outlier boxplot and a Student's t quantile-quantile plot

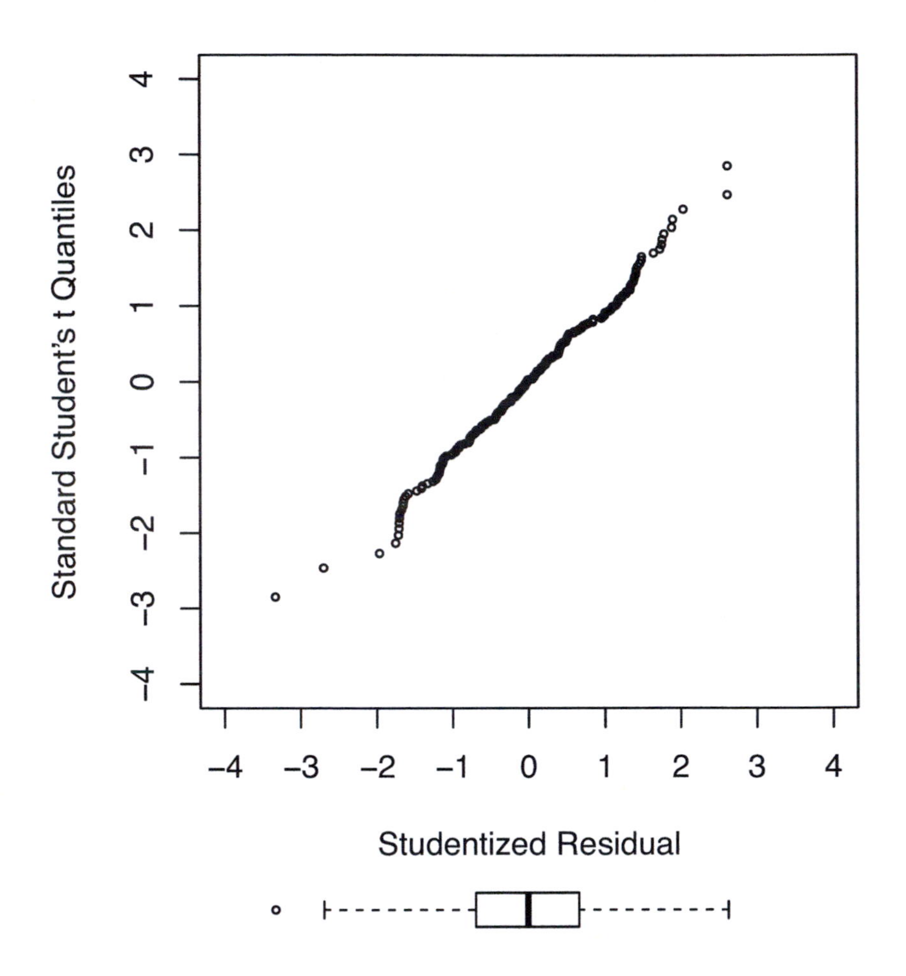

Figure 11.16 *Student's t quantile-quantile plot and outlier boxplot of studentized residuals from modeling mid-offspring height as a linear function of mid-parent height for Galton's Records of Family Faculties*

for the studentized residuals from modeling mid-offspring height as a linear function of mid-parent height for Galton's Records of Family Faculties. With 203 degrees of freedom there is not much difference between Student's t-distribution and the standard normal distribution.

Note that the residuals fall nearly in a straight line in the Student's t quantile-quantile plot in Figure 11.16. This is consistent with the condition that random error be normally distributed. Figure 11.17 gives a scatterplot of studentized residuals versus mid-parent height. There are no apparent patterns or outliers.

The studentized residuals depicted in Figures 11.16 and 11.17 were produced

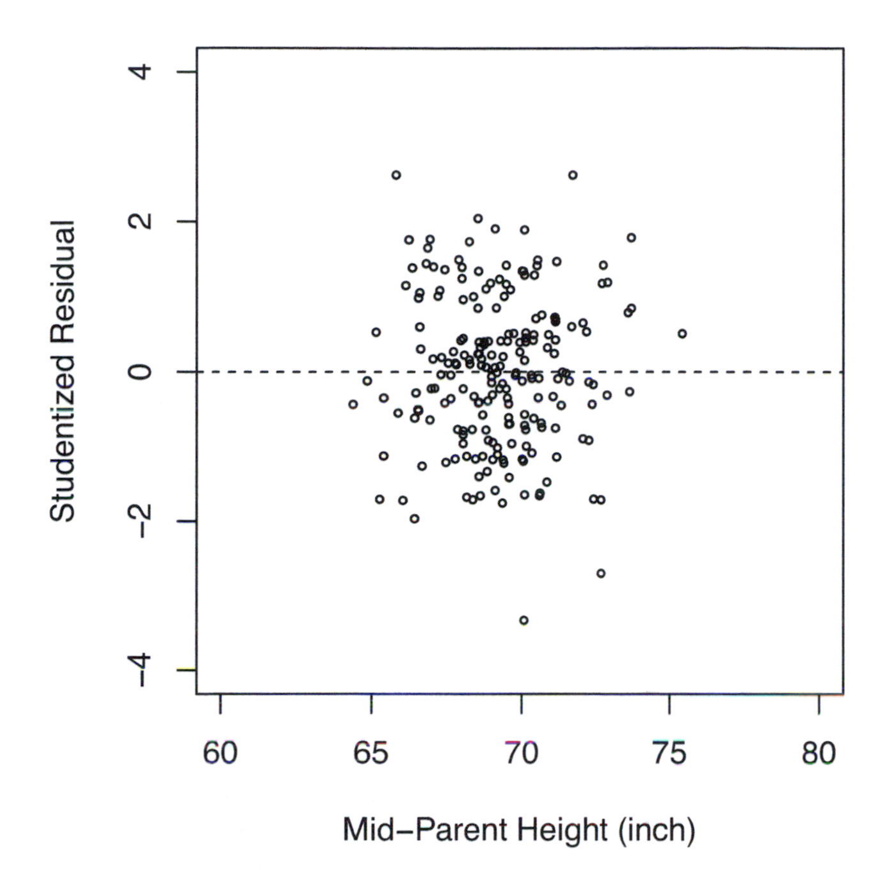

Figure 11.17 *Scatterplot of studentized residuals versus mid-parent height*

by the R function `rstudent` after using the function `lm` to fit the simple linear regression model.

Although the standardized and studentized residuals depend on leverage, they are generally used for identifying outliers based on values of the response variable. The next few measures are intended to identify outliers based on values of an explanatory variable.

11.5.5 Depicting Leverage

The following facts can also be established regarding the diagonal elements of the hat matrix in the case of simple linear regression:

$$0 \leq h_{ii} \leq 1,$$
$$\textstyle\sum_{i=1}^{n} h_{ii} = p,$$

where $p = 2$ for simple linear regression.

A leverage value h_{ii} is:

- large if it is more than twice as large as its average value of p/n, that is, if

$$h_{ii} > 2\frac{p}{n};$$

- moderately large if $0.2 \leq h_{ii} \leq 0.5$; and
- is very large if $h_{ii} > 0.5$.

Typically, one plots leverage h_{ii} against i in a scatterplot with the explanatory variable along the horizontal axis and adds reference lines to aid in determining points for which the explanatory variables have high leverage. Observations with large leverages are usually denoted as being *influential observations.*

An example of a *leverage plot* is given in Figure 11.18. Note that there are 19 high leverages but no moderately or very large leverages. What is somewhat disconcerting about Figure 11.18 is the quadratic pattern with respect to mid-parental height.

The explanation for the pattern in the leverage plot of Figure 11.18 is that in a linear regression with only one explanatory variable, leverage is a quadratic function of the distance of the explanatory value from its mean. After all, the principle of least squares is being used to estimate the regression coefficients.

When looking at a leverage plot one also looks for significant gaps. The family with the highest leverage, separated from the rest, is the family with the greatest mid-parent height. This is family #1, which is a tall family indeed. The father is 6 feet 6.5 inches. His wife is 5 feet 7 inches. Their son is 6 feet 1.2 inches (Galton, or his observer, was being quite precise for this measurement). Their eldest daughter is 5 feet 9.2 inches and their other two daughters are each 5 feet 9 inches. There is no compelling reason, however, to remove this family from the analysis.

11.5.6 Depicting DFFITS

A useful measure of influence is given by

$$(DFFITS)_i = \frac{\hat{y}_i - \hat{y}_{i(i)}}{\sqrt{MSE_{(i)}h_{ii}}}. \tag{11.23}$$

The letters DF in the acronym stand for the Difference in Fitted values. It can be shown that this measure of influence for the ith observation can be computed from the original model fit with the formula:

$$(DFFITS)_i = t_i\sqrt{\frac{h_{ii}}{1 - h_{ii}}}, \tag{11.24}$$

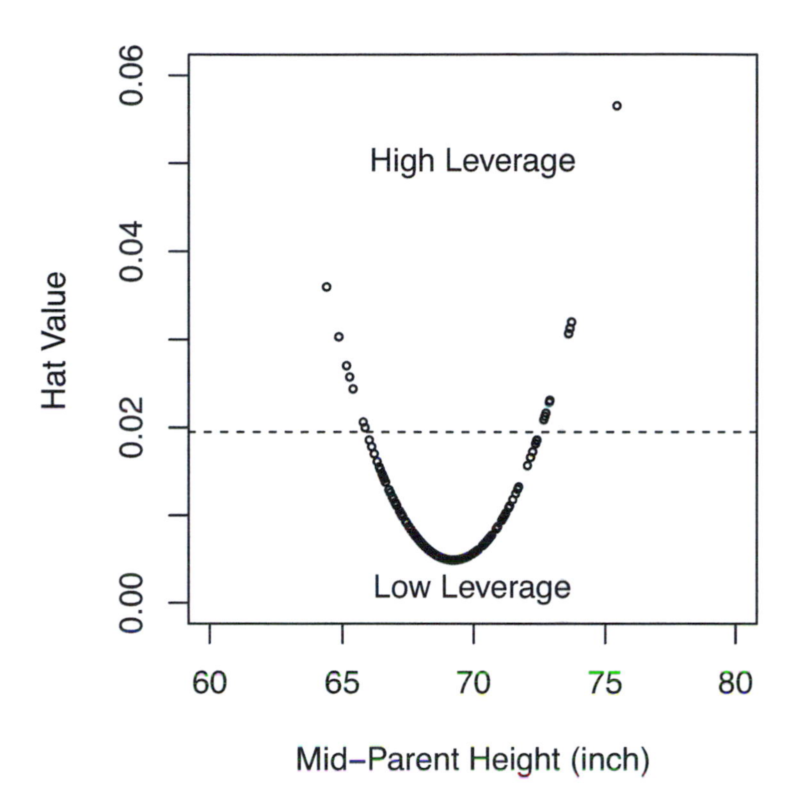

Figure 11.18 *Leverage plot for a model of mid-offspring height as a linear function of mid-parental height for Galton's Records of Family Faculties*

which depends on the studentized residual t_i and the leverage h_{ii}.

According to Neter *et al.* [86], a guideline for identifying influential observations is when the absolute values of $DFFITS$ exceed 1 for small to medium data sets and $2\sqrt{p/n}$, where $p = 2$ for simple linear regression, for large data sets when applying a simple linear regression model.

Figure 11.19 presents $DFFITS_i$ as a function of mid-parent height. According to the criterion that $DFFIT_i > 2\sqrt{2/n}$, 12 families are considered to be of high leverage. There are no other patterns discernible.

The $DFFTITS$ values plotted in Figure 11.19 were obtained from the R function `dffits`.

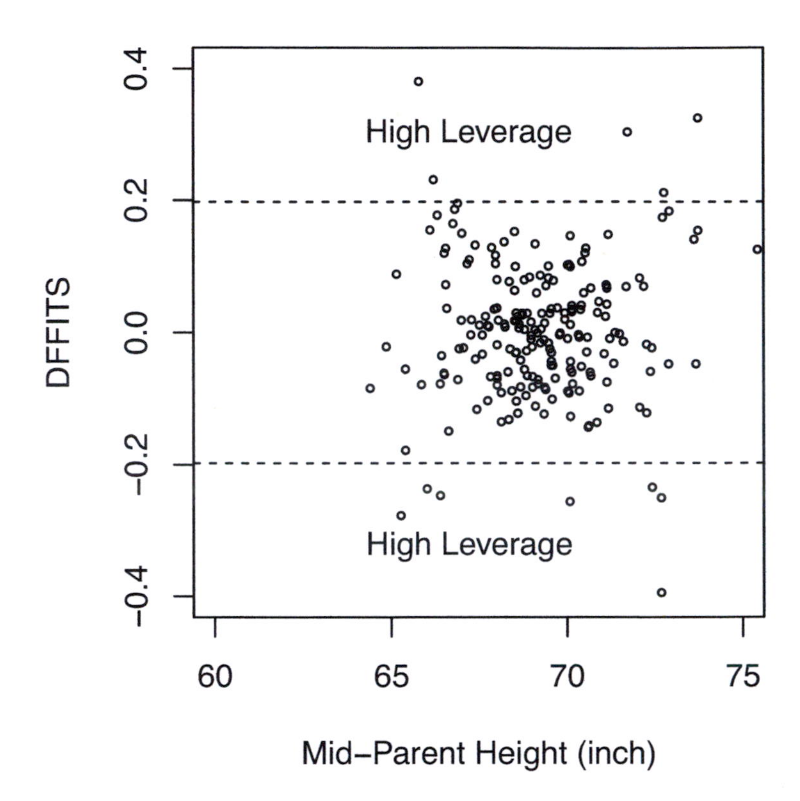

Figure 11.19 *Scatterplot of $DFFITS$ versus mid-parent height for Galton's 205 families when modeling mid-offspring height as a linear function of mid-parent height*

11.5.7 Depicting $DFBETAS$

An important consideration is the amount of influence each observation has on the estimate of the slope and intercept parameters. This can be estimated with the influence measure

$$(DFBETAS)_{k(i)} = \frac{\hat{\beta}_k - \hat{\beta}_{k(i)}}{\sqrt{MSE_{(i)}c_{kk}}} \tag{11.25}$$

for $k = 0$ or 1 where c_{kk} is the $(k+1)$th diagonal element of the matrix $(\mathbf{X}^T\mathbf{X})^{-1}$. Neter *et al.* [86] recommend considering an observation influential if the absolute value of $DFBETAS$ exceeds 1 for small to medium data sets and $2/\sqrt{n}$ in large data sets.

Figure 11.20 presents the $DFBETAS$ in the top panel for the intercept and

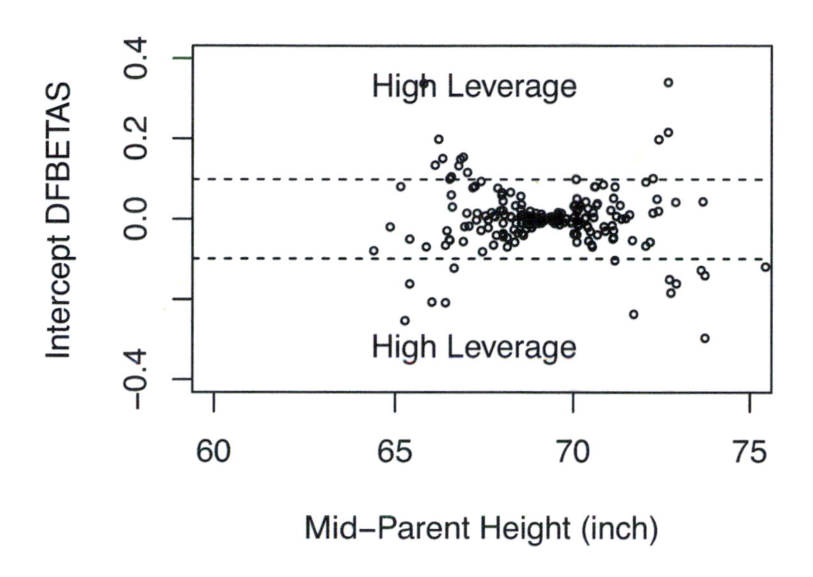

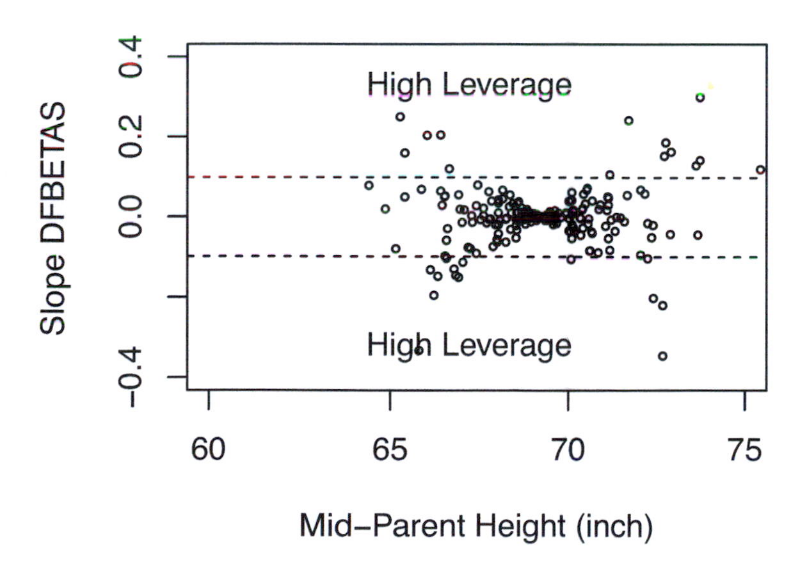

Figure 11.20 *Scatterplots of DFBETAS versus mid-parent height for Galton's 205 families when modeling mid-offspring height as a linear function of mid-parent height*

in the bottom panel for the slope. A common pattern to both scatterplots is that the high influential values appear at extremes from the sample mean of mid-parent heights. This is neither unusual nor a major concern.

The $DFBETAS$ plotted in Figure 11.20 were obtained from the R function `dfbetas` after fitting the linear model with the function `lm`.

11.5.8 Depicting Cook's Distance

A measure of the aggregate of the differences between the fitted value when all observations are used and the fitted value when the ith observation is deleted is given by *Cook's distance measure*

$$D_i = \frac{\sum_{j=1}^{n} [\hat{y}_j - \hat{y}_{j(i)}]^2}{pMSE} \tag{11.26}$$

where $p = 2$ for the simple linear regression model. This measure was introduced by Dennis Cook [27] in 1977. An expression available for determining Cook's distance measure without calculating the least squares estimates of the regression coefficient as each observation i is deleted is given by

$$D_i = \frac{e_i^2}{pMSE} \left[\frac{h_{ii}}{(1 - h_{ii})^2} \right]. \tag{11.27}$$

So Cook's distance D_i is a nonlinear function of both the residual e_i and the leverage h_{ii} of the ith observation.

While D_i does not follow the F distribution, conventional wisdom relates the value D_i to the F distribution with p numerator and $n-p$ denominator degrees of freedom. The F-transform of Cook's distance measure is given by finding the percentile of D_i from the $F_{p,n-p}$ distribution:

$$Q_i = F_{p,n-p}(D_i). \tag{11.28}$$

The ith observation is judged to have little influence if the percentile value is less than about 10%, according to Cook [27]. On the other hand, if the percentile is near 50 percent, or more, the observation is judged by Cook [27] to have a major influence.

Note that it can also be established that Cook's distance measure is coincidentally an overall measure of the combined impact of the ith observation on all of the estimated regression coefficients.

The F-transform of Cook's distance for mid-offspring modeled as a linear function of mid-parent height is given in Figure 11.21. None of the observations appear to be influential as the distance is less that 10%.

The R function `cooks.distance` was used to determine the values plotted in Figure 11.21.

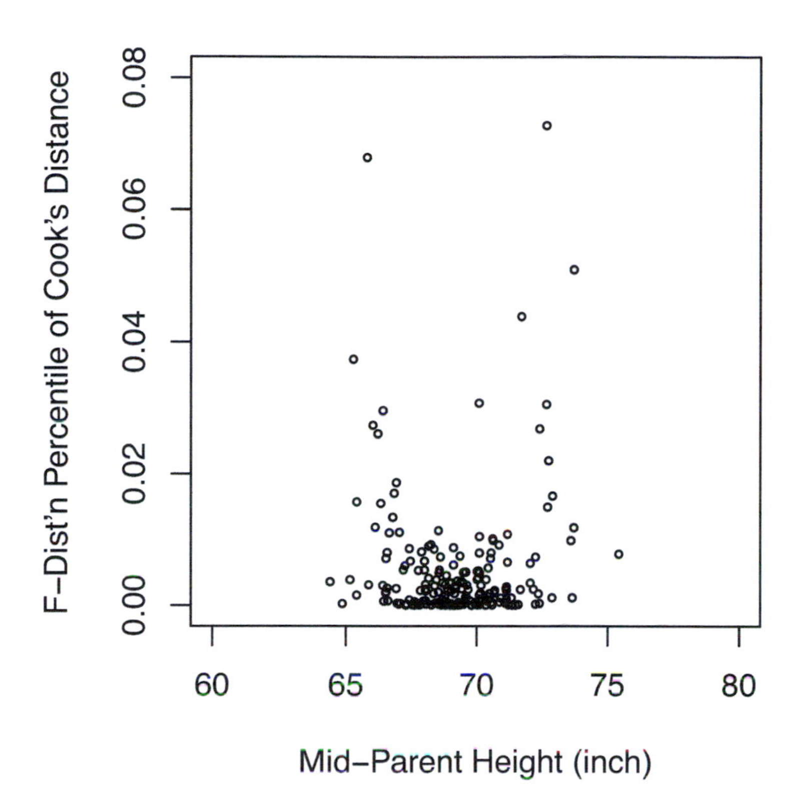

Figure 11.21 *Scatterplot of the F-transform of Cook's distance versus mid-parent height for Galton's 205 families when modeling mid-offspring height as a linear function of mid-parent height*

11.5.9 Influence Plots

Some statistical software packages provide influence plots that combine a number of influence measures at once. The *bubble plot* in Figure 11.22 was produced by the R function `influencePlot` in the package `car`. The function call is given below for the linear model of mid-offspring height on mid-parent height stored in the list `molm`.

```
influencePlot(molm,xlim=c(-0.01,0.06),
ylim=c(-4,4),labels="")
```

The horizontal axis is for leverages, or *hat values* as they are sometimes called. The vertical axis is for studentized residuals. The areas of the circles are

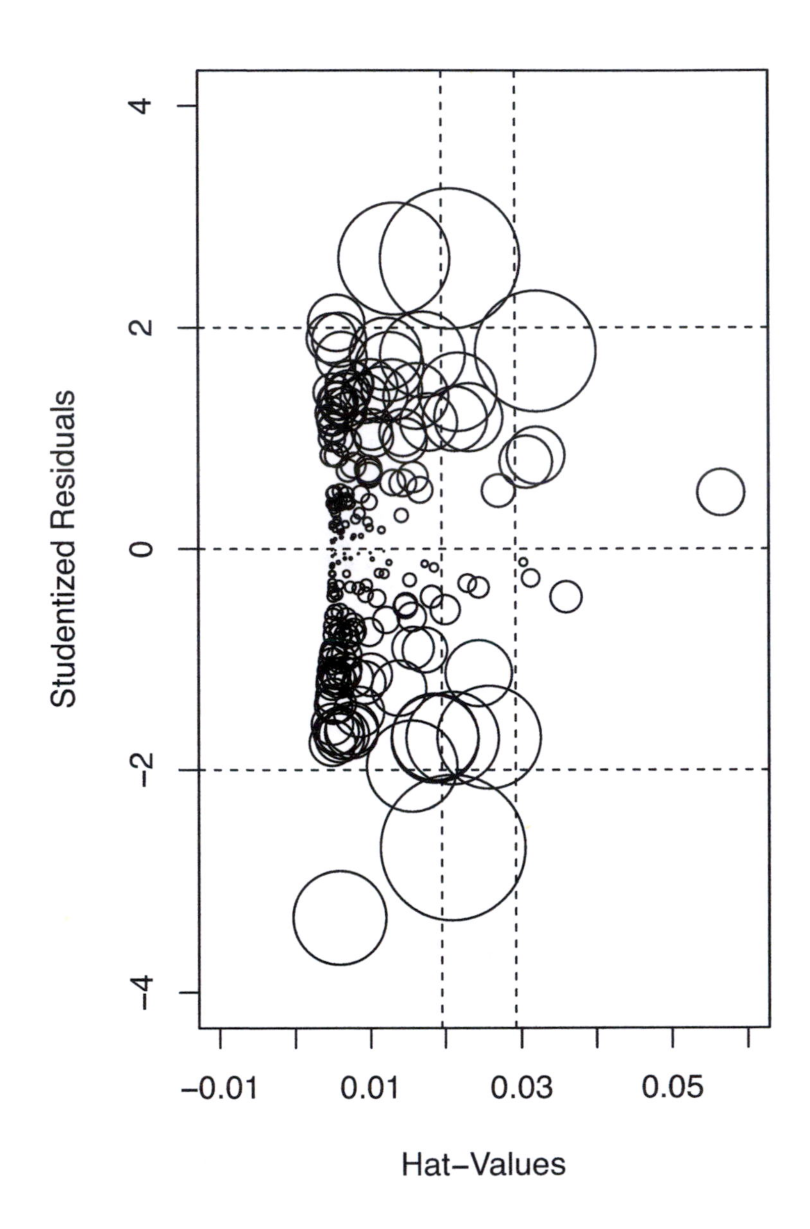

Figure 11.22 *Influence plot for the linear model of mid-offspring height on mid-parent height (the area of the circles is proportional to Cook's distance)*

proportional to Cook's distances. The circles are drawn in black, instead of gray, if Cook's distance is among the top 5% of observed values.

Notice that horizontal reference lines have been drawn at -2, 0, and 2. Vertical reference lines correspond to twice and thrice the average hat value.

There are no observations in the upper left or lower left regions that correspond to influential values of both the studentized residuals and leverage. The unremarkable gray circle at the far right but in the interval $[-2, 2]$ for the studentized residuals is family #1, which has the largest mid-parental height in Galton's sample.

All in all, Galton's data collectively appear not to give cause for concern with respect to influential observations. This concludes the re-analysis of Galton's original data concerning the heritability of stature in humans with the simple linear regression model. Little did Galton know what he started. But his start did come with a graphical presentation and graphical analysis of his data.

11.6 Conclusion

Graphical displays are the cornerstone of modern regression analysis. They can be used to discover patterns in the data and illustrate relationships among variables. They can confirm or negate model assumptions. Graphical displays can detect errors in data, such as an outlier due to a coding error. Graphical displays can be used to assess the adequacy of the fitted model and even suggest remedial action.

Graphical displays for regression analysis can be classified into one of two broad categories: graphs executed before fitting the model to the data, and graphs executed after.

Graphical displays before fitting the data include graphs depicting the distribution of the response and explanatory variables. These can be stemplots, dotplots, boxplots, quantile-quantile plots, and kernel density estimates for each variable separately. For the joint distribution of the response variable and the explanatory variable, a bagplot or a plot depicting a bivariate kernel density estimate can be used. The choice of graphical displays will be governed by the technical sophistication of the intended viewer.

At the point of fitting the model, a scatterplot is typically executed with the least-squares regression line added. If overlapping ordered pairs is a problem, a sunflower plot can be used. There is a wide choice of diagnostic plots after the fitting is done. If a viewer is comfortable with the scatterplot of the response and explanatory variables, then scatterplots of residuals can be presented even if the viewer has no previous exposure.

The residuals ought to be plotted against the explanatory and response variables and any other potential explanatory variables if possible. There ought to be one plot, and preferably more, depicting the distribution of the residuals with the choice, or choices, limited only by the anticipated technical sophistication of the viewer. These diagnostic plots need be examined for patterns or outliers. The lack of either suggests a good fit of the model. The presence of either suggests another round of modeling.

The use of graphical displays to identify influential observations is another kettle of fish. This material is sufficiently technical not to find its way into introductory statistics textbooks. Discussions with respect to outliers, based on residual analysis, tend to be straightforward. The issue of influence is a bit murkier.

As a result of a pattern in the plot of leverage versus mid-parental height, it was discovered that Galton ordered the families in his notebook according to mid-parental height. To make fuller use of the influence measures, more choices for the horizontal axis ought to have been made than just mid-parental height. If something had turned up, a graphical display would have been presented.

With respect to selection among the different measures of influence, including the different standardizations of the residuals, personal preference seems to be the rule.

Returning to the topic of the case study presented in this chapter, it is quite something that the first regression data set from the nineteenth century is available for a full analysis as if new. Comparing what Galton was able to do and what can be done, because of Galton, is an interesting historical study.

With respect to the representativeness of Galton's sample from 1884, Galton [53] lists the names of the 84 prize winners from 1884, with their community name or London street address, in his book *Natural Inheritance* published in 1889. The addresses appear to broadly cover Great Britain. The plotting of the addresses on a map is left for another time.

Galton stayed true to his promise of 500 pounds Sterling in total prizes for contributors. He paid prizes of 7 pounds Sterling to 40 contributors and 5 pounds to 44 contributors. Among the prize winners were a few Anglican clergy or their wives, a knight of the realm, a Lieutenant Colonel, and presumably the wife of the Governor of Her Majesty's Camden Road Prison in London. To use an English turn of phrase, the 205 families appear to be well bred, that is to say, upper middle and upper class.

The net result of Galton's [50, 51, 54] work on the heritability of stature with his original data is that the mean stature of adult children is related to the mid-parental stature by a ratio of 2:3. That is, before the principle of least squares was available, Galton estimated the slope coefficient to be $2/3$. It is not at all clear in his writings how Galton came to this estimate. His approach, in retrospect, was more advanced than Pearson's a decade later. Galton worked with medians of ranges of values of height and used these values to determine a line of best fit. It is somewhat ironic that Galton's robust method pre-dated the least-squares method. Robust regression analysis became a hot area of research nearly a century after Galton's pioneering work on the topic when the circuits in computers were replaced by solid-state devices with the requisite increase in processing speed needed for robust regression methods.

For the sake of interest, with the mid-parent heights paired with the 934 offspring heights, the least-squares estimate of the slope is 0.7132 for predicting offspring height as a function of mid-parent height. Because the ordered pairs of 934 points do not constitute a simple random sample, it is not possible to provide the standard error for this estimate. The value Galton found was 2/3. The value Dickson found at Galton's behest was $2 \times 6/17.6 \approx 0.6818$. With the 205 families assumed to be from a simple random sample, the point estimate of the slope is 0.6703 and a 95% confidence interval is [0.5573, 0.7833]. Galton's and Dickson's results, both reported in 1885 by Galton [50], are quite close to that using the method due to Pearson [91] as reported in 1896. The genetic theory equating this slope with heritability h^2 had to wait until Fisher [37] in 1918. The calculation of the confidence interval had to wait until the publication in 1925 by Fisher [39] of the small-sample statistical theory for Student's t-distribution.

The number of loci affecting stature in human populations is still unknown. The degree of polymorphism for each loci is also unknown. From Galton's own data, the point estimate of heritability, on a scale from zero to one, for height is $h^2 = 0.6703$. This estimate represents the proportion of the phenotypic variance due to the additive effect of one or more genes.

An alternative approach to the problem of dependencies among Galton's 934 pairs of offspring and mid-parents was taken by Pearson [91] more than a decade later in 1896. Pearson regarded the nuclear families as a simple random sample and did not combine parents' heights into a single mid-parental value. Instead, with Galton's data, Pearson calculated all sample product-moment correlations: father and mother; father and son; father and daughter; mother and son; mother and daughter; brother with sister; brother with brother; and sister with sister. He treated his pairs as independent for each correlation when estimating its standard error. This limitation was not overcome until 1987 when Keen [74] wrote down the likelihood function for nuclear families and found the maximum likelihood estimates of the familial correlations and their standard errors by numerical calculation for two, or more, quantitative traits (possibly correlated).

The solution to the problem for extended families had to wait until 2003 when Keen and Elston [75] derived robust asymptotic sampling theory. As previously noted, Pearson [91] was concerned about whether the calculation of the sample correlation between fathers and mothers should be weighted according to the number of husband and wife pairs or by the number of their offspring. In essence, Pearson [91] proposed weighted regression analysis as an alternative solution to the problem. Keen and Elston [75] returned to the approach of

Pearson [91] allowing any weighting scheme in their robust asymptotic solution 119 years after Galton starting collecting data for his Records of Family Faculties.

Although somewhat stimulating on its own merits, there wasn't too much excitement on account of interesting features in the residual and influence plots. The case study of the next chapter will relieve any boredom on this account.

11.7 Exercises

1. The data summary in Table 11.1 of the heights of 928 offspring with their mid-parents was not, in fact, depicted by Galton in any of his writings since obtaining the data in 1884. Nor was it depicted in this chapter. The scatterplot of Figure 11.2 and the sunflower plot of Figure 11.3 were done with Galton's own original data from the Records of Family Faculties.
   (a) Graph the contents of Table 11.1 in a scatterplot with the plotting symbols reporting the count for each cell.
   (b) Graph the contents of Table 11.1 in a sunflower plot.
   (c) Which of the two graphical displays from parts (a) and (b) do you prefer? Discuss.
2. Refer to Exercise 1. Add a 95% constant probability ellipse to each of the two plots produced for parts (a) and (b) in Exercise 1 using Galton's own original data on heights for each of the adult offspring and their mid-parental heights. Examine the fit of the ellipse from the original data to Galton's own tabular summarization. Discuss.
3. In the Conclusion section, it was noted that the preparation for a simple linear regression analysis ought to include depicting the marginal distributions of the two variables of interest and their joint distribution. Yet this was not done with the case study for this chapter.
   (a) Produce a boxplot for mid-parental height. Discuss.
   (b) Produce a boxplot for mid-offspring height. Discuss.
   (c) Produce a bagplot for the joint distribution of heights of mid-parents and mid-offspring. Discuss.
4. In the Conclusion section, it was noted that the preparation for a simple linear regression analysis ought to include depicting the marginal distributions of the two variables of interest and their joint distribution. Yet this was not done with the case study for this chapter.
   (a) Produce a kernel density estimate for mid-parental height. Discuss.
   (b) Produce a kernel density estimate for mid-offspring height. Discuss.
   (c) Produce a graphical display of the bivariate kernel density estimate for the joint distribution of heights of mid-parents and mid-offspring. Discuss.

5. In the Conclusion section, it was noted that the preparation for a simple linear regression analysis ought to include depicting each of the two variables of interest in a quantile-quantile plot. This was not done with the case study for this chapter.
   (a) Produce a normal quantile-quantile plot for mid-parental height. Discuss.
   (b) Produce a normal quantile-quantile plot for offspring height. Discuss.
   (c) Although not explicitly considered in this chapter, it is not a bad idea to consider whether the joint distribution of a response and an explanatory variable is bivariate normal. Show how this can be done graphically and discuss what you see.
   (d) Comment on whether bivariate normality is a requirement under the simple linear regression model.
6. In Figure 11.11, each residual was plotted against the family's location in the sequence of families as recorded in Galton's notebook. At no point in the discussion in this chapter was there a comment noting that a pattern was found relating to family sequence. It is recommended that semistudentized residuals and other measures of influence be plotted against sequence observation number. Yet this was not done. Do this. Report in 250 words or less on what patterns were discerned.
7. Galton's coding scheme for his 205 families used numbers 1 through 204, only. A family labeled 136A was inserted between families 135 and 136. Galton's coding scheme for family numbers is not obvious. But it is non-random.
   (a) Plot mid-parental height against family sequence in Galton's notebook. Join the points with line segments for visual effect.
   (b) Plot mid-offspring height against family sequence in Galton's notebook. Join the points with line segments for visual effect.
   (c) Using the plots for parts (a) and (b), develop a coding scheme and then check it against the original data. State Galton's coding scheme.
   (d) What implications are there for the least-squares fit of the simple linear regression model for mid-offspring height as a function of mid-parental height? Is there any cause for concern?
8. Andrews and Herzberg [4] produced a book published in 1985 that consists only of multiple data sets and their stories. The data sets can be downloaded from the StatLib archive at http://lib.stat.cmu.edu/datasets/Andrews. For this exercise, consider the urine data in their Table 44.1. Consider only the controls without calcium oxalate crystals in their urine. Exercises in the previous chapter delved into depicting the marginal and joint distributions of urea and calcium concentrations.
   (a) Model urea concentration as a simple linear function of calcium concentration. Report the estimates and two-sided p-values for the regression coefficients.

(b) Produce a scatterplot with the least-squares regression line. Be sure the fitted line is banked to 45° as recommended by Cleveland [22].

(c) Conduct a residual analysis. Report your conclusions in 250 words or less with accompanying graphical displays.

9. Refer to the previous exercise.

   (a) Conduct a diagnostic analysis of the influence measures. Report your conclusions in 250 words or less with accompanying graphical displays.

   (b) Write a brief concluding paragraph summarizing both the residual and influence analyses.

10. Consider the urine data for Exercise 8. Focus only on the controls without calcium oxalate crystals in their urine. Urea concentration is thought to regulate osmolarity of urine. Model osmolarity in a simple linear regression model dependent only on urea concentration. Produce all necessary figures required for a 500-word discussion of the model fit.

11. Consider the urine data for Exercise 8. Focus only on the cases with calcium oxalate crystals in their urine. Urea concentration is thought to regulate osmolarity of urine. Model osmolarity in a simple linear regression model dependent only on urea concentration. Produce all necessary figures required for a 500-word discussion of the model fit.

12. Find a bivariate dataset of your choosing and do a complete linear regression analysis using everything you have learned so far in this book.

Chapter 12

Polynomial Regression and Data Smoothing: Graphical Displays

12.1 Introduction

For the case study for this chapter, the data source will be the research trawl on the freshwater lake known as Längelmävesi near the city of Tampere in Finland, as reported by Brofeldt [14] in 1917. Each fish caught was weighed, and measurements taken of width, height, length. Three different measurements were taken of length:

- from the tip of the nose to the beginning of the tail;
- from the tip of the nose to the notch in the tail; and
- from the tip of the nose to the end of the tail.

Catch-and-release regulations for sport anglers usually specify minimum lengths, and sometime maximum lengths, according to length measured from the tip of the nose to the notch in the tail. This avoids problems associated with the accuracy of weigh scales carried by sports fishers. Rulers or tapes for measuring length are less prone to error.

Fisheries researchers, on the other hand, carry accurate weigh scales with them. Moreover, species biomass for a given lake is measured in kilograms. Fisheries regulations are drafted based on management of biomass. Decisions to preserve breeding stock are based on size as measured by mass of a fish, not its length.

The problem for the fisheries regulator is different from the typical situation in engineering for the so-called *calibration problem*, which seeks to substitute a cheaper measure for a more expensive one. In engineering the cheaper measurement is made and then converted to what could be expected to be the more expensive measurement. In the setting of fisheries management, determining the mass of fish is more expensive than a measurement of length. But

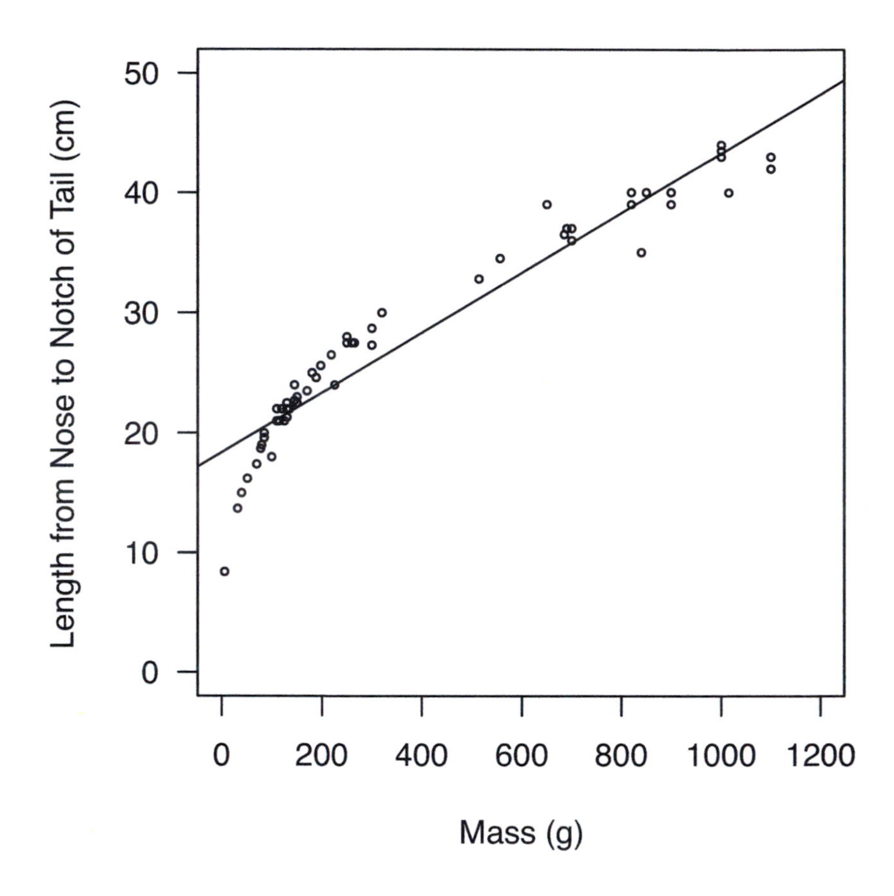

Figure 12.1 *Scatterplot of 56 perch with a least-squares regression line modeling length as a function of mass*

the mass of fish doesn't need to be estimated by the sports angler. It is much easier if the catch-and-release regulations refer only to length.

Catch-and-release regulations generally are species specific. For this case study, the national fish of Finland has been selected: the perch (*Perca fluviatis*). If the only tool one has is a hammer, then everything looks like a nail. Figure 12.1 depicts a scatterplot for a simple linear regression model of perch length, in centimeters, as a function of mass, in grams. Length is measured from tip of the nose to the notch in the tail. Evidently, the fit is so bad that further diagnostic plots are not necessary. But the shape of the dependency suggests that the addition to the model of terms associated by mass squared or mass cubed, or both, might do the trick. This is the topic of the next section.

Regression Coefficient	Estimate	Standard Error	Two-Sided p-Value
β_0	1.225×10^{1}	6.413×10^{-1}	$< 2 \times 10^{-16}$
β_1	8.322×10^{-2}	6.025×10^{-3}	$< 2 \times 10^{-16}$
β_2	-1.010×10^{-4}	1.273×10^{-5}	1.60×10^{-10}
β_3	4.693×10^{-8}	7.560×10^{-9}	9.03×10^{-9}

Table 12.1 *Summary of results for the fit of a polynomial regression model to third order of tip-to-notch length as a function of weight in 56 perch*

12.2 Learning Outcomes

When you complete this chapter, you will be able to do the following.

- Know the definition of the polynomial regression model, be able to fit it to data, and produce any and all residual plots, including an influence plot.
- Understand the mathematical model for spline approximation and be able to use the function `smooth.spline` in the R package `stats` to fit a spline approximation to data.
- Know the definition of the locally weighted (LOWESS) polynomial regression model and be familiar with the `stats` functions `lowess` and `loess` for fitting this model to data.
- Be able to depict LOWESS smoothing for bivariate data using either the `graphics` package or `ggplot2` package.
- Be able to use the R function `locpoly` in the package `KernSmooth` to do local kernel regression smoothing of bivariate data.

12.3 The Polynomial Regression Model

An order $p-1$ *polynomial regression model* is given by

$$y_i = \beta_0 + \beta_1 x_i + \beta_2 x_i^2 + \cdots + \beta_{p-1} x_i^{p-1} + \epsilon_i \tag{12.1}$$

where y_i is the response variable and x_i is the explanatory variable for $i = 1, 2, \ldots, n$. The random errors $\{\epsilon_i\}$ are assumed to be independently normally distributed with mean 0 and variance σ^2. The parameters $\{\beta_0, \beta_1, \beta_2, \ldots, \beta_{p-1}\}$ are the *regression coefficients.* Note that there are p regression coefficients in total as the sequential numbering begins with zero for the intercept. All p regression coefficients are typically unknown and are estimated from a simple random sample of ordered pairs $\{(x_i, y_i)\}$.

The top panel in Figure 12.2 is a scatterplot of tip-to-notch length versus weight for 56 perch with a cubic polynomial model. The estimated coefficients and their standard errors are listed in Table 12.1. Also listed in Table 12.1 are

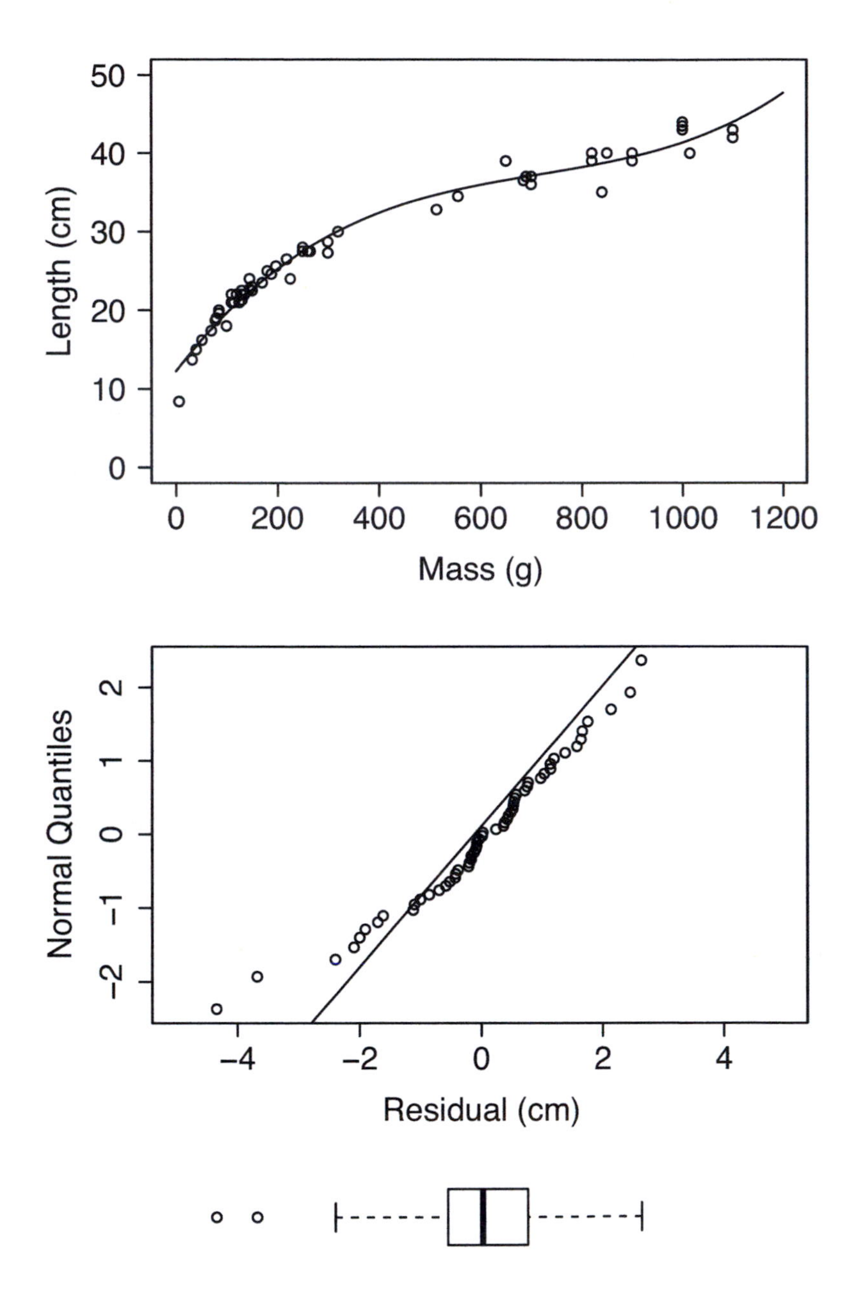

Figure 12.2 *The top panel is a scatterplot of 56 perch with a least-squares fit of a third-order polynomial modeling length as a function of mass, the middle panel is a scatterplot of residuals versus weight, and the bottom panel is an outlier boxplot of the residuals*

p-values for the corresponding hypothesis test that each regression coefficient is zero against a two-sided alternative.

Not reported in Table 12.1 are the following summary statistics. The residual standard error for the fit is 1.401 on 52 degrees of freedom. The squared multiple correlation $R^2 = 0.9772$. The p-value for the F-statistic is less than 2.2×10^{-16}, which indicates that the fit is statistically significant.

The normal quantile-quantile plot of residuals, the middle panel in Figure 12.2, suggests that the normal assumption for the model appears to be reflected in the data. The outlier boxplot of the residuals, the bottom panel of Figure 12.2, appears to be symmetric, as would be expected for a normal distribution. The two outliers on the left of the outlier boxplot correspond to fishes #104 and #143.

Fish #104 is the runt of the perch catch with a length of only 7.5 cm. Fish #143 has been encountered in an analysis in a previous chapter as an outlier. It was found to have had 6 roach in its stomach. Both fish are reasonable biological specimens and there is no compelling reason to remove them from the analysis.

As a further check on the residuals, they have been plotted against the response variable of length in the top panel of Figure 12.3. The residuals are plotted against the explanatory variable of mass in the bottom panel. Although the bimodality of size for this predator is apparent, there are no apparent patterns in the residuals themselves. This concludes the residual analysis for the cubic polynomial model.

Three measures of influence are depicted in the bubble plot of Figure 12.4. There is nothing really remarkable upon which to comment regarding influential points.

All in all, the polynomial fit for determining length from the mass of perch has worked out quite well. This is not always the case for polynomial regression. Were the interest in the original concept of the calibration, that is, requiring sports anglers to measure the length of their catch and then use a calculator to estimate the mass with regression coefficients for a cubic polynomial conversion as supplied by the Finnish fisheries service, the outcome would not have been nearly as good.

Figure 12.5 displays the third-order polynomial regression model for mass as a function of fish length from the tip of the nose to the notch in the tail. The fit is particularly bad for fish less than 10 cm in length. Without even plotting the residuals against mass and length, it is apparent in Figure 12.5 that there will be patterns. This is revealed as expected in Figure 12.6, which displays a scatterplot of the residuals against length and mass.

In both panels of Figure 12.6, as either length or mass increase, the amount of spread in the residuals fans out. But this should come as no surprise given

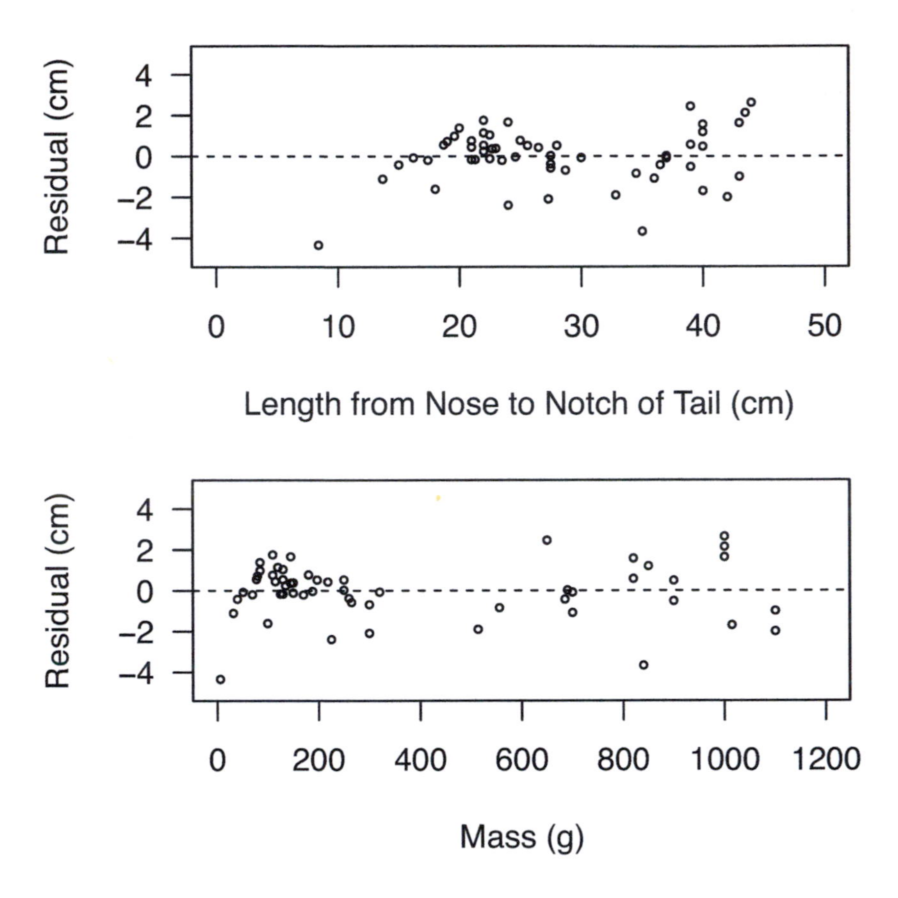

Figure 12.3 *The top panel is a scatterplot of residuals versus length, the bottom panel is a scatterplot of residuals versus mass, for length as a cubic polynomial function of mass*

the bimodality detected in the two-dimensional kernel density estimate in Figure 10.23 (and also depicted in the stereographic pair in the cover art for this second edition). The observation in Chapter 10 was that the major mode depicted prey and the minor mode predator size-ranges of this single species of perch. This spread in variance from prey to predator, also known as heteroscedasticity, cannot be fixed by adding more polynomial terms to a unimodal model.

When polynomial regression works, it works well. Lower order polynomial models work best in a local area over a small interval along the horizontal axis. Globally, over the full range of the data, polynomials typically do not work consistently well. Alternative local polynomial models are discussed in the next two sections.

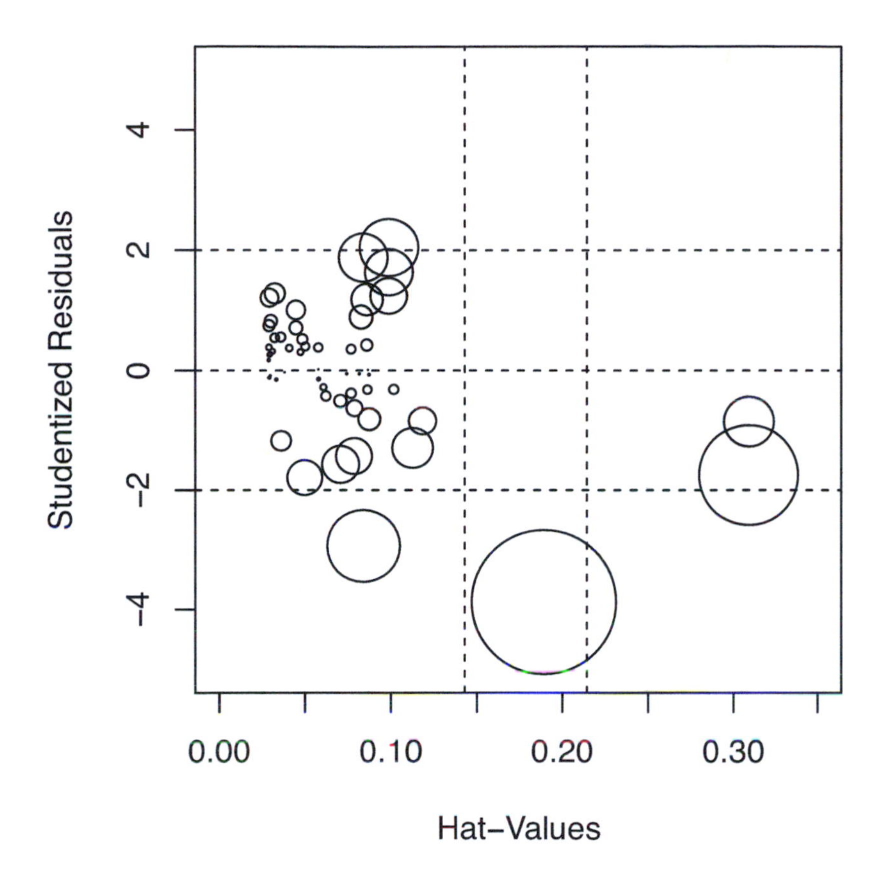

Figure 12.4 *Influence plot for least-squares fit of a cubic polynomial for tip-to-notch length as a function of mass (the area of each circle is proportional to Cook's distance, and the black circles correspond to the 5 largest values for Cook's distance; otherwise, the circles are gray)*

12.4 Splines

An alternative to using kernel density estimation for fitting a smoothed histogram to data or fitting a straight line or other simple curve through data is to use a spline approximation. *Splines* were jointed rubber, wooden, or metal strips that were previously used in engineering and physics practice to draw or interpolate the shape of an object between predetermined points called *knots*. This is now done on a computer with a variety of methods available for representing splines quite quickly.

Spline approximation is a three-stage process:

(i) determine the knots;

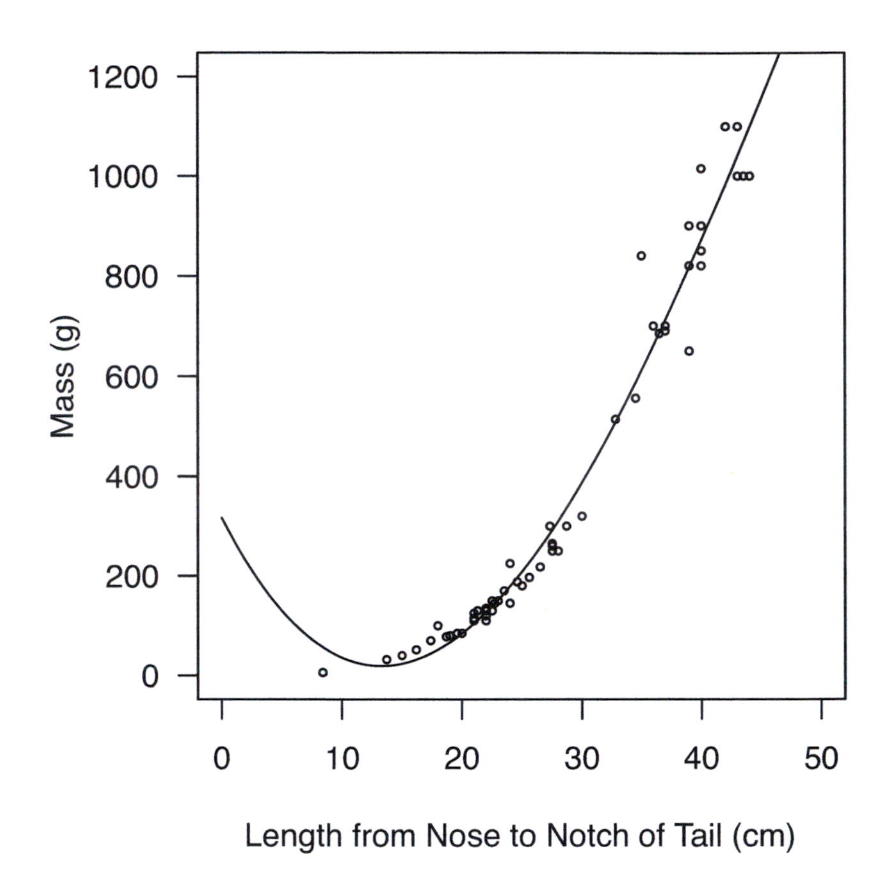

Figure 12.5 *Scatterplot of mass versus length with least-squares fit of a cubic polynomial for mass as the response variable and length as the explanatory variable*

(ii) find the approximating equation that fits the data reasonably well according to some criterion; and

(iii) evaluate the interpolation function at the required points.

There are two additional considerations: the smoothness or stiffness of the curve passing through the data, and whether the spline curve must pass through the knot points. Typically, in statistical applications, since there can be more than one ordinate for a given value on the abscissa, especially if replication is involved, there is no requirement for the spline function to pass through knot points.

If the knot points for a spline are taken to be $\{x_i\}$ for $i = 0, 1, 2, \ldots, k+1$ and a local basis function $b_i(x)$ is nonzero only if $x_{i-1} \leq x \leq x_i$, then the spline

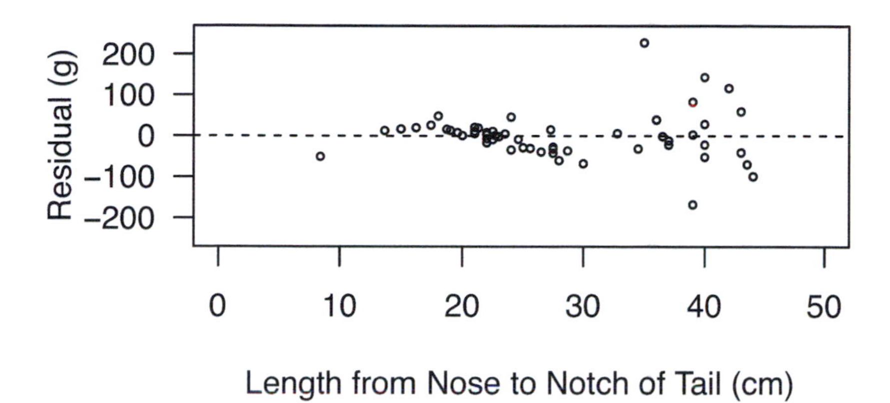

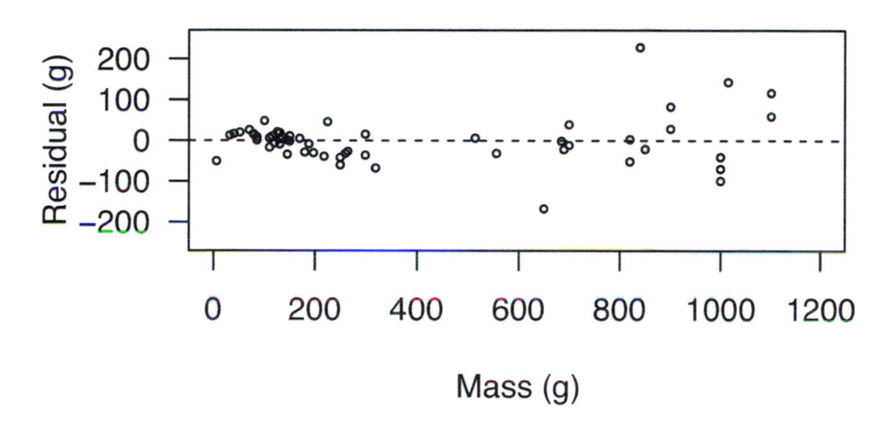

Figure 12.6 *The top panel is a scatterplot of residuals versus length, the bottom panel is a scatterplot of residuals versus mass, for mass as a cubic polynomial function of weight*

function through the data $h(x)$ can be represented as

$$h(x) = \sum_{i=1}^{k+m} a_i b_i(x) \tag{12.2}$$

where the $\{b_i\}$ are polynomials of some order $m = p - 1$. The most frequent choice of m is 3. It is typical to require the pieces to join smoothly so that the derivatives up to order $m - 1$ must be continuous at the knots. If one sets $m = 1$ and requires only that the spline function is continuous, then the spline becomes a piecewise linear function through the data.

The criterion in the R function `smooth.spline` for finding a cubic smoothing spline that bests fits the data is finding the function $\hat{h}$ that minimizes the

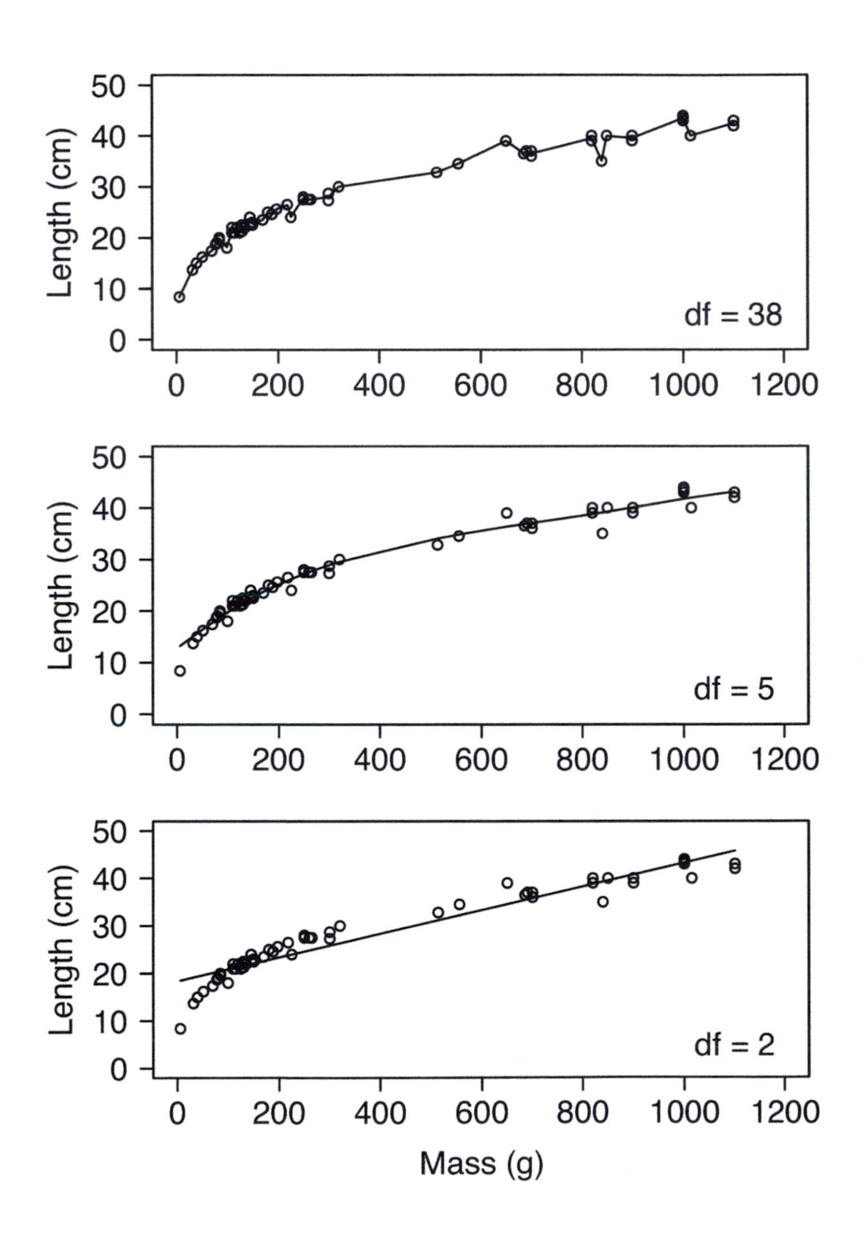

Figure 12.7 *Scatterplots of nose-to-notch length versus mass with cubic splines (the degrees of freedom associated with each spline is indicated in the lower right of each panel)*

Polynomial Regression Sum of Squares

$$PRSS = \sum_{i=1}^{N} [y_i - h(x_i)]^2 + \lambda \int [h''(t)]^2 \, dt \tag{12.3}$$

at the N unique values of the explanatory variable x_i. The smoothing parameter λ controls the trade-off between fidelity to the data and smoothness. It is possible to specify values of λ for the R function `smooth.spline`. Alternatively, the degree of smoothness in the fit produced by `smooth.spline` can be controlled by the degrees of freedom

$$2 \leq df \leq k + 2. \tag{12.4}$$

Three examples of cubic splines fitted through data are presented in Figure 12.7 for perch length from the tip of the nose to the notch in the tail as a function of mass. The splines were produced by the R function `smooth.spline` with the degrees of freedom as indicated in each panel in Figure 12.7.

The top panel in Figure 12.7 with the jagged spline has the greatest stiffness. That is to say, the spline in the top endeavors to get as close as possible to each point. The bottom panel has the loosest fit: a straight line in effect. The most eye-appealing spline of the three is depicted in the middle panel. It takes a certain amount of trial and error before finding a value for the degrees of freedom that produces an eye-appealing spline.

The smallest value for the degrees of freedom for the R function `smooth.spline` is 2. The largest is the number of unique horizontal values. There are 41 unique values for mass, so the maximum number of degrees of freedom is 41.

Because there was a global fit problem with a cubic polynomial for mass as a function of length, cubic splines were used in Figure 12.8 for this reversal of axes compared to Figure 12.7. Note that the three local cubic spline fits depicted in Figure 12.8 appear to be fine.

Because there are only 39 unique values of nose-to-notch length among the 56 perch, the maximum degrees of freedom for mass as a function of length is 39, as indicated in the top panel of Figure 12.8.

Splines were originally used by scientists and engineers when graphical displays of mathematical functions were done by hand with pen and ink. Computation was time intensive with the use of tables of logarithms, or slide rules, prior to electronic calculators. Several points would be calculated, and the draftsman, or draftswoman, would fill in the gaps with a few well-placed splines. The method of splines is not ideally suited for a statistical situation with multiple ordinates for given values of the abscissa. The method of the next section was developed more along the lines of statistical practice.

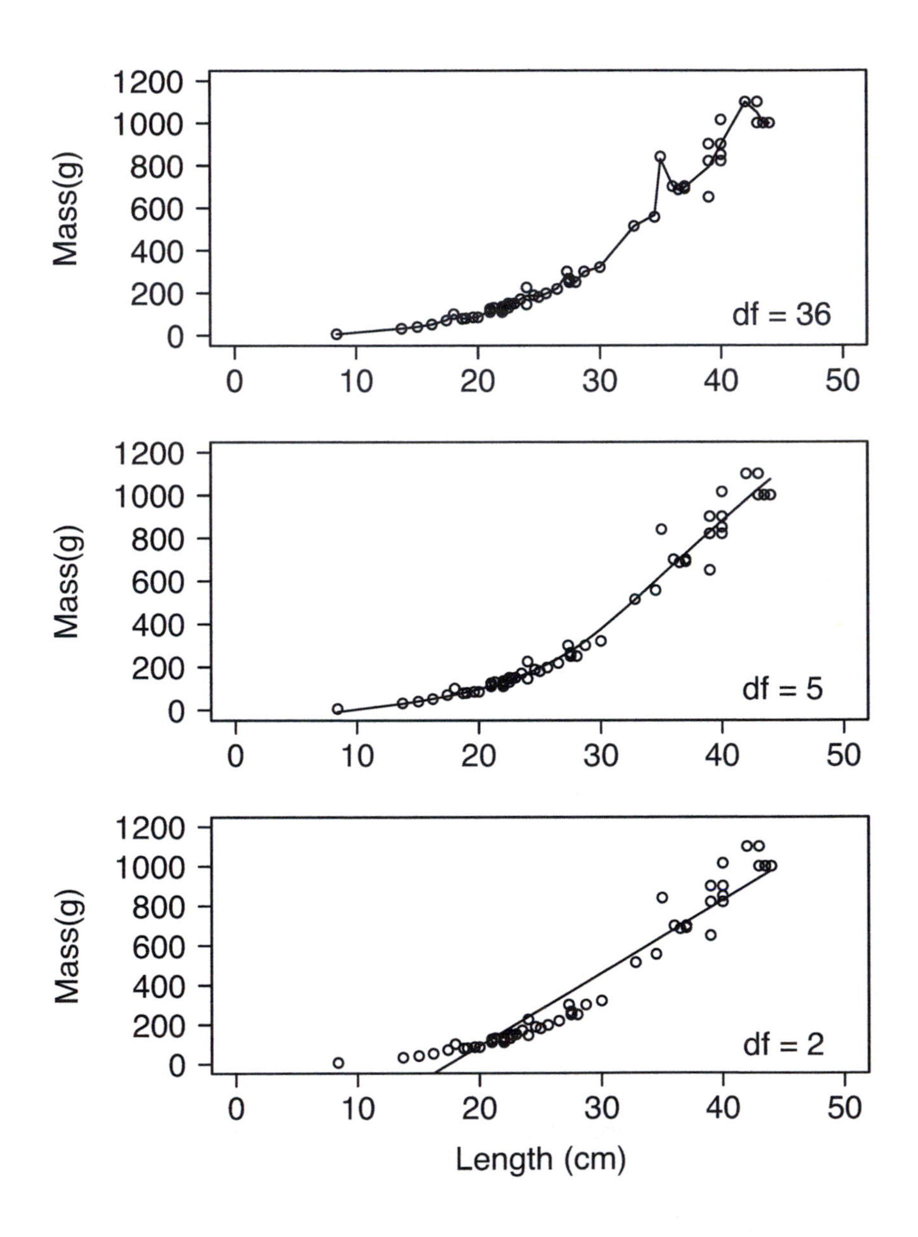

Figure 12.8 *Scatterplots of mass versus nose-to-notch length with cubic splines (the degrees of freedom associated with each spline is indicated in the lower right of each panel)*

12.5 Locally Weighted Polynomial Regression

The gist of smoothing is to find some smooth function g for ordered pairs $\{(x_i, y_i)\}$ of data such that

$$y_i = g(x_i) + \epsilon_i \tag{12.5}$$

to good approximation where the $\{\epsilon_i\}$ are random variables with a mean 0 and a constant variance σ^2. A *locally weighted regression smoother* g estimates the response variable at a point x by computing a weighted average of all those values y_j in the sample that have explanatory values x_j close to x. Moreover, the weights depend smoothly on this closeness. For the situation of a *locally weighted polynomial regression smoother*, the function g must be a polynomial.

A *weight function* W must have the following properties:

- $W(x) > 0$ for $|x| < 1$ (nonnegativity);
- $W(-x) = W(x)$ (symmetry);
- W is a nonincreasing function for $x \geq 0$ (nonincreasing); and
- $W(x) = 0$ for $|x| \geq 1$ (nullity outside the interval $[-1, 1]$).

For each x_i, the weights $w_k(x_i)$ are defined, for all x_k with $k = 1, 2, \ldots, n$, using the weight function W centered at x_i and scaled so that the point at which W first becomes zero is at the rth nearest neighbor of x_i.

If h_i denotes the distance from x_i to the rth nearest neighbor of x_i, then

$$w_k(x_i) = W\left(\frac{x_k - x_i}{h_i}\right). \tag{12.6}$$

As with the kernel in kernel density estimation, there are many different possible choices for the weighting function W. In 1979, Cleveland [20] proposed the following two choices. *Tukey's biweight function* is defined by

$$B(x) = \begin{cases} (1 - x^2)^2 & \text{if } |x| < 1, \\ 0 & \text{if } |x| \geq 1. \end{cases}$$

The *tricube function* is defined by

$$C(x) = \begin{cases} (1 - |x|^3)^3 & \text{if } |x| < 1, \\ 0 & \text{if } |x| \geq 1. \end{cases}$$

As an alternative to specifying r to determine the nearest neighbor distance, Cleveland [20] defines the *span* f such that $0 < f \leq 1$ and $f = r/n$.

Cleveland's [20] algorithm for locally weighted polynomial regression and robust locally weighted polynomial regression is as follows.

1. For each i calculate the weighted least-squares estimates $\{\hat{\beta}_0(x_i), \hat{\beta}_1(x_i), \ldots, \hat{\beta}_{p-1}(x_i)\}$ of the regression coefficients in a polynomial of degree $m = p-1$ using weight $w_k(x_i)$ for ordered pair (x_k, y_k). That is, the values $\left\{\hat{\beta}_j(x_i)\right\}_{j=1}^{p-1}$ are the values of $\{\beta_j\}$ that minimize

$$\sum_{k=1}^{n} w_k(x_i)\left(y_k - \beta_0 - \beta_1 x_k - \cdots - \beta_{p-1}x_k^{p-1}\right)^2. \tag{12.7}$$

   The smoothed value of the response variable at x_i is the fitted value of this regression:

$$\hat{y}_i = \sum_{j=0}^{p-1} \hat{\beta}_j(x_i)x_i^j. \tag{12.8}$$

   This completes the algorithm for *locally weighted polynomial regression.*

2. Let the residuals from the current fit be given by

$$e_i = y_i - \hat{y}_i. \tag{12.9}$$

   Let $\tilde{e}$ denote the median of the absolute deviations $\{|e_i|\}$. Define *robustness weights* by

$$\delta_k = B\left(\frac{e_k}{6\tilde{e}}\right). \tag{12.10}$$

3. Compute a new fitted value $\hat{y}_i$ for each i by fitting an mth degree polynomial using weighted least squares with weight $\delta_k w_k(x_i)$ at (x_k, y_k) for $k = 1, 2, \ldots n$.
4. Cycle through steps 2 and 3 a total of t times. The final $\{\hat{y}_i\}$ are the *robust locally weighted polynomial (LOWESS) regression* fitted values.

Figure 12.9 adds a LOWESS curve to the scatterplot of nose-to-notch length versus mass for the 56 perch caught during a research trawl of Längelmävesi. The curve was computed using the R function `lowess` in the `stats` package with the following default values: the degree of the polynomial is $m = 1$; the neighborhood fraction $f = 2/3$; the number of iterations $t = 3$; and the tricube function C was used to create the weights $w_k(x_i)$. The R code for producing Figure 12.9 is as follows.

```
plot(Perch$Weight,Perch$Length2,
ylab="Length from Nose to Notch of Tail (cm)",
xlab="Mass (g)",ylim=c(0,50),xlim=c(0,1200),
las=1,cex=0.5)

lines(lowess(Perch$Weight,Perch$Length2,delta=0))
```

An option in the function `lowess` to speed up computation was not used. Instead of computing the local polynomial fit at each data point, it is not

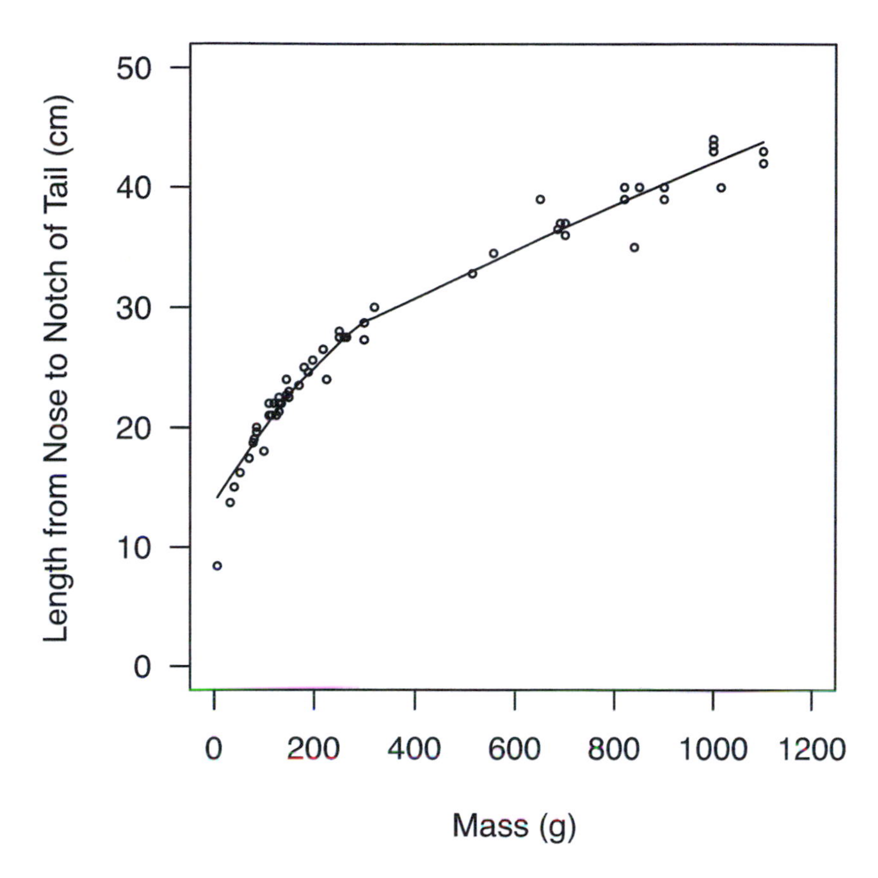

Figure 12.9 *Scatterplot of 56 perch with a locally weighted polynomial regression (LOWESS) curve modeling length as a function of mass*

computed within a distance `delta` of the last computed point with linear interpolation being used to fill in the fitted values for the skipped points. The default value for the `lowess` function parameter `delta` is the range of the explanatory variable x divided by 100. In setting `delta=0` for producing Figure 12.9, the decision has been made not to speed up computations as they occur within a blink of an eye for the observations on 56 perch.

Figure 12.10 is the `ggplot2` version of Figure 12.9 but with 95% confidence intervals added. The code for generating Figure 12.10 is as follows.

```
figure<-ggplot(Perch,aes(x=Weight,y=Length2)) +
geom_point(shape=21) +
geom_smooth(method="loess") +
theme(axis.title=element_text(size=12),
axis.text=element_text(size=12)) +
```

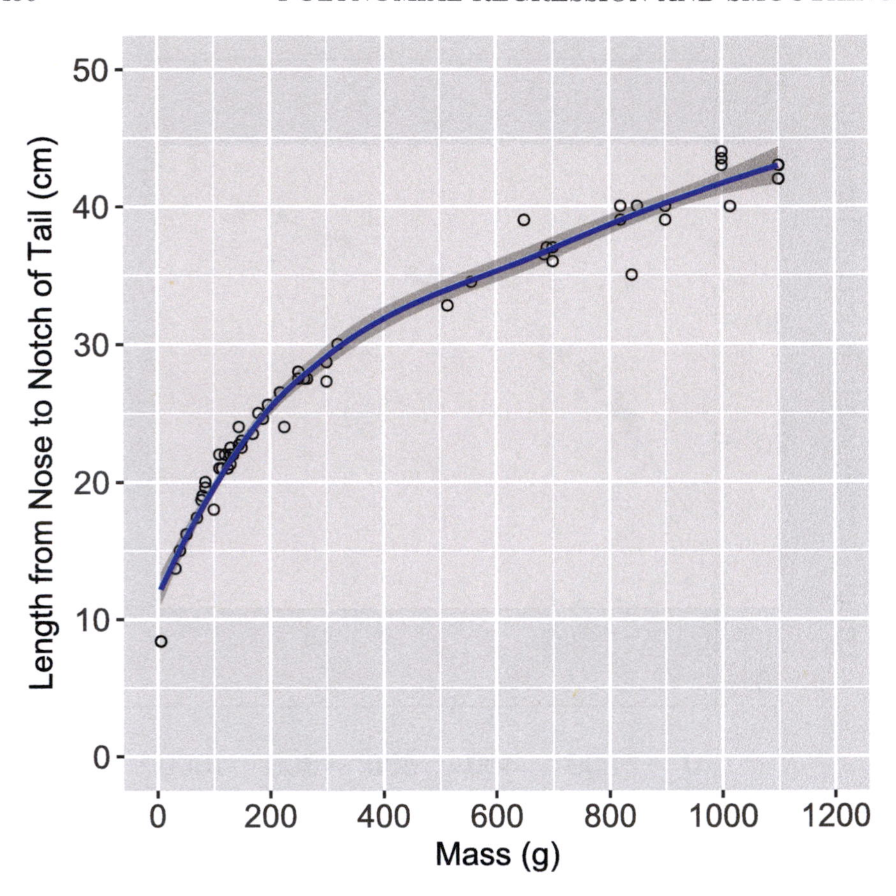

Figure 12.10 *Scatterplot of 56 perch with a locally weighted polynomial regression (LOWESS) curve modeling length as a function of mass with 95% confidence interval using* ggplot2

```
scale_x_continuous(breaks=seq(0,1200,200),
limits=c(0,1200)) +
scale_y_continuous(limits=c(0,50)) +
labs(x="Mass (g)",
y="Length from Nose to Notch of Tail (cm)")

print(figure)
```

Because reliance is made on all the defaults save font size for the axis titles and labels, this is about as parsimonious as one gets when using `ggplot2`. Notice that the LOWESS curve is blue and the 95% confidence interval is a light semi-transparent overlay.

A close comparison of the LOWESS curves in Figures 12.9 and 12.10 reveals

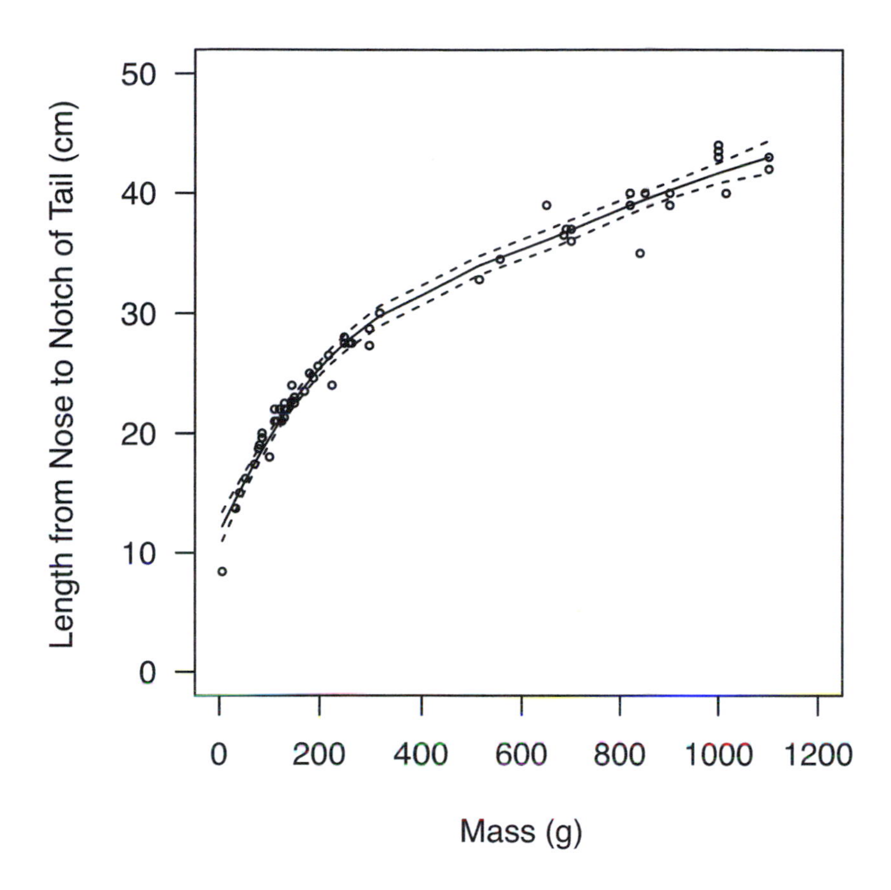

Figure 12.11 *Scatterplot of 56 perch with a locally weighted polynomial regression (LOWESS) curve modeling length as a function of mass with 95% confidence interval*

that the curves are in fact not the same. The reason for this is that the two figures were produced by two different implementations of locally weighted polynomial regression in the `stats` package in R. The following code using the `stats` and `graphics` packages reproduces a black-on-white version of `ggplot2` Figure 12.10 in Figure 12.11.

```
plot(Perch$Weight,Perch$Length2,
ylab="Length from Nose to Notch of Tail (cm)",
xlab="Mass (g)",ylim=c(0,50),xlim=c(0,1200),
las=1,cex=0.5)

uwt<-data.frame(
Weight=unique(sort(Perch$Weight)))
a<-predict(loess(Length2   Weight,Perch),
```

```
newdata=uwt,se=TRUE)
cv<-qt(0.025,df=a$df)

lines(uwt$Weight,a$fit)
lines(uwt$Weight,a$fit-cv*a$se.fit,lty=2)
lines(uwt$Weight,a$fit+cv*a$se.fit,lty=2)
```

This is one of the rare instances in which the `ggplot2` code requires fewer characters. There are a few things happening in the code above. To plot correctly, the explanatory variable `Weight` is sorted and the non-unique values must be removed and the resulting numerical vector stored in a new `data.frame`. The `stats` function `loess` is called to do the locally weighted polynomial regression, which is passed as an argument to the `predict` function to produce the list `a` that contains the prediction and its standard error for each sorted, unique value of `Weight` in the `newdata` list `uwt`. The variable `df` produced by `predict` and stored in the list `a` is an estimate of the effective degrees of freedom used in estimating the residual scale, intended for use with t-based confidence intervals. The `stats` function `qt` is then called with the degrees of freedom (which are not necessarily integer) to find the corresponding critical value from Student's t distribution to be used for an approximate 95% confidence interval. The concluding three calls to the `graphics` function `lines` overlay the fitted values, lower confidence limits, and upper confidence limits, respectively. In `ggplot2`, this all happens by issuing the command `geom_smooth(method="loess")`, which is much easier.

The function `loess` is more recent than `lowess`. The function `loess` can do more than `lowess` as `loess` can accommodate more than one response variable, produce standard errors for predicted values, accept prior weights, estimate the equivalent number of parameters in the resulting model, and it accepts a formula rather than vectors or matrices. On the other hand, `lowess` is easier to use, executes faster, and can succeed when `loess` fails. So it is expected that both `lowess` and `loess` will be around for some time in the `stats` package.

It is possible to tweak the function `loess` to produce the same results as `lowess`. The two functions count iterations differently. The iterations are the same when `iter` is set to one more in `loess`. The default span f in `lowess` is `f = 2/3` while the default in `loess` is `span = 0.75`. To get the same result, pick a common value for the span f. By far and away the major difference between the two functions is that `lowess` only does linear local regression, whereas the default for `loess` is quadratic. But this can be overcome in `loess` by setting and passing `degree = 1`. Setting `delta` to a non-zero value in `lowess` was formerly used to speed up computation by using linear interpolation to fill in for the skipped points. This has not been needed as much as it was ten years before the printing of the First Edition of this book. Setting `delta= 0` results in a local polynomial fit at each data point.

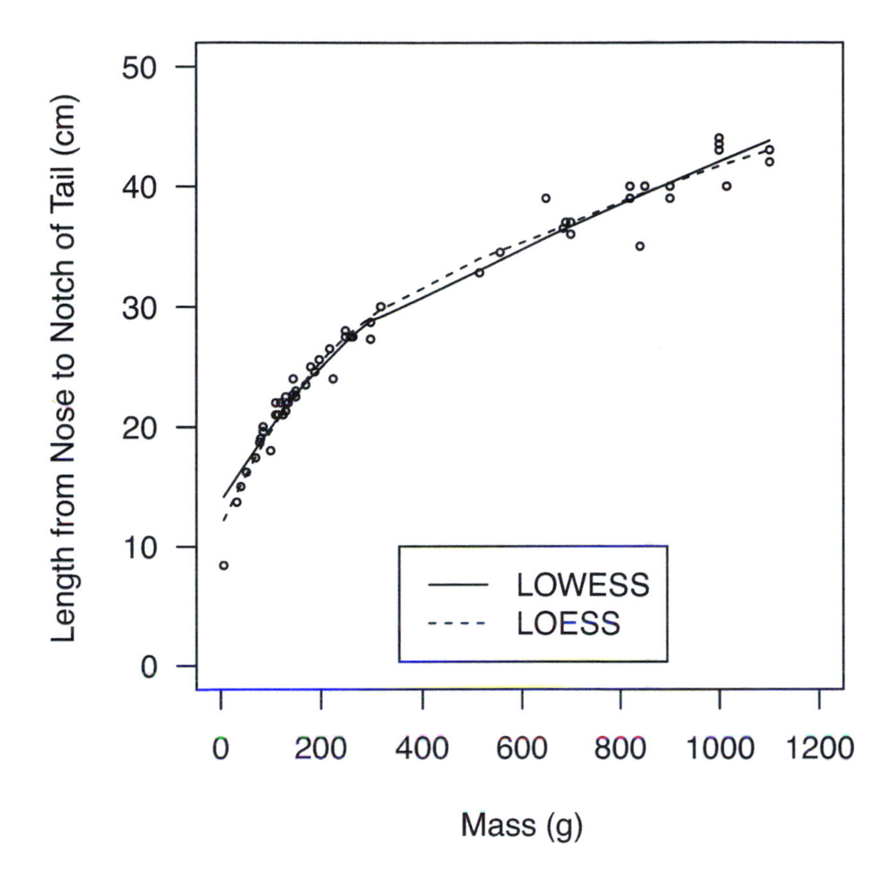

Figure 12.12 *Scatterplot of 56 perch with a locally weighted polynomial regression curve modeling length as a function of mass using* `stats` *package functions* `lowess` *and* `loess`

Figure 12.12 shows just how close the locally weighted polynomial regression curves get for the perch data with the functions `lowess` and `loess`. Unfortunately, `lowess` is not an option for `geom_smooth` or its nearly synonymous alternative `stat_smooth` in `ggplot2`.

Figure 12.13 compares the locally weighted linear regression model with the globally fitted cubic polynomial for length as a function of mass for the 56 perch caught in the research trawl on Längelmävesi. The two models are reasonably close in their fitted values, except for the mid-range of fish mass between 300 and 700 grams.

Cleveland [20] makes recommendations concerning choices for the degree m of the polynomial, the weighting function W, the number t of iterations, and the fraction f of observations to be included for each local regression.

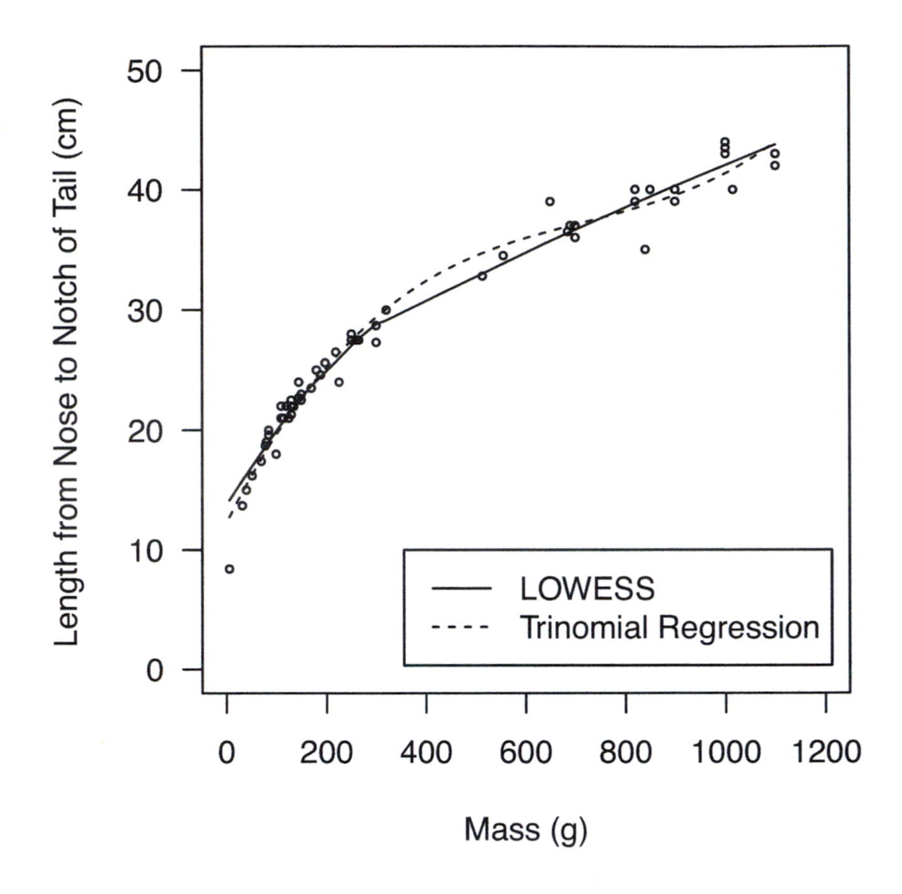

Figure 12.13 *Scatterplot of different models for length versus mass for 56 perch*

Cleveland [20] believes that "Taking $m = 1$ should almost always provide adequate smoothed points and computational ease." In fact $m = 1$ is the only choice available in the R function `lowess` and, based upon experience, this choice appears to work well.

Cleveland [20] noted that the choice of the tricube function C for the weight function W enhances the approximation of a chi-square distribution to an estimate of the error variance and provides adequate smoothing in most situations. The R function `lowess` only supports the tricube function.

Cleveland [20] reported that experimentation with a large number of real and artificial data sets indicated that two iterations ($t = 2$) should be adequate for almost all situations. The default in R function `lowess` is $t = 3$.

Cleveland [20] noted that increasing f tends to increase the smoothness of the smoothed points. The goal is to select a value for f that is as large as

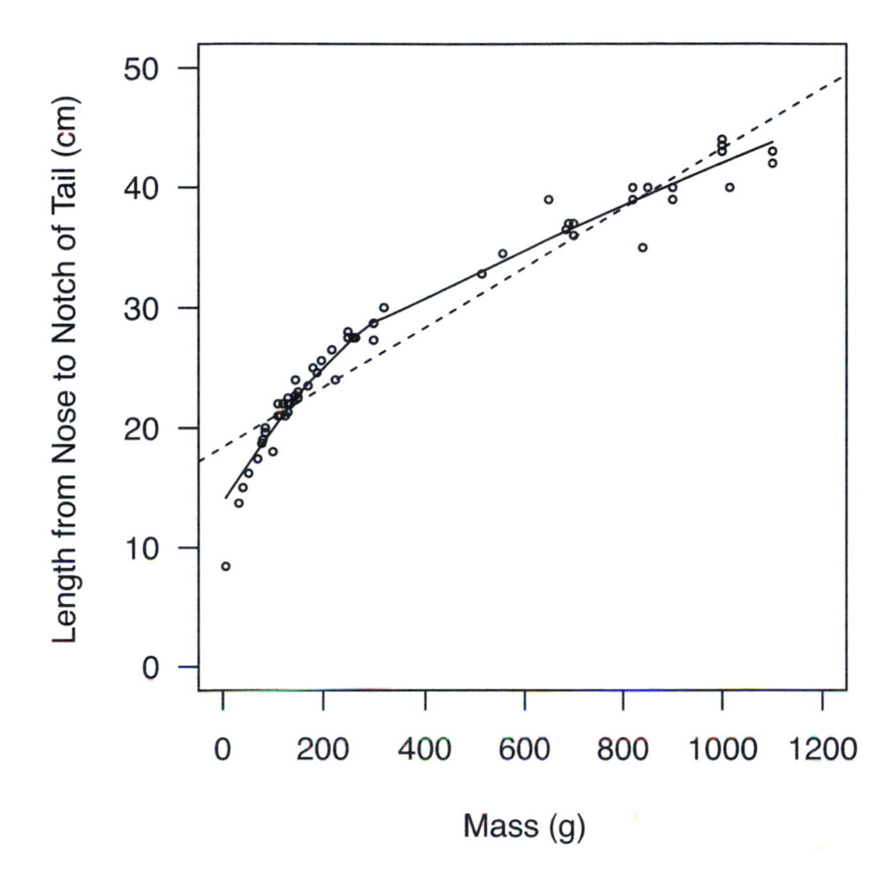

Figure 12.14 *Scatterplot of different models for length versus mass for 56 perch (the dashed line is that of the simple linear regression model, and the solid curve is LOWESS)*

possible to minimize the variability of the smoothed points without distorting the value of the pattern in the data. He recommended choosing f in the range 0.2 to 0.8 with $f = 0.5$ being a reasonable starting value. The default in the R function `lowess` is $f = 2/3$ but $f = 3/4$ in `loess`. This is consistent with the spirit of Cleveland's [20] recommendation.

The thesis behind Cleveland's [20] presentation of the theory of robust locally weighted regression is that "The visual information on a scatterplot can be greatly enhanced, with little additional cost, by computing and plotting smoothed points." This has been depicted in Figure 12.9 for perch length as a function of weight.

The LOWESS curve was added to the scatterplot of Figure 12.14 for comparison with the dashed line representing the least-squares fit of the simple

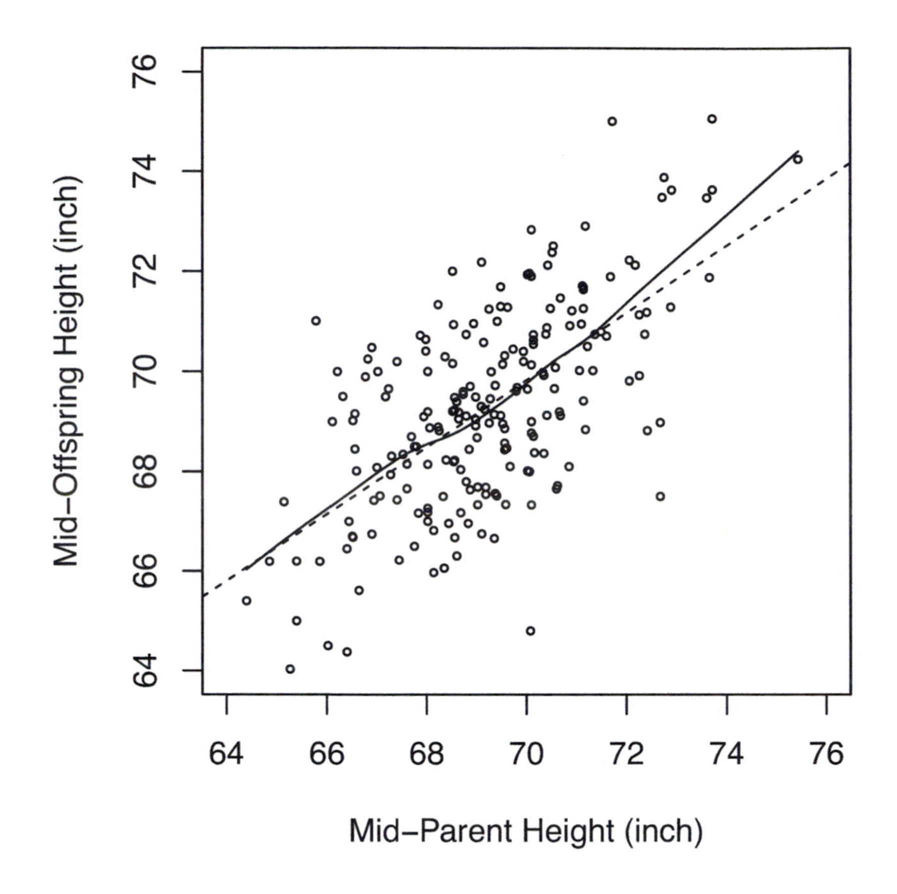

Figure 12.15 *Scatterplot of heights of mid-parents and mid-offspring*

linear regression model. It is there for diagnostic purposes. It is a visual aid for checking on whether there could be a better model than simple linear regression.

Figures like Figure 12.14 often appear without a key for the dashed line and solid curve in either an inset containing a legend or comments in the figure's caption, as in the scatterplot of Figure 12.15 of mid-offspring height versus mid-parental height for the 205 families collected by Galton in 1884 in his Records of Family Faculties. If the intended viewer, or viewers, is not familiar with this convention, then do supply the verbal cues.

The fidelity of the simple linear regression model for Galton's data is enhanced by the presence of the LOWESS curve and its proximity to the least-squares regression line.

Although not a formal hypothesis test in itself, the addition of the LOWESS

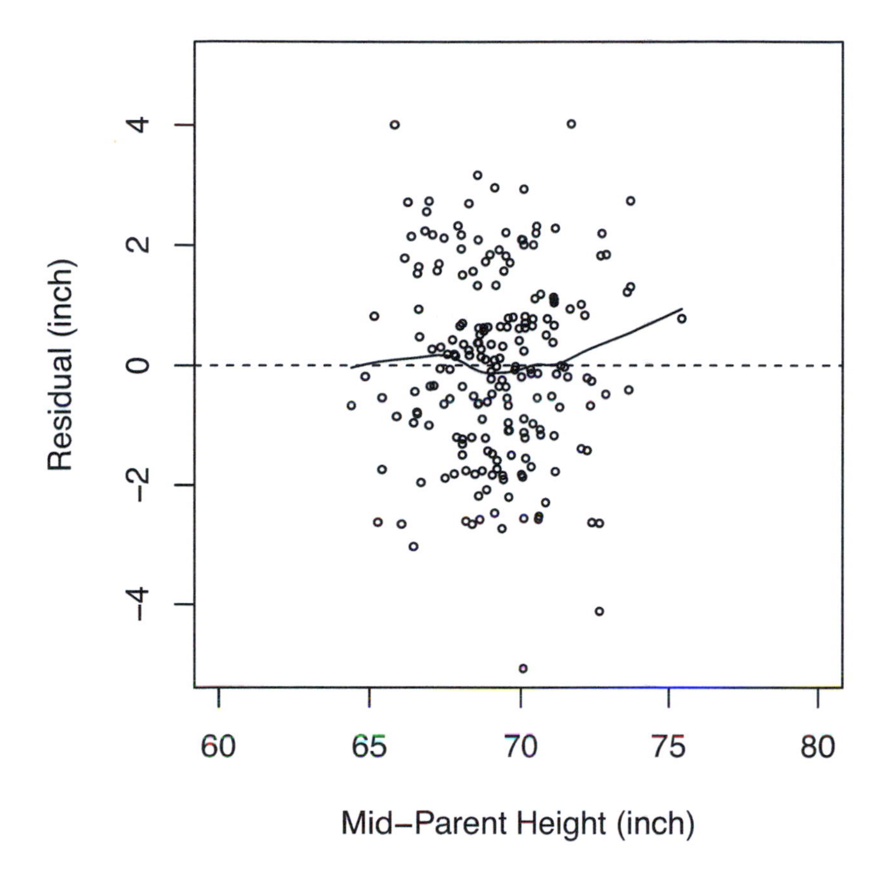

Figure 12.16 *Scatterplot of residuals of mid-offspring height versus heights of mid-parents*

curve to Figure 12.15 provides a visual comparison between the null hypothesis of the simple linear regression model and the broader model of locally weighted robust linear regression.

There is also a role for LOWESS in diagnostic plots. A single example is given in Figure 12.16 for residuals from the fit of the linear regression model of mid-offspring height as a function of mid-parental height. Human eyes can sometimes miss a pattern deviating from the reference line of zero residuals.

The addition of a LOWESS curve to a residual plot can help focus attention to where it is needed. In Figure 12.16 all looks well except perhaps for the upper extreme of mid-parental height.

The potential influence of family #1 with the tallest mid-parental height has been noted previously. But its impact is not thought to be unduly influential and so family #1 has been retained in the final analysis.

12.6 Conclusion

When used to smooth a scatterplot, the locally weighted polynomial regression model of Cleveland [20] is called a *scatterplot smoother* or a *regression smoother*.

The letters LOWESS represent the phrase LOcally WEighted Scatterplot Smoother. Examples of LOWESS curves have been given in Figures 12.9 through 12.16, inclusive. In their textbook of 1995, Wand and Jones [131] refer to Cleveland's [20] regression smoother as a *local polynomial regression smoother*. In so doing, they highlight the fact the weight functions used by Cleveland are, in fact, polynomial kernels.

The difficulties associated with kernel selection and bandwidth previously discussed in the context of kernel density estimation also apply to LOWESS, which implicitly uses a rectangular kernel. The R function `locpoly` is in the `KernSmooth` package designed to accompany Wand and Jones [131]. The function `locpoly` offers the choice of other kernels. Figure 12.17 compares the LOWESS smoother with a first-degree polynomial smoother that uses a Gaussian kernel. The difference between these two local linear regression estimators is slight.

Also depicted in Figure 12.17 is another local polynomial regression estimator with degree $m = 0$ and a Gaussian kernel. This smoother is also known as the *Nadaraya-Watson estimator* after its independent co-discoverers [85, 132] in 1964.

The Nadaraya-Watson estimator in Figure 12.17 does not appear to do as well as either local linear kernel regression smoother. This is consistent with comments by Cleveland [20] that degree $m = 1$ seemed to do better than $m = 0$.

This concludes the discussion of two variables, continuous or otherwise.

12.7 Exercises

1. Consider the data on the pike (*Esox lucius*) species caught during a research trawl on the Finnish lake Längelmävesi near Tampere. Pike is an important gamefish for anglers. This apex-predator species is ubiquitous in freshwater lakes throughout the northern hemisphere. Using appropriate graphical displays, show that the data do not support a model of nose-to-notch length as a linear function of mass.
2. Refer to the previous exercise concerning the species pike.

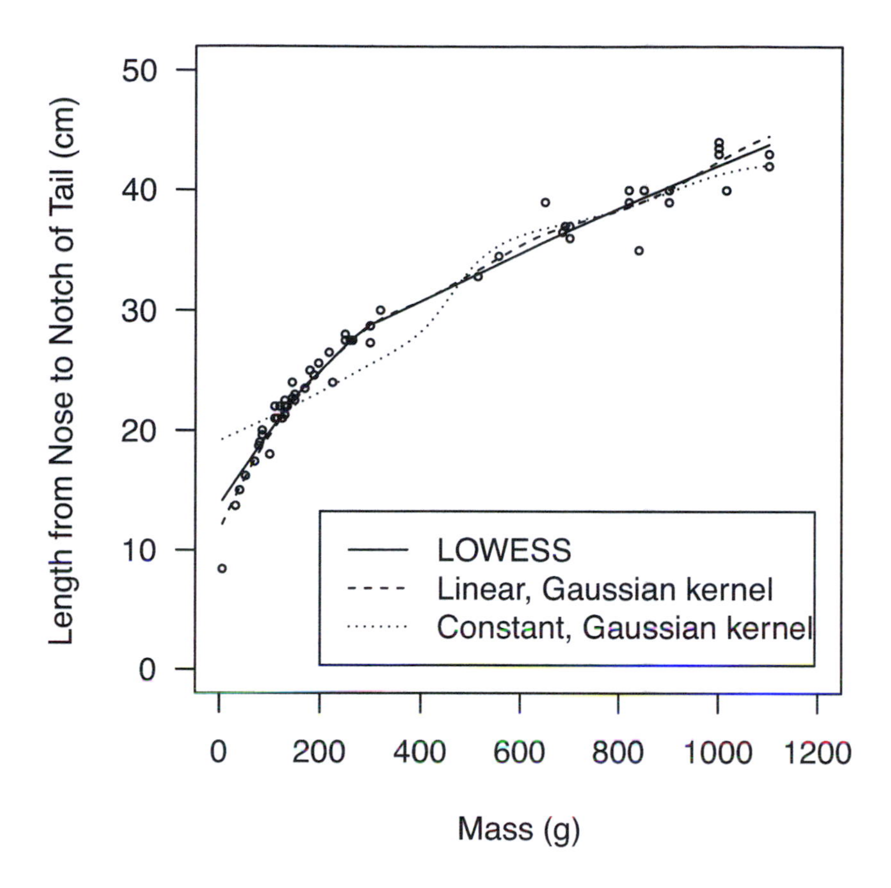

Figure 12.17 *Scatterplot of residuals of nose-to-notch length versus mass for 56 perch with different polynomial regression kernels*

(a) Fit a cubic polynomial regression for nose-to-notch length as the response variable and mass as the explanatory variable. Summarize and comment on the results.

(b) Conduct a regression analysis for the fit of part (a). Briefly comment.

(c) Conduct an influence analysis for the fit of part (a). Briefly comment.

(d) Summarize in 250 words or less the results of your diagnostic analysis from parts (b) and (c). Comment on the viability of modeling the length of pike as a cubic polynomial function of mass.

(e) Can you reasonably expect to set catch-and-release regulations for pike based upon the model fit of part (a)? Discuss.

3. Table 12.2 reports numbers of fatalities as reported by the *2005 and 2006 Canadian Motor Vehicle Traffic Collision Statistics*, published by Trans-

Year	Number of Fatalities	Licensed Drivers
1986	4,063	16,226
1987	4,283	16,927
1988	4,154	17,155
1989	4,238	17,592
1990	3,963	17,718
1991	3,690	18,090
1992	3,501	18,465
1993	3,615	18,843
1994	3,263	19,243
1995	3,351	19,327
1996	3,062	19,964
1997	3,033	20,148
1998	2,911	20,744
1999	2,984	20,934
2000	2,927	20,593
2001	2,776	20,879
2002	2,932	21,163
2003	2,768	21,436
2004	2,722	21,673
2005	2,905	21,937
2006	2,889	22,278

Table 12.2 *Fatalities as reported by the 2005 and 2006 Canadian Motor Vehicle Traffic Collision Statistics (published by Transport Canada)*

port Canada. Let the number of fatalities be the response variable. Let the year be the explanatory variable.

(a) Create a scatterplot of the fatalities versus year.

(b) Fit a simple linear regression model. Summarize the results.

(c) Perform a complete diagnostic analysis. Summarize the results.

(d) Based upon your answers to parts (b) and (c), does the simple linear regression model explain the pattern in the numbers of fatalities? Discuss.

4. Table 12.2 reports numbers of fatalities as reported by the *2005 and 2006 Canadian Motor Vehicle Traffic Collision Statistics*, published by Transport Canada. Let the number of fatalities be the response variable. Let the year be the explanatory variable. (Suggestion: recode the year by subtracting 1985 from each value.)

(a) Create a scatterplot of the fatalities versus year. Add a LOWESS curve to the scatterplot.

(b) Fit a fourth-degree polynomial regression model. Summarize the results.

(c) Perform a complete diagnostic analysis. Summarize the results.

(d) Based upon your answers to parts (a) and (b), does the quartic polynomial regression model explain the pattern in the numbers of fatalities? Is it possible to use instead a polynomial regression model of lower degree? Discuss.

5. Refer to Table 12.2. The previous exercise required that the number of fatalities be the response variable and time be the explanatory variable. Given that Canada has universal health insurance, it is not unreasonable to model the number of fatalities as a function of year because healthcare for mortally injured patients incurs real healthcare costs. The third column in Table 12.2 reports estimates for years 1986 to 2006 of the number of licensed Canadian drivers. Create a fatality rate index by dividing the number of fatalities in each year by the number of licensed drivers. Use the fatality rate index as the response variable. Keep time as the explanatory variable.

   (a) Create a scatterplot of the fatalities versus year.

   (b) Fit a simple linear regression model. Summarize the results.

   (c) Perform a complete diagnostic analysis. Summarize the results.

   (d) Based upon your answers to parts (b) and (c), does the simple linear regression model explain the pattern in the numbers of fatalities? Discuss.

6. Refer to the previous exercise. Keep fatality rate index as the response variable and time as the explanatory variable. (Suggestion: recode the year by subtracting 1985 from each value.)

   (a) Create a scatterplot of the fatalities versus year. Add a LOWESS curve to the scatterplot.

   (b) Fit a fifth-degree polynomial regression model. Summarize the results.

   (c) Perform a complete diagnostic analysis. Summarize the results.

   (d) Based upon your answers to parts (a) and (b), does the quartic polynomial regression model explain the pattern in the numbers of fatalities?

   (e) Suggest possible explanations for the decrease in the fatality rate over time. Could it be that the increase in younger drivers is resulting in a decrease in motor vehicle collisions, and thus fatalities, because there are greater numbers of drivers with faster reflexes as time goes by? Or, is the decrease in fatality rates associated with the reduction in smoking rates in Canada and so fewer drivers are distracted with cigarettes while smoking? Or are all Canadian drivers becoming more cautious because of concerns about rationing of healthcare due to cost limits in their universal health insurance system? Discuss.

7. Refer to Table 12.2. Take the number of fatalities in each year as the response variable and year as the explanatory variable.

   (a) Create a scatterplot of the number of fatalities versus year.

   (b) Add a LOWESS curve to the scatterplot of part (a).

   (c) Add a spline curve to the scatterplot of part (a).

   (d) How does the spline function perform compared to the LOWESS curve?

8. Refer to Table 12.2. Create a fatality rate index by dividing the number of fatalities in each year by the number of licensed drivers. Use the fatality rate index as the response variable. Keep time as the explanatory variable.
   (a) Create a scatterplot of the rate of fatalities versus year.
   (b) Add a LOWESS curve to the scatterplot of part (a).
   (c) For each year in Table 12.2 starting with 1990, calculate a 5-year moving average using the year and the previous four. Plot this moving average on the scatterplot of part (a) and connect the points with a spline function.
   (d) How does the spline function fitted to the moving average perform compared to the LOWESS curve?
9. Consider the data on the pike caught during a research trawl on the Finnish lake Längelmävesi near Tampere. Consider nose-to-notch length as the response variable and mass as the explanatory variable.
   (a) Create a scatterplot of the nose-to-notch length versus mass.
   (b) Add a LOWESS curve to the scatterplot of part (a).
   (c) Add the least-squares fit of a cubic polynomial model to the scatterplot of part (a).
   (d) Load the `KernSmooth` package and use the function `locpoly` to fit a third-degree locally weighted polynomial to the data with a rectangular, so-called `box` kernel.
   (e) Compare the fitted cubic polynomial curve, the LOWESS curve, and the third-degree locally weighted polynomial. Do you agree with Cleveland's [20] comment that first order is good enough? Discuss.
10. Discuss whether adding the LOWESS curve to a scatterplot ought to be a mandatory convention when considering a linear or polynomial model.
11. In Figure 12.2 the LOWESS curves produced by the `stats` functions `lowess` and `loess` curve are different. Make the necessary changes when calling `loess` to obtain the same curve as `lowess` when modeling the length of perch from the nose to the notch of the tail as a function of mass.
12. Write a `ggplot2` extension package to implement `lowess` as an alternative to `loess` in `geom_smooth`.

Chapter 13

Visualizing Multivariate Data

13.1 Introduction

With the advent of computer graphics, visualization of multivariate data has become possible. With the ubiquity of personal computers, visualization of multivariate data is no longer confined to research departments within corporations, government agencies, or universities.

Development of visualization tools for multivariate data is an active area of research for chemists, physicists, engineers, computer scientists, and statisticians. Many of these disciplines have peer-reviewed journals specifically for the purpose of conveying breaking news regarding new software and hardware. The overarching name for this topic of research is *scientific visualization.*

The application of scientific visualization tools in marketing research is an integral part of *data mining.* Statistically trained workers search mountains of data for pockets of market share to mine for profit. The data typically come from a variety of sources that are merged together to form one large data set.

In this chapter, graphical displays are presented for three variables and then more. Discussion begins with the simplest setting of three discrete random variables and a well-known example.

13.2 Learning Outcomes

When you complete this chapter, you will be able to do the following.

- Be able to draft a thermometer chart using the function `symbols` in the `graphics` package in R.
- Be able to draft a three-dimensional bar chart using the function `cloud` in the `lattice` package in R and incorporate this style of chart in a trellis display.
- Be able to use the `graphics` function `plot` to produce a superposed scatterplot.

- Be able to use the `lattice` function `cloud` to produce a superposed three-dimensional scatterplot.
- Be able to draft a superposed scatterplot matrix in R with the function `pairs` in the `graphics` package and the function `ggpairs` in the `ggplot2` extension package `GGally`.
- Be able to produce a parallel coordinates plot using the function `parallelplot` in the `lattice` package in R.
- Be able to draft Chernoff faces using the function `faces` in R's `aplpack` but understand the severe limitations associated with how the data are displayed.
- Be able to use the function `stars` in the `graphics` package to create star plots and rose (or coxcomb) plots to depict multi-continuous variables for one or more discrete variables.
- Be knowledgeable about and be able to produce a residual scatterplot matrix and a leverage scatterplot matrix using the function `scatterplot.matrix` in the `car` package.
- Review the use of the `influencePlot` function in the `car` package to produce an influence plot but in the context of a multiple linear regression model instead of a simple linear regression model as in the previous chapter.
- Be knowledgeable about and be able to produce a partial-regression scatterplot matrix using the function `avPlots` in the `car` package.
- Be knowledgeable about and be able to produce a partial-residual scatterplot matrix using the function `crPlots` in the `car` package.
- Whenever color is used in the charts and plots discussed in this chapter, be able to use the function `brewer.pal` in the `RColorBrewer` package and the function `daltonize` in the package `LSD` to produce color palettes friendly to individuals with color vision deficiencies. (More information about the types and prevalences of color vision deficiencies as well as the `RColorBrewer` and `LSD` packages is given in Appendix A.)

13.3 Depicting Distributions of Three or More Discrete Variables

13.3.1 The Sinking of the Titanic

On the night of 14 April 1912, the ocean liner Titanic struck an iceberg. Within three hours early on 15 April 1912, the Royal Mail Steamer (RMS) Titanic sank with 1503 lives lost. Only 337 bodies were ever recovered with 119 buried at sea. The Titanic had a lifeboat capacity of 1178 despite being able to carry 3547 people.

On Monday 22 April 1912, the Lord Chancellor Robert, Earl Loreburn, appointed Charles Bigham, Lord Mersey of Toxteth, as Wreck Commissioner.

Type of Individual

Class	**Men**		**Women**		**Children**	
	Carried	**Saved**	**Carried**	**Saved**	**Carried**	**Saved**
First	173	58	144	139	5	5[a]
Second	160	13	93	78	24	24
Third	454	55	179	98	76	23
Crew	875	189	23	21	0	0

Table 13.1 *Passenger and crew numbers (excerpted from Appendix 1 of the British Wreck Commissioner's Inquiry [13])*

[a]One first-class child Helen Loraine Allison was lost with her parents Hudson Joshua Creighton Allison and Bessie Waldo Allison of Montréal. Her infant brother Hudson Trevor Allison survived the sinking.

Lord Mersey, at the time of appointment, was President of the Probate, Divorce, and Admiralty Division of the High Court. According to a compilation done for the British Wreck Commissioner's Inquiry [13], a total of 2206 souls were embarked. Others report the number on board when the iceberg was struck as 2201, 2223, and 2240.

The loss of life as a result of the sinking is variously estimated as 1490, 1503, and 1517. The authors of the Report of the British Wreck Commissioner's Inquiry noted the following factors contributing to the loss of life.

- The Titanic was traveling at excessive speed in waters where icebergs were known to be present.
- The design of watertight compartments was inadequate.
- There were not enough lifeboats.
- The crew manning the lifeboats were not adequately drilled.
- The Marconi radio sets for transmitting and receiving Morse code were not manned on a 24-hour basis on the ships involved in the rescue and this led to delay in aid arriving on scene. (The SS Californian was within 8 nautical miles of the Titanic when it sank but ignored signal distress rockets while its radio operator slept.)

The Report of the British Wreck Commissioner's Inquiry was released on 30 July 1912, just three and one-half months after the sinking. It made specific recommendations to correct these deficiencies. Too late for the victims of the Titanic.

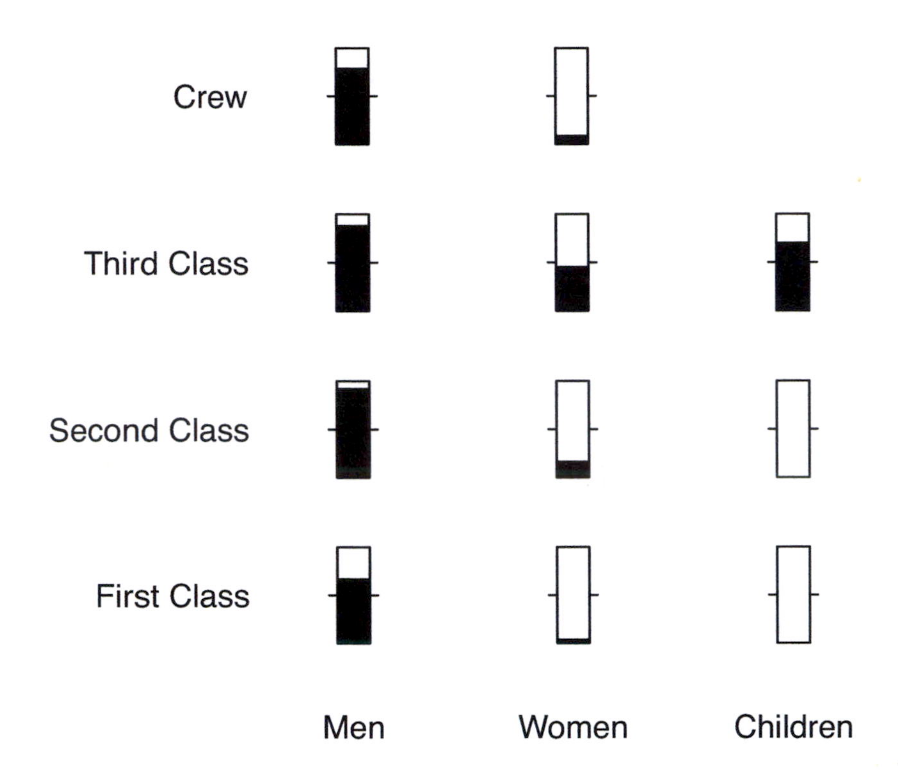

Figure 13.1 *Thermometer chart of survivors (white) and dead (black) among those carried on the ocean liner Titanic*

13.3.2 Thermometer Chart

Table 13.1 reports the number carried and the number that survived of men, women, and children among the three ticket classes and the crew. Table 13.1 is excerpted from Appendix 1 of the British Wreck Commissioner's Inquiry [13]. It is a tabular summary of the testimony on the second day of the inquiry by the Attorney-General for the United Kingdom, The Right Honourable Sir Rufus Isaacs. A pictorial representation of Table 13.1 is given in Figure 13.1.

Questions have been raised as to whether passengers and crew were evacuated to the lifeboats on the basis of "women and children first." It has been suggested that social class was the determining factor for survival. Before this can be examined, it is necessary to analyze the number of people with respect to age, gender, and class.

Table 13.1 reports the number of men, women, and children among the three ticket classes and the crew, as compiled by the British Wreck Commissioner's Inquiry [13]. According to Table 13.1, a total of 2206 souls were embarked.

Sometime after the Report of the British Wreck Commissioner's Inquiry was released, it was determined that a two-year-old girl, Helen Loraine Allison, among the first class passengers was lost. Although the body of her father Hudson Joshua Creighton Allison was recovered, her body and that of her mother Bessie Waldo Allison were not. Her infant brother Hudson Trevor Allison survived the sinking in the arms of his nursemaid, Alice Catherine Cleaver. Because it is known that Loraine's name was left off the list of embarked passengers, the amended number of souls carried in Table 13.1 ought to be adjusted to report 2207. But it shall be left as reported in testimony by the British Attorney-General.

The number of persons aboard a vessel should be determined by a census not a random sample. Ignore any possible undercount or errors in recording and treat Table 13.1 as being fully accurate and without error.

The number of survivors for each two-way classification in Figure 13.1 is proportional to the white area enclosed by each bar. The number of dead for each two-way classification in Figure 13.1 is proportional to the black area enclosed by each bar.

Figure 13.1 was produced by the R function symbols, in part, by the following R code.

```
RMSTitanic<-array(
data=c(173,58,144,139,5,5,160,13,
93,78,24,24,454,55,179,98,76,23,
875,189,23,21,0,0),
dim=c(2,3,4),dimnames=list(c("Carried","Survived"),
c("Men","Women","Children"),
c("1st","2nd","3rd","Crew")))

SurviveTitanic<-RMSTitanic[2,,]/RMSTitanic[1,,]

x<-rep(1:3,4)
y<-c(rep(1,3),rep(2,3),rep(3,3),rep(4,3))

symbols(x=x,y=y,thermometers=cbind(rep(1,12),
rep(3,12),c(1-SurviveTitanic)),
inches=0.5,xaxt="n",yaxt="n",bty="n",
xlab=" ",ylab=" ")

axis(1,1:3,tick=FALSE,
labels=c("Men","Women","Children"))
axis(2,1:4,tick=FALSE,
```

```
labels=c("First Class","Second Class",
"Third Class","Crew"),las=1)
```

The R array `RMSTitanic` contains the summary table of the number of souls carried and those who survived for each of the three classes of passengers and the crew. (There were no children among the crew.) The R array `SurviveTitanic` is calculated to report the proportion that did survive.

The bars in Figure 13.1 for each of the two-way classifications were produced with the option `thermometers`. Thus Figure 13.1 is referred to as a *thermometer chart*. The vectors `x` and `y` are set to the horizontal and vertical coordinates, respectively, for displaying the thermometers. In the call to the `graphics` package function `symbols`, the parameter `thermometer` is set to a matrix with three columns. The first two columns give the width and height of the thermometers. The third column is set to the proportion in each two-way classification that did not survive. The setting `inches=0.5` produces symbols, which are thermometers in Figure 13.1, that are 0.5 inch in height.

The default numerical axes produced by the plotting routine `symbols` have been switched off with the parameters `xaxt="n"` and `yaxt="n"`. The labels for both axes in the two-way classification are added to Figure 13.1 by two calls to the lower-level plotting function `axis`. Note that the labels have been plotted without tick marks by setting `tick=FALSE` in each call of `axis`.

The argument `bty="n"` switches off the rectangular box that would have surrounded the plot in Figure 13.1 by default. The bounding box was considered to be an unnecessary distraction in this plot. By default, the detracting labels "x" and "y" would have been added to Figure 13.1 if not for the arguments `xlab=" "` and `ylab=" "` being passed to the plotting function `symbols`.

Note that the height of the black in the thermometers for Figure 13.1 portrays the proportion that did not survive for each cross-classification. Each thermometer is a framed rectangle with tick marks on either side halfway up the vertical sides. Hence, Cleveland [21] refers to a thermometer chart as a *framed-rectangle graph*.

Note that a thermometer can be used to depict two categorical outcomes for a single discrete variable or the value of one quantitative variable. This makes it quite a handy icon.

From Figure 13.1, it evidently was not "women and children first." The survival rate for first-class adult men exceeds that for third-class children. Among the adult men, the crew fared better than those in the second and third classes.

Titanic collided with the submerged spar of an iceberg at 11:40 p.m. ship's time. Most of the children aboard would have been in their berths. The third-class cabins were situated at the lowest levels of the great steamship but only the aftmost part of G deck was below the waterline.

Locked barriers were in place to keep the third-class passengers away from contact with other passengers. One role of the Titanic was to transport emigrants from Europe and these individuals would require health checks, and possibly quarantine on arrival at the new world port of destination.

Neither the Captain nor the Chief Officer gave orders to unlock the third-class barriers. Although 176 out of 709 third class passengers survived, the 75% fatality rate could be attributed to the locked barriers. Class distinction apparently did play a role in survival on board the Titanic.

With respect to Figure 13.1, the icon for each combination of class and type of individual is essentially a stacked bar chart. A stacked bar plot can easily depict more than two classifications. With respect to efficiency, a better choice would be a side-by-side bar chart. This way there would at least be a common axis for each row of bar charts.

13.3.3 Three-Dimensional Bar Chart

Given that there are two explanatory variables for survival in Table 13.1, a reasonable approach would be to take the definition for the bar chart and add a dimension. This is done in Figure 13.2 with a *three-dimensional bar chart*.

The proportion of survivors for each two-way classification in Figure 13.2 is depicted by a line segment, also called a *needle*. Each needle has been capped at the top by an asterisk. This helps because the survival rate for men in second and third class is so low.

Because needles have been used instead of blocks, there is not an issue related to comparing the volume. Figure 13.2, however, does not present a common baseline for comparison of the lengths of the needles. There is also the issue of contending with three-dimensional perspective depicted on a two-dimensional image. It is concluded then that the three-dimensional bar chart of Figure 13.2 is not as efficient as the thermometer plot of Figure 13.1.

Figure 13.2 was produced by the function `cloud` in the R package `lattice` as follows.

```
figure<-cloud(prop.table(Titanic, margin = 1:3),
layout=c(1,2,1),par.settings=par.set,
type=c("p","h"),
strip=strip.custom(strip.names=TRUE),
scales=list(arrows=FALSE,distance=2),
panel.aspect = 0.7,zlab = "Prop.")[,2]

print(figure)
```

The `list` passed to the parameter `screen` is the orientation in degrees relative to the x, y, and z axes for the impression of three-dimensional perspective.

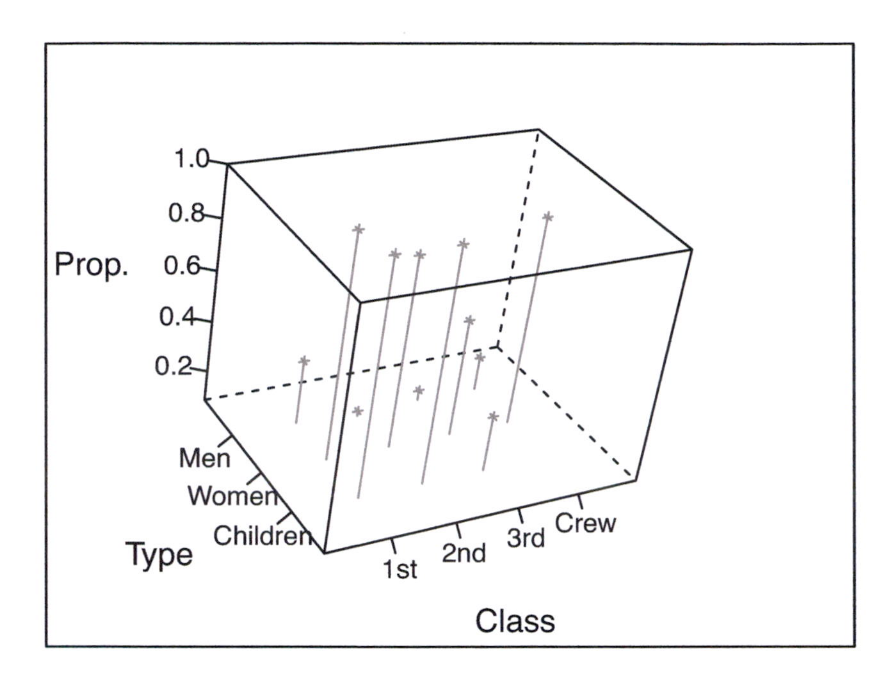

Figure 13.2 *Three-dimensional bar chart reporting proportion of survivors among those carried on the ocean liner Titanic*

13.3.4 Trellis Three-Dimensional Bar Chart

There was a second tabulation of the numbers of the passengers carried and perished at the sinking of the Titanic that was published in the Report of the British Wreck Commissioner's Inquiry.

The hearings began on 2 May 1912. The tabulation reported in Table 13.1 was presented on the second day of hearings. The Board of Trade, of Great Britain, worked with the White Star Line to produce a more accurate accounting of those carried and those who perished. On the thirty-fourth day of hearings, of thirty-six days of hearings, Attorney-General Isaacs stated that 2201 were aboard at the time of collision. The revised tabulation relating to this number is Appendix 10 to the Report and is presented in Table 13.2.

The results in Table 13.2 incorporate corrections that were not available until the last three days of the inquiry in Great Britain. But note that the death

	Type of Individual							
	Males				Females			
	Adult		Child		Adult		Child	
Class	**Carried**	**Saved**	**Carried**	**Saved**	**Carried**	**Saved**	**Carried**	**Saved**
First	175	57	5	5	144	140	1	1[a]
Second	168[b]	14	11	11	93	80	13	13
Third	462	75	48	13	165	76	31	14
Crew	862	192	0	0	23	20	0	0

Table 13.2 *Passenger and crew numbers (excerpted from Appendix 10 of the British Wreck Commissioner's Inquiry [13])*

[a]One first-class child Helen Loraine Allison was lost with her parents Hudson Joshua Creighton Allison and Bessie Waldo Allison of Montréal. Her infant brother Hudson Trevor Allison survived the sinking.

[b]There is a transcription error in Appendix 10. The number of second class adult male passengers carried ought to be 168 as tabulated above not 175 as recorded in Appendix 10.

of two-year-old first-class passenger Loraine Allison is still unreported. So the total number carried needs to be adjusted from 2201 to 2202.

The eight bandsman aboard the Titanic are reported as second-class adult males in Table 13.2. All eight members of the Ship's Orchestra perished. Only three of their bodies were identified with two being buried in separate cemeteries in Halifax. The body of Bandmaster Wallace Henry Hartley was repatriated to Colne, Lancashire, for burial. The musicians were all employees of the Liverpool firm C. W. and F. N. Black. They were not employees of the White Star Line.

Contrary to popular myth, they played ragtime tunes on deck, not the hymn *Nearer, My God, to Thee* while the lifeboats were lowered. The English idiomatic phrase *and the band played on* is, however, attributable to them. It is more subtle than another English idiom: *like rearranging deckchairs on the Titanic.*

Note that there are four variables in Table 13.2: class (including crew), gender, age (adult or child), survival (saved or not). This table can also be obtained from the data set `Titanic`, which is one of the standard `datasets` available with the R statistical software package. As a convenience, this data set can also be downloaded from the website for this book.

The four variables are displayed in the *trellis three-dimensional bar chart* of Figure 13.3. As with Figure 13.2, the function `cloud` from the R package `lattice` was used to generate Figure 13.3. But it is only with Figure 13.3 that one can appreciate the power of Trellis graphics. About the same amount of effort was required to produce either Figure 13.2 or Figure 13.3 as can be seen in the following code that produced Figure 13.3.

```
par.set<-list(plot.line=list(col=gray(0)),
plot.symbol=list(col=gray(0),cex=2.0),
strip.background=list(col=gray(c(0.9,1))),
box.3d=list(col=gray(.7)))

figure<-cloud(prop.table(Titanic, margin = 1:3),
layout=c(1,2,1),par.settings=par.set,type=c("p","h"),
strip=strip.custom(strip.names=TRUE),
scales=list(arrows=FALSE,distance=2),
panel.aspect = 0.7,zlab = "Prop.")[,2]

print(figure)
```

The list `par.set` passed as an argument to the parameter `par.settings` when calling the `lattice` function `cloud` gives the parameters for setting up the trellis display in Figure 13.3.

In comparing adults in the top panel of Figure 13.3 to children in the bottom

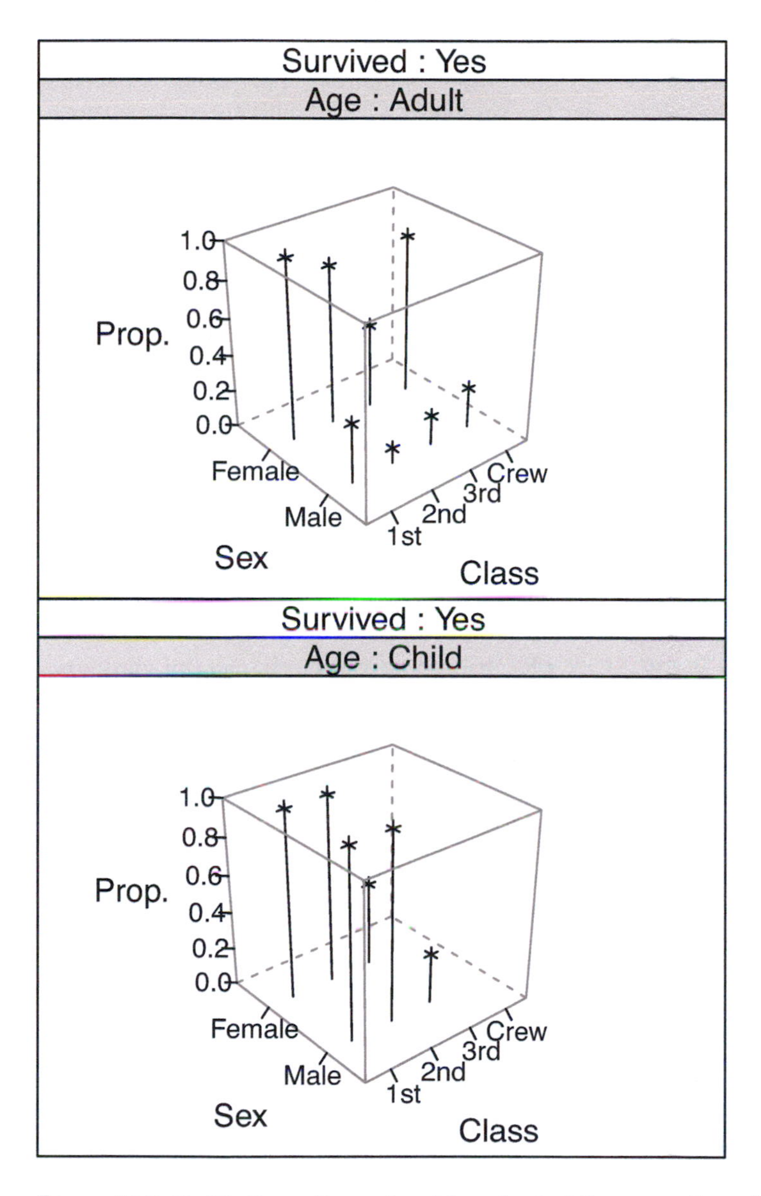

Figure 13.3 *Trellis three-dimensional bar chart of the proportion of survivors on the Titanic*

panel, it is clear that the survival rates for adult males was poor compared to the women and children. With a bit of effort, the anomalously higher survival rate for first-class men becomes apparent. As does the poor survival rate for children, male or female, in third class compared to first or second class.

Additional variables for a trellis three-dimensional bar chart would be accommodated with additional panels. The problems with a trellis three-dimensional bar chart are no different than for a three-dimensional bar chart: the lack of a common horizontal reference, scale for at least one of the variables, and presence of three-dimensional perspective contribute to requiring much more effort and time for studying these types of graphical displays.

13.4 Depicting Distributions of One Discrete Variable and Two or More Continuous Variables

13.4.1 Anderson's Iris Data

Edgar Anderson was an American botanist who made important contributions to botanical genetics. In 1922 he was awarded a D.Sc. in agricultural genetics. In 1929 he received a fellowship to undertake studies at the John Innes Horticultural Institute in Great Britain. There he worked with Cyril Dean Darlington, Ronald Aylmer Fisher, and James Burdon Sanderson Haldane.

Darlington was a cytogeneticist who discovered the mechanism of chromosomal crossover. Fisher and Haldane, together with Sewall Wright, founded the discipline of population genetics. Fisher, Haldane, and Wright were later to be awarded the Darwin Medal of the Royal Society. Anderson, Fisher, and Haldane were among twenty individuals awarded the Darwin-Wallace Medal, in silver, by the Linnean Society of London in 1958.

One of Anderson's research topics was on the development of techniques to quantify geographic variation in the flowering plant *Iris versicolor*. His article "THE IRISES OF THE GASPE PENINSULA" was published in 1935 in the *Bulletin of the American Iris Society*. In this article, Anderson [3] discusses three species of *Iris*: *setosa*, *versicolor*, and *virginica*.

The following year, the *Annuals of Eugenics*, now known by the more politically correct name *Annuals of Human Genetics*, published a paper by Fisher that presented the theory of linear discriminant analysis. The data chosen by Fisher [42] to illuminate this theory were those gathered by Anderson on the Gaspé Peninsula of Québec. In Table 1, Fisher [42] listed the observations of 50 flowers each for the three species of *Iris*. The data have come erroneously to be known as *Fisher's iris data*. They are, in fact, *Anderson's iris data*. Fisher [42] gave full credit to Edgar Anderson as the source of the data by writing the following.

> Table I shows measurements of the flowers of fifty plants each of the two

> species *Iris setosa* and *I. versicolor*, found growing together in the same colony and measured by Dr. E. Anderson, to whom I am indebted for the use of the data.

Although not reproduced here, Table I also contains measurements for 50 plants of the species *Iris virginica* that were collected from a separate colony.

As to why an article on statistical methodology might appear in a genetics journal, there is the obvious connection that Fisher was a member of the *Eugenics Society.* But this would overlook the discussion in Fisher [42] concerning allopolyploidy. As reported by Fisher, Anderson confirmed that *Iris setosa* is a diploid species with 38 chromosomes, *Iris virginica* is tetraploid, and *Iris versicolor* is hexaploid. *Iris versicolor* was conjectured by Fisher [42] to be a polyploid hybrid of the two other species. If the genetics are additive, then the values for *Iris versicolor* ought to be intermediate between the other two species with twice as much difference between *Iris setosa* compared to *Iris virginica.* Keep this in mind when looking at the graphical displays.

With respect to the geographical origin of Anderson's data, the Gaspé Peninsula is on the south shore of the Gulf of Saint Lawrence in Canada's province of Québec. *Iris versicolor* is the official flower of the Province of Québec. The common name for *Iris setosa* is the *beachhead iris.* The common name for *Iris virginica* is the *Virginia iris* for it is native to the state of Virginia and the eastern half of North America.

Although all three species of iris have blue flowers, the common name for *Iris versicolor* is the *harlequin blueflag*, or, simply, *blueflag.* The provincial flag for Québec is called the blue flag: it is blue with a white central cross and a white fleur-de-lis in each quadrant. Some would argue that the fleur-de-lis on this flag is not the *Iris pseudacorus*, or yellow flag, of the Old World and an enduring symbol of France, but the *Iris versicolor* of the New World and New France.

Anderson's iris data have become, over time, an important data set not just for teaching linear discriminant analysis but also for illustrating useful graphical displays of three and more dimensions. It is available as a data example for most, if not all, statistical packages. This section will make extensive use of Anderson's data, with thanks to Fisher for publishing the actual measurements.

13.4.2 The Superposed Scatterplot

In general, if each ordered pair of continuous random variables can be identified with a third discrete random variable, then a different marker or color (or a combination of the two) can be assigned for each category. This allows for:

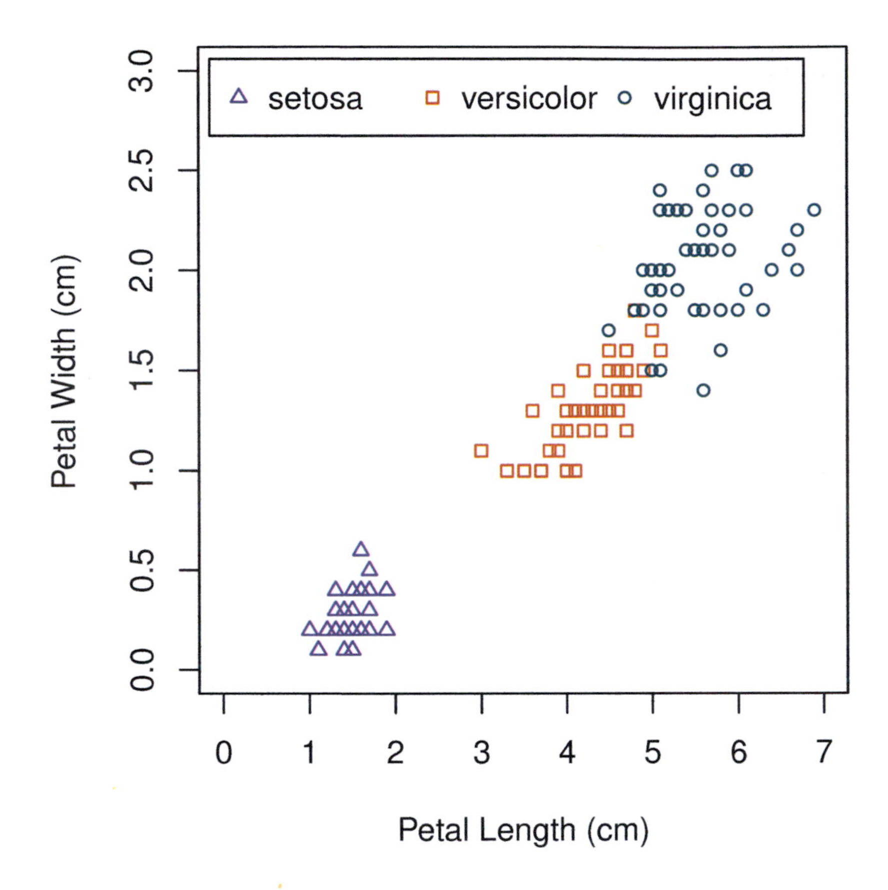

Figure 13.4 *Superposed scatterplot for three species of iris*

- superposing (overlaying) more than one scatterplot; and
- searching for distinguishing features, or so-called clusters, among the different categorical levels with respect to the random variables plotted along the horizontal and vertical axes.

Figure 13.4 presents a *superposed scatterplot* with petal length along the horizontal axis and petal width along the vertical axis. A different color has been used to depict each species of iris. Just as importantly, for individuals with color vision deficiencies, a different symbol has been used as well. The R code for producing Figure 13.4 is as follows.

```
colpal<-rev(brewer.pal(3,"Dark2"))

plot(iris3[,,1][,3],iris3[,,1][,4],cex=0.8,
pch=2,col=colpal[1],
```

```
xlim=c(0,7),ylim=c(0,3),xlab="Petal Length (cm)",
ylab="Petal Width (cm)")

points(iris3[,,2][,3],iris3[,,2][,4],pch=0,
col=colpal[2],cex=0.8)

points(iris3[,,3][,3],iris3[,,3][,4],pch=1,
col=colpal[3],cex=0.8)

legend(x=-0.15,y=3.06,
legend=c("setosa","versicolor","virginica"),
horiz=TRUE,pch=c(2,0,1),
col=colpal,pt.cex=0.8)
```

Preceding the call to the `graphics` function `plot` is a call to the `RColorBrewer` package function `brewer.pal` to find three colors in the `Dark2` palette deemed to be color-vision-deficient-friendly for a qualitative variable. The array `iris3` passed to the function `plot` is part of the R `datasets` package that is included in the base installation of the R statistical software system. The call to the function `plot` in the R `graphics` package produces the plot of the purple triangles for the species *setosa*. The following call to the function `points` then adds the buff squares for the species *versicolor*. The second call to `points` adds the blue-green circles for the observations of petal length and width for the species *virginica*.

An examination of Figure 13.4 reveals distinct clusters for the three species of iris with respect to petal length and width. As suggested by Fisher [42], the cluster for *Iris versicolor* is closer to that for *Iris virginica* with just a little overlap.

13.4.3 The Superposed Three-Dimensional Scatterplot

Anderson also measured sepal length of each flower. The answer to the question of what to do when there are three variables is given in Figure 13.5. Note the symbols for each species are two dimensional but they are plotted in a three-dimensional perspective. The R script, in part, for producing the legend and lower panel in Figure 13.5 is as follows.

```
colpal<-rev(brewer.pal(3,"Dark2"))

trellis.par.set(list(superpose.symbol=list(pch=c(24,22,21),
col=colpal)))

figure<-cloud(Sepal.Length   Petal.Length*Petal.Width,
groups=Species,data = iris,screen = list(z =20,x=-70,y=3),
```

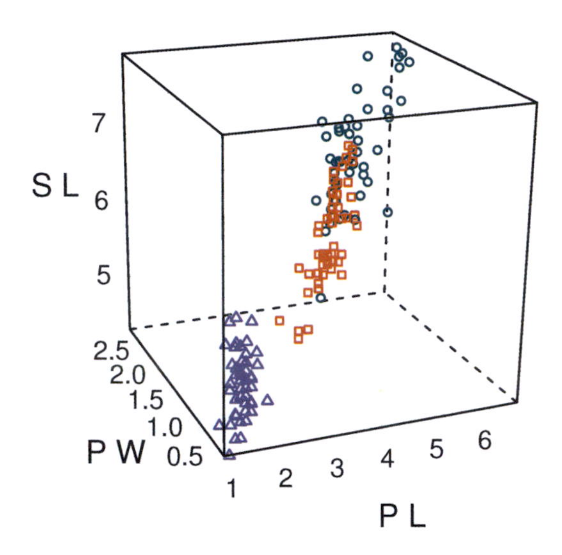

Figure 13.5 *Superposed three-dimensional scatterplot for three species of iris (S L denotes sepal length, P L denotes petal length, and P W denotes petal width, and all units are in centimeters)*

```
par.settings=list(axis.line=list(col="transparent")),
key=list(text=list(c("setosa","versicolor","virginica"),
cex=rep(1,3)),
space="top",columns=3,between=1,
points=list(cex=rep(0.5,3),pch=c(2,0,1),col=colpal)),
cex=0.5,zlab="S L",ylab="P W",xlab="P L",
clip=list(panel="off"),
scales=list(arrows=FALSE,distance = 1.2))

print(figure)
```

The call to the utility function `trellis.par.set` sets the shape and gray-tone of the symbols for each species of iris. Within the call to the `lattice` function `cloud`, which does the actual plotting of the data, the viewer's eye is set at the coordinates passed to the argument `screen`. The directions for creating the legend are in the list passed to the argument `key`.

Labeling of three-dimensional scatterplots can be awkward so the abbreviations of S L, P W, and P L have been used for sepal length, petal width, and petal length, respectively. The decision has also been made not to place tick marks on the three axes. The latter represents a trade-off between clarity and efficiency.

With respect to sepal length, petal width, and petal length, the cluster for *Iris setosa* is quite distinct from those for the other two species. The nature of the overlap between *I. setosa* and *I. virginica* can perhaps be better appreciated with the three-dimensional scatterplot in Figure 13.5 compared to the two-dimensional scatterplot in Figure 13.4.

The maximum number of visual dimensions available is three. For practical purposes, we can only examine three-dimensional slices or projections of data of higher dimension. Three dimensions can be depicted on a two-dimensional surface (paper, movie screen, or a video display monitor) and viewed by:

- stereographic pairs of images and special glasses;
- IMAX technology and special glasses;
- the use of motion, especially rotation;
- viewing a three-dimensional perspective rendering with two eyes and a good imagination, or if this fails;
- covering one eye and viewing a three-dimensional perspective rendering with one eye and relying on vergence to get the illusion of depth.

If you have difficulty getting the sense of perspective with the three-dimensional scatterplot in Figure 13.5, turn your eyes away from the Figure and look ahead. Cover one eye and after 10 seconds worth of time for adjustment, look again at the three-dimensional scatterplot in Figure 13.5. The three-dimensional scatterplot should now be in perspective.

The package `rgl` is a three-dimensional real-time rendering system for the statistical software system R. Among other things, its function `scatter3d` can be used to create a three-dimensional scatterplot, which the user can then rotate and zoom in on points of interest. The `rgl` function `indentify3d` can be used with a rectangular lasso to identify data points of interest. The R package `Rcmdr` provides a convenient spreadsheet-based front-end for `scatter3d` and `indentify3d`.

Given that Anderson measured petal length and width, it is not reasonable to expect that he stopped at measuring sepal length. He did measure sepal width. The next issue to address is how to visualize four continuous variables for each of the three species of iris. Static three-dimensional scatterplots are not quite up to the task. The workload is less if the static graphic display executed in two dimensions is portraying just two dimensions.

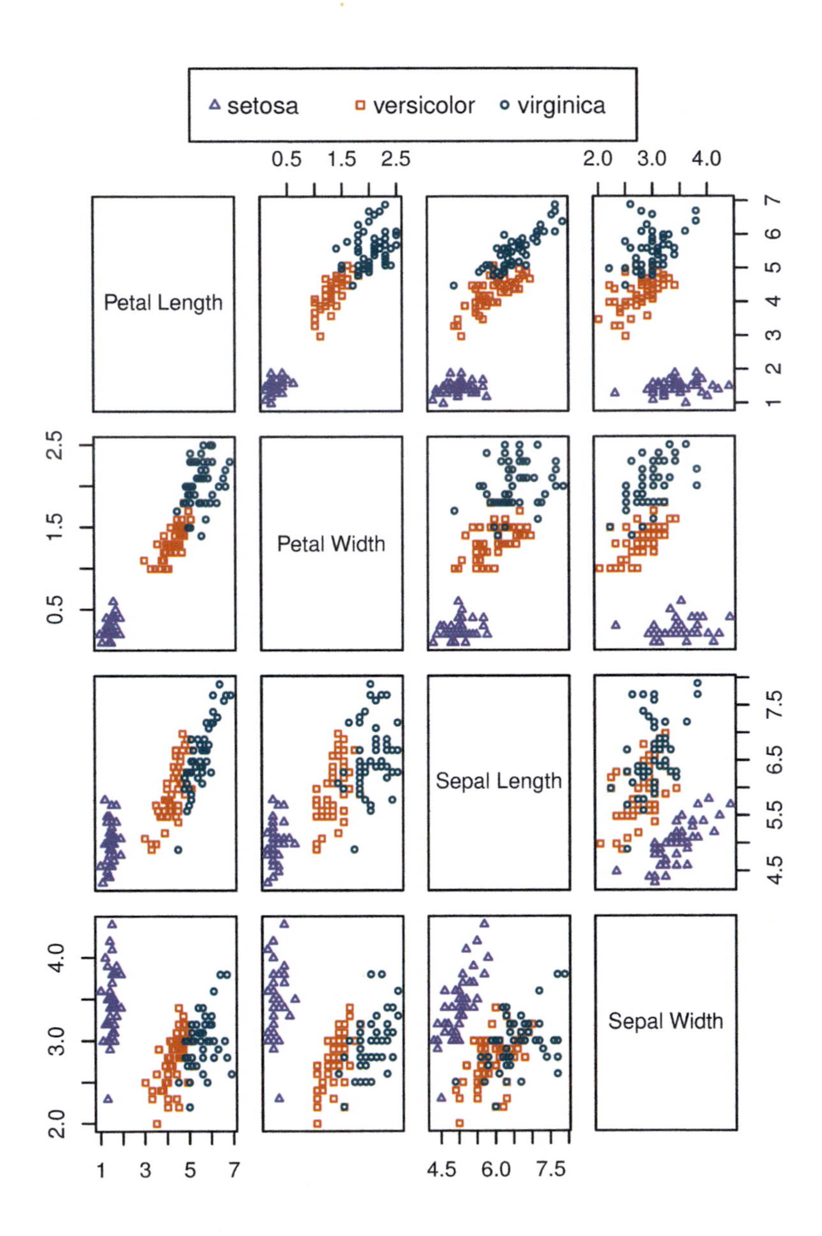

Figure 13.6 *Scatterplot matrix for all three species of iris in Anderson's data (all units are centimeters)*

13.4.4 The Scatterplot Matrix

If there are more than two quantitative variables, it is still possible to use scatterplots by plotting all possible scatterplots. These are known as either *casement plots* or a *scatterplot matrix*. The acronym *SPLOM* is used for a ScatterPLOt Matrix. An example of a scatterplot matrix for Anderson's data is given in Figure 13.6. This graphic was created by the following R code.

```
colpal<-rev(brewer.pal(3,"Dark2"))

pairs(iris[c(3,4,1,2)], main = " ",
cex=0.6,pch = c(2,0,1)[unclass(iris$Species)],
col = colpal[unclass(iris$Species)],
labels=c("Petal Length","Petal Width",
"Sepal Length","Sepal Width"))

legend(x="top",legend=c("setosa","versicolor",
"virginica"),cex=0.75,x.intersp=0.5,bty="o",
inset=-0.025,col=colpal,pch=c(2,0,1),pt.cex=0.5,
horiz=TRUE)
```

The legend has been inset in the top margin, where a title could have been placed, as there is no other room for it.

Note the redundant scatterplots with the axes interchanged on either side of the main diagonal. The redundant patterns may be omitted leaving either an upper or lower triangular presentation. Figure 13.7 deletes the scatterplots below the main diagonal by adding the code `lower.panel=NULL` in the call to the function `pairs`.

Figures 13.6 and 13.7 both have been produced by the function `pairs` in the standard R `graphics` package. Technically, both figures are *superposed scatterplot matrices*.

Figure 13.8 is a version of the upper scatterplot produced by the function `ggpairs` in the `ggplot2` extension package `GGally`. The code for producing Figure 13.8 is as follows.

```
colpal<-rev(brewer.pal(3,"Dark2"))

figure<-ggpairs(iris,aes(color=Species),
axisLabels="none",
columns=1:4,legend=c(1,1),
diag=list(continuous="densityDiag"),
upper=list(continuous="points",combo="blank"),
lower=list(continuous="blank",combo="box")) +
theme(legend.position="bottom")
```

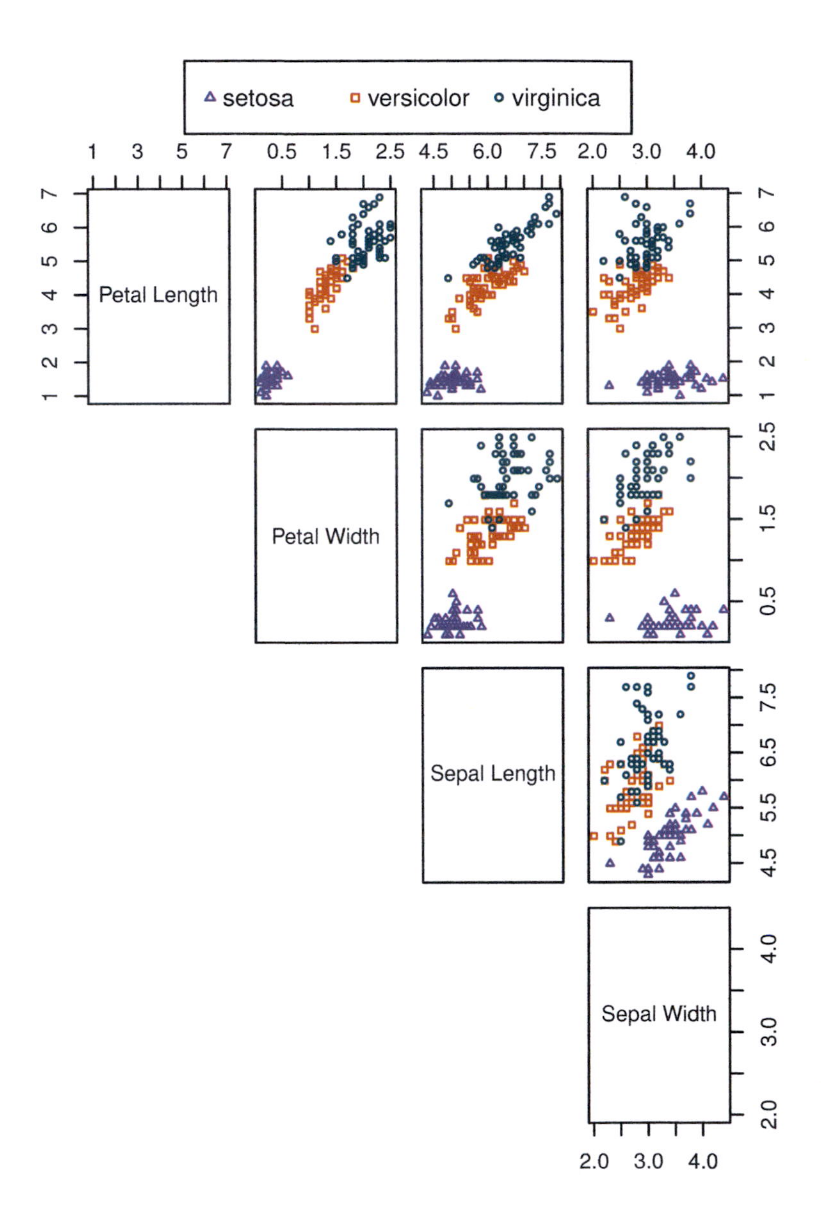

Figure 13.7 *Upper scatterplot matrix for all three species of iris in Anderson's data (all units are centimeters)*

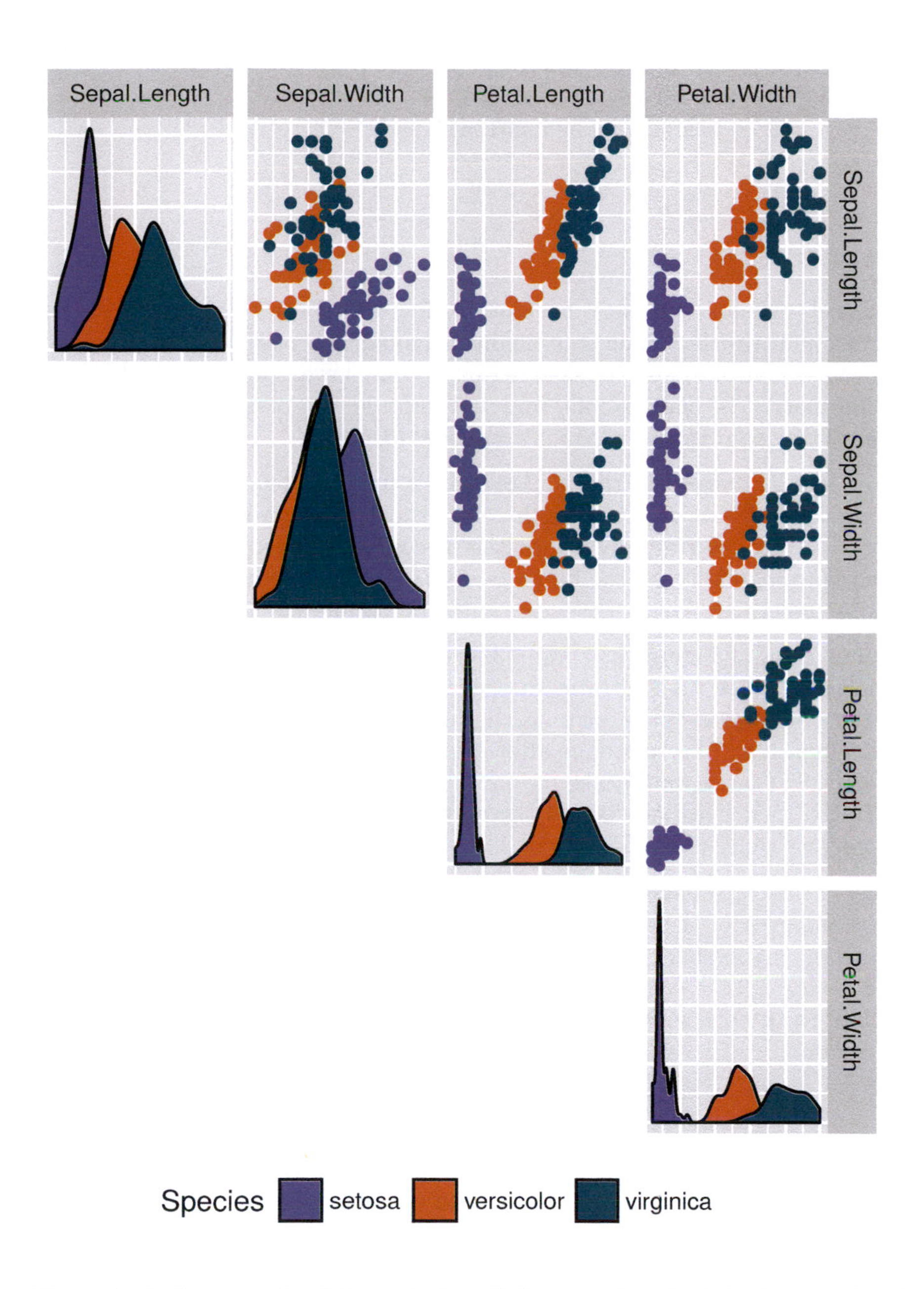

Figure 13.8 *Upper scatterplot matrix for all three species of iris in Anderson's data (all units are centimeters) using* `ggplot2` *and* `GGally`

```
for(i in 1:figure$nrow) {
for(j in 1:figure$ncol){
figure[i,j] <- figure[i,j] +
scale_fill_manual(values=colpal) +
scale_color_manual(values=colpal)
}
}

print(figure)
```

A kernel density estimate is plotted for all species for each of the lengths and widths along the main diagonal by passing the argument `diag=list(continuous="densityDiag")`. The lower portion of the scatterplot matrix is blanked out and not plotted by passing the argument `lower=list(continuous="blank",combo="blank")`.

Again, a call is made to the `RColorBrewer` function `brewer.pal` to obtain three colors friendly for those with color vision deficiencies. But in order for these colors to be properly used, it is necessary after calling the function `ggpairs` to manually set the fill and outline for the plotted symbols for each density and scatterplot in the figure with the double `for` loops.

A comparison of Figures 13.7 and 13.8 reveals that tick marks and their numerical labels are missing from the `ggpairs` version in Figure 13.8. In fact, these were suppressed by passing the argument `axisLabels="none"` when calling `ggpairs`. This was done after some experimentation with `ggpairs`. It was decided that it would take too much effort otherwise to remove the overprinting of tick mark labels and the cutting off of the variable names along the top and right side of Figure 13.8.

In any of the three figures, there is clear separation in all scatterplots between *setosa* and each of the other two species. The overlap between *versicolor* and *virginica* is greatest for the two variables sepal length and sepal width. It is clear from all three scatterplot matrices that *Iris versicolor*, although intermediate to the other two species, more closely resembles *Iris virginica* than *Iris setosa*.

13.4.5 The Parallel Coordinates Plot

Figure 13.9 presents a *parallel coordinates plot* for the three species of iris in Anderson's data. Each of the 150 plants measured by Anderson is represented by a piecewise linear curve. Colors have been used to some effect to distinguish among the 50 different plants in the panel for each species.

Scales are established for the variables on parallel line segments. Each scale is determined from the minimum and the maximum of each continuous variable across all values of the discrete variable. The points corresponding to a single observation are plotted and then joined with line segments. Once all data points have been plotted, the graphic can be studied to compare location and spread.

Figure 13.9 was produced by the following call to the function `parallel` in the R package `lattice`.

```
figure<-parallelplot(~iris[c(2,1,4,3)] | Species, iris,
col=colpal)

print(figure)
```

Depending on the order of the variables, patterns may be discerned. Looking at Figure 13.9, it is apparent that *virginica* and *versicolor* have a similar pattern—as would be expected. There is an increase in spread for sepal length and sepal width compared to petal length and petal width if looking down the panels for both species.

The general sweep for the species *Iris setosa* is essentially the mirror image of the other two species. The impression from Figure 13.9 is that the beachhead iris is the smallest of the three by every measure except sepal width. Moreover, the variability progressively increases through petal length, petal width, sepal length, and finally sepal width, for which the beachhead iris has greater spread than blueflag or the Virginia iris.

13.4.6 The Trellis Plot

Consider again the data obtained from a research trawl on the Finnish lake Längelmävesi around 1917. Sports fishers do have the means to measure height and width, as well as length. With respect to approximating mass, the mass of a rectangular solid box with known constant density ρ and volume V is given by

$$m = \rho V. \tag{13.1}$$

If the length of the rectangular solid box is ℓ, its width is w and its height is h, then

$$m = \rho \ell w h. \tag{13.2}$$

Taking the logarithm, to base ten, of this equation yields

$$\log(m) = \log(\rho) + \log(\ell) + \log(w) + \log(h). \tag{13.3}$$

Allowing for a little inaccuracy in the model by adding in a few fudge factors, the result is

$$\log(m) = \log(\rho) + \beta_1 \log(\ell) + \beta_2 \log(w) + \beta_3 \log(h). \tag{13.4}$$

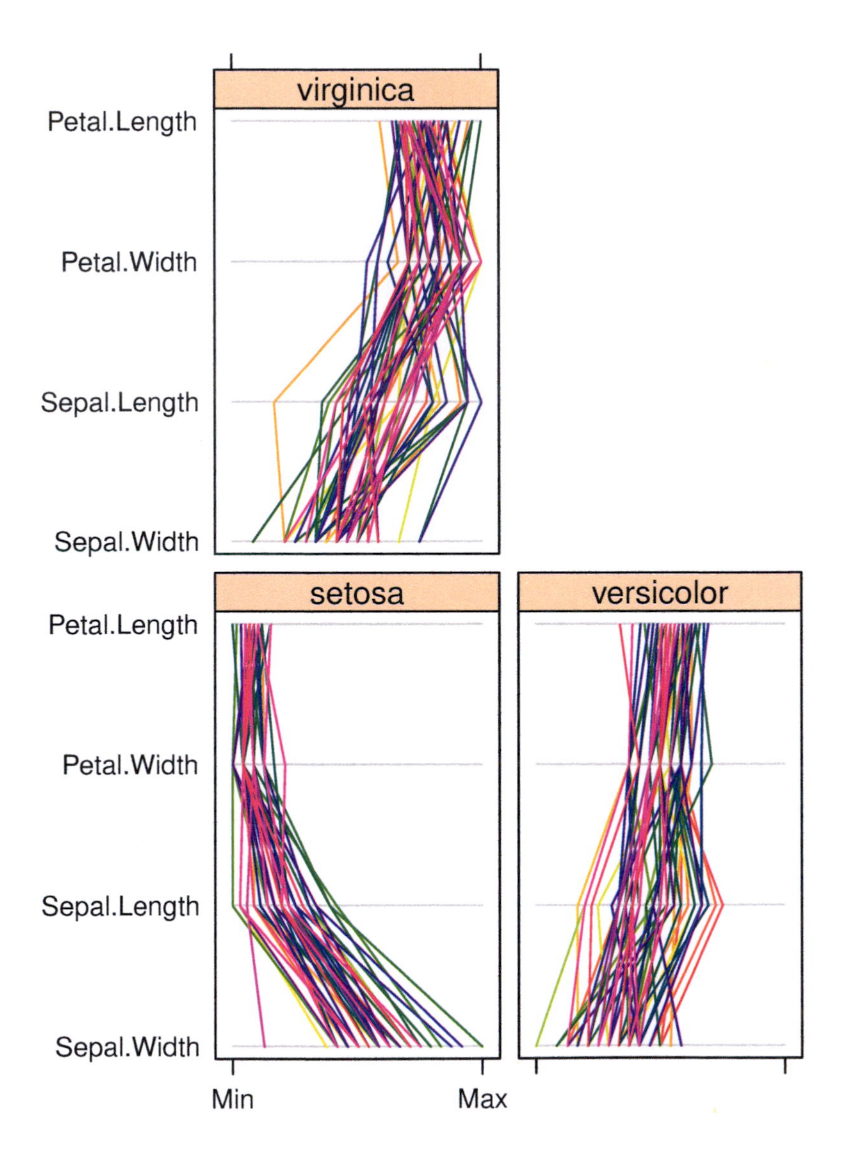

Figure 13.9 *Parallel coordinates plot for the three species of iris in Anderson's data (all units are centimeters)*

Fish mass m is known to follow a long-tailed distribution by and large. It can be modeled by a log-normal distribution. That is, $\log(m)$ is approximately normally distributed. As William Cleveland [21] wrote, "Use a logarithmic scale when it is important to understand percent change or multiplicative effect." Basic, perhaps oversimplified, physics of biological organisms justifies taking logarithms of the response and explanatory variables in this case study. Each species is indicated by a different colored plotting symbol.

It would be worthwhile to know from the data for 7 species caught during the research trawl on Längelmävesi, whether a linear model involving the logarithm of each of nose-to-notch length, width, and height can be used to estimate the logarithm of mass for most species. Figure 13.10 uses a *trellis plot* to display all the log-transformed variables for each species. The R script for creating Figure 13.10 is as follows.

```
colpal<-rev(brewer.pal(7,"Dark2"))

fc<-Fishcatch

fc$Species<-ordered(fc$Species,
levels=c(5,7,6,4,2,1,3),
labels=c("Smelt","Perch","Pike","Silver Bream",
"Ide","Bream","Roach"))

attach(fc)

width<-equal.count(log(Width_pc*Length3/100),3,1/2)
height<-equal.count(log(Height_pc*Length3/100),3,1/2)

figure<-xyplot(log(Weight) ~ log(Length2) | width*height,
groups=Species,auto.key=list(columns=3),aspect=1,
par.settings = simpleTheme(pch=0:6,col=colpal),
xlab = "log(Length)",ylab = "log(Mass)",
strip=strip.custom(var.name=c("log(Width)","log(Height)")))

print(figure)

detach(fc)
```

The R list `Fishcatch` is saved into the list `fc`. The species' labels in `fc` are re-ordered and changed from the integers from 1 to 7, inclusive, to their common names in English.

Using the R function `attach`, the list `fc` is attached to the R search path. This means that the database is searched by R when evaluating a variable, so objects in the database can be accessed by simply giving their names. When done, `detach(fc)` is executed to remove `fc` from the search path.

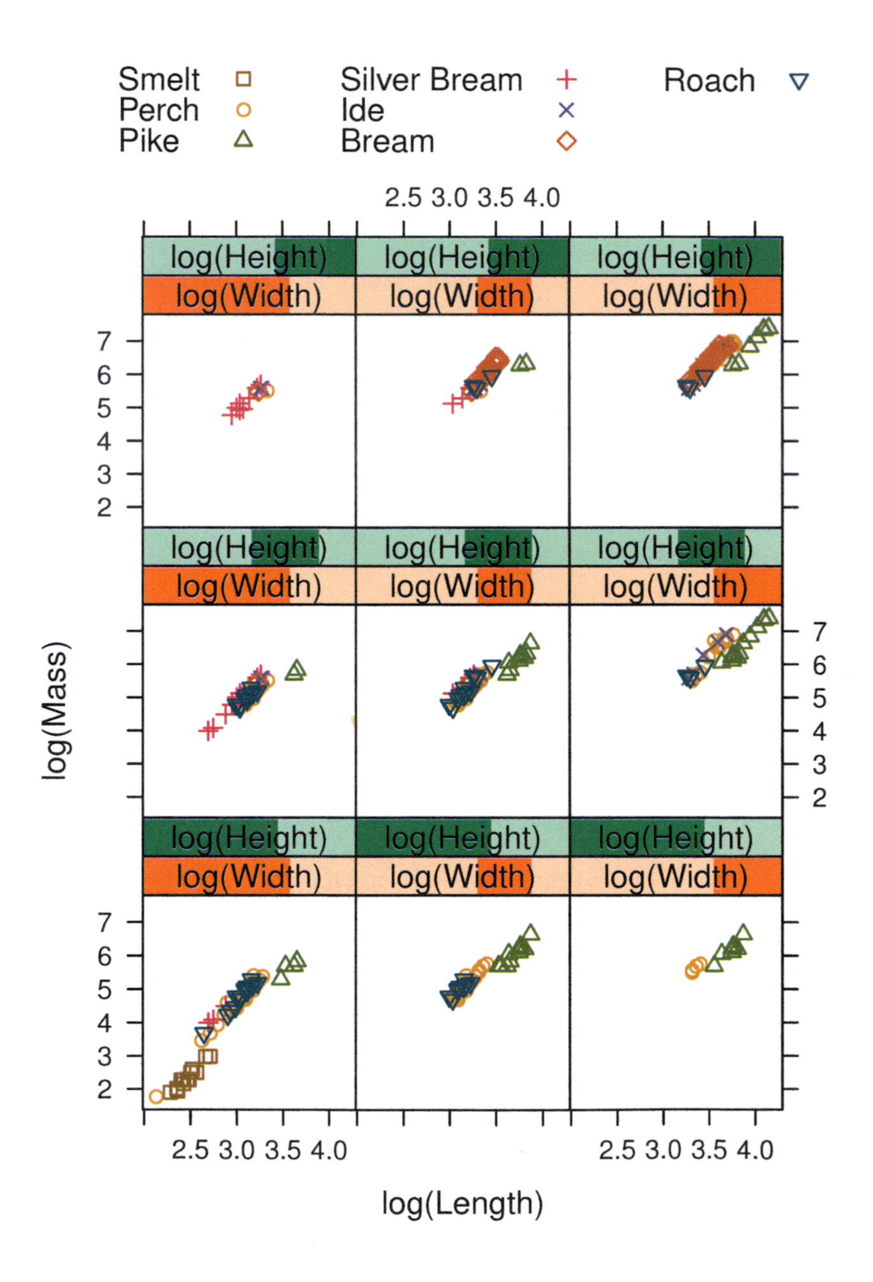

Figure 13.10 *Trellis display of the log-transform of variables measured in fish caught in 1917 during a research trawl on Längelmävesi in Finland*

For each of the nine panels in Figure 13.10, the logarithm of mass is displayed on the vertical axis and the logarithm of nose-to-notch length is displayed on the horizontal axis. The other two variables are displayed using *shingles*. Orange shading is used within the shingle for the logarithm of width. Green shading is used within the shingle for the logarithm of height.

The shading within each shingle denotes the range of values used to select observations in the panel immediately below it. Note that there is a degree of overlap for the ranges of values denoted by the color-shaded bands for each of log(Width) and log(Height). This is adjustable by the user when calling the function `xyplot` in the package `lattice` in the preceding R code. The `lattice` function `equal.count` was called before calling `xyplot` to create shingles for width and height. Note that three shingles were created for each of width and height with 50 percent overlap. These were stored in `shingle` data-structures `width` and `height`, respectively.

Examination of Figure 13.10 reveals strong linear patterns for each species. It would be nice if just one model could be used for all species, but one clearly notes that the species perch and pike are different from each other and the rest. Separate models for each species does not appear to be a bad idea from Figure 13.10. This will be pursued for perch later in this chapter.

Figure 13.10 gives a clear demonstration of the power of Trellis graphics. For more information on Trellis graphics consult the user's manual by Becker and Cleveland [9]. For more examples of Trellis graphics consult Becker, Cleveland, and Shyu [10] or Cleveland's [23] book *Visualizing Data*.

13.5 Depicting Observations of Multiple Variables

13.5.1 OECD Healthcare Service Data

The Organisation for Economic Co-operation and Development (OECD) is an international organization of 30 countries that accept the principles of representative democracy and free-market economy. Most members are high-income developed nations. It originated with the Marshall Plan for the reconstruction of Europe after the Second World War. Membership was later extended to non-European states. Its Statistics Directorate is one of the world's premier statistical agencies.

The OECD offers the most comprehensive source of comparable statistics on health and healthcare systems in OECD countries. Cross-country comparisons of national health systems play an important role in informing healthcare policy in each of the member nations. The Statistics Directorate compiles these figures for the OECD Directorate of Employment, Labour, and Social Affairs.

The 2009 version of the OECD health data information system includes 1200

Country	MD per 1,000	Nurses per 1,000	MD Graduates per 1,000 MD	Nursing Graduates per 1,000 Nurses	Total Beds per 1,000	Acute Beds per 1,000	Psychiatric Care Beds per 1,000	MRI Units per Million	CT Scanners per Million	Total Expend. per capita US (PPP)
Australia	2.68	10.11	32.1	29.7	3.9	3.6	0.4	3.7	45.2	2865
Austria	3.45	7.13	61.4	49.7	7.7	6.2	0.5	15.9	29.2	3392
Canada	2.12	8.49	25.9	37.2	3.4	2.9	0.3	4.9	10.7	3220
Czech Rep.	3.52	8.11	23.4	62.8	7.6	5.4	1.1	2.8	12.6	1387
Denmark	2.99	14.05	66.9	26.5	3.8	3.1	0.7	10.2	14.6	3055
France	3.39	7.46	17.3	43.3	7.5	3.8	1.0	3.1	7.4	3115
Germany	3.39	9.70	31.6	30.1	8.6	5.9	0.5	6.6	14.6	3160
Hungary	3.34	5.78	33.3	76.0	7.8	5.5	0.4	2.6	6.8	1305
Italy	4.19	6.71	27.4	23.0	4.0	3.4	0.1	14.1	26.3	2399
Korea	1.57	3.80	55.6	73.3	7.4	6.4	0.7	11.1	31.5	1155
Netherlands	3.60	9.39	29.0	38.8	4.5	3.1	1.3	6.2	7.1	3310
Slovak Rep.	3.06	6.32	34.6	96.6	6.9	4.9	0.9	3.7	10.2	1058
Spain	3.37	7.44	28.8	48.1	3.4	2.6	0.5	7.7	13.3	2126
UK	2.30	10.30	34.9	31.6	3.9	3.0	0.8	5.0	7.0	2557
USA	2.39	10.28	26.5	24.3	3.3	2.8	0.3	26.6	32.2	6194

Table 13.3 *OECD healthcare statistics for 2004 as reported in the report:* OECD Health Data 2009 *[88]*

series. Table 13.3 reports a subset of complete data for 15 member nations for 2004 from *OECD Health Data 2009* [88].

The 2009 health data reports 11 statistics pertaining to healthcare resources. Missing from Table 13.3 is the number of mammographs per million and radiation therapy equipment per million population. Too many countries were missing information for these two variables. Missing from Table 13.3 are countries with missing data for one or more of the other nine variables. Added to Table 13.3 is the total health expenditure per capita in US dollars adjusted for purchasing power parity (which uses long-term currency exchange rates).

The graphical displays illustrated in this section are intended for situations as given in Table 13.3 in which there are a dozen or so variables but not more than four dozen or so observations. There are various different types of plots available for displaying multivariate data in several dimensions. All of these plots rely on the use of icons or so-called *glyphs*. The next graphical display is considered a classic.

13.5.2 Chernoff's Faces

Figure 13.11 uses *Chernoff's faces* to depict the OECD healthcare service data from 2004. In 1973, Herman Chernoff [19] used facial features in a sketch diagram to depict measurements made on 87 fossil specimens. In his original proposal as many as 18 variables could be depicted.

Chernoff's idea regarding the faces is that humans are very good at recognizing individuals based on subtle differences in their facial features. The extension of this is that we should, therefore, be able to recognize items based on differences in their factors if they are converted to facial characteristics. To do this, it is important to know which variable maps to which facial feature.

Figure 13.11 was produced using the function `faces` in the R package `aplpack`. It does not use the same variable mapping, or even the same facial features, as that of Chernoff's [19] original plotting convention. See Table 13.4 for the mapping for the variables in the 2004 OECD data. Notice that the first 5 variables are repeated in sequence as the 16 facial features exceeded the 10 variables.

The single line of R code for creating Figure 13.11 with faces arranged in 5 rows and 3 columns is as follows.

```
faces(OECD[,1:10],nrow.plot=5,ncol.plot=3,face.type=0)
```

There is a tendency when using Chernoff's faces to ascribe human emotions based on the features. This is considered a serious drawback to using Chernoff's faces. There is also the problem that subtle changes in small features, such as around the eyes, can draw far more attention than other parts of the

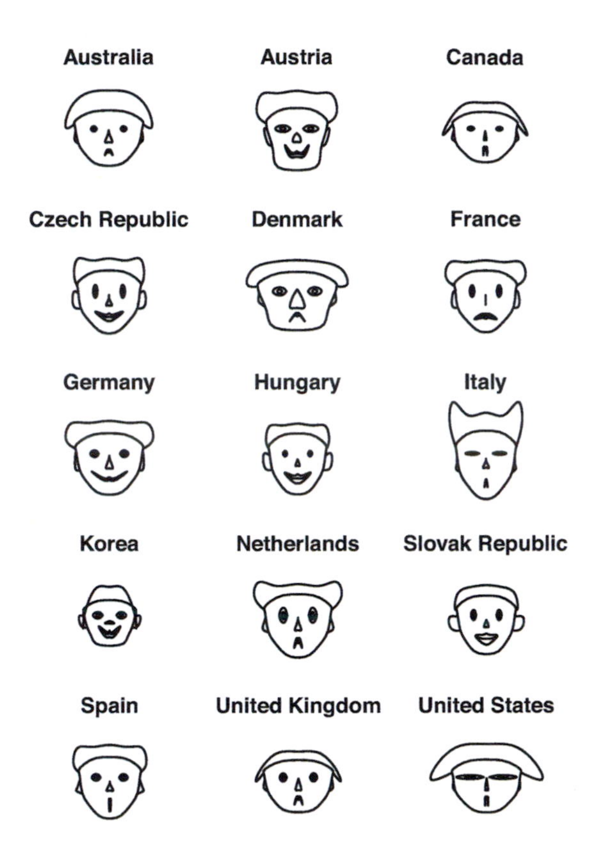

Figure 13.11 *Chernoff's faces depicting 2004 OECD heathcare service data*

face. Chernoff's original plotting convention mapped variables to the slant and size of eyebrows. These two features tended to draw more attention than they deserved. They are not listed in Table 13.4 and so are not available in the R function `faces`.

Looking at Figure 13.11, one might be tempted to conclude that the residents of the German-speaking nations of Austria and Germany are quite happy with their healthcare system or that residents of the English-speaking nations of Australia, Canada, the United Kingdom, and the United States are quite glum. The curve of the smile measures the number of acute care beds. It just so happens that the number of acute care beds in hospitals in each of the two

Index	Facial Feature	OCED Healthcare Service Variable
1	Height of face	Practicing physicians per 1,000 population
2	Width of face	Practicing nurses per 1,000 population
3	Shape of face	Medical graduates per 1,000 practicing physicians
4	Height of mouth	Nursing graduates per 1,000 practicing nurses
5	Width of mouth	Total hospital beds per 1,000 population
6	Curve of smile	Acute care beds per 1,000 population
7	Height of eyes	Psychiatric care beds per 1,000 population
8	Width of eyes	MRI units per million population
9	Height of hair	Computed Tomography Scanners per million population
10	Width of hair	Total expenditure on health per capita (US$ purchasing power parity)
11	Styling of hair	Practicing physicians per 1,000 population
12	Height of nose	Practicing nurses per 1,000 population
13	Width of nose	Medical graduates per 1,000 practicing
14	Width of ears	Nursing graduates per 1,000 practicing
15	Height of ears	Total hospital beds per 1,000 population

Table 13.4 *Chernoff facial feature and corresponding OECD healthcare statistics (if any)*

German-speaking nations is nearly double that of any one of the countries known by the acronym CANUKUSA.

One is tempted to make comparisons with respect to styling of hair and both nose and ear shape, but a quick check of Table 13.4 reveals the first 5 healthcare variables have been assigned to these.

Figure 13.12 is a representation of Figure 13.11 with color added. The code for obtaining Figure 13.12 is as follows.

```
numcolors<-20
col_nose<-daltonize(rainbow(numcolors),cvd="p",
```

Figure 13.12 *Color Chernoff's faces depicting 2004 OECD heathcare service data*

```
show=FALSE)[[2]]
col_eyes<-daltonize(rainbow(numcolors,start=0.6,end=0.85),
cvd="p",show=FALSE)[[2]]
col_hair<-daltonize(terrain.colors(numcolors),cvd="p",
show=FALSE)[[2]]
col_face<-daltonize(heat.colors(numcolors),cvd="p",
show=FALSE)[[2]]
col_lips<-daltonize(rainbow(numcolors,start=0,end=0.2),
cvd="p",show=FALSE)[[2]]
col_ears<-daltonize(rainbow(numcolors,start=0,end=0.2),
cvd="p",show=FALSE)[[2]]
```

```
faces(OECD[,1:10],nrow.plot=5,ncol.plot=3,face.type=1,
ncolors=numcolors,col.nose=col_nose,col.eyes=col_eyes,
col.hair=col_hair,col.face=col_face,col.lips=col_lips,
col.ears=col_ears)
```

The function `daltonize` in the package `LSD` has been used to produce color palettes enhanced for viewing by persons with protanopic color vision deficiencies for the default nose, eye, hair, face, lip, and ear color palettes in the function `faces`. For painting elements of a face, colors are found by averaging sets of variables: indices 7 and 8 for the irises in the eyes; indices 1, 2, and 3 for the lips; indices 14 and 15 for the ears; indices 12 and 13 for the nose; indices 9, 10, and 11 for the hair; and indices 1 and 2 for the face. It is very difficult to surmise what the impact of these complex averages is on perception of the data. Visual interpretation of colored Chernoff faces is further impacted when the facial features are too small for colors to be discerned in some of the faces, as in Figure 13.12. The coloring of faces is fanciful but not useful.

Hair width denotes the total healthcare expenditures per capita in 2004 by each country. The United States ought to appear outstanding in this regard as it spends nearly twice that of its nearest rivals per capita. But this difference is muted with Chernoff's faces. Of course, this variable could be exchanged with the variable for the feature of facial height. But this only serves to highlight the importance of the subjective choices for facial feature assignment.

Caution in the use of Chernoff's faces must be exercised. Chernoff's faces are less likely to be understood by the uninitiated as well as being difficult to explain. It is doubtful that a room full of administrators would react patiently to an explanation of how the facial features represent the variables of interest.

It is inconceivable that either Figure 13.11 or 13.12 would be presented in a briefing on comparative healthcare expenditures to the leaders of the 30 member nations of the OECD. This serves to highlight the fact that Chernoff's faces are not designed so much as a tool for summarizing data as they are for analyzing data. The iconographic plot of the next section is a better tool.

13.5.3 The Star Plot

The *star plot* of Figure 13.13 uses the icon of a star to depict the values of 10 healthcare service variables for a given member nation. The distance along each of 10 rays emanating from the center of origin for each country is used to denote a scaled value. If the variable is at the minimum for all observations for that variable, then the length of the ray is zero. Note that the ray for the number of practicing physicians per 1,000 population for Korea is not drawn because its length is zero. The length of the ray for this variable is at its maximum for Italy, which has 4.19 practicing physicians per 1,000 population, the largest of any of the nations depicted in Figure 13.13.

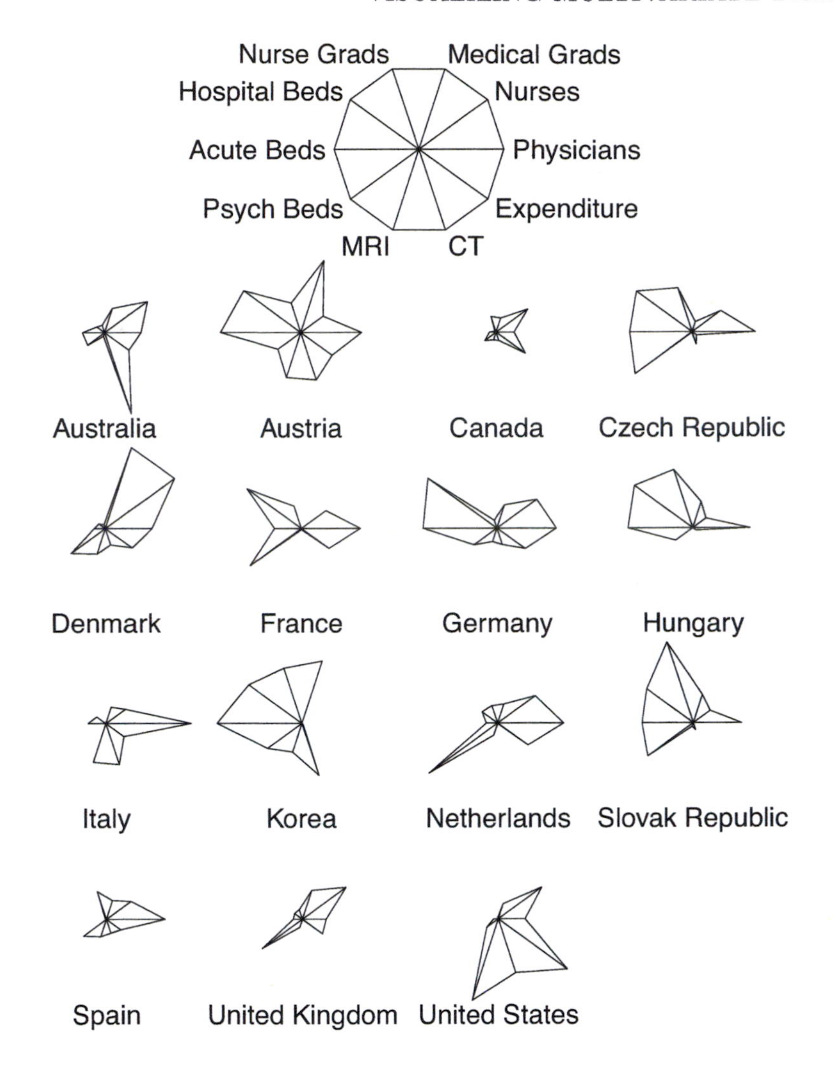

Figure 13.13 *Star plot depicting 2004 OECD healthcare service data*

Note the key at the top of Figure 13.13 that indicates the angular location for each of the 10 healthcare service variables. A star plot tends to look more objective and less trivial than a plot of Chernoff's faces. A simple key indicating the angle for each variable in a star plot is quicker to consult than a table that takes up three-quarters of a page. The large total healthcare expenditure per capita for the United States is more easily discerned in the star plot than in the plot of Chernoff's faces.

A star plot is not intended for a quick glance. It is a tool for data analysis. But it takes much more effort than a corresponding table of numbers. In searching for similar stars, one finds that the Netherlands and the UK are alike. The CANUKUSA nations actually are not all that similar to each other. Neither are the German-speaking nations of Austria and Germany.

Austria and Hungary have gone their separate ways, as have the Czech and Slovak Republics with respect to investments in their national healthcare systems. For a country to be middle of the road for all ten variables, one would need to see a star with rays of nearly equal length. Not one of the 15 countries in Figure 13.13 fits this bill.

The one country in Figure 13.13 that is remarkable for being the most unremarkable is Canada. The country with the smallest of the stars is Canada, which also tends to avoid the large imbalances in some of the variables seen in the 14 other countries.

When it comes to healthcare spending and investing in health professional training, Canadians come across as uniformly more tightfisted than even the Dutch. This is interesting because, contrary to popular belief, Canada does not have a single national healthcare authority. Instead, each of Canada's ten southern provinces and three northern territories have their own healthcare system. The federal government in Canada is the major contributor to these programs but is not the sole source of funding.

The star plot of Figure 13.13 was generated by the R function `stars`, which is in the standard `graphics` package. The R code that produced Figure 13.13 is as follows.

```
nations<-unlist(OECDHealth2004[1],use.names=FALSE)
categories<-c("Physicians","Nurses","Medical Grads",
"Nurse Grads","Hospital Beds","Acute Beds","Psych Beds",
"MRI","CT","Expenditure")

OECD<-matrix(unlist(OECDHealth2004,use.names=FALSE)[16:165],
nrow=15,ncol=10,dimnames=list(nations,categories))

stars(OECD[,1:10],key.loc=c(6,11.35),cex=0.8,
flip.labels=FALSE)
```

The R function `unlist` is used to create a vector variable from a list structure. This is done initially to create a character vector for the names of the nations. It is used again to create a matrix of expenditures from the OECD data stored in the list `OECDHealth2004`. On calling the function `stars`, the argument `key.loc=c(6,11.35)` is used to set the location of the legend. The argument `cex=0.8` reduces the size of the character labels. The function `stars` can flip labels alternatively from below star to above star. This is turned off by setting `flip.labels=FALSE`.

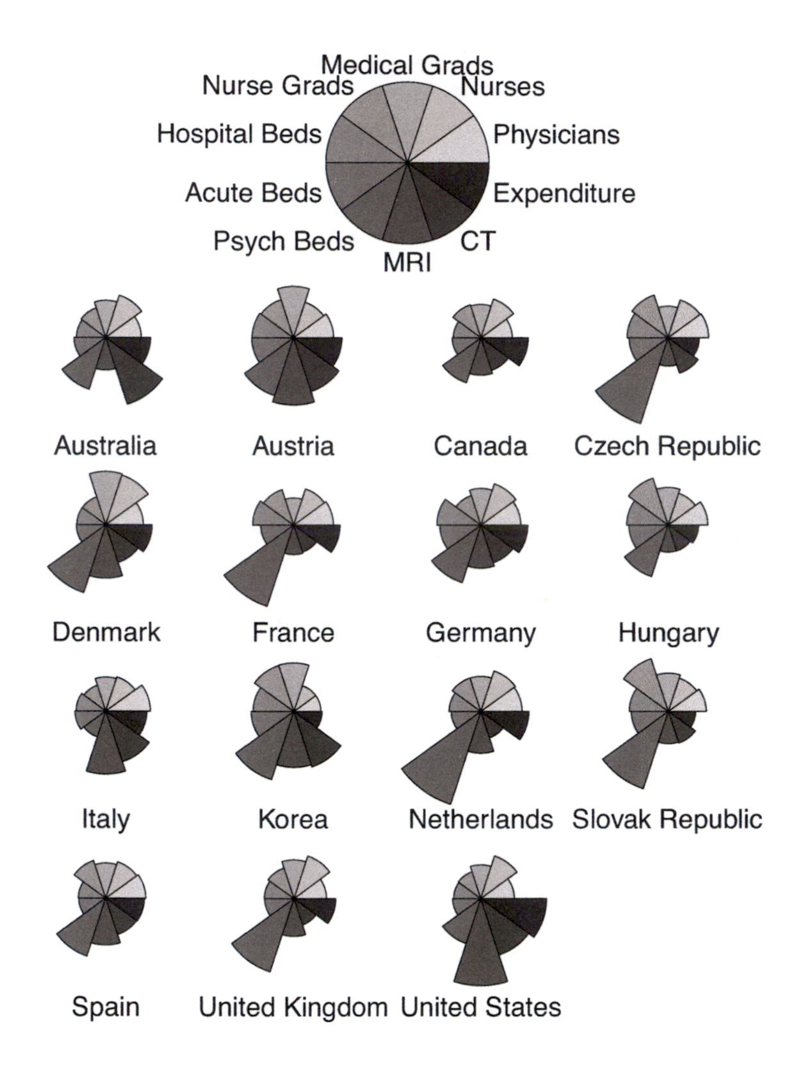

Figure 13.14 *Rose plot depicting 2004 OECD healthcare service data*

13.5.4 The Rose Plot

The *rose plot* of Figure 13.14 is a variation on Florence Nightingale's rose (or coxcomb) diagram introduced in Chapter 6. Each variable in the rose plot is identified with a sector, whereas for the star plot each variable is identified with a ray.

Care has been taken in the following call of the function `stars` to draft Figure 13.14 so that the area of each sector, and not its radial, is proportional to the value of the variable.

```
stars(sqrt(OECD %*% diag(1./apply(OECD,2,min)))/3,
scale=FALSE,key.loc=c(6,11.35),cex=0.8,draw.segments=TRUE,
col.segments=gray.colors(10)[10:1],key.xpd=TRUE,xpd=TRUE,
key.labels=categories,flip.labels=FALSE)
```

It is not unusual when comparing a rose plot to a star plot of the same data that different features in each are perceived. In the star plot of Figure 13.13 the focus is on line segment length with zero length corresponding to a minimum across the 15 countries. In the rose plot of Figure 13.14 the focus is on segment area, which necessarily cannot be started at zero.

Something leaping out in the rose plot is the greater investment per capita by the United States and Austria compared to all other countries in Magnetic Resonance Imaging (MRI) equipment.

Perhaps more outstanding is the greater number of psychiatric care beds for all countries in comparison to Australia, Canada, Italy, and the United States. Could Australia, Canada, and the United States be benefactors of the soothing influence of the wide open spaces? Could Italy be reaping the benefit of the Mediterranean diet? It is the time-honored question of nature versus nurture.

13.6 The Multiple Linear Regression Model

13.6.1 *Definition*

Consider the *multiple linear regression model* :

$$y_i = \beta_0 + \beta_1 x_{1\,i} + \beta_2 x_{2\,i} + \cdots + \beta_{p-1} x_{p-1\,i} + \epsilon_i \qquad (13.5)$$

where y_i is the response variable and x_{ij} are the explanatory variables from n observations and ϵ_i are independently normally distributed random errors with mean 0 and variance σ^2. Note carefully that there are p regression coefficients: $\{\beta_0, \beta_1, \beta_2, \ldots, \beta_{p-1}\}$.

Note that the simple linear regression model is a special case of formula (13.5) with $p = 2$ and only one explanatory variable. The polynomial regression model of order $m = p - 1$ is a special case obtained with just one explanatory variable x and $x_i = x^i$. Note the model is still linear when $x_i = x^i$. This is because the term *linear* refers not to how the variables enter the model in formula (13.5) but how the regression coefficients $\{\beta_0, \beta_1, \beta_2, \ldots, \beta_{p-1}\}$ enter the model.

The matrix notation for the multiple linear regression can be obtained from the discussion in Section 11.4.2 on matrix notation for the simple linear model by merely adding an additional column to the matrix $\mathbf{X}$ corresponding to each additional explanatory variable in the model. That is, the matrix of explanatory variables is given by

$$\mathbf{X} = \begin{bmatrix} 1 & x_{1\,1} & \cdots & x_{p-1\,1} \\ 1 & x_{1\,2} & \cdots & x_{p-1\,2} \\ \vdots & \vdots & \ddots & \vdots \\ 1 & x_{1\,n} & \cdots & x_{p-1\,n} \end{bmatrix}. \tag{13.6}$$

Arrange the regression coefficients in a column vector as follows:

$$\boldsymbol{\beta} = \begin{bmatrix} \beta_0 \\ \beta_1 \\ \vdots \\ \beta_{p-1} \end{bmatrix}. \tag{13.7}$$

Let the column vector of response variables be given by

$$\mathbf{y} = \begin{bmatrix} y_1 \\ y_2 \\ \vdots \\ y_n \end{bmatrix}. \tag{13.8}$$

Let the column vector of random errors be given by

$$\boldsymbol{\epsilon} = \begin{bmatrix} \epsilon_1 \\ \epsilon_2 \\ \vdots \\ \epsilon_n \end{bmatrix}. \tag{13.9}$$

The multiple linear regression model in matrix notion is

$$\mathbf{y} = \mathbf{X}\boldsymbol{\beta} + \boldsymbol{\epsilon}. \tag{13.10}$$

The *ordinary least-squares* estimator of $\boldsymbol{\beta}$ is

$$\hat{\boldsymbol{\beta}} = (\mathbf{X}^T\mathbf{X})^{-1}\mathbf{X}^T\mathbf{y}. \tag{13.11}$$

The column vector of *fitted values* is thus

$$\hat{\mathbf{y}} = \mathbf{X}(\mathbf{X}^T\mathbf{X})^{-1}\mathbf{X}^T\mathbf{y}. \tag{13.12}$$

The *hat matrix* $\mathbf{H} = [h_{ij}]$ is defined to be

$$\mathbf{H} = \mathbf{X}(\mathbf{X}^T\mathbf{X})^{-1}\mathbf{X}^T. \tag{13.13}$$

The column vector of *residuals* is given by

$$\mathbf{e} = (\mathbf{I} - \mathbf{H})\mathbf{y}. \tag{13.14}$$

The leverage is measured by the entries $\{h_{ii}\}$ along the main diagonal of the hat matrix $\mathbf{H}$.

The diagnostic plots presented in Chapter 11 for the simple linear regression model also apply to the multiple linear regression model, except that there are more of them because there are now two or more explanatory variables. There will be a few new diagnostic plots presented in this chapter because they are not really needed until there is more than one explanatory variable.

There is a critical assumption for both the simple and multiple linear regression models involving the random errors $\{\epsilon_i\}$. Its discussion was deferred from Chapter 11 until this chapter because only the theory of the multiple linear regression model affords a solution to the problem. The critical assumption is that the variance of the $\{\epsilon_i\}$ is a constant σ^2. If this is the case, then the variance is said to be *homoscedastic*. If this is not the case, then the variance is said to be *heteroscedastic*.

The *scedasticity* of the random errors is assessed by examining the residuals

$$e_i = y_i - \hat{y}_i \tag{13.15}$$

where the fitted values are obtained from

$$\hat{y}_i = \hat{\beta}_0 + \hat{\beta}_1 x_{1i} + \hat{\beta}_2 x_{2i} + \cdots + \hat{\beta}_{p-1} x_{p-1\ i} \tag{13.16}$$

after the least-squares estimates $\{\hat{\beta}_0, \hat{\beta}_1, \hat{\beta}_2, \ldots, \hat{\beta}_{p-1}\}$ have been determined from the data. Often it is possible to remove distinct patterns in the spread of the residuals by adding more explanatory variables to the model.

13.6.2 Modeling Perch Mass

An example for the general linear regression model is taken from the data collected on perch by a research trawl on the Finnish lake Längelmävesi around 1917. Sports fishers required to report fish sizes cannot be expected to provide accurate weigh scales for determining the mass of their catch. As noted in the discussion in the previous chapter on polynomial regression, catch-and-release regulations are typically set based upon fish length from tip of the nose to the notch in the tail.

Sports fishers do have the means to measure height and width, as well as length. The following model for mass has already been noted as a possibility in this chapter:

$$\log(m) = \log(\rho) + \beta_1 \log(\ell) + \beta_2 \log(w) + \beta_3 \log(h). \tag{13.17}$$

Regression Coefficient	Estimate	Standard Error	Two-Sided p-Value
Intercept	-2.3165	0.3981	2.28×10^{-7}
log(Length)	1.6197	0.2265	2.84×10^{-9}
log(Width)	0.8226	0.2167	3.86×10^{-4}
log(Height)	0.5622	0.1803	2.96×10^{-3}

Table 13.5 *Summary of results for the fit of a multiple linear regression model for the logarithm of mass of 56 perch*

This model is not at all different from formula (13.5) in the definition of the multiple linear regression model.

The least-squares estimates of the regression coefficients are summarized in Table 13.5. Each of the regression coefficients is significantly different from zero.

The overall F-statistic is highly significant with $p < 2.2 \times 10^{-16}$. The squared multiple correlation $R^2 = 0.994$ is quite high. The estimate of the error standard deviation σ is 0.08673 on 52 degrees of freedom. On the surface, the fit looks good. But the devil could be in the details. The next step is a detailed analysis of the residuals.

13.6.3 Residual Scatterplot Matrix

The *residual scatterplot matrix* for the multiple linear model of the log-transform of mass as a function of the log-transform of each nose-to-notch length, width, and height is given in Figure 13.15. Note that for efficiency's sake, only the lower triangular portion of the scatterplot matrix has been plotted.

Along the main diagonal are normal quantile-quantile plots for each variable. The multiple linear regression does not require that the response and explanatory variables follow a multivariate normal distribution. Although marginal normality for each variable is not a guarantee of multivariate normality, this is usually checked. All that is required under the multiple linear regression model is that the residuals are normally distributed. This does appear to be the case in the lower right panel of Figure 13.15.

The logarithms of the response and explanatory variables do not appear to be normal according to the other plots along the main diagonal. The bimodal nature in each of these variables is evident in the normal quantile-quantile plots. It is a relief to see that there is no evidence of more than one mode for the residuals.

Note that if the residuals are plotted as the last variable, then the residual plots

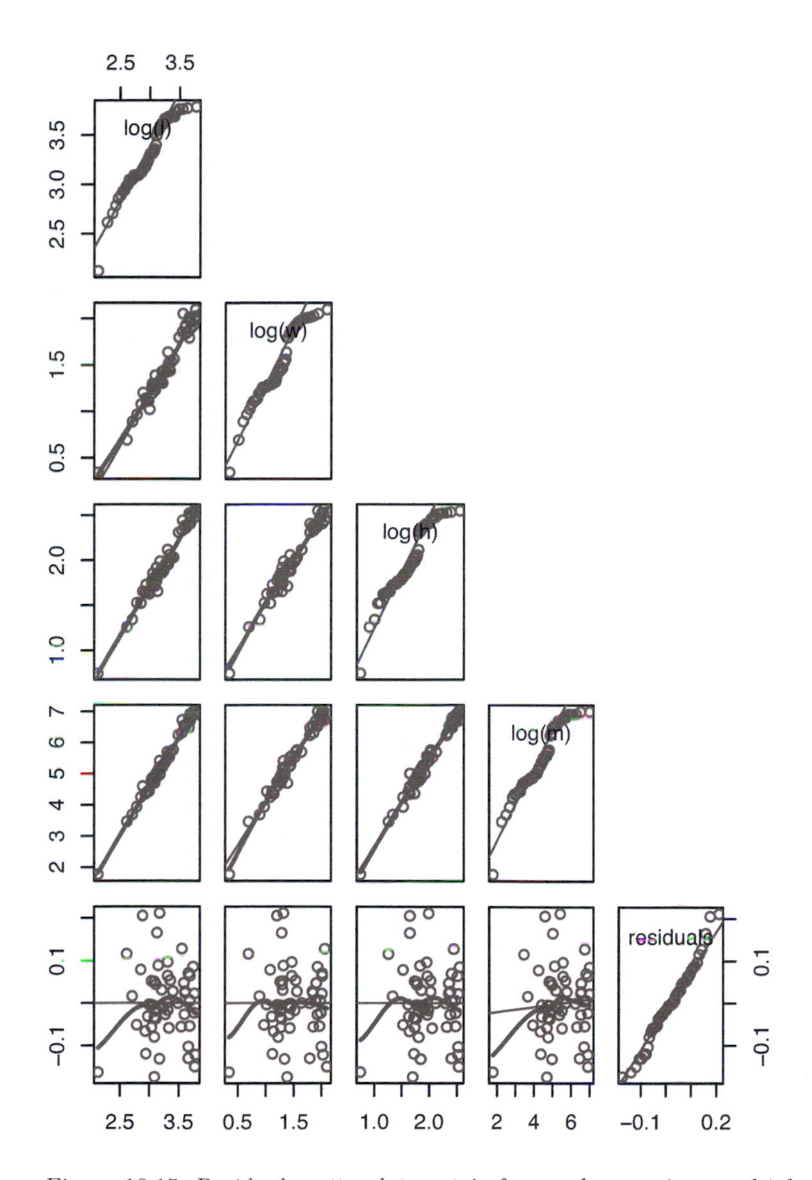

Figure 13.15 *Residual scatterplot matrix for perch mass in a multiple linear model with length, width, and height of fish (a normal quantile-quantile plot is displayed for each variable along the main diagonal)*

on the lowest line of panels will be plotted in the conventional arrangement of the residuals along the vertical axis and the response or an explanatory variable along the horizontal matrix. Note that all the scatterplots in Figure 13.15 have a dotted simple least-squares regression line and a LOWESS curve.

The least-squares regression lines will have a slope and intercept both set to zero when the residuals are plotted against the explanatory variables. This is seen in Figure 13.15. Based on the scatter of points and the proximity of the LOWESS curve to the horizontal reference line, there appears to be no sign of a pattern, or heteroscedasticity, for the residuals when plotted against each of the explanatory variables. The outcome is similar for the residuals plotted against the response variable of the log-transform of mass.

If the response variable is chosen to be the second-to-last variable, then the second-to-last row of scatterplots will depict the response variable along the vertical axis and each of the explanatory variables along the horizontal axis. The linear fit with each explanatory variable is so strong, the LOWESS curve is difficult to distinguish in the scatterplots for each of the three explanatory variables.

The remaining three scatterplots are on the second and third rows from the top in the scatterplot matrix of Figure 13.15. There are no residuals displayed in these scatterplots. What can be discerned from these three scatterplots is that the three explanatory variables are highly mutually linear with each other. But not so much as to be a concern with modern linear algorithms used to solve for the least-squares estimates of the regression coefficients.

It goes without saying that the size of the plots in Figure 13.15 are about the size of a thumbprint. If there is detail attracting attention in any one plot that requires magnification, then by all means either zoom in on the computer screen or produce a new larger plot specifically for that feature.

There is one data point in each of the plots in Figure 13.15, save the normal quantile-quantile plot of the residuals, that is attracting attention. This is the smallest perch in the catch with a mass of 5.9 grams. Although it does not stray far from the least-squares regression line in the top six scatterplots, it might be influential.

The fitting of the linear model and the plotting of Figure 13.15 was done by the following R script.

```
Perch$Mass<-Perch$Weight
Perch$Length<-Perch$Length2
Perch$Width<-Perch$Width_pc*Perch$Length3/100
Perch$Height<-Perch$Height_pc*Perch$Length3/100

lmlog<-lm(log(Mass) ~ log(Length)+log(Width)+log(Height),
data=Perch)
```

```
Perch$res<-residuals(lmlog)
yxr<-cbind(log(Perch$Length),log(Perch$Width),
log(Perch$Height),log(Perch$Mass),Perch$res)

scatterplotMatrix(yxr,diagonal="qqplot",
reg.line=lm,spread=FALSE,smoother=loessLine,
var.labels=c("log(l)","log(w)","log(h)","log(m)",
"residuals"),col=gray(0.5),upper.panel=NULL)
```

The plotting of the scatterplot matrix is done by the function `scatter-plot.matrix` of the `car` package that was written to accompany the book *An R and S-PLUS Companion to Applied Regression* written by John Fox [43] in 2002. Although this function just sets up a call to the function `pairs` in the `graphics` package, it is a convenient and quicker alternative to doing it yourself. The normal quantile-quantile plot along the main diagonal for each variable was selected by the argument `diagonal="qqplot"`.

13.6.4 Leverage Scatterplot Matrix

Figure 13.16 presents a *leverage scatterplot matrix* for the perch model. To plot leverage along the vertical axis in a scatterplot for the response variable, and each of the explanatory variables, only requires that leverage be the last variable. The following R script calculates the leverage (`hatvalues`) from the list `lmlog` produced by the function `lm` that fits the linear model. Then Figure 13.16 is produced by the function `scatterplot.matrix` of the `car` package.

```
Perch$hat<-hatvalues(lmlog)
yxr<-cbind(log(Perch$Length),log(Perch$Width),
log(Perch$Height),log(Perch$Mass),Perch$hat)

scatterplotMatrix(yxr,diagonal="density",
reg.line=lm,spread=FALSE,smoother=loessLine,
var.labels=c("log(l)","log(w)","log(h)","log(m)",
"leverage"),col=gray(0.5),upper.panel=NULL)
```

Note that the argument `diagonal="density"` produces a kernel density estimate for each variable along the main diagonal.

Given that $p = 4$ for the multiple linear regression model, leverage h_{ii} for the ith observation is judged large if $h_{ii} > 2(p/n) \approx 0.071$. Twenty observations meet this criterion. But only 3 observations meet the criterion of being moderately large ($0.2 \geq h_{ii} < 0.5$). And none meet the criterion for very large. The largest leverage is 0.2719 and this for fish #104, the smallest fish.

Rather than depicting for a second time normal quantile-quantile plots along

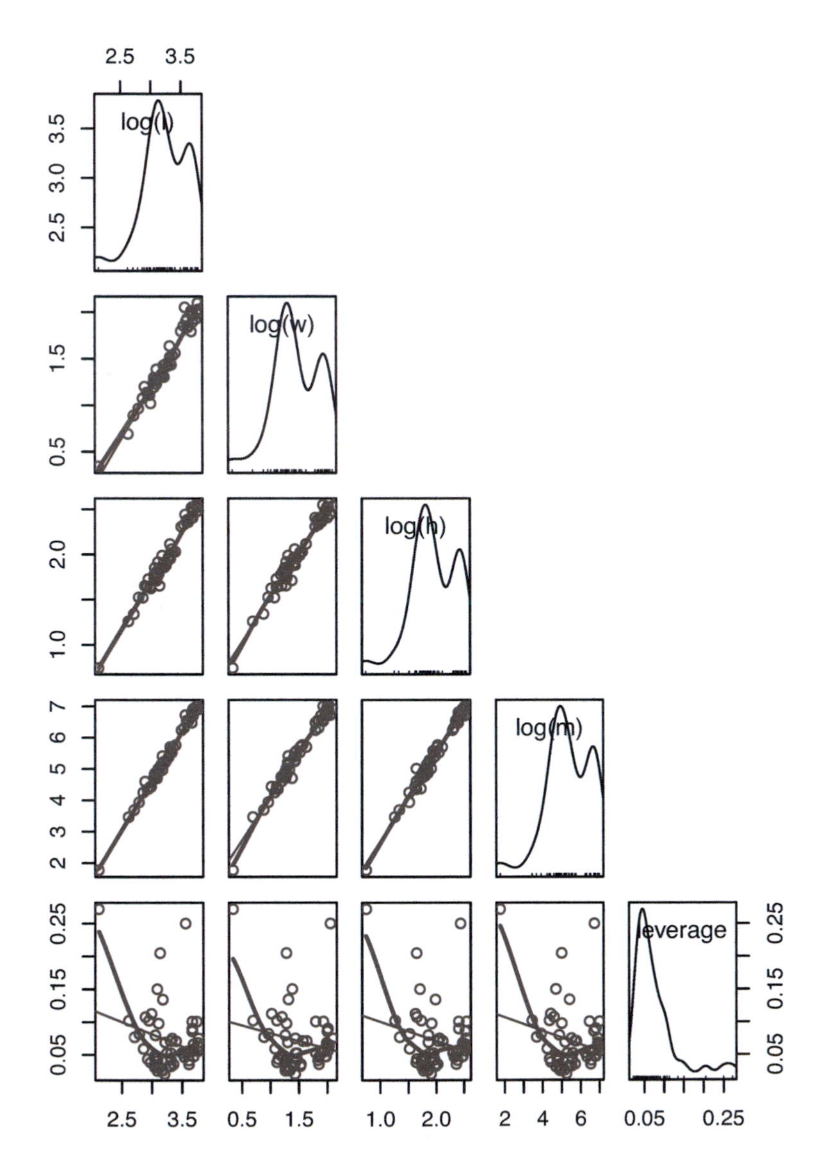

Figure 13.16 *Leverage scatterplot matrix for perch mass in a multiple linear model with length, width, and height of fish (a kernel density estimate is overlaid on a barcode plot for each variable along the main diagonal)*

the main diagonal of a scatterplot matrix, Figure 13.16 presents marginal kernel density estimates superimposed on barcode plots. The bimodality for the marginal distribution of each of the response and explanatory variables is clear. Leverage is not expected to be normally distributed. The lack of evidence for two or more modes for leverage, however, is welcome.

13.6.5 *Influence Plot*

The influence for the multiple linear regression model of the mass of perch is given in Figure 13.17. This figure is produced by the two lines of R code below which call the function `influencePlot` from the package `car`.

```
influencePlot(lmlog,ylim=c(-4,4),xlim=c(0,0.35),labels="")
```

The only fish with a studentized residual with an absolute value greater than two and a leverage greater than three times the average leverage $(3(p/n) = 3(4/56) \approx 0.2143)$ is Fish #104, the smallest fish. With respect to leverage, the 56 perch are otherwise unremarkable.

13.6.6 *Partial-Regression Scatterplot Matrix*

Let $e_i^{(j)}$ denote the residual from a multiple linear regression model of y on the subset of explanatory variables with the variable x_j removed. For example,

$$\begin{aligned} e_i^{(1)} &= y_i - \hat{y}_i^{(1)} && (13.18) \\ &= y_i - \left[\hat{\beta}_0^{(1)} + \hat{\beta}_2^{(1)} x_{2i} + \cdots + \hat{\beta}_{p-1}^{(1)} x_{p-1\ i}\right] && (13.19) \end{aligned}$$

where the ordinary least-squares regression coefficient estimates $\{\hat{\beta}_i^{(1)}\}$ are found without the explanatory variable x_1 in the model.

Let $f_i^{(j)}$ denote the residual from a multiple linear regression model of x_j on the subset of explanatory variables with the variable x_j removed. For example,

$$\begin{aligned} f_i^{(1)} &= x_i - \hat{x}_i^{(1)} && (13.20) \\ &= x_i - \left[\hat{\alpha}_0^{(1)} + \hat{\alpha}_2^{(1)} x_{2i} + \cdots + \hat{\alpha}_{p-1}^{(1)} x_{p-1\ i}\right] && (13.21) \end{aligned}$$

where the ordinary least-squares regression coefficient estimates $\{\hat{\alpha}_i^{(1)}\}$ are found without the explanatory variable x_1 among the explanatory variables.

The residuals $e_i^{(1)}$ and $f_i^{(1)}$ have a couple of convenient properties. Firstly, the least-squares estimate of the slope for the simple linear regression of $e_i^{(1)}$

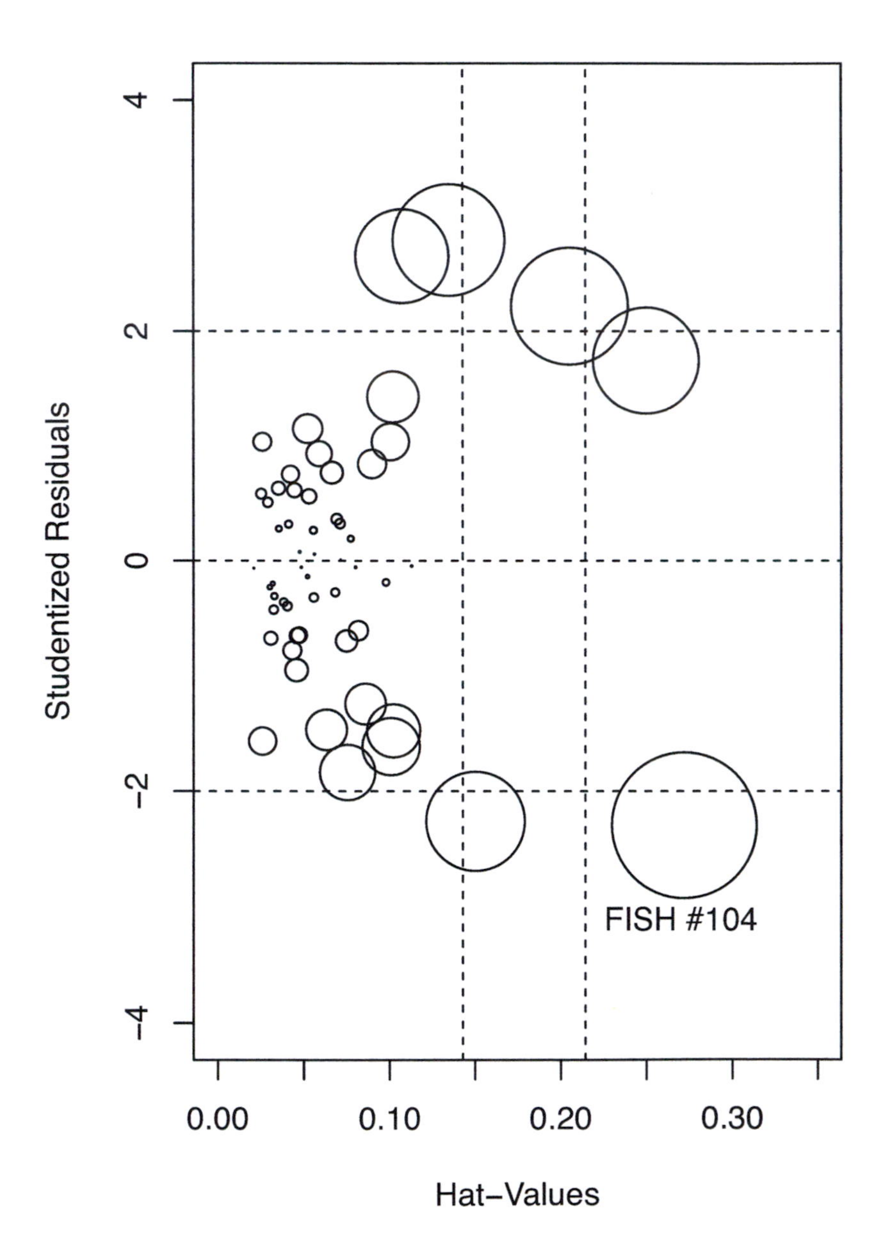

Figure 13.17 *Influence plot for the multiple linear regression model of the mass of perch (the area of each circle is proportional to the Cook's distance for the observation at the circle's center; and vertical reference lines are drawn at twice and thrice the average leverage value)*

on $f_i^{(1)}$ is the ordinary least-squares estimate $\hat{\beta}_1$ from the multiple linear regression for the full model with all explanatory variables. Lastly, the residuals satisfy the following equation

$$e_i^{(1)} = \hat{\beta}_1 f_i^{(1)} + e_i. \tag{13.22}$$

Plotting $e_i^{(1)}$ against $f_i^{(1)}$ permits examination of influence on $\hat{\beta}_1$ and also provides an impression of the precision of the estimate $\hat{\beta}_1$. This can also reveal nonlinearity and suggest whether a relationship is monotone.

Figure 13.18 presents a scatterplot matrix of $e_i^{(j)}$ against $f_i^{(j)}$ for all $j = 0, 1, 2, 3$ in the multiple linear regression model for the mass of perch. These plots are called *partial-regression plots* or *added-variable plots*. Figure 13.18 was produced by the following two lines of R code below, which make a call to the `avPlots` function in the `car` package.

```
avPlots(lmlog,intercept=TRUE,col="black",
col.lines="black",lwd=1,grid=FALSE,main="")
```

The argument `intercept=TRUE` overrides the default `intercept=FALSE` that causes `avPlots` to produce an added-variable plot for the intercept term in the statistical model. The argument `grid=FALSE` suppresses the (background) grid in the plot.

The added-variable plots in Figure 13.18 give no reason for concern, although the fit with the log-transform of the nose-to-notch length appears tighter than that for the other two explanatory variables.

13.6.7 Partial-Residual Scatterplot Matrix

Define the partial residual for the jth explanatory variable of the ith observation as

$$g_i^{(j)} = e_i + \hat{\beta} x_{i\,j}. \tag{13.23}$$

What this does is add back the linear component of the partial relationship between the response variable y and the explanatory variable x_j to the least-squares residuals. The resulting partial residual $g_i^{(j)}$ can include an unmodeled nonlinear component. One looks for this in a plot of $g_i^{(j)}$ versus x_j. This plot is called a *partial-residual plot* or a *component-plus-residual plot*. A partial-residual plot is done for all explanatory variables, that is, $j = 1, 2, \ldots, p-1$.

Figure 13.19 is a scatterplot matrix depicting a component-plus-residual plot for each of the explanatory variables. In each plot is a least-squares regression line and a LOWESS curve. By the tight clustering around the least-squares regression line for each plot and the tight proximity of the LOWESS curve to

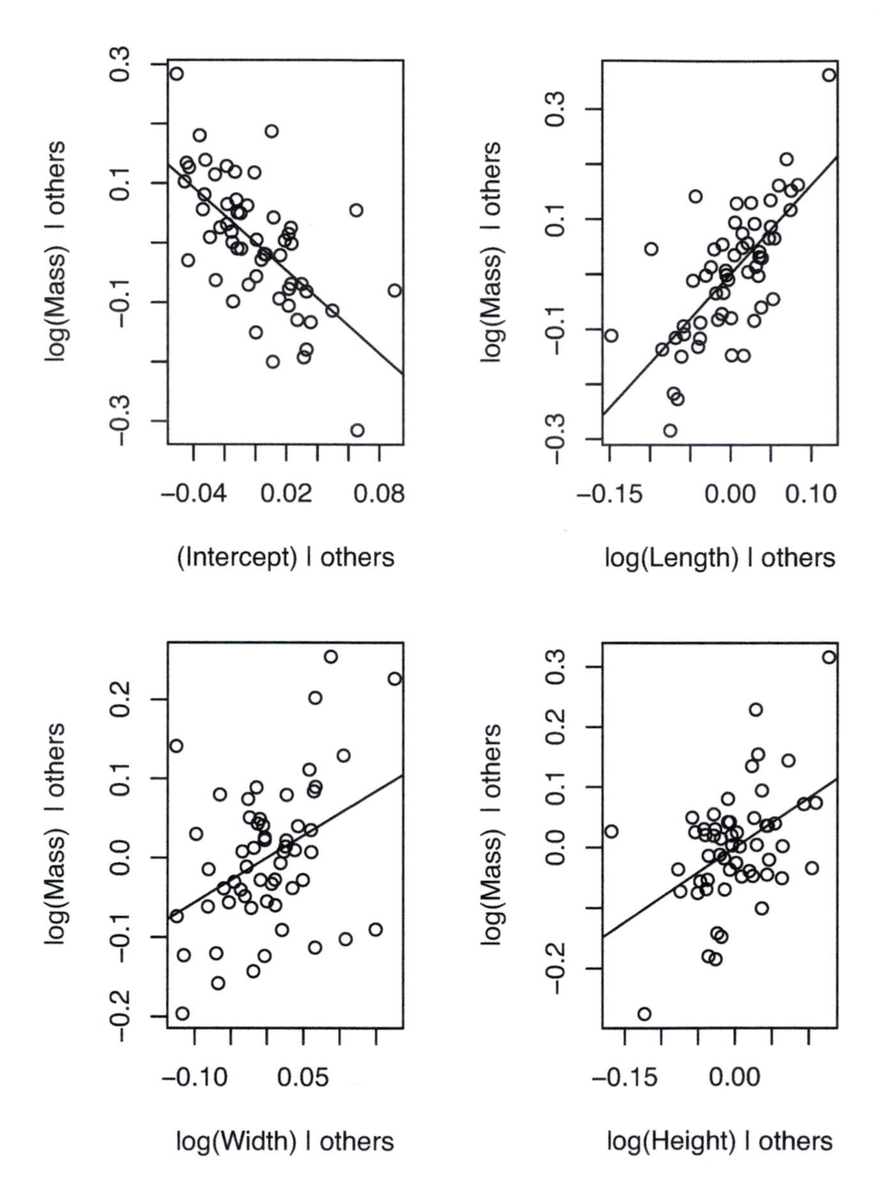

Figure 13.18 *Partial-regression scatterplot matrix for the multiple linear regression model for perch mass*

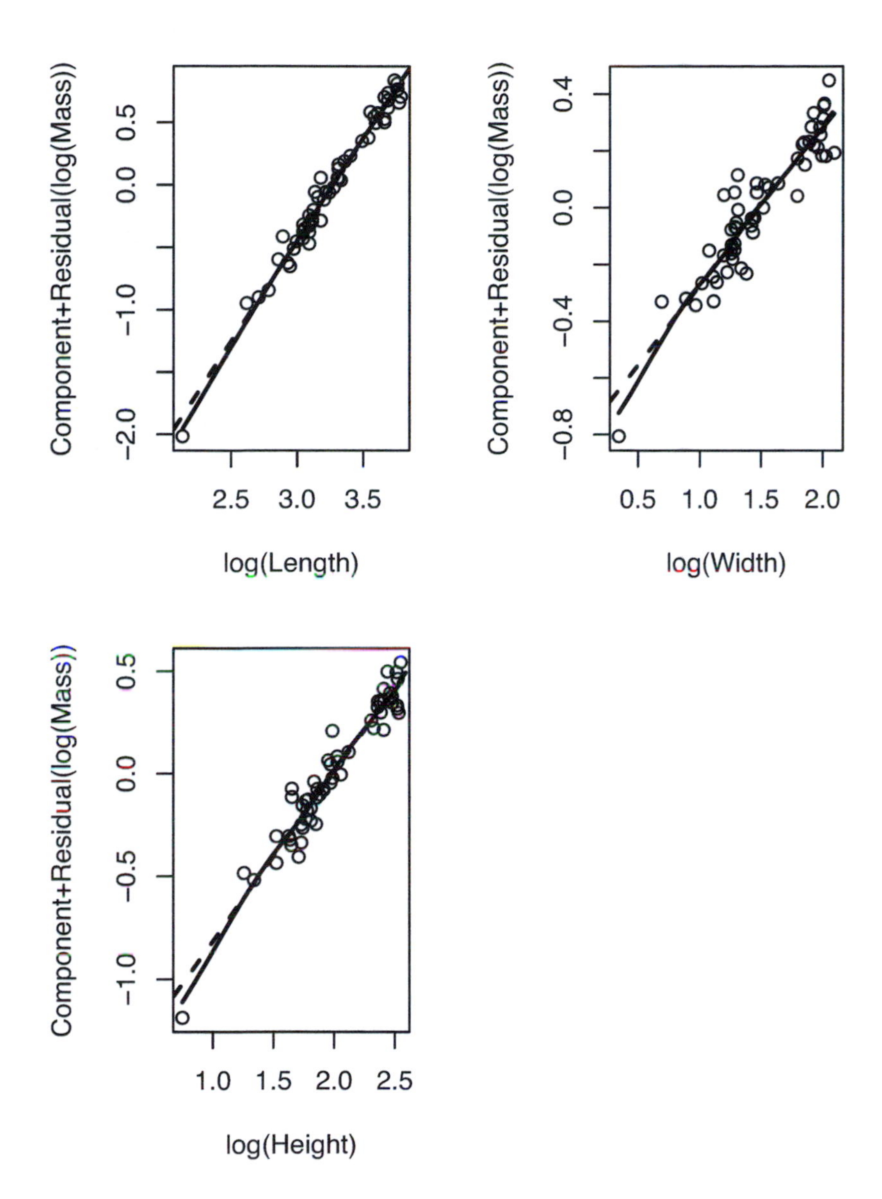

Figure 13.19 *Partial-residual scatterplot matrix for the multiple linear regression model for perch mass*

the least-squares regression line, there appears to be no unmodeled nonlinear effect.

Figure 13.19 was produced by the following two lines of R code that make a call to the `crPlots` function in the `car` package.

```
crPlots(lmlog,col="black",line=TRUE,smoother=loessLine,
col.lines=c("black","black"),lwd=2,grid=FALSE,main="")
```

The only cause for concern in Figure 13.19 is the aforementioned Fish #104, the smallest perch. It is exerting a downward force on the fit of the least-squares regression line.

13.6.8 Summary of the Model for Perch Mass

With respect to the sample of 56 perch caught in the research trawl on Längelmävesi, the following model for weight appears to be quite adequate based on the graphical regression diagnostics performed:

$$\begin{aligned} \log(\text{Mass}) &= -2.165 + 1.6197\ \log(\text{Length}) \qquad (13.24) \\ &\quad + 0.5622 \log(\text{Width}) + 0.8226\ \log(\text{Height}). \qquad (13.25) \end{aligned}$$

A simpler representation of this result is

$$\text{Mass} = 0.004825\ \text{Length}^{1.6197}\ \text{Width}^{0.5622}\ \text{Height}^{0.8226}. \qquad (13.26)$$

No outliers were found although one influential point was identified through a number of different diagnostic plots relating to influence and leverage. The influential fish was #104 and was the smallest of the perch at 5.6 grams, barely a minnow. The residuals appeared to be normally distributed and no patterns were found.

The square of the correlation coefficient was found to be 99%. It is believed that the model fits the data well and that it would be entirely reasonable for sports fishers to make measurements in centimeters of length, height, and width of each catch and then convert these measurements to mass in grams with formula (13.26) above. In all likelihood, however, the formula (13.26) would only be of scientific interest to a fisheries biologist.

13.7 Conclusion

Many of the components in this chapter have been presented in detail in earlier chapters. Normal quantile-quantile plots, kernel density estimates, bar charts and plots, rose diagrams, these are not new to this chapter. But what is new is the use of color, symbol, and spatial layout to put these simple graphical displays to use in settings of three or more variables.

The ability to visualize multiple dimensions lies at the core of statistical modeling. Assembling this information in a time-efficient manner is essential. The majority of the graphical displays of this chapter are not so much intended to be presented to an audience; they are a bit too technical for that purpose. These graphical displays are intended to play an active role in the process of data analysis. Star plots, parallel coordinates plots, and shingled trellis plots are great for highlighting what there really is to see in the data themselves.

The scatterplot matrix lends itself initially to exploration. But after the model is fitted, extensive use is made again of the scatterplot for diagnostic purposes to see whether the model really captures the essence of variability in the data.

With the advances in computer technology resulting in a virtual explosion of data for just about any topic of research interest, skill in the use of the tools for depicting multivariate data is critical. A good understanding of the basic principles of data display, gleaned in simpler settings with one or two variables, needs to be part of the process of developing this skill.

13.8 Exercises

1. Consider the numbers carried and those who survived the collision of the Titanic with the iceberg on 14 April 1912 as given by the second set of tabulation for the Report of the British Wreck Commissioner's Inquiry [13]. These are given in Table 13.2. Sometimes with discrete variables it is possible to take two variables and, by creating a new single discrete variable, reduce the workload. Combine gender and age to produce a new discrete variable to replace them. Let the values of the new variable be: men, women, boys, girls.
   (a) Produce a thermometer chart like Figure 13.1 to depict the values in Table 13.2.
   (b) Are there any significant differences between Figure 13.1 and your answer to part (a)? Is there any evidence to suggest that British Attorney-General Rufus Isaacs attempted to whitewash findings of the criminally negligent treatment of third-class passengers?
2. Consider the numbers carried and those who survived the collision of the Titanic with the iceberg on 14 April 1912 as given by the second set of tabulation for the Report of the British Wreck Commissioner's Inquiry [13]. These are given in Table 13.2.
   (a) Retain all four variables from Table 13.2 and create a thermometer chart to depict survival as a function of the other three discrete variables.
   (b) Based on your answer to part (a), was the rate of survival for boys markedly better than for girls? Discuss.
3. In the superposed scatterplot of Figure 13.4 there is overlap of repeated data points that cannot be discerned. This lack of clarity can be addressed

with a superposed sunflower plot. Produce such a plot for Anderson's data for the variables of petal length and width.

4. Use a real-time three-dimensional scatterplot tool in R to explore Anderson's iris data. Summarize your findings in 500 words or less.
5. Explore the mass, length, width, and height of the 7 species of fish from the research trawl in 1917 on Längelmävesi in Finland. Ignore gender but remove any fish with missing observations for these three variables.
   (a) Use a real-time three-dimensional scatterplot tool in R to explore mass, length, width, and height.
   (b) After your exploration in part (a), produce four superposed three-dimensional scatterplots viewed from angles that best represent each of three of the four variables.
   (c) Produce a superposed scatterplot matrix for the data.
   (d) Summarize your findings in 500 words or less.
6. Explore the `ggplot2` extension package `ggChernoff` for mapping multivariate data. Is `geom_chernoff` as capable as `faces` in `aplpack`? Discuss.
7. Explore the use of the functions `coord_polar` and `geom_bar` to produce a `ggplot2` version of the rose (coxcomb) plot of Figure 13.14.
8. Using the data from the research trawl on Längelmävesi, model bream (*Abramis brama*) mass in a multiple linear model as a function of nose-to-notch length, width, and height without any transformations of the response and explanatory variables.
   (a) Create a scatterplot matrix of bream mass, length, width, and height. Briefly comment on what you see.
   (b) Report the least-squares estimates of the regression coefficients.
   (c) Summarize the overall fit of the model.
   (d) Summarize your findings in 500 words or less.
9. Refer to Exercise 8.
   (a) Create a residual scatterplot matrix. Briefly comment.
   (b) Create a leverage scatterplot matrix.
   (c) Create an influence plot for the model.
   (d) In 250 words or less, comment on the plots created for parts (b) and (c).
10. Refer to Exercises 8 and 9.
   (a) Produce a partial-regression scatterplot matrix for the multiple linear regression model of Exercise 6.
   (b) Produce a partial-residual scatterplot matrix for the multiple linear regression model of Exercise 6.
   (c) In 250 words or less, comment on the adequacy of the model in light of your answers to parts (a) and (b) of this exercise.

11. Refer to Exercises 8, 9 and 10. Find the optimal Box-Cox transformation for each of the response and explanatory variables before fitting the multiple linear regression model for bream mass.
12. Using the data from the research trawl on Längelmävesi, model pike (*Esox lucius*) mass in a multiple linear model as a function of nose-to-notch length, width, and height. Use any transformation as appropriate.

Part VI

Appendices

Appendix A

Human Visualization

A.1 Introduction

Human visual perception is complex. Analysis of this process draws upon the sciences of physics and physiology. Research into human visualization is conducted by scientists in the disciplines of ophthalmology, neurology, psychology, computer science, and statistics. Research is also conducted by human-factor engineers and practitioners of the fine, graphic, and printing arts. Because graphical images are omnipresent in our society, it can safely be said that issues concerning human visualization are more than multidisciplinary; they are in fact *omnidisciplinary* and involve a broad mixture of physical and biological theory.

A.2 Learning Outcomes

When you complete this appendix, you will be able to do the following.

- Appreciate the history of the development of the discipline of optics.
- Understand the basics of the thin lens theory of optics.
- Recognize the basic colors and understand the formula relating wavelength and frequency for the light spectrum.
- Know the anatomy of the human eye.
- Appreciate how color is perceived.
- Appreciate Weber's, Stevens's, and the Gestalt Laws in characterizing the perception of graphical objects as well as Kosslyn's image processing model.
- Comprehend the hierarchy of tasks of graphical perception from easiest to hardest.

A.3 Optics

A.3.1 Introduction

The underlying physical theory of light and light transmission is known collectively as *optics*. The list of contributors to this body of scientific theory is luminary and includes the greatest scientific minds of all the ages: Pythagoras, Aristotle, Euclid, Claudius Ptolemy, Sir Roger Bacon, Leonardo da Vinci, Galileo Galilei and contemporaries Johannes Kepler and Willebord Snell, René Descartes and contemporary Sir Robert Hooke, Sir Isaac Newton and contemporary Christian Huygens, Michael Farraday and contemporary Augustin Jean Fresnel, James Clerk Maxwell, Albert Einstein and contemporaries Max Karl Ernst Ludwig Planck, Niels Henrik David Bohr, and Erwin Shrödinger.

Roger Bacon (1215–1294) is credited with the idea for using lenses to correct vision. Leonardo da Vinci (1452–1519) is credited with the conception of the *camera obscura* and surely must be the grandfather of photography. We owe much to the Dutch opticians for many optical devices we take for granted. Masters of the art of lens grinding for the purpose of creating corrective lenses for spectacles, in their spare time, turned to tinkering. Hans Lippershey (1587–1619) applied for the first patent of a refracting telescope and his fellow countryman Zacharias Janssen (1588–1632) is credited with inventing the compound microscope. Willebrord Snell (1591–1626) of the University of Leiden is credited with empirically discovering the law of refraction that bears his name.

At the age of 23 years, Isaac Newton (1642–1727) wrote: *"I procured me a triangular glass prism to try therewith the celebrated phenomena of colours."* And then showed that white light was composed of all the colors of the rainbow. With that the topic of color achieved importance within the body of optical theory. He later developed the reflecting telescope in 1668 because he erroneously believed that the chromatic aberration in refracting telescopes was not correctable.

But of what is light composed? Newton believed that light was corpuscular. Sir Robert Hooke (1635–1703) believed that light was a rapid vibratory motion: a wave. The debate has raged ever since these two immovable bodies collided. Newton maintained that the corpuscles of light associated with the various colors excited the ether into characteristic vibrations with the sensation of red corresponding to the longest vibration of the ether, and violet the shortest. The ether was postulated to be a medium that permeated the universe.

Michael Farraday (1791–1867) established the interrelationship between electromagnetism and light when he discovered through an experiment that the polarization direction of a light beam could be altered by a strong magnetic field. James Clerk Maxwell (1831–1879) concluded that *light was an electromagnetic disturbance in the form of waves* and established a set of equations that bears his name for describing electromagnetic and light waves.

Max Planck (1858–1947) in 1900 attacked the complacency of the wave theory with his theory of quantum mechanics. There was a return to the corpuscular theory of light in which light consists of discrete particles, called quanta, of energy in the *theory of special relativity* published in 1905 by Albert Einstein (1879–1955). These particles are now called *photons*. With this theory, Einstein debunked the idea of an all-pervading ether with his *special theory of relativity* and showed that the speed of light is the same in every inertial frame of reference. But he also showed that the path of a particle of light could be deflected by a gravitational field.

In 1913, Niels Bohr (1885-1962) was able to predict the waves of the emission spectrum of light with the change in orbitals of the electron in the hydrogen atom. Quantum mechanics went on to establish that a particle of momentum p had an associated wavelength λ such that

$$p\lambda = h \tag{A.1}$$

where $h = (6.6256 \pm 0.0005) \times 10^{-34}$ Joule-second. This is called Planck's constant. With this equation the wave-particle dichotomy for light becomes the wave-particle duality. But the light photon is a very special particle: it has zero rest mass. Collide an electron with a positron and obtain two photons with all mass being converted to light energy.

Einstein later went on in 1915 to describe the phenomenon of *gravitational redshift of a star*, whereby the photon having escaped the gravitational pull of the star will have a lower frequency, by his *general theory of relativity*. The magnitude of the gravitational redshift compared to the Doppler redshift from our sun (which rotates about once every 24.7 days) is of the same magnitude for light emitted at the sun's equator and received at the earth's surface. The prediction of the amount of deflection of the path of a photon in the presence of a gravitational field was further refined from the special theory of relativity in the general theory of relativity.

It should be noted that the theory of quantum mechanics, which underpins our concept of light, was developed at the turn of the twentieth century and depends on our notions of probability theory. That the theories of quantum mechanics, probability, and statistics marched forward hand in hand, so to speak, is remarkable. Here we are using light to construct statistical graphics and our theory for light is essentially a statistical one. The circularity of this is glaring.

A.3.2 Geometrical Optics

The theory of optics is conventionally divided into a number of areas. One of those areas is concerned with the application of geometrical methods: *geometrical optics*. There are a number of interesting issues in this area but we shall

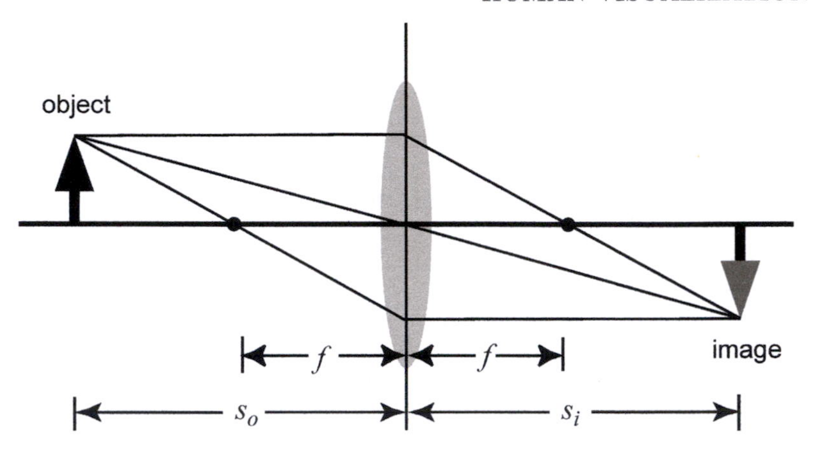

Figure A.1 *Object, image, and focal length location for a thin lens*

confine ourselves to one issue. The human eye is a complex optical device. The simplest component of the human eye, at least from a theoretical perspective concerning its optical properties, is the lens.

Assuming that the thin lens theory will match fairly well the function of a human lens, consider the following basic elements of thin lens theory. A property of any thin lens is that it can be characterized by something called the focal length, which can be determined empirically. Let the focal length be denoted by f. Consider an object located a distance s_o along the axis of the lens from the plane of the lens. The distance of the image of this object behind the lens is denoted by s_i and must satisfy the Gaussian lens equation:

$$\frac{1}{f} = \frac{1}{s_o} + \frac{1}{s_i}. \tag{A.2}$$

For a visual depiction of the geometry of this equation, see Figure A.1.

Letting y_o denote the height of the object above the axis of the lens and y_i the height of the image, the *lateral* or *transverse magnification* of a thin lens is given by

$$M_T = \frac{y_i}{y_o}. \tag{A.3}$$

The term *magnification* is a bit of a misnomer because an image could very well be smaller than the object as is certainly the case with the human eye.

Notice also from Figure A.1 that the sign of y_i will be negative as the image is inverted. Thus a positive M_T denotes an erect image and a negative M_T denotes an inverted image. Clearly, the human lens must produce inverted images smaller than the objects we see on the retina. The images we perceive as right-side up are actually upside down on the inside of our eyes.

Generally, we do not refer to the focal length of the human eye but rather to the power. The power p of a thin lens is given by

$$p = \frac{1}{f}. \tag{A.4}$$

If the focal length is measured in meters, then the unit of measure for power is the *diopter*. Prescriptions for corrective lenses are given in units of diopters.

A.3.3 The Light Spectrum

As light is a form of electromagnetic radiation, a comparison of the visible light spectrum with the entire range of the electromagnetic spectrum reveals a very small part of the whole. Light occurs in the narrow band of frequencies from 3.84×10^{14} Hz to 7.69×10^{14} Hz. The relationship between frequency ν and wavelength λ is given by

$$\nu = \frac{c}{\lambda} \tag{A.5}$$

where the constant $c = 299,792.4574 \pm 0.0012$ km/s is the speed of light.

The orange-red line of krypton (isotope 86) occurs at a frequency of approximately 400 MHz, which corresponds to a wavelength of 605.7802105 nm. It is the metric standard of length: 1,650,763.73 times this wavelength is defined to be one meter. Notice that light provides us with our standards for both length and time.

The equation (A.5) is valid only in a vacuum. Christian Huygens (1629–1695) correctly concluded that light slows down on entering more dense media giving rise to refraction as exemplified by the Gaussian equation for a thin lens. Another point to note is that light waves are transverse, whereas sound waves in air are longitudinal.

The color spectrum and approximate boundaries for the colors of red, orange, yellow, green, blue, and violet are given in Figure A.2. The values for the wavelengths and frequencies are valid in a vacuum.

A.4 Anatomy of the Human Eye

The human eye is a sensory organ that perceives and responds to light. The eye is similar in structure and function to a camera. The eye uses a single lens to focus a picture on a surface called a *retina*. The retina contains densely packed light-sensitive photoreceptor cells. An image of the external object being viewed is absorbed by the eye in a point-by-point fashion and is then

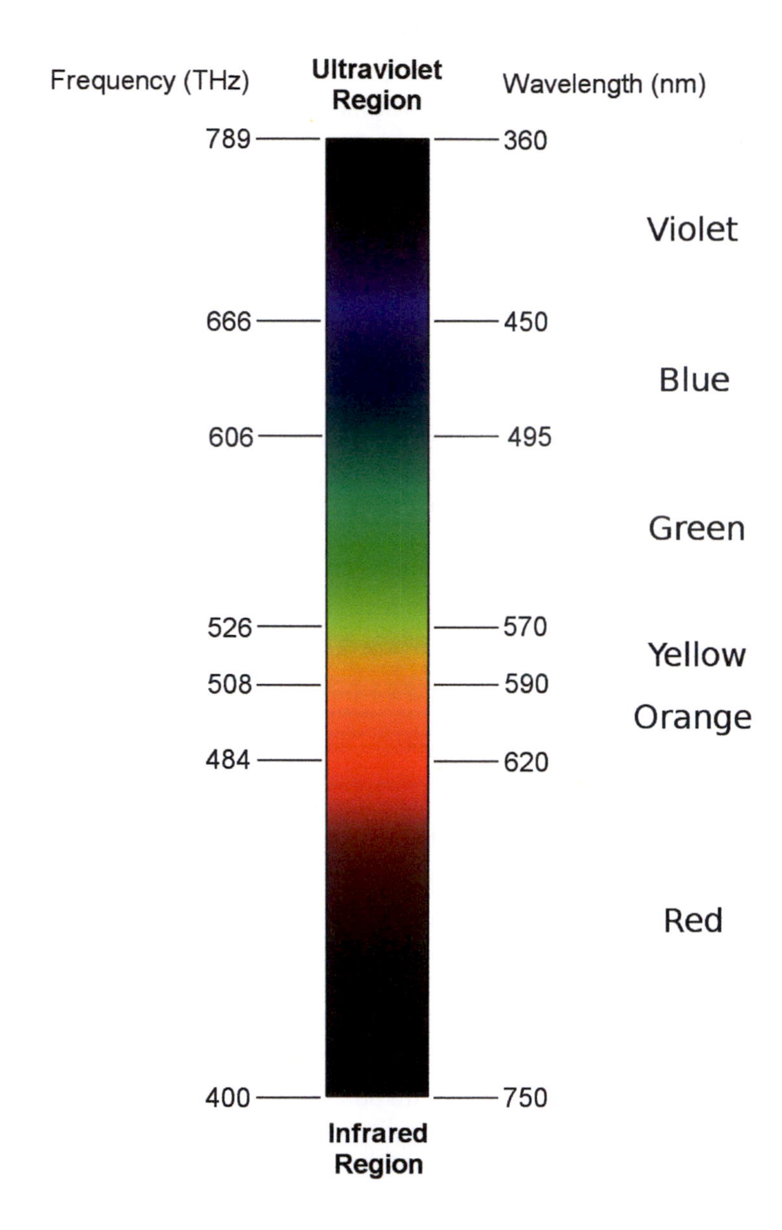

Figure A.2 *Light spectrum and approximate color boundaries for the colors of red, orange, yellow, green, blue and violet (from the bottom) by frequency and wavelength in a vacuum (Note that 1 TeraHertz [THz]* $= 10^{12}$ *Hertz and 1 nanometer [nm]* $=$ 10^{-9} *m. Adapted from a public domain image courtesy of Wikimedia Commons.)*

transmitted to the brain via the optic nerve, where the complex process of perception occurs. Information from the eye, like a piece of a puzzle, is analyzed in the brain and fitted into meaningful forms.

The eyeball is housed within a protective bony socket called the orbit. Each eye is suspended within its orbit by supporting tissue and is surrounded by a cushion of fat and blood vessels and motor and sensory nerves, including the optic nerve. There are six small muscles attached to each eye that mediate coordinated movement of both eyes. The eyelids provide some protection in front of the eye and control the amount of light that is permitted to enter. They also serve to keep the surface of the eye lubricated by spreading the tear fluid with each blink, as well as an oily fluid produced by the Meibomian glands present within the lid. Tear fluid is produced by lacrimal glands near the outer portion of each eyebrow. Fluid is collected and drained through tiny canals within the upper and lower lids near the nose. The tears eventually flow into the nasal passages and are swallowed.

The adult human eye is roughly spherical, with a diameter of approximately 2.5 cm (1 in). The wall of the eye is composed of three coats.

- The outermost coat supplies the basic support of the eye and gives it shape. It is divided into the *cornea* and the *sclera*. The cornea is the front portion of the outer coat and is the transparent surface membrane situated in front of the lens. Behind the cornea is the sclera, the strong white coat of the eye. The muscles that move the eyeball are attached to the sclera.
- The middle coat is composed of three regions. In the front is the *iris* that imparts a distinctive color to an eye. The iris contains the central opening: the *pupil*. The pupil varies in size with the intensity of ambient light and is controlled by a sphincter and a dilator muscle within the iris. Suspended behind the pupil is the lens. The iris merges with the ring-shaped *ciliary muscles* that can change the lens shape via thin fibers that attach to the lens. The ciliary muscles can change the thickness of the lens to provide a variation in power from about 45 diopters to 59 diopters. The ability to change lens thickness diminishes with age. Behind the ciliary muscles is the *choroid*, which carries blood vessels to and from the eye. It has a heavy melanin pigmentation to prevent light scattering within the eye.
- The innermost coat is the retina, which lies behind the lens. The retina is composed of approximately 120 million photoreceptor cells known as *rods* for low light, achromatic vision; and approximately 6 million photoreceptor cells known as *cones* for high-acuity color vision. The retina also contains the *optic disc* (blind spot), which is the junction of nerve fibers passing to the brain. The blood vessels supplying the inner aspect of the eye enter the eye through the optic disc and then branch out over the retina.

The rods and cones convert light energy into nerve impulses. The rods contain the visual pigment rhodopsin, or visual purple. They also are important for motion detection in addition to their use in low-light-level vision. The rods are so sensitive that the human eye can detect a match struck at night at a distance of 1500 meters. Cones are responsible for color vision and for the apprehension of bright images.

The greatest concentration of cone cells is found in a tiny depression in the center of the retina called the *fovea.* Only cones are present there, rods are absent; so color vision is the most acute at the fovea.

Among mammals, only humans and primates possess color vision. Only birds have color perception superior to humans. Honey bees can be trained to distinguish colors but are color-blind to red. Some fish have color vision but members of the shark and skate family do not.

Not all animals share with humans the ability to fuse visual information from separate eyes into a single, comprehensive, three-dimensional picture. *Binocular vision*, as it is called, requires eyes be in the front and not the side of the head. Members of the whale and dolphin species lack binocular vision but some species of dogs, such as the poodle, have it. Even among humans, just 95% of the population can perceive depth from binocular disparity, that is, when two eyes see a slightly different view of the world. The depth information we get from binocular disparity is relative, not absolute. Distances greater than 1 meter are underestimated while distances less than 1 meter are overestimated. What is impressive is that we can distinguish stereo depth differences smaller than the width of a single photoreceptor cell.

Accommodation cues signaled by physical sensations in the muscles of the eye can detect differences in depth of greater than 0.5 meter but the relationship is not linear.

Vergence is another oculomotor cue to depth that occurs when we look at a nearby object with both eyes as the eyes turn in ever so slightly so that the lines of sight converge at a point on the surface of an object. Experimental evidence has shown that viewers can judge depth based on vergence cues up to a depth of 6 meters. Even more impressive is the fact that vergence responses can be evoked from perspective images viewed with a single eye. Moreover, we perceive more depth in a perspective plot if we look at it with one eye. When encountering difficulty coming to terms with a perspective image on a two-dimensional surface, try covering one eye.

The chemical process that transforms light into nerve impulses is basically similar in all land vertebrates and marine fishes. Of the 100,000 genes in the human genome, 50,000 are thought to be important for sensory awareness. Therefore, some of that genetic material that we humans have for sight must be in common with these other species.

Nerve fibers from the retina come together in the optic nerve, which relays visual information to the brain. Each optic nerve has about 1.2 million nerve fibers.

If a bright light is shone into the eye, the pupil will immediately constrict. This is the *light reflex* that protects the retina from light that is too bright.

The lens system, which focuses the light rays, is composed of the cornea and the lens. The process by which the lens focuses on objects at different distances from the eye is called *accommodation*. The lens is fairly flat when a distant object is viewed but thickens or curves outward as the object moves closer. Lens shape is controlled by the ciliary muscles. A blurred image causes reflex impulses to the ciliary body that promote contraction or relaxation until the image is in focus.

When the eye is focused on an object at 14 m, objects at 7 m to infinity are in focus—this is the *depth of focus*. Visual acuity depends on the lens system and retina shape so people wear spectacles to correct for a mismatch between the power of the lens and the range of the object as well as variations in retina shape *(astigmatism)*. Object focus is not usually considered as a design factor in static 2D plots. But it is a factor in dynamic 3D plots as shall be seen later.

Many people suffer from errors in refraction. *Myopia* (nearsightedness) causes distant objects to appear blurred. *Hyperopia* (farsightedness) causes near objects to appear blurred. *Presbyopia*, which arises after the age of 43 years, causes the lens of the eye to harden so it loses its ability to change shape and near objects do not appear in focus. Presbyopia can correct myopia and as some people reach middle age their vision corrects to normal. *Strabismus* denotes misalignment of the eyes resulting in the inability of the eyes to simultaneously view an object. *Photophobia* denotes excessive sensitivity to light. It may result from certain eye diseases or disorders of the central nervous system.

The right eye is connected to the left half of the brain and the left eye is connected to the right half of the brain. Both halves of the picture are seen right-side up, though the retinas receive inverted images. The optic nerves cross one another at the *optic chiasm*, a major nerve junction. The nerves then continue backward to enter an area on the underside of the brain called the *lateral geniculate body*, which partially processes the data before passing it to the *occipital (visual)* cortex in the back of the brain.

The eyes are never at rest. Eyes are held in balance by three pairs of antagonistic muscles. Instability in this balance can cause a continuous small-amplitude tremor. The *moiré effect* occurs when regularly spaced high-contrast lines cause the image to appear to vibrate because of one such tremor. *Saccades* are rapid intermittent jumps of the eye that occur when reading. *Pursuit* refers to the eye movement that occurs when watching a moving object at a fixed distance. *Convergence* refers to the eye movement that occurs when watching a moving object that changes distance.

A.5 The Perception of Color

Humans can identify up to 8 million shades and tints of color and 10 million gradations of light intensity. This is achieved with the three types of cones:

- those sensitive to blue light (optimal absorption at a wavelength of 440 nm);
- those sensitive to green light (optimal absorption at a wavelength of 535 nm); and
- those sensitive to red light (optimal absorption at a wavelength of 570 nm).

Relative eye sensitivity is the greatest at the center of the visible region: 555 nm to be specific. Light of this wavelength is perceived as yellow-green by people without color vision deficiencies.

Mixtures of color can produce the same perceived color as monochromatic light. Also, photoreceptor cells decrease response under conditions of continuous stimulation. For example, if one looks at a blue saturated field that changes to white some time later, then one will briefly see the complementary color red because the relative response to blue has been reduced.

Studies in statistical graphics have shown that a few groups in a scatterplot can be effectively distinguished using color but that color is a poor choice for representing one or more continuous variables because humans do not order color along a linear scale. The spectral, or rainbow, ordering of color by wavelength is effective only to the extent that the ordering is learned through repeated use. An alternative to spectral ordering is one based on color saturation. On the other hand, the grayscale provides a consistent and effective ordering. Thus, color ought to be the last resort for representing a continuous variable.

An additional consideration in the use of color in graphs is color blindness in the viewers, perhaps even in the creators. Total color blindness (anomalous monochromatism) is extremely rare. It is due to a rod or cone defect. A person with red-blindness (*protanopia*) cannot distinguish the colors green and red. A person with green-blindness (*deuteranopia*) cannot see the green part of the spectrum. A person with blue-blindness (*tritanopia*) cannot distinguish the colors blue and yellow. Collectively protanopia, deuteranopia and tritanopia are forms of partial color blindness known as *anomalous dichromacy.*

Degenerative diseases or injury to the optic nerve or retina may also result in color blindness, usually to the colors yellow and blue. Color-vision testing cards and disks can determine the type and severity of the defect.

It is good practice to avoid putting a region in red adjacent to a region in green to avoid confusion for individuals with protanopia and deuteranopia and likewise a region in yellow adjacent to a region in blue for individuals with tritanopia.

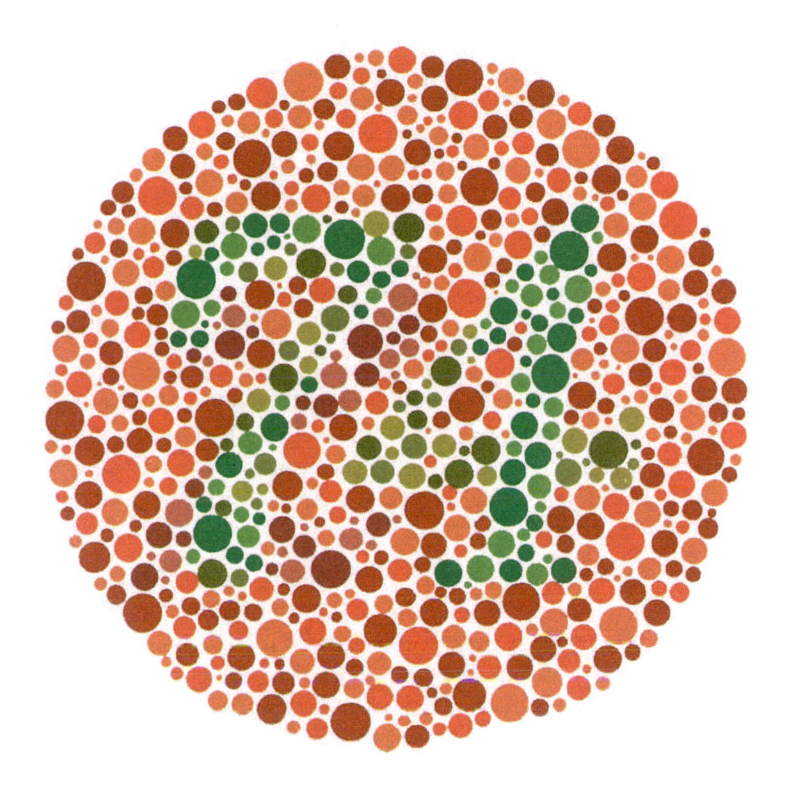

Figure A.3 *Example of an Ishihara color test plate (What number is visible? See text for discussion. Public domain image courtesy of Wikimedia Commons.)*

Color blindness is much more prevalent in men than women because color blindness is predominantly a sex-linked recessive disorder. Inherited color-blindness is almost always caused by a genetic defect on the long arm of the X chromosome. It occurs primarily in men, who have only one X chromosome. About 8 percent of all men have color-deficient vision. Women are often protected from the flawed gene because they have two X chromosomes. However, studies have shown small defects in the color vision of women who carry the gene. The prevalence of both forms of red-green color blindness is about 2.3 percent in males and 0.03 percent in females. The prevalence of tritanopia is about 0.001 percent in males and 0.03 percent in females. Complete color blindness (achromatopsia) is rare: occurring once in every 10,000,000 males or 10,000,000 females. The overall prevalence of color blindness in men is 4%

while the prevalence of color blindness in women is only 0.4%. That 1 in 25 males have some form of color blindness is a substantial rate of disability.

In addition to color blindness, approximately 1.3 percent of males and 0.02 per cent of females are red deficient (protanomaly). Approximately 5.0 percent of males and 0.35 per cent of females are green deficient (deuteranomaly). Approximately 0.0001 percent of both genders are blue deficient (tritanomaly). Collectively, protanomaly, deuteranomaly and tritanomaly are forms of partial color blindness known as anomalous trichromacy. Although anomalous trichromacy is less limiting than anomalous dichromacy, that more than 1 in 10 males have some form of color defective vision ought to be taken into account when producing color graphics, especially for the public.

One of the more popular methods of assessing color vision is through the Ishihara test for color perception. This was first published in 1917 by Dr. Shinobu Ishihara, an ophthalmologist and researcher at the University of Tokyo. The Ishihara color vision test consists of using a sequence of pseudo-isochromatics plates. The so-called *Ishihara plates* each contain a circle of dots appearing to be randomized in color and size.

An example of an Ishihara plate is given in Figure A.3. This single plate can be used to test whether an individual has either achromatopsia or has one of anomalous dichromacy or anomalous trichromacy. Viewers of Figure A.3 with achromatopsia will see no number at all. Viewers of Figure A.3 with either anomalous dichromacy or anomalous trichromacy may see the number 21. More Ishihara plates are needed to determine the type of color vision defect and the severity of it. Viewers of Figure A.3 with normal color vision ought to find the number 74 to be clearly visible. Versions of the full Ishihara color blindness test can be found on the Internet.

There are a few R packages that can be used to allow users with normal color vision to see how individuals with different forms will perceive different colors. One of these packages is `dichromat`. Let us start with a color topographical plot in Figure A.4 of the Maungawhau volcano in New Zealand that uses the standard terrain color palette in R. Figure A.5 was produced with the `dichromat` function in the package of the same name showing to someone with normal color vision how someone with protanopia perceives Figure A.4. As to how someone with deuteranopia or tritanopia perceive Figure A.4, see Figures A.6 and Figure A.7, respectively, for simulations designed for someone with normal color vision.

An interesting question is whether individuals with normal vision, protanopia, deuteranopia, or tritanopia all extract the same information from the original plot in terrain colors for Figure A.4. Conventional wisdom would assert that individuals with any form of anomalous dichromacy extract less information from Figure A.4 than individuals with normal color vision. Individuals with some form of anomalous trichromacy lie in between and would be expected to extract more information than individuals with any anomalous dichromacy.

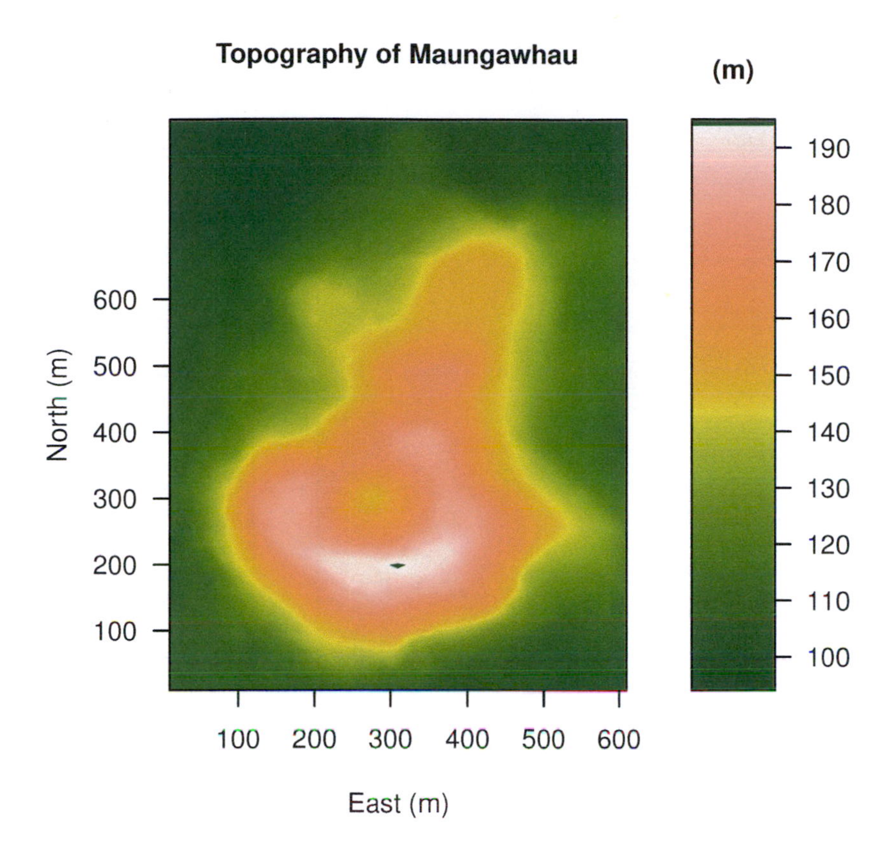

Figure A.4 *Color topographic map of the Maungawhau volcano in New Zealand in standard terrain colors*

Without being able to give normal color vision to individuals with either anomalous dichromacy or anomalous trichromacy, it is difficult to test these hypotheses. What can be done, however, is to ask someone with normal color vision to examine Figures A.4 and A.7 and ask whether the same features are equally perceived. Or whether some features are better perceived in the figures simulating color blindness.

Studies of individuals with color deficient vision have shown their visual perception to be different from those whose color vision is not deficient. Individuals with color deficient vision see lines and contours on the edges of objects more prominently than individuals with non-deficient color vision who rely instead on changes and transitions in colors. Although color deficient vision can be a barrier to a number of careers including aircraft pilot and bridge-keeping officer on board a ship, color blind individuals were recruited during

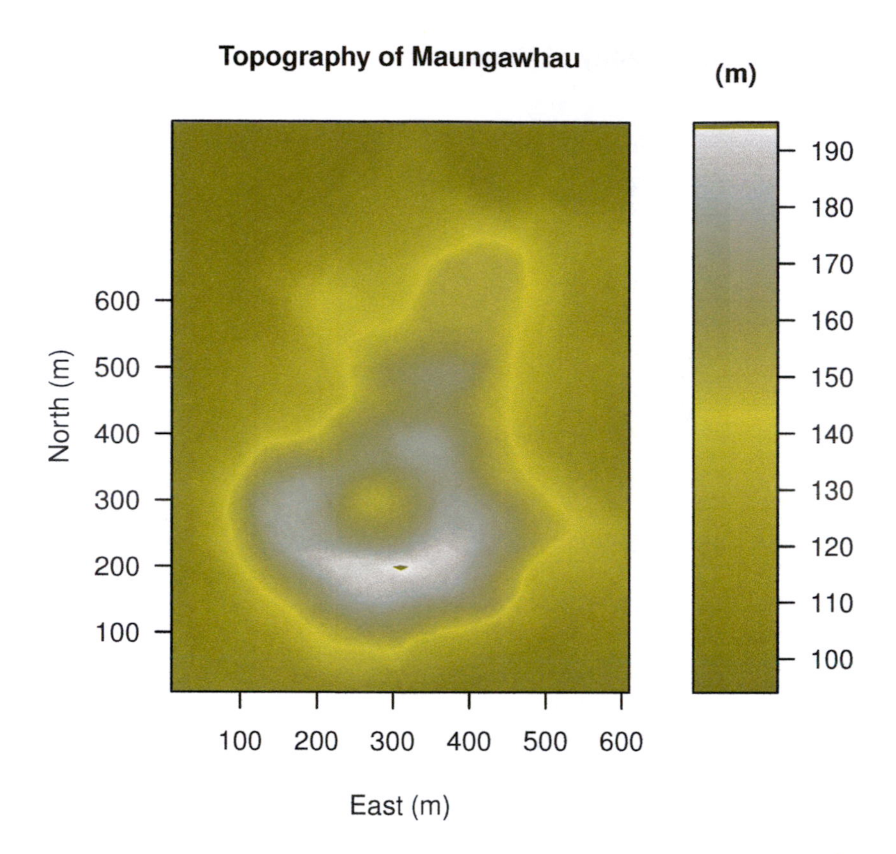

Figure A.5 *Color topographic map of the Maungawhau volcano in New Zealand in standard terrain colors as seen by someone with protanopia*

the Second World War as bombardiers on night bombers and as ship gunnery officers because they could visually cue on the darkened cities and the geometrical camouflage on warships, respectively, that their color-sighted colleagues would miss.

Other standard color palettes are available in R other than just terrain. Figure A.8 depicts the elevation of the Maungawhau volcano in suburban Auckland using the cyan-magenta color palette. For many with normal color, the cyan-magenta color palette of Figure A.8 renders the higher elevations more clearly than the terrain color palette of Figure A.4. This begs the question as to whether there are better choices for color palette for individuals with normal color vision.

It is important to be conscious of the *blue problem*. Note that

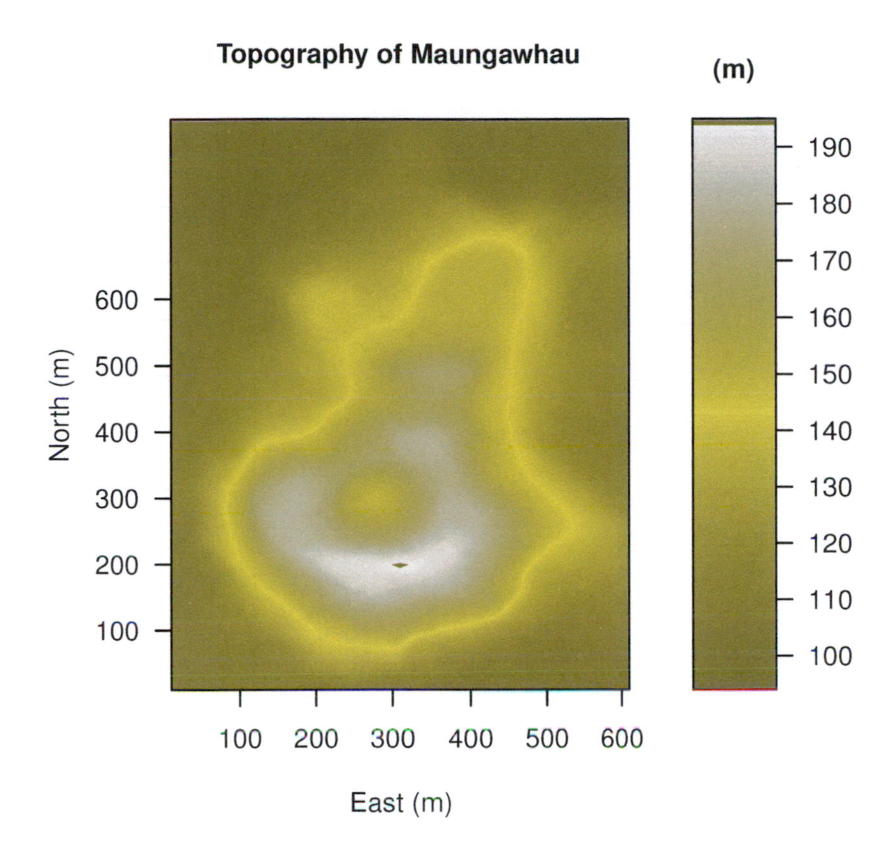

Figure A.6 *Color topographic map of the Maungawhau volcano in New Zealand in standard terrain colors as seen by someone with deuteranopia*

- blue cones are much more sparsely distributed than the red and green cones;
- there are no blue cones in the center of focus; and
- blue cones focus differently compared to red and green cones.

Consequently, blue color is fine for a background but never use blue for text for it will render it hard to read. Note that a cyan-magenta color is inherent in Figure A.7, which is intended to simulate for individuals with normal vision the blue-yellow color blindness of individuals with deuteranopia. Given the blue problem for individuals with normal vision, it is worthwhile to compare Figure A.7 with Figure A.8, as the latter uses the standard cyan-magenta color palette in R, and ask individuals with normal vision which of the color palettes in the two figures is preferred for revealing features in the topography of Maungawhau. This begs the question as to whether there could be a choice

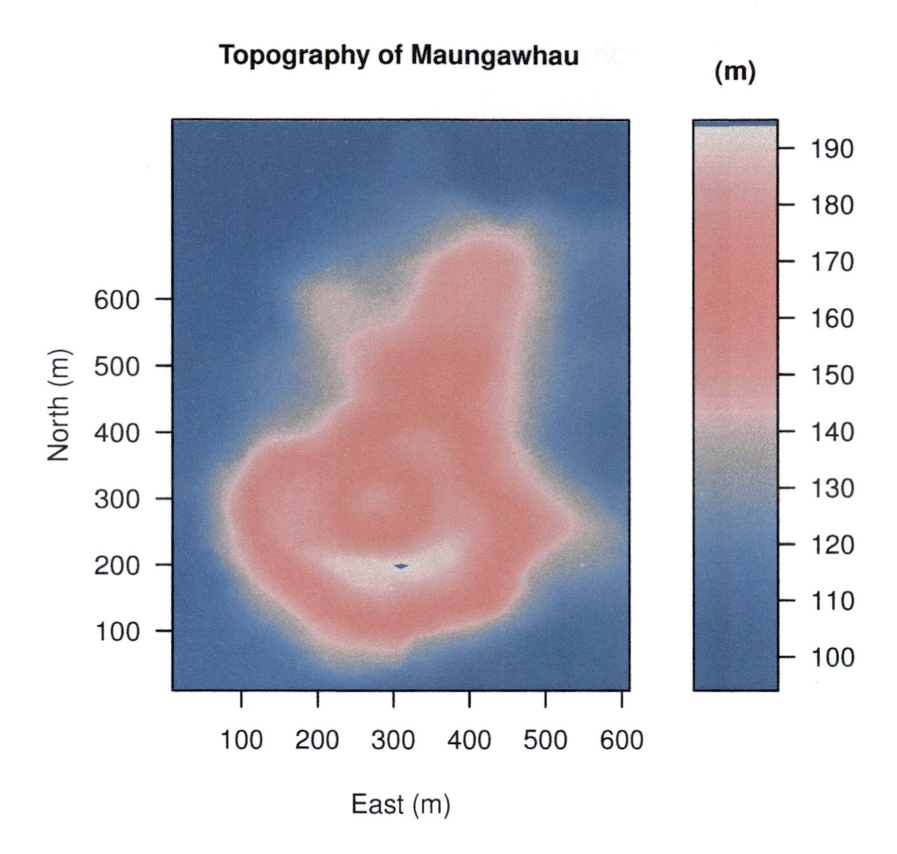

Figure A.7 *Color topographic map of the Maungawhau volcano in New Zealand in standard terrain colors as seen by someone with tritanopia*

of color palette that would place individuals with normal vision and those with tritanopia on a near equal footing.

Other packages exist in R to simulate color deficient vision. One such package is `LSD` that includes the function `daltonize`, which is named after the chemist John Dalton who was the first scientist to describe color blindness in detail in 1798 after he discovered that both he and his brother were affected by deuteranopia. The function `daltonize` in the `LSD` package was used to produce Figure A.9, which shows nine colors in the standard topographic color pallet in R and a simulation for an individual with normal color vision of how this would appear to a deuteranope. But the function `daltonize` does more. It implements the Daltonization Algorithm from the Internet site www.dalton.org that compensates for color vision deficiency by shifting wavelengths away from the portion of the spectrum invisible to either a deuteranope, a protanope, or

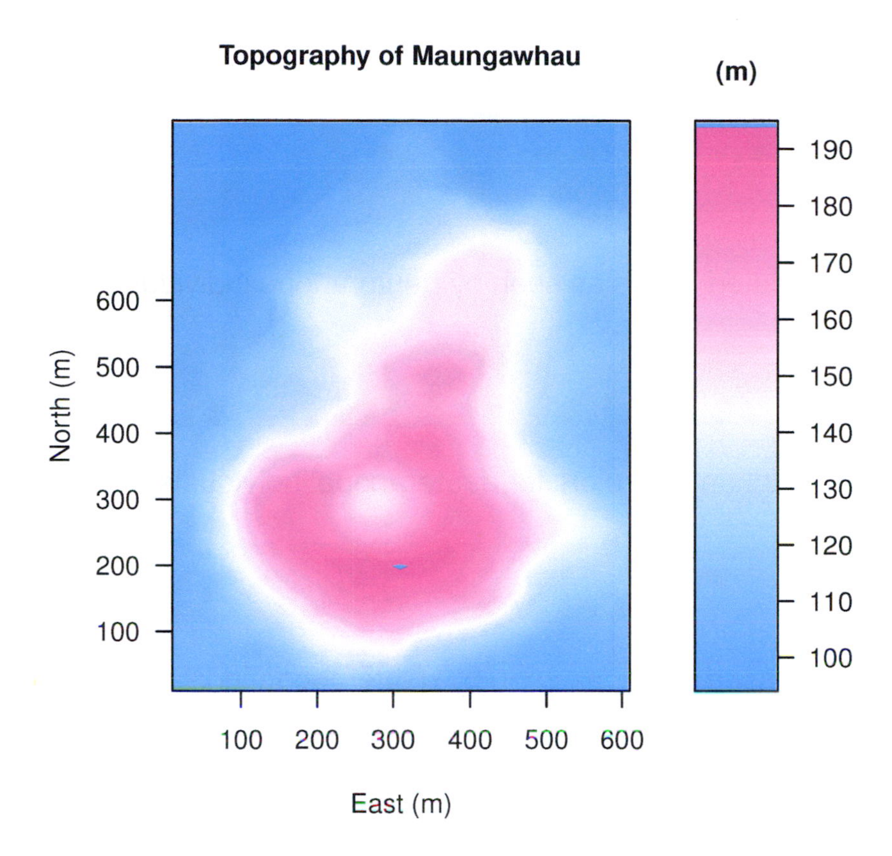

Figure A.8 *Color topographic map of the Maungawhau volcano in New Zealand in the standard cyan-magenta color palette*

a tritanope. Of course, all three cannot be done simultaneously. Figure A.9 shows an alternative selection of enhanced colors for a deuteranope and a simulation of what these colors would look like for a deuteranope.

Figure A.10 displays nine colors in cyan-magenta color palette in R, a simulation for a individual with normal color vision of how these appear to a individual with tritanopia, a suggestion for an enhanced version according to the Daltonization Algorithm, and how these enhanced colors appear to a tritanope. An interesting question for an individual with normal color vision is whether the actual enhanced colors and the depiction of how a tritanope would see them are equally effective despite being different colors.

Another package in R that can assist with the selection of color schemes for an audience with color blind members is `RColorBrewer`. This package provides palettes in R from the `ColorBrewer` software package, which at the time of

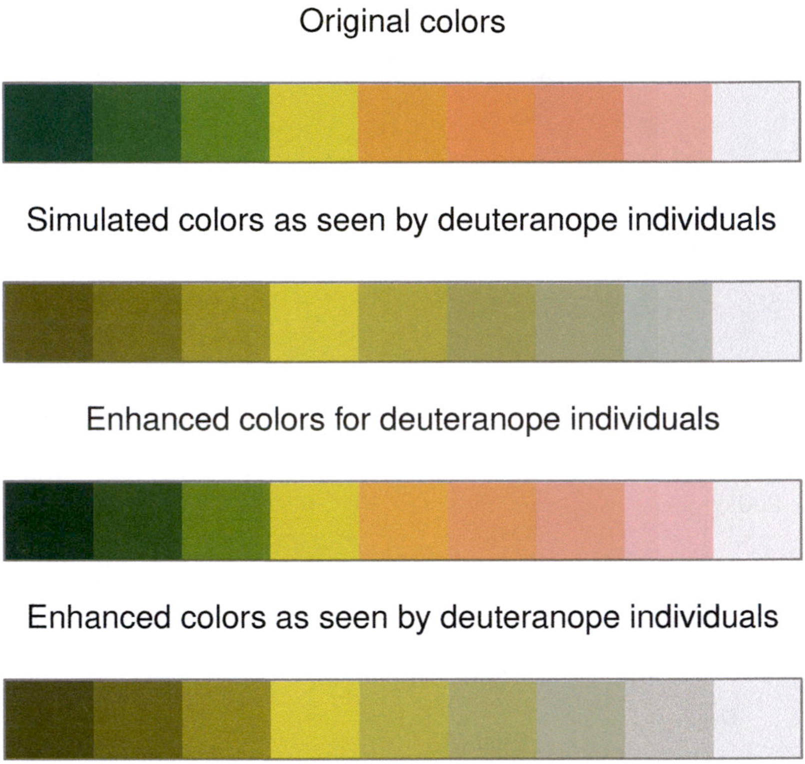

Original colors

Simulated colors as seen by deuteranope individuals

Enhanced colors for deuteranope individuals

Enhanced colors as seen by deuteranope individuals

Figure A.9 *The terrain color palette in R and* `daltonize` *enhanced terrain palette colors in R as seen by individuals with normal color vision and individuals with deuteranopia*

writing is in its second edition and available from the website colorbrewer2.org. Cynthia Brewer and colleagues have developed a number of palettes for use in cartography for depicting variables for geographical areas. Separate collections of palettes have been developed for sequential data, diverging data, and qualitative data. The software on the colorbrewer2.org website allows a user to evaluate the robustness of these palettes. Options are available to screen for palettes thought to be photocopy safe, print friendly, and color blind safe.

Sequential palettes were chosen for `ColorBrewer` on the basis of being best suited for ordinal data progressing from low to high with lighter colors for low and darker colors for high. Diverging palettes put equal emphasis on mid-range values and extremes with lighter colors in the mid-range and darker colors at the two extremes with contrasting hues. Qualitative palettes are intended for

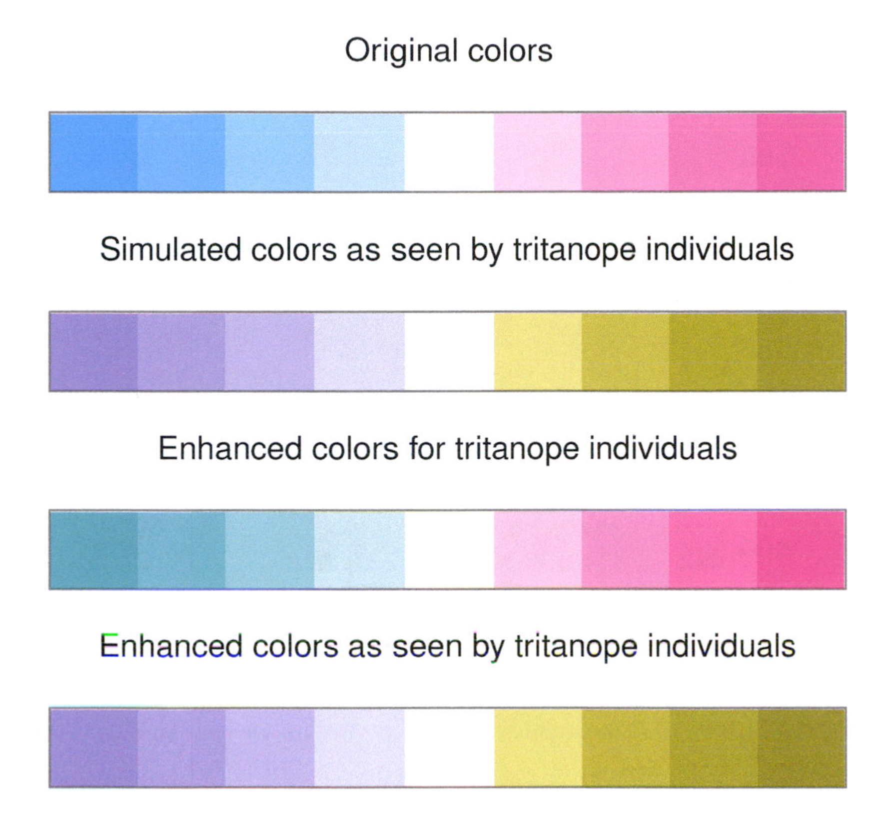

Figure A.10 *The cyan-magenta color palette in R and* `daltonize` *enhanced cyan-magenta palette colors in R as seen by individuals with normal color vision and individuals with tritanopia*

categorical data with a nominal scale. Figure A.11 depicts the eight categorical palettes available in `RColorBrewer` together with their slight cryptic names. Figure A.12 depicts the subset of three categorical palettes available in `RColorBrewer` thought to be color-blind-friendly together with their slight cryptic names. The function `brewer.pal` in the package `RColorBrewer` will also recommend a sub-palette if you require fewer colors but the minimum is at least three.

None of the previous discussion of color choice is relevant to individuals with achromatopsia. Figure A.13 attempts to place everyone with normal color vision or some form of color blindness on level footing by using not fifty but one hundred shades of gray from pure white to solid black to depict the elevations of Maungawhau (Mount Eden). As previously noted, color blind and color

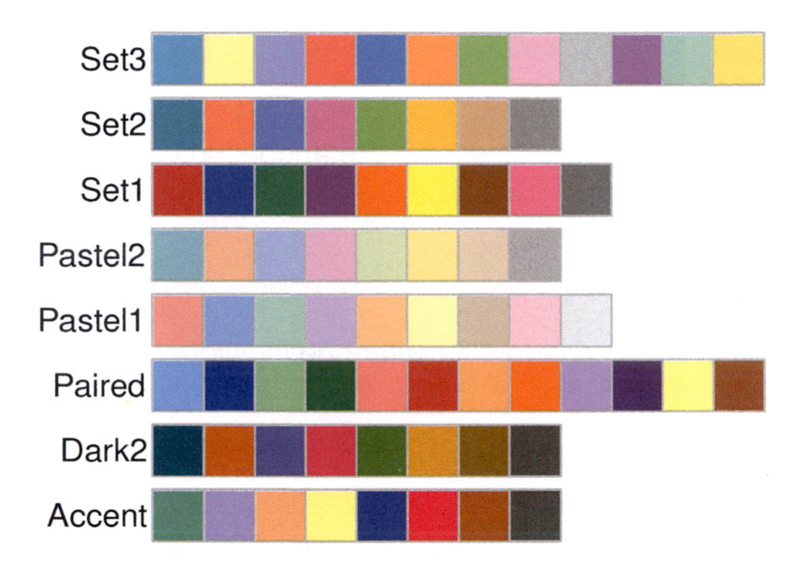

Figure A.11 *All the qualitative palettes available in* `RColorBrewer` *and their names*

vision deficient individuals may arguably be more adept at distinguishing shades of gray. Figure A.13 has been produced with a linear scale of gray. For an individual with normal trichromatic vision, a better approach is to calculate grayscale values so that they have the same relative luminance as in the original color space as is done for photographic images. This process is inherently nonlinear.

Nevertheless, with a few simple principles to bear in mind, the use of color in statistical graphics can be very effective either for parts of the chart itself or as colored text. Color can be used to enliven, to decorate, or to represent or imitate reality. Remember to accommodate the viewers, especially those who might be color blind or color vision deficient. Avoid red-green contrasts. Avoid blue-yellow contrasts. Avoid dark saturated reds. Use high contrast and avoid small bits of color in order to avoid acuity problems. Above all else, duplicate information carried by color. Understand the media being used and the intended audience. Test a graphic design before going to print or making a presentation. When in doubt, be conservative. If all this is done, then the use of color will enhance the graphical display.

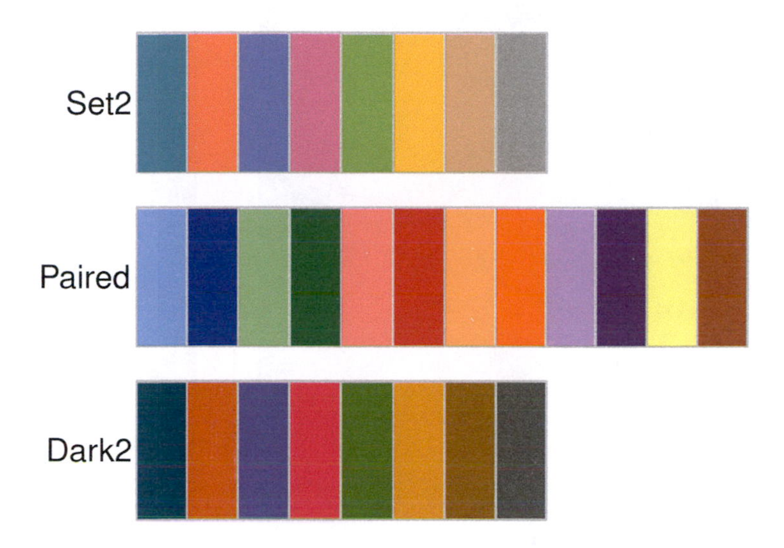

Figure A.12 *Subset of qualitative palettes available in* `RColorBrewer` *thought to be color-blind-friendly and their names*

A.6 Graphical Perception

The preceding sections in this appendix have encompassed physics, medical science, and psychology insofar as psychophysics pertains to the perception of color. In this section we continue with the results of psychological experimentation as it pertains to the perception of black-and-white images on a graphical chart.

A.6.1 Weber's Law

Weber's law was formulated in the nineteenth century by the famed psychophysicist E. H. Weber and represents one of the fundamental and controversial laws of human perception. Let x denote the length of a line segment. Let $w_p(x)$ be a positive number such that a line segment of length $x + w_p(x)$ is detected with probability p of being longer than the line segment of length x. Weber's law states that for fixed probability p:

$$w_p(x) = k_p x \tag{A.6}$$

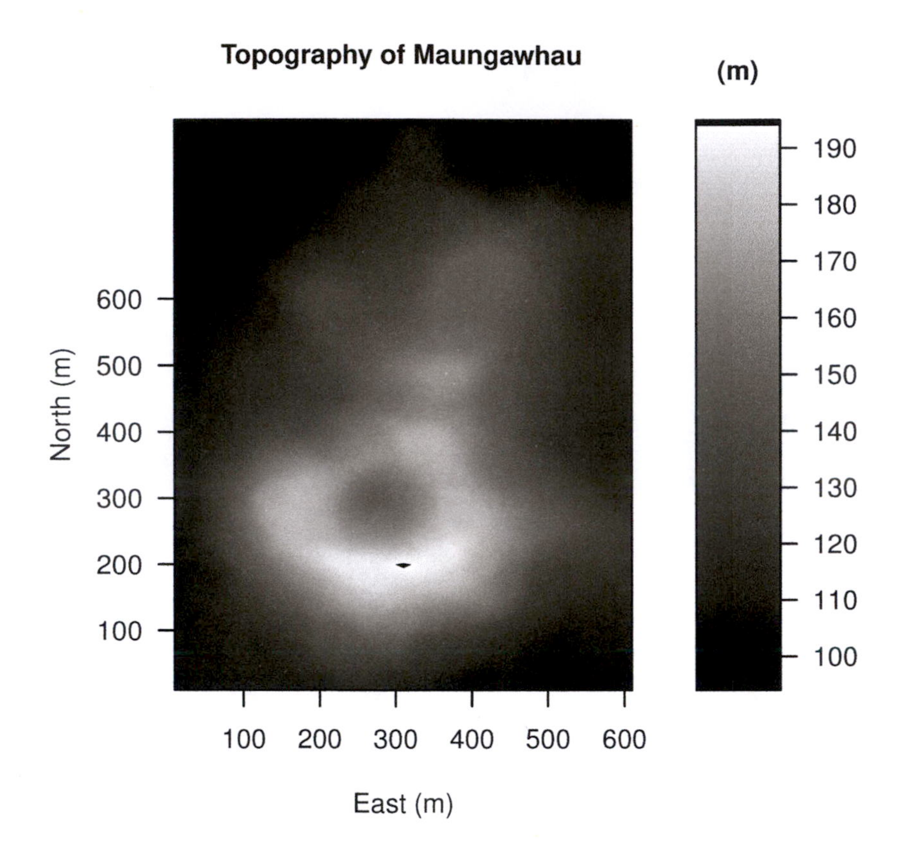

Figure A.13 *Topographic map of the Maungawhau volcano in New Zealand in grayscale*

where the constant k_p is not a function of x. Moreover, this law appears to work well for many other perceptual judgments including position, length, and area.

One implication of Weber's law is that it is easier to detect a difference between two line segments 2.0 cm and 2.5 cm in length than between two line segments 50.0 cm and 50.5 cm in length because the percentage difference of the former is 25% while the percentage difference of the latter is only 1%.

Another implication is that it is easier to detect differences along identical non-aligned scales than length judgments. Consider the two solid black rectangles in Figure A.14. It is difficult to determine which rectangle, A or B, is longer. That is because the difference in length is slight. Add a bounding box to form a framed rectangle each of equal length for each bar, as is done in Figure A.15, and it is readily apparent that rectangle A is shorter. Our eyes cue visually

Figure A.14 *Bars and Weber's Law (Which of bar A or B appears larger? Consult the text.)*

on the differences in the size of the white bars atop the solid black bars A and B in accordance with Weber's law.

This has implications for the drafting of any rectangular figure that depicts statistical information. Notice that most of the figures in this book are set in fully framed displays. In some instances this feature can be switched off by overriding a default setting but in other instances this is difficult or impossible to do. There are some who do not like frames or bounding boxes for axes. But evidently by design or accident, many developers of statistical software believe in Weber's law and provide these features because in many instances the frames and bounding axes cannot be switched off.

Figure A.16 represents a variation on Figure A.14 with the base of bars A

Figure A.15 *Framed rectangles and Weber's Law (Which of bar A or B appears larger? Consult the text.)*

and B set on an invisible common horizontal axis in the lower portion of the figure. It is quite evident that bar B is larger than bar A. Moreover, this task is more quickly accomplished in Figure A.16 with the common axis than with the framing rectangles in Figure A.15. So we now have a hierarchy developing with respect to tasks relating to graphical perception.

A.6.2 Stevens's Law

Another psychophysical law is given to us by S. S. Stevens in 1975. S. S. Stevens is the same Stevens who gave us the scales of measurement discussed

Figure A.16 *Bars set on a common axis and Weber's Law (Which of bar A or B appears larger? Consult the text.)*

in the previous chapter. Suppose a collection of graphical objects are given to a group of test subjects and they are asked to judge the magnitudes of some attribute of the objects; for example, area. *Stevens's law* states that a person's perceived scale is

$$p(x) = \alpha x^{\beta} \tag{A.7}$$

where x is the magnitude of the attribute. Evidently, the perceived scale is a power of the original magnitude so Stevens's law is also known as *Stevens's power law.*

Many experiments have been conducted to estimate the power parameter β. For length judgments, estimates of β are generally in the range 0.9 to 1.1. For area judgments, estimates of β are generally in the range 0.6 to 0.9. For volume judgments, estimates of β are generally in the range 0.5 to 0.8.

Because β is nearly one for length comparisons, there is little bias in length judgments. For area, the power of the judgments is less than one. Consider a pictograph with 3 icons with areas 2, 4, and 8 in square units. Assume $\beta = 0.7$. In Chapter 2 we asserted that the lie factor for a comparison of areas 4 and 2 would be equal to a ratio of $\sqrt{(4/2)} \approx 1.41$. By Stevens's law, it follows that this is an underestimate as the perceived ratio is $(4/2)^0.7 \approx 1.62$. And it gets worse. The lie factor for a comparison of areas 2 and 8 would be equal to a ratio of $\sqrt{(8/2)} = 2$. By Stevens's law, the perceived ratio is $(8/2)^0.7 \approx 2.64$. The increase in the perceived ratio of areas is disproportionately higher on comparing the smallest area with the largest area rather than the smallest area with the middle-sized area. The situation is even worse for volume comparisons.

So based on Stevens's law and experimental results we have obtained the following ordering of tasks from most accurate to least accurate:

1. length judgment;
2. area judgment; and
3. volume judgement.

Moreover, length judgments are unbiased while area and volume judgments are biased.

A.6.3 The Gestalt Laws of Organization

As noted in the first chapter, visual information is stored as perceptual units in short-term memory. The nature of these units was studied by the Gestalt psychology school founded in 1912 with the peak of its activity during the 1920s. Their experiments began with the premise that we do not see every little dot or smudge in an image but rather the patterns they form. Four of these laws relevant to statistical graphics are as follows.

Continuation: Marks such as "- - - - - - - - - -" are seen as a single perceptual unit not 10 distinct ones.

Proximity: Marks such as "xxx xxx" are seen as two groups, whereas, marks such as 'xx xx xx" are seen as three.

Similarity: "ooo|||" is seen as two groups because marks that have similar shapes or orientations are grouped together.

Common Fate: Marks or lines going in the same direction are grouped together.

The concept underlying the gestalt phenomenon is that we perceive the whole as greater than the sum of the parts. The recommendations for duplicating messages coded in color in statistical graphics by other methods is intended to enhance gestalt and could be considered an application of the law of common fate.

A.6.4 Kosslyn's Image Processing Model

Assuming that the visual buffer of short-term memory, as depicted in Figure A.15, functions as if it were a coordinate space but is constrained by limited resolution causing contours to become obscured if the object is too small, S. M. Kosslyn produced a model based upon four premises.

Premise 1. Images are transformed a portion at a time, either in parallel or serially.

Premise 2. The image processing system is inherently noisy like all physical systems.

Premise 3. There is a variance around and proportional to the distance a portion is moved in an image.

Premise 4. Procedures that realign portions of an image are effective only with relatively minor amounts of deformation.

These four premises can be taken as guiding principles in the production of statistical graphics.

A.7 Conclusion

To pull the discussions of this appendix together with those of Chapter 1, note the following hierarchy of eleven visual perception tasks in increasing order of difficulty as follows:

1. judgment of position along a common scale; judgment of position along framed rectangles of similar size;
2. judgment of position along non-aligned scales;
3. judgment of length;
4. judgment of direction; or angle;
5. judgment of area;
6. judgment of volume; or curvature;
7. judgment of grayscale shading; and
8. judgment of color saturation.

Notice that judgment of position and judgment of length are differentiated. In case of disagreement with this decision, this judgment can be subsumed with the previous three.

Concerning angle judgments, in the nineteenth century it was established that acute angles are underestimated while obtuse angles are overestimated. Experiments a century later showed also that angles whose bisectors are horizontal tend to be seen as larger than those whose bisectors are vertical. In view of these empirical results, judgments of direction or angle are ranked as less accurate than judgments of length.

The discussion of Weber's law and Stevens's law is adapted from Cleveland (1985, pp. 241–245) [21]. The importance of discussing these two laws of psychophysics derives from comments made by Kosslyn [77], as was the need to discuss the Gestalt laws. Even more impressive is the discussion concerning the relationship between image and mind in the monograph by Kosslyn [78].

For those readers interested in learning in depth about experiments conducted by statisticians with respect to graphical perception, consult Cleveland and McGill [25] and Simkin and Hastie [111]. For details concerning perception experiments for the boxplot, consult Stock and Behrens [116].

For readers who do not believe that bounding frames are important or are unnecessary, consult Cleveland and McGill [25]. Virtually all of the statistical graphical displays in Cleveland and McGill [25] come with bounding boxes for the axes or are set in rectangular frames—presumably on the assumption of the validity of Weber's law.

Cleveland [21] lists ten elementary tasks, and liberties have been taken in subdividing one of these tasks into two for a total of eleven tasks. The task of judgment of position along a common scale has been split into judgment of position along a common scale and judgment of position along framed rectangles of similar size.

For those interested in learning more about optics or reviewing what they may have forgotten, see Hecht and Zajac [63]. For those interested in learning about the eye and the interaction between eye and brain, consult Gregory [56].

In summary, by considering the guidance given in the preceding list of eleven perceptual tasks and giving due consideration to the range of human abilities in the perception of color, it is possible to create effective statistical graphics. Also required is taking the time to test graphical displays on unsuspecting colleagues, a random selection drawn from the intended audience, or both. With patience for executing redrafts as necessary; success can be anticipated in using the chosen graphics medium or media.

A.8 Exercises

1.(a) What is the power p in diopters of one of your own corrective lenses or one of a family member or friend?

(b) Take your answer to part (a) above and find the focal length f of the thin corrective lens.

(c) With your answer to part (b), for an object 6 meters away from the thin lens, find the distance of the image s_i from behind the lens along the axis of the lens.

(d) With your answer to part (b), for an object 10 meters away from the thin lens, find the distance of the image s_i from behind the lens along the axis of the lens.

2. Find the formula relating the transverse magnification of a thin lens to its focal length. Using your answers to Exercise 1 above, do the following.
   (a) For an object 6 meters away from the thin lens, find the transverse magnification.
   (b) For an object 10 meters away from the thin lens, find the transverse magnification.
3. The double concave shape of the thin lens depicted in Figure A.1 is no longer used in modern corrective lenses to avoid being brushed by eye lashes. Study a modern corrective lens and describe its shape.
4. Draw a sketch of the human eye showing major structures.
5. Check your library, medical school, visit an ophthalmologist or optometrist, or go online to the internet for tests of color vision and check your own color acuity.
6. For the 5-day period from Monday to Friday, clip or copy the articles from a newspaper or online news outlet that have graphs that display data.
   (a) Note the use of color in graphics.
   (b) Comment on whether color is used to express information or merely to be eye-catching for each of the graphs.
   (c) Look for deficiencies, such the moiré effect or color for a continuous variable, in the use of color.
   (d) Suggest how the use of color in the graphs could be improved.
7. Review one issue of a journal in your field for the graphics that are used.
   (a) Comment on whether color is used to express information or merely to be eye-catching for each of the graphs.
   (b) Look for deficiencies, such the moiré effect or color for a continuous variable, in the use of color.
   (c) Suggest how the use of color in the graphs could be improved.
8. Use the package `dichromat` in R to see how someone with protanopia might perceive Figure A.5. Discuss.
9. Use the package `dichromat` in R to see how someone with protanopia, deuteranopia, or protanopia would perceive the color-blind-friendly Color-Brewer palettes in Figure A.12. Discuss.
10. Living among us are people with four cones rather than the usual three. Although a trichromat can distinguish nearly ten million tints of color, a tetrachromat can distinguish about one hundred million tints of color with each tint fracturing into a hundred more subtle tints for which there are no names and no paint swatches. Quite likely, a tetrachromat would have no way of knowing that their vision goes far beyond the accepted limits

of human vision. What would color look like through the super vision of a tetrachromat's eyes? Only a tetrachromat would know.

(a) Have there been any documented cases of human tetrachromacy?

(b) Goldfish and zebrafish are examples of species with cones for red, green, blue, and ultraviolet life. What is the estimate for the prevalence of tetrachromacy in humans?

11. For the 5-day period from Monday to Friday, clip with scissors or electronically cut the articles from a newspaper or online news outlet that have graphs that display data. Discuss these articles with respect to the hierarchy of the eleven visual tasks.

12. For one issue of a journal in your field, photocopy or electronically cut the articles that have graphs that display data. Discuss these articles with respect to the hierarchy of the eleven visual tasks.

Appendix B

Color Rendering

B.1 Introduction

Colored graphics are omnipresent on computer monitors, tablets, cellular phones, and printed media. It could well be thought that the science and art behind the processes for screen and printed display of colored images is rather trivial but nothing could be further from the truth. Different color spaces have been developed for different media and even by different manufacturers for the same media. Think of the different types of paper used in newspapers and magazines, and by photocopiers. The appearance of these papers can be dull or bright. The reaction of colored inks to the various different types of paper is different. All these factors need to be managed. With respect to computer monitors, tablets and cellular phones, the various manufacturers have not adopted a common color space. Very few appreciate that the colors of the same image can appear differently based on the manufacturer. This appendix explores these issues.

B.2 Learning Outcomes

When you complete this appendix, you will be able to do the following.

- Appreciate the history of the development of the theory of color.
- Understand the philosophy underlying the RGB and XYZ color spaces of 1931.
- Comprehend the difference between lightness and value in the HSL and HSV color spaces.
- Understand the basics of the CIELAB and CIELUV color models of 1976.
- Understand the rationale for additive and subtractive color models.
- Appreciate the effort in developing the International Color Consortium profiles for creating and re-creating color images on different paper stocks.
- Know that the sRGB color gamut is the principal color model in R.

- Be able to convert colors to and from sRGB and other color spaces in R.
- Be able to save color images from R in .pdf and .eps files using the CMYK color space.

B.3 RGB and XYZ Color Spaces

In 1931, the Commission Internationale d'Éclairage (CIE), or International Commission on Lighting, released two color spaces. The first of two color spaces released in 1931 was CIE RGB derived from the addition of the colors red, green, and blue. See Figure B.1 for an illustration of the RGB process using a Venn diagram. Figure B.2 depicts the average red ($\bar{r}$), green ($\bar{g}$), and blue ($\bar{b}$) color matching functions adopted for the CIE 1931 RGB color space. The vertical axis of Figure B.2 gives the average amounts of primaries needed in human experiments to match the monochromatic test color at the wavelength on the horizontal axis. Note where the wavelength interval for the color matching function for red is negative—this is the blue-green part of the visible spectrum.

There is a problem with the RGB process that arises when judging relative luminance, or brightness. Greens are perceived as brighter than reds or blues of equal power. The CIE XYZ color space was released in 1931 to address this problem. The distribution of cones in the human eye is concentrated on the retina directly behind the lens in an area known as the *fovea centralis*. Because of this, tristimulus perception of color depends on the observer's field of view. The CIE 1931 XYZ color space was created to represent an average human chromatic response within a two-degree arc inside the fovea centralis. The CIE 1931 XYZ color space model defines Y as luminance, Z as roughly equal to blue stimulation, and X is a linear combination of cone response curves. CIE also uses the (lowercase) variables x, y, and z that are normalized versions of X, Y, and Z, and are defined as follows:

$$\begin{aligned} x &= \frac{X}{X+Y+Z}, \\ y &= \frac{Y}{X+Y+Z}, \\ z &= \frac{Z}{X+Y+Z}. \end{aligned}$$

Figure B.3 depicts the average $\bar{x}$, $\bar{y}$, and $\bar{z}$, color matching functions adopted for the CIE 1931 RGB color space versus wavelength above an illustration of the visible color spectrum. A comparison of the average amounts of constituents to achieve a given wavelength of color in the visible spectrum for the CIE 1931 RBG color space model in Figure B.2 with the CIE 1931 XYZ color space model in Figure B.3 shows that the subtractive amount of red in the

Figure B.1 *RBG Additive color space (Public domain image courtesy of Wikimedia Commons.)*

RBG color space is replaced by a second minor (positive) mode for $\bar{x}$ in the XYZ color space.

See Figure B.4 for a depiction in chromaticity variables x and y of all the colors visible to the average person. The outer curved boundary in the figure is the wavelength in nanometers. Although the CIE 1931 XYZ color space can represent all visible colors, the CIE 1931 RGB color gamut does not. The inner triangle in Figure B.4 depicts the more restricted CIE 1931 RGB color gamut as a subset of the CIE 1931 XYZ color space. Also in Figure B.4 is an open circle labeled E, which depicts the *standard illuminant*—theoretical source of white light. Illuminant E has equal tristimulus values in the CIE 1931 XYZ color space. Standard illuminants serve as theoretical references for color.

The colors in Figure B.4 are simulated regardless of which media the viewer is using. Only recent technological developments can reproduce all the visible

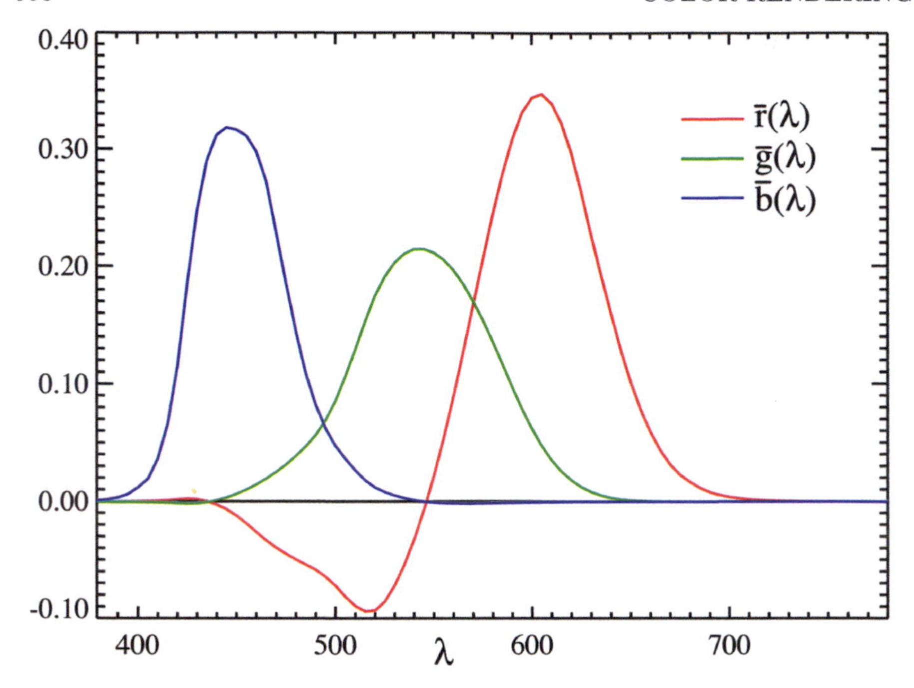

Figure B.2 *CIE 1931 RGB average color matching functions for red, green, and blue (See discussion in text for details. Public domain image courtesy of Wikimedia Commons.)*

colors that humans can see—and which are in the CIE 1931 XYZ color space. In fact, technology does not reproduce all the colors in the full CIE 1931 RBG color gamut. For electronic media displays, the limitations are due to software and hardware.

R statistical software implements the 8-bit standard RGB (sRGB) color palette, which has 16,777,216 colors. Recall from Appendix A that a human can identify up to 8 million tints of color and 10 million gradations of light intensity. So the sRBG color palette is not too shabby. As depicted in Figure B.5, sRGB is a subset of the CIE 1931 RGB color gamut, which itself is a subset of the CIE 1931 XYZ color space, which can represent all colors visible to the average human. Also depicted in Figure B.5 is the Adobe RGB 1998 color gamut, which is more comprehensive than sRGB but still falls short of the CIE 1931 RGB and XYZ color spaces. Not all electronic displays or media can support the Adobe RGB 1998 color gamut.

An improvement on the 8-bit sRGB color gamut is 48-bit scRGB, which has been available since the Microsoft Windows 7 operating system. For all practical purposes, scRGB can represent all colors visible to the average human. (The initials "sc" in scRGB actually stand for nothing.) The scRGB color

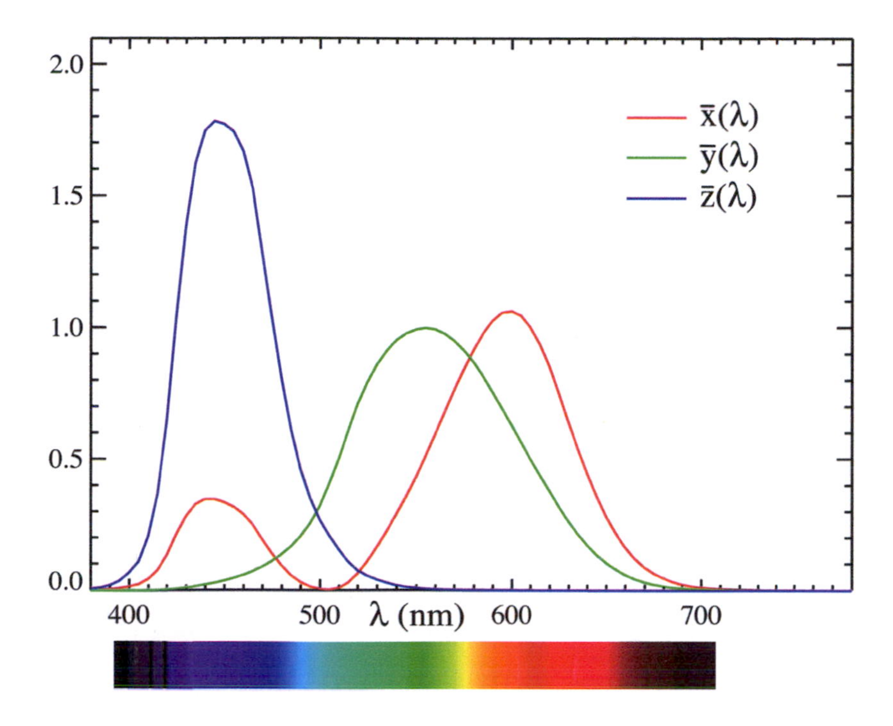

Figure B.3 *CIE 1931 XYZ average color matching functions (See discussion in text for details. Public domain image courtesy of Wikimedia Commons.)*

gamut has backward compatibility with the sRGB color gamut but the cost is that approximately 80% of the scRGB color gamut consists of colors not visible to the average human. Monitor hardware that fully implements the Windows Color System can emit all the colors the average human eye can perceive. Do note that Apple RGB is distinct from sRGB and scRGB.

Note from Figure B.5 that Adobe software on a computer running a Microsoft Windows 7 or newer operating system with the Windows Color System will deliver more colors than in sRGB but fewer than in scRGB. Also depicted in Figure B.5 are the color gamuts for ProPhoto RGB developed by Kodak and ColorMatch RGB, which was developed by Radius for its ColorMatch RGB line of graphics monitors for professional printers at a time when the available sRGB color monitors were thought to have too small of a color gamut for proofing of CMYK on offset printing presses.

The standard illuminant for CIE 1931 RGB and XYZ color spaces is E, which is also known by the alias D65. This standard illuminant is also used for sRGB, scRGB, Adobe RGB 1998, and Apple RGB. High-definition television (HDTV), ultra-high-definition television (UHDTV), and the PAL/SECAM

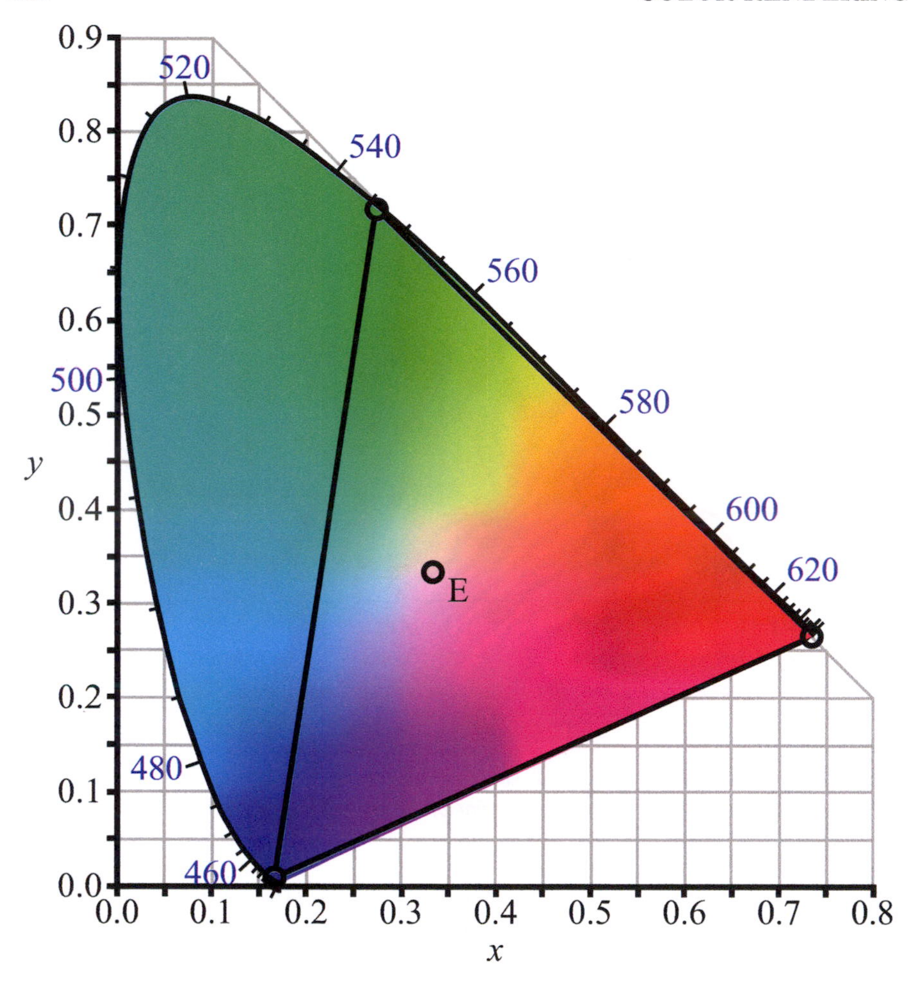

Figure B.4 *CIE 1931 xy chromaticity diagram of the XYZ color space with the RGB color space depicted as the inner triangle (Public domain image courtesy of Wikimedia Commons.)*

(1970) and NTSC (1987) visual recording standards have also adopted the D65 white point. Japanese NTSC (1987) uses D93 reflecting different tastes. ProPhoto RGB, ColorMatch RGB and the newer Adobe Wide Gamut RGB have adopted D50 as the white point. The D series of illuminants were developed to represent natural daylight and although difficult to reproduce artificially they are easily mathematically modeled. When multiplied by 100, the digits in the D series are approximately equal to the correlated color temperature (CCT) of a radiating black body. D50 corresponds to horizon daylight and D65 to noon or overhead daylight.

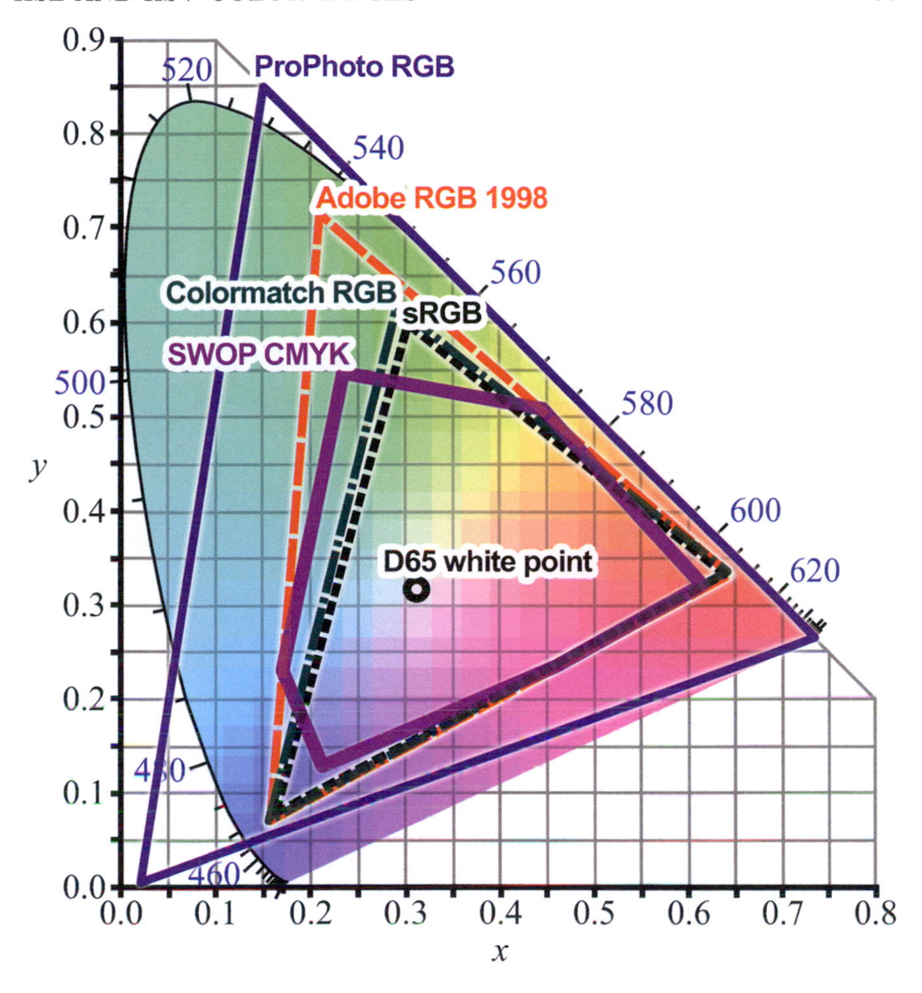

Figure B.5 *Comparison of some RGB color spaces with a CMYK color space on the CIE 1931 xy chromaticity diagram (Public domain image courtesy of Wikimedia Commons.)*

B.4 HSL and HSV Color Spaces

HSL (hue, saturation, lightness) and HSV (hue, saturation, value) were developed by computer graphics researchers as alternative representations of the CIE 1931 RGB color space to better fit human perception of color. HSL and HSV color models both use a cylindrical geometry with linear mixtures between adjacent pairs of additive primary and secondary colors: red, yellow, green, cyan, blue, and magenta. Colors of each hue are arranged in a radial slice in both the HSL and HSV color spaces around a central axis of neutral colors ranging from white at the top, down to black at the bottom. The HSL

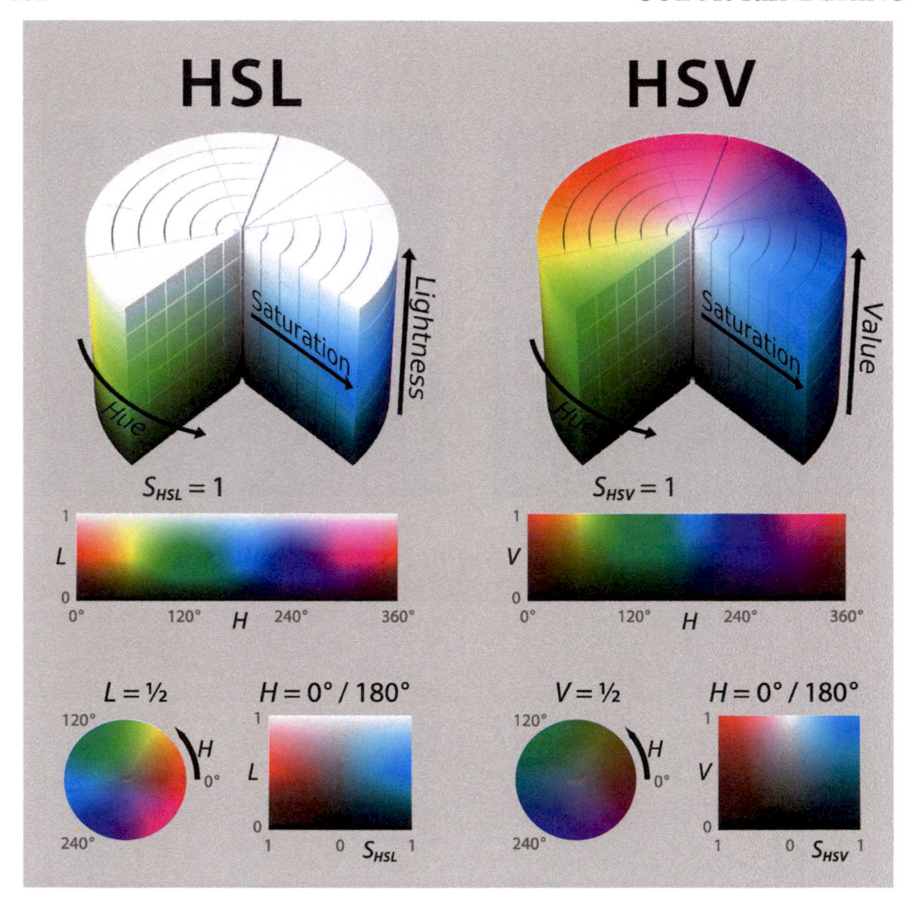

Figure B.6 *HSL and HSV color spaces (Public domain image courtesy of Wikimedia Commons.)*

color space places fully saturated colors around a circle at a lightness value of 0.5 on a scale from 0 (black) to 1 (white). On the other hand, the HSV color space uses saturation to model various shades of brightly colored pigments with values describing the mixture of the pigments with varying amounts of black or white. Saturated colors have a lightness of 0.5 in HSL and a value of 1 in HSV. See Figure B.6 for a comparison of the HSL and HSV color spaces in their cylindrical coordinate systems.

B.5 CIELAB and CIELUV Color Spaces

The CIEL*a*b* 1976 (CIELAB) color space has asterisks in it to distinguish it from the Lab color space due to Richard Hunter in 1948. Hunter's Lab color scale was conceived on the basis of providing a scale that could be more easily comprehended than the CIE 1931 XYZ color scale. In the Lab color space, "L" is for lightness, "a" for red-green, and "b" is for blue-yellow. Hunter's Lab coordinates are created by a square-root transformation of the CIE 1931 XYZ color space, whereas the CIELAB coordinates use a cube-root transformation.

The CIELAB color space exceeds the color gamuts of the CIE 1931 RBG color space and the versions of the CMYK color space available in color printing. The basic concept behind both Hunter's Lab and the CIELAB color spaces is to refine the CIE 1931 XYZ color space so that the result is more perceptually uniform, that is, the same amount of change in color value should be perceptually the same regardless of color. In this way, the Lab and CIELAB color spaces are intended to approximate human vision. The CIELAB 1976 color space is copyright-free, license-free, and in the public domain. But much of CIELAB coordinate space falls outside of normal human color vision.

Although the CIELAB 1976 color space is mathematically defined, there are no simple formulas for conversion to and from either CIE 1931 XYZ or CIE 1931 RGB.

The CIEL*u*v* 1976 (CIELUV) color space was created in 1976 as a perceptually uniform transformation of the CIE 1931 XYZ color space that would be easier to compute than the CIELAB 1976 color space. CIELUV and CIELAB were adopted simultaneously by CIE in 1976 when no clear consensus could be reached to select one over the other. CIELUV is used extensively in computer gaming applications.

In 2002, CIE released the CIECAM02 color appearance model to address shortcomings in both CIELAB and CIELUV. The two major components of CIECAM02 are a chromatic adaptation transform and equations for relating luminance, lightness, colorfulness, chroma, saturation and hue. The Windows Color System uses CIECAM02 for visible colors.

B.6 HCL Color Space

The HCL color space uses hue, chroma, and luminance. It too was designed to better represent human perception of color than the CIE 1938 XYZ color space. It is believed by some to better present data by avoiding the bias due to varying saturation. The HCL color model is based on the CIELAB color model of 1976. HCL retains the luminance (L) axis of the CIEL*u*v* 1976 color space but transforms uv to polar coordinates in which the distance from the origin is *chroma* (which represents color) and the phase angle is hue. Because steps of equal size in the HCL color space correspond roughly to equal perceptual changes in color, HCL can be thought of as a perceptually uniform transformation of the HSV color space.

B.7 CMYK Color Space

The abbreviation CMYK stands for the first letters of the colors cyan, magenta, and yellow and the letter k in black. The CMYK color space can be thought to represent the idiom: "What you see, you don't always get." Examining Figure B.5 reveals a CMYK color gamut close to but somewhat different from the sRGB gamut and quite a bit less than 48-bit scRBG. The CMYK color gamut is achieved through the four-color process used on paper. CMYK is a subtractive color as illustrated in Figure B.7. With respect to the idiom, the issue is that RGB color spaces are not used in color printing on paper. So what is seen on an electronic device in an RGB palette is not reproduced on paper. Experience typically shows that the CMYK printed color image looks darker and less vibrant than the image on an electronic device. A workaround for this has been the development of the CcMmYK color model that adds light cyan and light magenta to CMYK to lighten the image and remove the harsh half-toning dot appearance. The newer inkjet color printers use the CcMmYK color model.

An even newer CMYK color gamut for high-quality, sheet-fed offset printing is the extended CMYK or XCMYK 2017 color space developed by Idealliance and finalized after testing in the United States, Canada, Pakistan, Malaysia, Singapore, Hong Kong, and China.

Depicted in the xy chromaticity diagram of the CIE 1931 XYZ color space of Figure B.5 is the SWOP CMYK color gamut. SWOP is the abbreviation for *Specifications for Web Offset Publications.* SWOP used to be the abbreviation for an organization as well as its specifications produced for improving the consistency and quality of materials produced by professional printers in the United States. Inks used by professional printing firms conforming to the SWOP specification are called *SWOP inks*, which are similar but not identical to ISO standard 2486-1:2006. There are also SWOP specifications for proofs because the printed colors in proofs are not necessarily the same as in the final product, as was discovered by the author in the preparation of the first edition of this book, because of the use of different paper products and different printers for the different processes of proofing and production.

There are also other SWOP standards for ICC profiles used in color management. An ICC profile is essentially a dataset that characterizes a device either for color input or color output according standards of the International Color Consortium (ICC). Profiles define a color mapping between a device source or target color space and a *profile connection space* (PCS). The PCS is either the CIE 1931 XYZ color space or the CIEL*a*b* 1976 color space (CIELAB). Adobe image management software, such as Photoshop, Illustrator, and Acrobat, allow for the use of different ICC profiles.

While color management on devices using the Apple and Windows operating systems is well established, the systems ColorSync and Windows Color System, are different. Operating systems, such as Linux, that use the X Window

Figure B.7 *CMYK subtractive color space (Public domain image courtesy of Wikimedia Commons.)*

System for graphics can also use ICC profiles but the process is less mature being coordinated through OpenICC at freedesktop.org and using the *Little CMS* (LCMS) open source color management system.

ICC profiles arguably exist for two purposes: to reproduce on paper as accurately as possible RGB images viewed on a monitor, and to view color images on an RGB monitor as they would appear printed on paper by a CMYK process. The optimal selection of an appropriate ICC is dependent on the type of printer or printing press being used and the form and quality of paper chosen.

Offset lithography is one of the most common methods for mass production of colored printed materials. Image quality by offset printing is not as good as photogravure or rotogravure, which both involve etching an image onto a carrier. The older computer-to-film offset printing method uses four separate films, one for each of cyan, yellow, magenta, and black, to burn an image onto a lithographic plate. The more modern computer-to-plate offset printing method bypasses film and instead uses desktop publishing software to directly prepare the lithographic printing plate. In modern lithographic production, a chemical coating is applied either to a plastic or metal plate. The plate can then either be used to print the image directly on paper, or it can be *offset* by transferring the image from the plate onto a flexible sheet for printing.

Form of paper refers to whether the source is a reel for *web* printing or stacked sheets of a pre-cut size. Quality can refer to whether the stock source is gloss or matte with various gradations for each. Paper acquires a glossy appearance through the use of coatings. Choices of paper and make of printing press are regionally dependent. So while the United Kingdom and Europe may make use of the same set of ICC profiles, other ICC profiles are more popular in the United States and Canada. The ICC CMYK profile used to produce the printed version of this second edition is *US Web Coated (SWOP) v2.* On the other hand, had printing been done on a sheet-fed press with Grade 1 coated paper, the better choice of CMYK color profile would have been *Coated GRACol 2006.* Had a reel of coated grade 3 paper been used, then *Web Coated SWOP 2006 Grade 3 Paper* would be the ICC profile of choice. If the press and type of paper are not known, then the conservative choice of CMYK profile would be ICC *Web Coated SWOP 2006 Grade 5 Paper.* For sheet-fed paper of Grade 1 or 2 (gloss or matte) in Europe, the preferred CMYK ICC profile would be *ISO Coated v2 (ECI).*

In addition to final production, the ICC profiles are used in the proofing process before final production runs. There are two major types of proof: soft proof and hard proof. *Soft proof* uses well-calibrated ICC profiles on a computer monitor and should be done under controlled and consistent viewing conditions. *Hard proof* requires a sample printed on paper and typically does not include a test copy produced by the printing press to be used in production because this is too expensive. Instead, an inkjet or thermal sublimation printer is used to produce what is hopefully a reliable and true color reproduction. Typically the same paper used in production is not used for the hard proof because of the technological differences between the proofing and production printers.

B.8 Displaying Color in R

Functions that manipulate color in R are found in the graphics devices package `grDevices`. These functions support both the `base` and `grid` graphics packages. The sRBG color space used by R is CIE standard RGB 61966. One version of this color space was released as IEC 61966-1-1:1999 in 1999; amended by CIE in 2003 and released as CIE 61966-2-1:1999/AMD1:1:2003; and then corrected in 2014 and subsequently released as CIE 61966-2-1:1999/COR1:2014.

The function `colors` in the package `grDevices` simply returns the names of colors in R. The function `palette` when called as `palette()` returns the current color palette in R. Passing a vector of colors to `palette` will set a new palette of colors. The following color palettes are available in R as functions capable of providing a vector of `n` contiguous colors, with `n` supplied as an argument: `cm.colors`, `gray`, `heat.colors`, `hsv`, `rainbow`, `terrain.colors`, and `topo.colors`. The function `adjustcolor` can be used to tweak colors by adjusting the coordinates r, g, and b (in sRGB space) and the opacity α. The function `col2rgb` converts an R color to the CIE 1931 RGB color space. The R color passed to either `col2rgb` or `adjustcolor` can be either a color name as listed by `colors()`, a hexadecimal string of the form `#rrggbb` or `#rrggbbaa` (with aa denoting the opacity α), or a positive integer `i` meaning `palette()[i]`.

The function `rgb` in `grDevices` accepts as arguments intensities between zero and their maximal values for the colors red, green, and blue; and optionally the opacity value α between 0 and 1, inclusive. The output from `rgb` is a hexadecimal number in the form `#rrbbgg` in sRGB or `#rrggbbaa` (with aa denoting the opacity α) if the opacity α was provided on the call to `rgb`. An opacity value of 0 corresponds to transparent while a value of 1 corresponds to opaque.

The function `hsv` in `grDevices` produces a vector of sRGB colors from hue, saturation, and value in the HSV color space. A vector of semi-transparent colors can also be obtained if the optional opacity argument α is passed. To reverse this operation, the function `rgb2hsv` in `grDevices` will convert a vector of colors in the sRGB color space into the HSV color space.

The function `hcl` in `grDevices` produces a vector of sRGB colors from hue, chroma, and luminance in the polar coordinate version of the CIELUV color space. Optionally, the opacity argument α can be passed to obtain semi-transparent colors. As some resulting sRGB values could lie outside the range of real colors, if an optional logical parameter `fixup` is set to `FALSE`, then the resulting color will be an `NA` value. The default value for `fixup` is `TRUE`, which results in a real color result.

The function `convertColor` in the package `grDevices` converts a matrix of colors from one color space to another. Possible color spaces are `"RGB"` (CIE 1931 RGB), `"XYZ"` (CIE 1931 XYZ), `"sRGB"`, `"Apple RGB"`, `"Lab"` (CIELAB 1976), and `"Luv"` (CIELUV 1976). The `Lab` and `Luv` color spaces in `convertColor` require the specification of an illuminant as a reference white color. Standard illuminants available for `convertColor` are `"A"`, `"B"`, `"C"`, `"D50"`, `"D55"`, `"D60"`, and `"D65"` (also known as `"E"`).

The function `make.rgb` in the package `grDevices` is used to define an RGB color space for use by `convertColor`. Arguments to `make.rgb` in addition to the intensities of the red, green, and blue colors, include the choice of standard illuminant, as described for `convertColor`, and gamma γ—both of which should be the same as that of the display monitor being used. Gamma, or more properly the gamma correction, is an operation to encode and decode luminance or tristimulus values for display monitors. Typically a value of γ less than one is used to encode images, whereas a value of γ in excess of one is used to decode images. Prior to the release of Apple OS X 10.6 in 2009, Apple monitors were encoded with a gamma of 0.55 and decoded with a gamma of 1.8, which is the value to be used when the `"Apple RGB"` color space is selected. The sRGB standard does not have a constant gamma but instead a more complicated function, although $\gamma = 2.2$ is a good constant approximation. However, when using the `"sRGB"` color space, set `gamma = "sRGB"` to obtain this complicated function that is part of the sRGB standard.

The function `colorRamp` in `grDevices` maps the interval $[0, 1]$ to colors using interpolation in a more sophisticated way than just calling the functions `cm.colors`, `heat.colors`, `hsv`, `rainbow`, `terrain.colors`, and `topo.colors`. Interpolation options for `colorRamp` are either linear or spline. Interestingly, the output from `colorRamp` is a function that can then be called to yield coordinates in either the CIE 1931 RGB color space or the CIE 1976 L*a*b* color space by setting the argument `space` to either `"rgb"` or `"Lab"`, respectively. Optionally, the opacity (alpha channel) values can be returned for the CIE 1931 RGB color space but only if the `space` parameter is not set. The documentation for `colorRamp` as of version 3.4.1 states that the conversion formulas do not appear to be accurate and so the color ramp values might not reach the extremes of CIELAB. The documentation recommends using the sequential and diverging ColorBrewer palettes available in the R package `RColorBrewer`.

Conversion between color spaces in R is also available through functions in the contributed package `colorspace`. The list of color models for `colorspace` is more comprehensive than for `grDevices`. Included in `colorspace` are the standard color models CIE 1931 RGB, HSV, HLS, CIE 1931 XYZ, CIELAB 1976, a polar coordinate version of CIELAB 1976 (which is called `polarLAB`), CIELUV 1976, and HCL (a polar coordinate version of CIELAB 1976 called `polarLUV` in `colorspace`). But also included in `colorspace` are the qualitative, sequential, and diverging color palettes of ColorBrewer based on HCL colors. There is a diverging color palette based on the HSV color space. There are also versions rainbow, heat, and terrain color palettes also based on the HCL color space.

Available as a `ggplot2` extension is the R package `colorplaner` that allows the visualization of an additional variable through a color aesthetic using the YUV color space. The choice of this color space is interesting as the YUV color space was developed to permit color display in a black-and-white television infrastructure. To the grayscale luminance variable (Y), the YUV color space adds color identification through two variables (U and V).

Note: semi-transparent colors with $0 < \alpha < 1$ are only supported on the following devices by the `grDevices` package: `pdf`, `windows`, and

`X11(type="cairo")`; and the associated bitmap devices `jpeg`, `png`, `bmp`, `tiff`, and `bitmap`. Semi-transparent colors are also supported by third-party devices such as in the packages `Cairo`, `cairoDevice`, and `JavaGD`. Most other graphics devices plot semi-transparent colors as fully transparent, usually with a warning on first use.

B.9 Saving Color Documents from R

The R package `grDevices` for graphics devices can save an image from an R graphics display window in a number of different widely used formats. Four of these formats are .bmp, .jpeg (or .jpg), .png, and .tiff. The names of the `grDevices` functions for so doing are the same as the format names. There are reasons for choosing one format over the others.

The .bmp format is standard on Windows and is supported by viewers in other operating systems. The default resolution in .bmp is 72 points per inch (ppi). While this is the size of text, the resulting image is a bit grainy. The parameter `res` on call to the function `bmp` can be set to a higher ppi.

The .jpeg format widely used for photographs is lossy. The `quality` parameter in the function `jpeg` has a default value of 75 measured in per cent. Smaller values will save storage space but at the expense of image degradation.

The .png format is lossless and is excellent for line diagrams and colored figures and can easily be converted into many bitmap formats. The default resolution on calling the function `jpeg` is 72 ppi but this can be changed by altering the value of `res`.

The .tiff (or .tif) format is another lossless format that stores RGB value uncompressed but this tends to yield rather large files compared to the other formats. Fortunately, the function `tiff` permits compression of the files produced with the parameter `compression`. Setting `compression = "none"` results in no compression. Other options are `"jpeg"`, `"lzw"`, `"lzw+p"`, `"rle"`, and `"zip+p"`.

The functions `bmp` and `png` will use a palette if there are fewer than 256 colors in the image but record 24-bit RGB otherwise.

All four functions allow for the `height`, `width`, and the `units` for same to be passed as arguments with a default of 480 for both `height` and `width` and pixels (`"px"` for the `units`). The name of the file is given in the argument `filename`. The background color (`bg`) can also be set different from the default value of `"white"`. The font family is passed to the argument `family` for which the default is Arial. The font size in big points is passed with the parameter `pointsize`. Colors are as in the monitor application for viewing the image. The parameter `type` when set to `""windows"` instructs that plotting be done by the Windows graphical display interface (GDI) and when set to `"cairo"'` uses X11's cairographic display interface.

The function png when used with the Windows GDI supports 16-bit transparent color backgrounds and records a 24-bit RGB otherwise. Setting windows= "cairo-png" when using the cairographics display interface will produce a larger 32-bit ARGB file.

Calls to any of the four functions on Windows will produce output from a hidden screen using Windows GDI calls. Calls on an X11 machine will likewise use a hidden X11 display. With an Apple operating system, these four functions will use the Quartz graphics system. For high-quality plots in a Windows environment, pass antialias = "cleartype" when calling any of these four functions.

Either pdf or dev.copy2pdf in the grDevices package can be used to produce a .pdf (or PDF) file. If using cairographics on X11 is enabled, a third option is the function cairo_pdf. (If using an Apple operating system, XQuartz may need to be loaded to access this third function.) Of the two functions pdf or dev.copy2pdf, most users will find dev.copy2pdf easier to use. The reasons for this are that the latter function inherits height, width, background color, and font family from the current device unless otherwise specified. The former function requires these to be manually specified. The .pdf files of all of the figures rendered in R in this book were saved to a file using dev.copy2pdf.

The functions postscript and dev.copy2eps can be used to save a graphic from a displayed window to a PostScript (PS or .ps) file or an Encapsulated PostScript (EPS or .eps) file, respectively. The passed parameters are the same for these two functions. Note that for postscript and dev.copy2eps the font family defaults to "Helvetica" as it also does for pdf and dev.copy2pdf. The paper size is passed to all four functions with the parameter paper. The default value for paper is "special", which simply means that the height and width of the current graphics window is used. A further setting is paper = "default", which is not the default but a direction that the paper size is taken from the environmental R variable papersize. Other options for paper size are "a4", "executive", "legal", and "letter" with "a4r" and "USr" for a rotated, or landscape, output.

A color model can be selected when calling each of pdf, dev.copy2pdf, postscript, and dev.copy2eps using the parameter colormodel. All four functions will accept "srgb", "gray", and "cmyk" as arguments for colormodel. The default color model for all four functions is "srgb", which is not ideal if printing the graphic on paper is desired, which requires in each instance setting colormodel = "cmyk". Note that "gray" converts any color but a pure gray to a luminance and is not the same as taking a so-called black-and-white photograph of a colored object.

For the functions postscript and dev.copy2eps, there are three more possible values for colormodel: "rgb", and "rgb-nogray", and "srgb+gray". The color models "rgb" and "rgb-nogray" have been provided for backward compatibility. The color model "rgb" uses uncalibrated RGB and corresponds to the color model of the same name in R prior to version 2.13.0. Plot files in `"rgb"` and `"rgb-nogray"` will be smaller and may render faster than `"srgb"`.

The choice of `"srgb+gray"` uses sRGB for colors but pure grayscale, including black and white, for grays. Although this can result in smaller files that are useful for some printers, there can be noticeable color gradients involving the grays.

All the figures in this book that were rendered in R were saved to an .eps file using `dev.copy2eps` in R with `colormodel = "cmyk"`. All other figures drafted in a different format were saved by Adobe Illustrator .pdf and .eps files with both sRGB and CMYK color spaces embedded: the former for monitor display and the latter if printed to paper.

B.10 Conclusion

It is altogether too easy to sit in front of a computer monitor looking at a graphical display with a complex application of color, to do a few points and clicks that produce a paper facsimile on a nearby color printer, and be entirely oblivious to the marvelous scientific and technological achievements needed to do all this. Ignorance is bliss. But when the printed product does not appear as intended and in fact appears to be doing something different with color when compared to the same image on the computer monitor, then it is necessary to dig a little deeper.

It is worthwhile to know that color appears on a computer monitor through an additive process and that color appears in print through a subtractive process. It is also worthwhile to know that color monitors and color printers cannot reproduce all the colors that a normal human can perceive. Due to the inherent process differences in reproducing color, there is never a perfect match of color palette between monitor and printer. Moreover, there is variation in the color spaces amongst different computer monitors and then again amongst different printers, either due to software, hardware, or both. A minimum core set of knowledge for statistical graphics should include knowing that computer monitors use an RGB color space and that printers use a CMYK color space.

To take full advantage of the features for rendering colors in R, it is necessary to know about other color spaces. To review, there are the color spaces CIE 1931 RGB, CIE 1931 XYZ, CIELAB 1976, and CIELUV 1976 developed and standardized by the Commission International d'Éclairage. Added to these by researchers interested in computer visualization are the HSL, HSV, and HCL color spaces. To complement these theoretical constructs of normal human color visualization are the following color gamuts created by manufacturers: sRGB, scRGB, Adobe RGB, Apple RGB, CMYK, CcMmYyK, and XCMYK to name but a few.

At an absolute minimum, an R user working with color in a statistical graphic needs to do just two things. The first is rather automatic: merely rely on computer software and hardware providers to render accurately the sRGB colorspace in R on a computer monitor. (Although recommended, it is not necessary to re-calibrate periodically the computer monitor's colors.) The second and last is not automatic but easy to do: when saving a color graphic to be printed afterward on paper, pass the argument phrase `colormodel = "cmyk"` so that the image is converted to the CMYK color space when saved to the file with an embedded file notation to the effect that the color space for the image is in fact CMYK.

B.11 Exercises

1. Although, in the discussion, the average color matching function $\bar{r}$ in the CIE 1931 RGB color space is compared with the average color matching function $\bar{x}$ in the CIE 1931 XYZ color space in Section B.3, the other two pairs of color matching functions are not. Do this.
2. Although the CIE 1931 XYZ color space can represent all colors that a normal human can perceive, the CIE 1931 RGB color space cannot. Examine the xy chromaticity diagram of Figure B.4 that compares these two color spaces. State which colors and, specifically, which wavelengths the CIE 1931 RGB color space does not capture.
3. Find out which RGB color space is implemented on the computer monitor you most frequently use. Determine whether this RGB color space is a subset of the CIE 1931 RGB color space or whether it contains all of the CIE 1931 XYZ color space and thus the full spectrum visibly perceptible by normal human vision.
4. Determine whether there are CIE standards for the HSL and HSV color spaces.
5. Examine Figure B.6, which depicts the HSL and HSV color spaces.
   (a) Explain what is being depicted in the two middle rectangles in this figure.
   (b) Explain what is being depicted in the two bottom circles in this figure.
   (c) Explain what is being depicted in the two bottom rectangles in this figure.
6. Consider the CIEL*a*b* 1976 (CIELAB) and CIEL*u*v* 1976 (CIELUV) color spaces.
   (a) State the reason for creating the CIEL*a*b* 1976 (CIELAB) color space.
   (b) State the reason for creating the CIEL*u*v* 1976 (CIELUV) color space.
   (c) State the name of the color space used in the Windows Color System. Can it depict all colors in the CIE 1931 XYZ color space?
7. Consider the HCL and HSV color spaces.
   (a) Determine whether the HCL color space can depict all colors in the CIE 1931 XYZ color space.

(b) Determine whether there are colors in the HCL color space that are not in the HSV color space.

8. The following relate to the color printer that you use most often.
   (a) Determine which version of the CMYK color space is used.
   (b) Determine the type and quality of paper that this printer typically uses. Does the paper have a matte or glossy appearance?
9. Use the `convertColor` function in the package `grDevices` to convert 10 colors from the `rainbow` palette in sRGB color space to the CIELAB 1976 color space and then back to the sRGB color space.
   (a) Compare the sRGB colors in `#rrggbb` format before and after the conversions.
   (b) Plot the 10-color rainbow palette before and after completing the conversion loop on a computer monitor and compare.
   (c) Print the 10-color rainbow palette before and after completing the conversion loop on paper and compare.
10. Explore reasons for using semi-transparent colors.
11. Which of the formats .bmp, .jpeg, .png, .tiff, .pdf, and .eps are you more likely to use and why?
12. Are you likely to use any color space other than `"srgb"` or `"cmyk"` when saving a color plot to an .eps file? Discuss.

Bibliography

[1] "graphicacy, n.". *The Oxford English Dictionary.* OED Online. Oxford University Press, Oxford, UK, ADDITIONS SERIES 1993, <http://dictionary.oed.com/cgi/entry/00293111>, 2nd edition, 1989.

[2] "numeracy, n.". *The Oxford English Dictionary.* OED Online. Oxford University Press, Oxford, UK, DRAFT REVISION Mar. 2004, <http://dictionary.oed.com/cgi/entry/00328393>, 2nd edition, 1989.

[3] E. Anderson. The irises of the Gaspé Peninsula. *Bulletin of the American Iris Society*, 59:2–5, 1935.

[4] D.F. Andrews and A.M. Herzberg. *Data: A Collection of Problems from Many Fields for the Student and Research Worker.* Springer, New York, NY, 1985.

[5] F.J. Anscombe. Graphs in statistical analysis. *American Statistician*, 27:17–21, 1973.

[6] F.J. Anscombe and J.W. Tukey. The estimation and analysis of residuals. *Technometrics*, 5:141–160, 1963.

[7] J.C. Baird. *Psychophysical Analysis of Visual Space.* Pergamon Press, Oxford, England, 1970.

[8] R.A. Becker, J.M. Chambers, and A.R. Wilks. *The New S Language.* Bell Telephone Laboratories, Inc., Murray Hill, NJ, 1988.

[9] R.A. Becker and W.S. Cleveland. *S-PLUS Trellis Graphics User's Manual.* MathSoft, Inc., Seattle, WA, and Bell Labs, Murray Hill, NJ, 1996.

[10] R.A. Becker, W.S. Cleveland, and M.-J. Shyu. The visual design and control of Trellis display. *Journal of Computational and Graphical Statistics*, 5:123–155, 1996.

[11] G. Blom. *Statistical Estimates and Transformed Beta Variables.* John Wiley and Sons, New York, NY, 1958.

[12] G.E.P. Box and D.R. Cox. An analysis of transformations. *Journal of the Royal Statistical Society, Series B*, 26:211–243, 1964.

[13] British Wreck Commissioner's Inquiry. *Report on the Loss of the "Titanic." (s.s.).* His Majesty's Stationery Office, London, England, 1912.

[14] P. Brofeldt. Bidgrag till kaennedom on fiskbestondet i vaara sjoear. laengelmaevesi. Finlands fiskeriet band 4, Meddelanden utgivna av fiskerifoereningen i Finland, Helsingfors, Finland, 1917.

[15] L.J. Brown and G. Hwang. How to approximate a histogram by a normal density. *American Statistician*, 47:251–257, 1993.

[16] D.A. Burn. *Designing Effective Statistical Graphics*, volume 9 of *Computational Statistics, Handbook of Statistics*, chapter 22, pages 745–773. Elsevier Science Publishers B. V., Amsterdam, The Netherlands, 1993.

[17] H. Cavendish. Experiments to determine the density of the earth. *Philosophical Transactions of the Royal Society of London*, 88:469–526, 1798.

[18] J.M. Chambers, W.S. Cleveland, B. Kleiner, and P.A. Tukey. *Graphical Methods for Data Analysis.* Wadsworth & Brooks/Cole, Monterey, CA, 1983.

[19] H. Chernoff. The use of faces to represent points in k-dimensional space graphically. *Journal of the American Statistical Society*, 68:361–368, 1973.

[20] W.S. Cleveland. Robust locally weighted regression and smoothing scatterplots. *Journal of the American Statistical Association*, 74:829–836, 1979.

[21] W.S. Cleveland. *The Elements of Graphing Data.* Wadsworth Advanced Books, Monterey, CA, 1985.

[22] W.S. Cleveland. A model for studying display methods of statistical graphics. *Journal of Computational and Graphical Statistics*, 2:323–343, 1993.

[23] W.S. Cleveland. *Visualizing Data.* Hobart Press, Summit, NJ, 1993.

[24] W.S. Cleveland, M.E. McGill, and R. McGill. The shape parameter of a two-variable graph. *Journal of the American Statistical Association*, 83:289–300, 1988.

[25] W.S. Cleveland and R. McGill. Graphical perception: Theory, experimentation, and application to the development of graphical methods. *Journal of the American Statistical Association*, 79:531–554, 1984.

[26] W.S. Cleveland and R. McGill. The many faces of a scatterplot. *Journal of the American Statistical Association*, 79:807–822, 1984.

[27] R.D. Cook. Detection of influential observation in linear regression. *Technometrics*, 79:15–18, 1977.

[28] H. Cramèr. *Mathematical Methods of Statistics.* Princeton University Press, Princeton, NJ, 1946.

[29] F.E. Croxton. Further studies in the graphic use of circles and bars: (ii) Some additional data. *Journal of the American Statistical Association*, 22:36–39, 1927.

[30] F.E. Croxton and H. Stein. Graphic comparison by bars, squares, circles, and cubes. *Journal of the American Statistical Association*, 27:54–60, 1932.

[31] F.E. Croxton and R.E. Stryker. Bar charts versus circle diagrams. *Journal of the American Statistical Association*, 22:473–482, 1927.

[32] D.P. Doane. Aesthetic frequency classifications. *American Statistician*, 30:181–183, 1976.

[33] W.C. Eells. The relative merits of circles and bars for representing component parts. *Journal of the American Statistical Association*, 21:119–132, 1926.

[34] V.A. Epanechnikov. Nonparametric estimation of a multidimensional probability density. *Theory of Probability and Its Applications*, 214:153–158, 1969.

[35] D.S. Falconer and T.F.C. Mackay. *Introduction to Quantitative Genetics*. Pearson Education, Harlow, UK, 1996.

[36] N.I. Fisher. *Statistical Analysis of Circular Data*. Cambridge University Press, Cambridge, UK, 1993.

[37] R.A. Fisher. The correlation between relatives on the supposition of Mendelian inheritance. *Transactions of the Royal Society of Edinburgh*, 52:399–433, 1918.

[38] R.A. Fisher. On the "probable error" of a coefficient of correlation deduced from a small sample. *Metron*, 1:3–32, 1921.

[39] R.A. Fisher. Applications of "Student's" distribution. *Metron*, 5:90–104, 1925.

[40] R.A. Fisher. *Statistical Methods for Research Workers*. Oliver and Boyd, Edinburgh, UK, 1925.

[41] R.A. Fisher. The arrangement of field experiments. *Journal of the Ministry of Agriculture of Great Britain*, 33:503–513, 1926.

[42] R.A. Fisher. The use of multiple measures in taxonomic problems. *Annals of Eugenics*, 7:179–188, 1936.

[43] J. Fox. *An R and S-PLUS Companion to Applied Regression*. SAGE Publications, Inc., Thousand Oaks, CA, 2002.

[44] D. Freedman and P. Diaconis. On the histogram as a density estimator: l_2 theory. *Zeitschrift für Wahrscheinichkeitstheorie und verwandte Gebiete*, 57:453–476, 1981.

[45] M. Frigge, D.C. Hoaglin, and B. Iglewicz. Some implementations of the boxplot. *The American Statistician*, 43:50–54, 1989.

[46] F. Galton. Family records [letter to the editor]. *The Times*, 9 January:10b, 1884.

[47] F. Galton. Prize records of family faculties [letter to the editor]. *The Times*, 19 May:9e, 1884.

[48] F. Galton. *Meteorographica, or Methods of Mapping the Weather.* Macmillan, London, UK, 1863.

[49] F. Galton. Record of family faculties [letter]. *Journal of the Royal Statistical Society*, 47:166, 1884.

[50] F. Galton. Presidential Address, Section H, Anthropology. *Report of the British Association for the Advancement of Science*, 55:1206–1214, 1885.

[51] F. Galton. Regression towards mediocrity in hereditary stature. *Journal of the Anthropological Institute of Great Britain and Ireland*, 15:246–263, 1886.

[52] F. Galton. Co-relations and their measurements, chiefly from anthropometric data. *Proceedings of the Royal Society*, 45:135–145, 1888.

[53] F. Galton. *Natural Inheritance.* Macmillan, London, UK, 1889.

[54] F. Galton and J.D. Hamilton Dickson. Family likeness in stature. *Proceedings of the Royal Society*, 40:42–73, 1886.

[55] B.D. Gaynor, J.D. Chidambaram, V. Cevallos, Y. Miao, K. Miller, H.C. Jha, R.C. Bhatta, J.S.P. Chaudhary, S. Osaki Holm, J.P. Whitcher, K.A. Holbrook, A.M. Fry, and T.M. Lietman. Topical ocular antibiotics induce bacterial resistance at extraocular sites. *British Journal of Ophthalmology*, 89:1097–1099, 2005.

[56] R.L. Gregory. *Eye and Brain: The Psychology of Seeing.* Princeton University Press, Princeton, NJ, 5th edition, 1997.

[57] P. Hall. Large sample optimality of least-squares cross-validation in density estimation. *Annals of Statistics*, 11:1156–1174, 1983.

[58] P. Hall and J.S. Marron. Estimation of integrated squared density derivatives. *Statistics and Probability Letters*, 6:109–115, 1987.

[59] J.A. Hanley. "Transmuting" women into men: Galton's family data on human stature. *The American Statistician*, 58:237–243, 2004.

[60] J.A. Hartigan and B. Kleiner. Mosaics for contingency tables. In W.F. Eddy, editor, *Proceedings of the 13th Symposium on the Interface Between Computer Science and Statistics*, New York, NY, 1981. Springer-Verlag.

[61] J.A. Hartigan and B. Kleiner. A mosaic of television ratings. *The American Statistician*, 38:32–35, 1984.

[62] T. Hayfield and J.S. Racine. Nonparametric econometrics: The np package. *Journal of Statistical Software*, 27:1–32, 2008.

[63] E. Hecht and A. Zajac. *Optics.* Addison-Wesley Publishing Company, Inc., Reading, MA, 1976.

[64] G.T. Henry. *Graphing Data: Techniques for Display and Analysis.* Sage Publishing, Inc., Thousand Oaks, CA, 1995.

[65] J.L. Hintze and R.D. Nelson. Violin plots: A box plot-density trace synergism. *The American Statistician*, 52:181–184, 1998.

[66] D.C. Hoaglin and B. Iglewicz. Fine-tuning some resistant rules for outlier labeling. *Journal of the American Statistical Association*, 82:1147–1149, 1987.

[67] J.L. Hodges and E.L. Lehmann. The efficiency of some nonparametric competitors of the t-test. *The Annals of Mathematical Statistics*, 13:435–475, 1956.

[68] D. Huff. *How to Lie with Statistics.* W. W. Norton, New York, NY, 1954.

[69] R.J. Hyndman and Y. Fan. Quantiles in statistical packages. *The American Statistician*, 50:361–365, 1996.

[70] Apple Computers Inc. *Apple Human Interface Guidelines: The Apple Desktop Interface.* Addison-Wesley, Reading, MA, 1987.

[71] SAS Institute. *SAS®9.2 Help and Documentation.* SAS Institute, Cary, NC, 2002–2009.

[72] W.G. Jacoby. *Statistical Graphics for Univariate and Bivariate Data.* Sage Publishing, Inc., Thousand Oaks, CA, 1997.

[73] M.C. Jones, J.S. Marron, and S.J. Sheather. A brief survey of bandwidth selection for density estimation. *Journal of the American Statistical Association*, 91:401–407, 1996.

[74] K.J. Keen. *Estimation of Intraclass and Interclass Correlations.* PhD thesis, University of Toronto, Toronto, Canada, 1987.

[75] K.J. Keen and R.C. Elston. Robust asymptotic sampling theory for correlations in pedigrees. *Statistics in Medicine*, 22:3229–3247, 2003.

[76] M. Kendall and A. Stuart. *The Advanced Theory of Statistics, Volume 1, Distribution Theory.* Macmillan Publishing Co., Inc., New York, NY, fourth edition, 1977.

[77] S. Kosslyn. Graphics and human information processing: A review of five books. *Journal of the American Statistical Association*, 80:499–512, 1985.

[78] S.M. Kosslyn. *Image and Mind.* Harvard University Press, Cambridge, MA, 1980.

[79] L. Lalanne. *Appendice À La Météorlogie De Kaemtz, traduite et annotée par M. Martins, Sur La Représentation Graphique Des Tableaux Météorologiques Et Des Lois Naturelles En Général.* Paulin, Paris, France, 1843.

[80] J.L. Lee and Z.N. Tu. A versatile one-dimensional distribution plot: The BLiP plot. *The American Statistician*, 51:353–358, 1997.

[81] L. Manchester. A technique for comparing graphical methods. *The Canadian Journal of Statistics*, 19:1–22, 1991.

[82] J.S. Marron. Discussion of: "Practical performance of several data driven bandwidth selectors" by Park and Turlach. *Computational Statistics*, 8:17–19, 1993.

[83] R. McGill, J.W. Tukey, and W.A. Larsen. Variations of box plots. *The American Statistician*, 32:12–16, 1978.

[84] P. Murrell. *R Graphics*. Chapman and Hall/CRC, Taylor and Francis Group, Boca Raton, FL, 2006.

[85] E.A. Nadaraya. On estimating regression. *Theory of Probability and its Applications*, 10:1189–1190, 1964.

[86] J. Neter, M.H. Kutner, C.J. Nachtsheim, and W. Wasserman. *Applied Linear Statistical Models*. Times Mirror Higher Education Group, Inc., Chicago, IL, fourth edition, 1996.

[87] F. Nightingale. *Notes on Matters Affecting the Health, Efficiency and Hospital Administration of the British Army*. Florence Nightingale, London, UK, 1858.

[88] Organisation for Economic Co-operation and Development. OECD Health Data 2009. http://www.oecd.org/health/healthdata, 2009. Accessed: 2009-08-13.

[89] K. Pearson. Contributions to the mathematical theory of evolution. [abstract]. ii. Skew variation in homogenous material. *Proceedings of the Royal Society of London*, 54:329–333, 1893.

[90] K. Pearson. Contributions to the mathematical theory of evolution. ii. Skew variation in homogenous material. *Philosophical Transactions of the Royal Society of London, A*, 186:343–414, 1895.

[91] K. Pearson. Mathematical contributions to the theory of evolution. iii. Regression, heredity and panmixia. *Philosophical Transactions of the Royal Society of London, A*, 187:253–318, 1896.

[92] K. Pearson. Data for the problem of the evolution in man. iii. On the magnitude of certain coefficients of correlation in man, &c. *Proceedings of the Royal Society of London*, 66:23–32, 1899.

[93] K. Pearson. Mathematical contributions to the theory of evolution. x. Supplement to a memoir on skew variation. *Philosophical Transactions of the Royal Society of London, A*, 197:443–459, 1901.

[94] K. Pearson. Mathematical contributions to the theory of evolution. xix. Second supplement to a memoir on skew variation. *Philosophical Transactions of the Royal Society of London, A*, 216:429–457, 1916.

[95] K. Pearson and A. Lee. On the laws of inheritance in man. i. Inheritance of physical characters. *Biometrika*, 2:357–462, 1903.

[96] W. Playfair. *The Commercial and Political Atlas.* Corry, London, UK, 1786.

[97] R Development Core Team. *R: A Language and Environment for Statistical Computing.* R Foundation for Statistical Computing, Vienna, Austria, 2007.

[98] A. Rhind. Tables to facilitate the computations of the probable errors of the chief constants of skew frequency distributions. *Biometrika*, 7:127–147, 1909.

[99] D.F. Roberts, W.Z. Billiewicz, and I.A. McGregor. Heritability of stature in a West African population. *Annals of Human Genetics*, 42:15–24, 1978.

[100] P.J. Rousseeuw and I. Ruts. The bagplot: A bivariate box-and-whiskers plot. Technical report, Universitaire Instelling Antwerpen, Antwerpen, Belgium, 1997.

[101] P.J. Rousseeuw, I. Ruts, and J.W. Tukey. The bagplot: A bivariate boxplot. *American Statistician*, 53:382–387, 1999.

[102] P.J. Rousseeuw, S. Van Aelst, and M. Hubert. Rejoinder to the discussion of regression depth. *Journal of the American Statistical Association*, 94:419–433, 1999.

[103] M.F. Schilling and A.E. Watkins. A suggestion for sunflower plots. *The American Statistician*, 48:303–305, 1994.

[104] C.F. Schmid. *Statistical Graphics.* John Wiley, New York, NY, 1983.

[105] D.W. Scott. On optimal and data-based histograms. *Biometrika*, 66:605–610, 1979.

[106] D.W. Scott. *Multivariate Density Estimation: Theory, Practice and Visualization.* John Wiley, New York, NY, 1992.

[107] D.W. Scott and G.R. Terrell. Biased and unbiased cross-validation in density estimation. *Journal of the American Statistical Association*, 82:1131–1146, 1987.

[108] S.J. Sheather. Density estimation. *Statistical Science*, 19:588–597, 2004.

[109] S.J. Sheather and M.C. Jones. A reliable data-based bandwidth selection method for kernel density estimation. *Journal of the Royal Statistical Society, Series B*, 53:683–690, 1991.

[110] B.W. Silverman. *Density Estimation for Statistics and Data Analysis.* Chapman & Hall/CRC, Boca Raton, FL, 1992.

[111] D. Simkin and R. Hastie. An information-processing analysis of graphical perception. *Journal of the American Statistical Association*, 82:454–465, 1987.

[112] R.D. Snee. Graphical display of two-way contingency tables. *The American Statistician*, 28:9–12, 1974.

[113] M.A. Stephens. Techniques for directional data. Technical Report 150, Department of Statistics, Stanford University, Palo Alto, CA, 1969.

[114] M.A. Stephens. EDF statistics for goodness of fit and some comparisons. *Journal of the American Statistical Association*, 69:730–737, 1974.

[115] S. Stigler. Do robust estimators work with real data? *The Annals of Statistics*, 5:1055–1098, 1977.

[116] W.A. Stock and J.T. Behrens. Box, line, and midgap plots: Effects of display characteristics on the accuracy and bias of estimates of whisker length. *Journal of Educational Statistics*, 16:1–20, 1991.

[117] Student. The probable error of a mean. *Biometrika*, 6:1–25, 1908.

[118] H.A. Sturges. The choice of a class interval. *Journal of the American Statistical Association*, 21:65–66, 1926.

[119] P. Tai, J. Tonita, E. Yu, and D. Skarsgard. Twenty-year follow-up study of long-term survival of limited-stage small-cell lung and overview of prognostic and treatment factors. *International Journal of Radiation Oncology, Biology, and Physics*, 56:626–633, 2003.

[120] J. Teghtsoonian. The judgment of size. *American Journal of Psychology*, 78:392–402, 1965.

[121] G.R. Terrell. The maximum smoothing principle in density estimation. *Journal of the American Statistical Association*, 85:470–477, 1990.

[122] G.R. Terrell and D.W. Scott. Oversmoothed nonparametric density estimates. *Journal of the American Statistical Association*, 80:209–214, 1985.

[123] E.R. Tufte. *The Visual Display of Quantitative Information.* Graphics Press, Cheshire, CT, 1983.

[124] E.R. Tufte. *Envisioning Information.* Graphics Press, Cheshire, CT, 1990.

[125] E.R. Tufte. *Visual Explanations: Images and Quantities, Evidence and Narrative.* Graphics Press, Cheshire, CT, 1997.

[126] J.W. Tukey. Some graphic and semigraphic displays. In T.A. Bancroft, editor, *Statistical Papers in Honor of George W. Snedecor*, pages 293–316. The Iowa State University Press, Ames, IA, 1972.

[127] J.W. Tukey. *Exploratory Data Analysis.* Addison-Wesley, Reading, MA, 1977.

[128] W.N. Venables and B.D. Ripley. *Modern Applied Statistics with S.* Springer, New York, NY, 4th edition, 2002.

[129] R. von Huhn. Further studies in the graphic use of circles and bars: (i) A discussion of the Eell's experiment. *Journal of the American Statistical Association*, 22:31–36, 1927.

[130] H. Wainer. *Visual Revelations: Graphical Tales of Fate and Deception from Napoleon Bonaparte to Ross Perot.* Springer-Verlag, New York, NY, 1997.

[131] M.P. Wand and M.C. Jones. *Kernel Smoothing.* Chapman and Hall, London, UK, 1995.

[132] G.S. Watson. Smooth regression analysis. *Sankhyā, Series A*, 26:101–106, 1964.

[133] H. Wickham. A layered grammar of graphics. *Journal of Computational and Statistical Graphics*, 19:3–28, 2010.

[134] H. Wickham. *ggplot2 Elegant Graphics for Data Analysis.* Springer-Verlag, Heidelberg, Germany, second edition, 2016.

[135] M.B. Wilk and R. Gnanadesikan. Probability plotting methods for the analysis of data. *Biometrika*, 55:1–17, 1968.

[136] L. Wilkinson. Dot plots. *The American Statistician*, 53:276–281, 1999.

[137] L. Wilkinson. *The Grammar of Graphics.* Springer-Verlag, New York, NY, first edition, 1999.

[138] L. Wilkinson. *The Grammar of Graphics.* Springer-Verlag, New York, NY, second edition, 2005.

[139] L. Wilkinson, D.J. Rope, D.B. Carr, and M.A. Rubin. The language of graphics. *Journal of Computational and Statistical Graphics*, 9:530–543, 2000.

Index